Boiler Operator's
Guide

Other McGraw-Hill Books of Interest

Boiler Operator's Guide

Anthony Lawrence Kohan

*Member of the American Society of
Mechanical Engineers
National Board Commissioned Inspector
(various state boiler inspector commissions)
Certified Safety Professional
Member of the National Society of
Professional Engineers
Member of the American Welding Society*

Fourth Edition

McGraw-Hill

New York   San Francisco   Washington, D.C.   Auckland   Bogotá
Caracas   Lisbon   London   Madrid   Mexico City   Milan
Montreal   New Delhi   San Juan   Singapore
Sydney   Tokyo   Toronto

Library of Congress Cataloging-in-Publication Data

Kohan, Anthony Lawrence.
    Boiler operator's guide / Anthony Lawrence Kohan.—4th ed.
        p.   cm.
    Includes index.
    ISBN 0-07-036574-1
    1. Steam-boilers—Handbooks, manuals, etc. I. Title.
TJ289.K55   1997
621.1'94—dc21   97-26076
                                                     CIP

McGraw-Hill

A Division of The McGraw-Hill Companies

    4 5 6 7 8 9 0   DOC/DOC   0 2 1 0

ISBN 0-07-036574-1

The sponsoring editor for this book was Harold B. Crawford, the editing supervisor was David E. Fogarty, and the production supervisor was Tina Cameron. It was set in Century Schoolbook by Estelita F. Green of McGraw-Hill's Professional Book Group composition unit.

Printed and bound by R. R. Donnelley & Sons Company.

This book is printed on recycled, acid-free paper containing a minimum of 50% recycled, de-inked fiber.

McGraw-Hill books are available at special quantity discounts to use as premiums and sales promotions, or for use in corporate training programs. For more information, please write to the Director of Special Sales, McGraw-Hill, Professional Publishing, Two Penn Plaza, New York, NY 10121-2298. Or contact your local bookstore.

Dedicated to plant boiler operators, jurisdictional and insurance company inspectors, repairers, installers, plant engineers, and managers involved with providing safe and reliable boiler construction, operation, maintenance, inspection, and repairs at a facility.

Special acknowledgment to Steve Elonka and John Beckert for their inspiration and encouragement.

Contents

Preface to the Fourth Edition

Heating, industrial process, institutional, and utility boiler plant operation continues to be affected by developments in electronic instrumentation and controls, which in turn produce more automatic operation. Regulatory requirements for controlling emissions and discharges are also affecting modern operations. This edition will emphasize some of these modern developments.

The fourth edition will continue to stress fundamental basic operating principles, as well as treating current regulatory requirements on emissions, construction and installation, maintenance and repair, safety controls and devices, and the application of the latest edition of the ASME Codes.

Other developments newly included or receiving expanded discussion are:

Potential impact on the Boiler Code of the ISO 9000 certification program

Low air-fuel ratio burning and NO_x control methods

Confined space testing per OSHA rules

Performance, efficiency testing, and related calculations

Cycling effects on previous base-load operated units and inspections required

Economic evaluations in repair or replace decisions

Combined-cycle cogeneration heat recovery steam generator's operation and inspection

ASME and National Board code revisions

Developments in sensors, transmitters, and actuators that are applicable to automatic operation

Boiler auxiliaries, water treatment chemistry, and chemical equations

Safety practices for the boiler plant

Fire prevention for the boiler or plant

Causes of boiler component failures

Practical questions and answers at the end of each chapter have been revised to reflect current boiler systems. However, some questions and answers on older boiler systems have been retained for those readers who must prepare for jurisdictional operator license examinations or for a National Board inspector's examination. As in previous editions, the problems follow the pattern of these examinations, stressing the practical applications and math skills usually required.

Many corporations and organizations have provided pictures and sketches as well as information on their products for this edition, and their assistance is gratefully acknowledged. Mention in particular is made of *Power* magazine, *Chemical Engineering* magazine, *Mechanical Engineering*, ASME Boiler Code, National Board of Boiler and Pressure Vessel Inspectors' Inspection Code, American Welding Society, National Fire Protection Association, Factory Mutual Engineering, Industrial Risk Insurers, Hartford Steam Boiler Inspection and Insurance Company, and various boiler manufacturers as well as the American Boiler Manufacturers Association. Credit for illustrations or pictures is provided in the text.

The author used due diligence and care in preparing the text, but assumes no legal liability for the information and accuracy contained therein or for the possible consequences of the use thereof. However, the author would sincerely appreciate being advised by readers of any errors or omissions so that necessary changes can be made in the text or illustrations.

Anthony Lawrence Kohan

Abbreviations
and Symbols

A or a	area
ASME	American Society of Mechanical Engineers
ASTM	American Society for Testing and Materials
AWS	American Welding Society
Bhn	Brinell hardness number
Btu	British thermal unit
C	carbon
C	coulomb
C	a constant
Ca	calcium
°C	degree Celsius (centigrade)
cm	centimeter
cm^3	cubic centimeter
CO	carbon monoxide
CO_2	carbon dioxide
Code, the	ASME Boiler and Pressure Vessel Codes
Cu	copper
D	diameter of a shell or drum
E	Young's modulus of elasticity = unit stress divided by unit strain (29 million for steel)
°F	degree Fahrenheit
Fe	iron
FS	factor of safety
ft	foot
gal	gallon
gr/gal	grain per gallon (concentration)

g/min	gallon per minute flow
H	hydrogen
H_2O	water
HAZ	heat-affected zone
HTHW	high-temperature hot-water system
hp	horsepower
hr	hour
HRT boiler	horizontal-return-tubular boiler
HS	water heating surface
ID	inside diameter
in.	inch
J	joule
k	a constant
kg	kilogram
kW	kilowatt
L	liter
l or L	length, in inches, unless otherwise specified
lb	pound
max	maximum
Mg	magnesium
$MgSO_4$	magnesium sulfate
mm	millimeter
Mn	manganese
NB	National Board of Boiler and Pressure Vessel Inspectors
N	nitrogen
NaOH	sodium hydroxide
$NaSiO_2$	sodium silicate
NDE	nondestructive examination
NDT	nondestructive testing
Ni	nickel
O	oxygen
OD	outside diameter
OH	hydroxide
oz	ounce
p	pitch, in inches, usually of a series of holes
P	maximum allowable working pressure
Si	silicon
SM boiler	scotch marine boiler

SiO_2	silica
SMAW	shielded metal arc welding
SO_4	sulfate
std	standard
t	thickness, in inches, unless otherwise stated
temp	temperature
TS	tensile strength
V	vanadium
VT boiler	vertical tubular boiler
W	watt
yd	yard
YP	yield point
%	percent
μm	micrometer (formerly micron)

Boiler Systems, Classifications, and Fundamental Operating Practices

Modern Operation and Responsibilities

Boiler plant operation, maintenance, and inspection requires the services of trained technical people because of the growth and technological development in new materials, metallurgical principles on why materials fail, welding in joining boiler components, and in repairs, sensor development which permits more automatic control, and finally the application of computers in tracking boiler operations and conditions.

Boilers are used at many different pressures and temperatures with large variations in output and different fuel-burning systems. Designers and fabricators apply heat transfer principles to design a boiler system but must also have broad technical skills in fluid mechanics, metallurgy, strength of materials to resist stress, burners, controls and safety devices for the boiler system, or as stipulated by Codes and approval bodies.

The skill and knowledge required of operators may vary because installations range from simple heating systems to integrated process and utility boiler systems. Operating controls can vary from manual to semiautomatic to full automatic. The trend is to automatic operation. However, experienced operators always study the boiler plant layout so that the components, auxiliaries, controls, piping, and possible emergency procedures to follow are thoroughly understood. The study should include a review of the fuel, air, water and steam and fuel-gas loops, and the assigned limitations each may have in operation.

Operators must be familiar with modern boiler controls that are based on an integrated system involving controlling:

1. Load flow for heat, process use, or electric power generation.
2. Fuel flow and its efficient burning.
3. Airflow to support proper and efficient combustion.
4. Water and steam flows to follow load.
5. Exhaust flow of products of combustion.

The highly automated plant requires the knowledge of how the system works to produce the desired results, and what to do to make it perform according to design. Manual operation may still be required under emergency conditions, which is why a knowledge of the different "loops" of a boiler system will assist the operator to restore conditions to normal much faster. With the advent of computers, if a boiler system is out of limits, skilled personnel must trace through the system to see if the problem is in the instruments or out-of-calibration actuators or if a component of the system has had an electrical or mechanical breakdown.

Fundamental operational responsibilities. Operators must be familiar with certain fundamentals that were commonly posted in the past, especially in manually operated systems. Among these were the following rules:

1. *Water level* maintenance and checking at least once per shift.
2. *Low water* and the actions required by the operator to minimize damage.
3. *Low water cutoff* testing to make sure it is functional, usually performed once per shift. This includes blowing down the float chamber or the housing in which the sensor is located, so it does not become obstructed from internal deposits.
4. *Gauge cocks* must be kept clean and dry. They should be tested once per shift in order to make sure that all connections to the water glass and water column are clear, and thus by testing gauge cocks, the true level in the gauge glass can be determined.
5. *Safety valves* should be tested at least once per month by raising the valve off the seat slowly. If the valve does not lift, it is an indication that rust or boiler compound is binding the valve and corrections or repairs are needed. The boiler should be secured, and not operated with a defective safety valve.

6. *Burners* should be kept clean and free of leaks with the flame adjusted so that it does not strike side walls, shells, or tubes. *Flame safeguards* should be checked every shift in order to make sure that they are functional and thus prevent a furnace explosion.

7. *Boiler internals* must be kept free of scale, mud, or oily deposits by proper water treatment and blowdown procedures in order to prevent overheating, bagged and buckled sheets, and the occurrence of a serious rupture or explosion.

8. The *outside* of the *boiler* should be kept clean and dry. Soot or unburned products should not be allowed to accumulate, as these will cause controls and actuators to bind and malfunction as well as causing corrosion to occur on the different parts of the boiler.

9. *Leaks* are a sign of distress on the boiler system and should be repaired immediately because of the possible danger involved, and also because they accelerate corrosion and grooving of system components that will result in forced shutdowns.

10. *When taking a boiler out of service,* do not accelerate the process by blowing off the boiler under pressure in order to prevent the heat of the boiler from baking mud and scale on the internal surfaces. Let the boiler cool slowly, then drain and thoroughly wash out the top and bottom parts of the internal surfaces.

11. *Dampers* should be kept in good condition in order to avoid unconsumed fuel from accumulating in the combustion chamber or furnace and cause a fire-side explosion. All connections and appurtenances should be kept in good working order to maintain efficient operation and also to prevent forced shutdowns.

12. *Idle boilers,* for any length of time, especially steel boilers, and if dry layup is to follow, should have their manholes and handholes removed, followed by thorough washing of the interior surfaces to remove scale and other contaminants. The boiler should be kept dry. (Later chapters will describe the methods used to keep a boiler dry.) Cast iron boilers are usually cleaned on the fire side, and kept layed up wet.

13. *Purging* should be thorough on any firing or restart in order to clear the furnace passages of any unconsumed fuel, and thus prevent a fire-side explosion.

14. *Preparing a boiler for inspections* per legal statute requires all critical internal surfaces to be made available for inspection (covered by later chapters). This requires manholes and handholes to be removed, with the boiler cooled slowly, and then cleaned internally and externally including fire sides of boiler components. All

valves should be tight in order to prevent any steam or water from backing into the idle boiler.

15. *Maintain boiler water treatment testing* and application of the treatment per guidelines established by water treatment specialists. This will assist in avoiding scale buildup and dissolved gases in the boiler water forming acids that can cause corrosion in the boiler system, and will also help maintain boiler efficiency.

16. *Maintain proper blowdown* in order to remove the sludge that may build up in the boiler water. Follow the recommendations of the water treatment specialist on frequency of blowdown and amount.

These fundamental responsibilities are important in maintaining a safe and efficient boiler plant and are considered minimum operator responsibilities. Later chapters will dwell on other features of boiler operation, maintenance, inspection, and repair.

New boiler installations, repairs, and retrofits. Experienced operators in high-pressure plants are also involved in bringing a new boiler into service by making sure that proper operating procedures are followed during preliminary and final checkouts of fuel burning equipment, fans, pumps, valves, controls, safety devices, and all components that may comprise the boiler system. Other activities on new boilers include cleaning out internal surfaces and boiling out and blowing out steam lines prior to final acceptance test runs. Also included in the acceptance procedure is hydrostatic testing, calibration of instruments and controls, safety valve testing, starting, testing, and making sure auxiliary boiler equipment performs per design. Output performance guarantees must be verified as well as stipulated efficiencies.

Skills updating. Operators of semiautomatic and automatic plants must continue to study the systems under their control, because of progress in controls and computer application as systems are more automated. Optimization of equipment performance is now considered a desirable goal in operation. This includes improving efficiency of operation, gains in environmental compliance, and the economic gains from better operation.

Computer application to energy systems now requires fewer people to operate a boiler system, but also requires more knowledge by the operator. For example, in a fully integrated boiler plant system, the operator is in a control room and is linked to the boiler, and perhaps generating equipment, by means of video displays that show different data by the operator's pushing the appropriate button on the computer. This can show the operator the status of each unit as respects load,

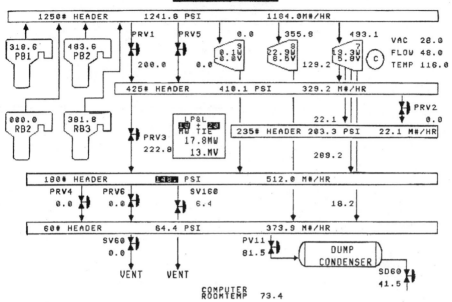

Figure 1.1 Printout of computer display of steam flow, pressure and turbo-generator loading, including outside power connection for a paper mill power plant.

pressure, and temperatures as shown in Fig. 1.1. The computer can be programmed for each subprocess to have startup and sequential shutdown features. There can be incorporated intelligent logic that can interrupt a starting sequence if conditions are not within set points. It is important for operators to be alert to new developments in the rapidly expanding on-line computer technology.

Jurisdictional operator licensing laws

Because of the inherent danger of explosions and fire that exists in a boiler system, many jurisdictions require boiler system operators to pass a written or oral examination provided the candidate also has appropriate experience under the supervision of another licensed operator. Figure 1.2 lists the jurisdictions that have operating engineers' licensing laws. Jurisdictional departments and street addresses for the licensing authorities are listed in the McGraw-Hill publication, *Plant Services and Operations Handbook* (Kohan, 1995).

Heat transfer and operation

A study of thermodynamics, vapor cycles, and basic heat transfer can assist boiler operation by instituting a program of heat tracing in

Jurisdiction	High-pressure boilers	Low-pressure boilers
U.S. cities and counties		
Buffalo, N.Y.	X	—
Chicago, Ill.	X	—
Dearborn, Mich.	X	X
Denver, Colo.	X	X
Des Moines, Iowa	X	X
Detroit, Mich.	X	X
E. St. Louis, Ill.	X	X
Kansas City, Mo.	X	X
Los Angeles, Calif.	X	X
Memphis, Tenn.	X	X
Miami, Fla.	X	X
Milwaukee, Wis.	X	X
New Orleans, La.	X	X
New York City, N.Y.	X	—
Oklahoma City, Okla.	X	X
Omaha, Neb.	X	X
St. Joseph, Mo.	X	X
St. Louis, Mo.	X	X
San Jose, Calif.	X	—
Spokane, Wash.	X	X
Tacoma, Wash.	X	X
Tampa, Fla.	X	X
Tulsa, Okla.	X	X
University City, Mo.	X	X
White Plains, N.Y.	X	—
Jefferson Parish, La.	X	X
St. Louis Co., Mo.	X	X
States		
Alaska	X	X
Arkansas	X	X
District of Columbia	X	X
Massachusetts	X	—
Minnesota	X	X
Montana	X	X
Nebraska	—	X
New Jersey	X	X
Ohio	X	X
Pennsylvania	X	X
Canadian provinces		
Alberta	X	X
British Columbia	X	X
Manitoba	X	X
New Brunswick	X	X
Newfoundland and Labrador	X	X
N.W. Territory	X	X
Nova Scotia	X	—
Ontario	X	X
Quebec	X	X
Saskatchewan	X	X
Yukon Territory	X	X

Note: Due to variations in the laws, it is necessary to check the jurisdiction for specific requirements on licensed operators.

Figure 1.2 Jurisdictions having operating engineer's licensing laws for boilers.

order to improve efficiency and track heat losses in boiler plant operation. A boiler is a heat transfer apparatus that converts fossil fuel, electrical, or nuclear energy through a working medium such as water, or organic fluids such as dowtherm, and then conveys this energy to some external heat transfer apparatus, such as is used for heating buildings or for process use. This energy may also be converted to produce power with mechanical drive steam turbines or with steam turbine generators to produce electrical power.

The flow of heat in a boiler can affect the efficiency of operation, and may even cause overheating problems, such as when scale is allowed to accumulate in tubes. The flow of heat can occur by *conduction, convection,* or *radiation,* and usually consists of all three inside a boiler.

Conduction is the transfer of heat from one part of a material to another or to a material with which it is in contact. Heat is visualized as molecular activity—crudely speaking, as the vibration of the molecules of a material. When one part of a material is heated, the molecular vibration increases. This excites increased activity in adjacent molecules, and heat flow is set up from the hot part of the material to the cooler parts. In boilers, considerable surface conductance between a fluid and a solid takes place, for example, between water and a tube and gas and a tube, in addition to conductance through the metal of a tube, shell, or a furnace.

While surface conductance plays a vital part in boiler efficiency, it can also lead to metal failures when heating surfaces become overheated, as may occur when surfaces become insulated with scale. The surface conductance when expressed in Btu per hour per square foot of heating surface for a difference of one degree Fahrenheit in temperature of the fluid and the adjacent surface, is known as the *surface coefficient* or *film coefficient.* Figure 1.3*a* shows stagnant areas near the tube where the film coefficient will reduce heat transfer.

The coefficient of thermal (heat) conductivity is defined further as the quantity of heat that will flow across a unit area in unit time if the temperature gradient across this area is unity. In physical units it is expressed as *Btu per hour per square foot per degree Fahrenheit per foot.* Expressed mathematically, the rate of heat transfer Q by conduction across an area A, through a temperature gradient of degrees Fahrenheit per foot T/L, is

$$Q = kA\frac{T}{L}$$

where k = coefficient of thermal conductivity.

Note that k varies with temperature. For example, mild steel at 32°F has a thermal conductivity of 36 Btu/(hr/ft²/°F/ft), whereas at 212°F it is 33.

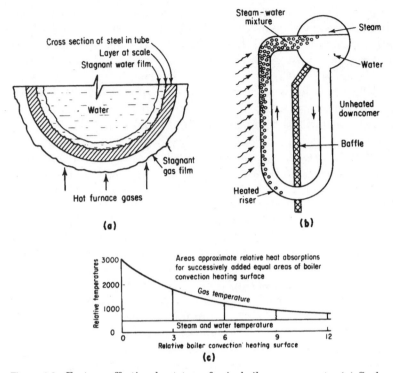

Figure 1.3 Factors affecting heat transfer in boiler components. (*a*) Scale and stagnant gas and water near the tube affect heat transfer across the tube. (*b*) Circulation depends on heated water rising while cooler water descends to replace it. (*c*) Adding boiler heating surface increases heat absorption but at a reduced rate.

Convection is the transfer of heat to or from a fluid (liquid or gas) flowing over the surface of a body. It is further refined into *free* and *forced convection.* Free convection is *natural* convection causing circulation of the transfer fluid due to a difference in density resulting from temperature changes.

For example, in Fig. 1.3*b* the heated water and steam rise on the left and are displaced by cooler (heavier) water on the right. This causes free convection of heat transfer between heat on one side of the U tube and cooler water on the other side. Actually, conduction has to take place first between the gas film and metal of the tube, then the water. But if the water did not circulate, eventually equal temperatures would result. Heat transfer would then cease.

Forced convection results when circulation of the fluid is made positive by some mechanical means, such as a pump for water or a fan for hot gases. The heat transfer by convection is thus aided mechanically.

Adding boiler surface may increase the heat absorption, but as shown in Fig. 1.3c, the temperature gradient will drop more and more. Then at some point the gain in efficiency will be far less than the cost of adding heating surface. Further, the mechanical power required for forced circulation will also increase with the addition of heating surface by convection.

The hydraulic circuit of a boiler consists of the paths of water flow created by the difference between heads of water and water-steam mixtures. Flow in tubes and risers is induced by the difference in density of water and water-steam mixtures. The heavier water will flow to the bottom as the lighter water-steam mixture rises in the boiler water-steam paths. The higher the steam pressure, the denser the steam becomes, which results in a loss of flow as the steam approaches water density. It is the reason that pumps are used to promote circulation in very high pressure boilers. Insufficient flows create inefficient use of heating surfaces, but can also result in tubes overheating due to water starvation.

Note that in Fig. 1.4a more tube area is required at lower pressure than higher pressure for the same circulation to exist. But the force producing circulation is less at high pressure than at low pressure. This involves the change in the specific weight of water and steam as pressures increase. The mixture actually weighs less in pounds per cubic foot at higher pressures. For example, in the sketch in Fig. 1.4b at the critical pressure (3206.2 psia), water and steam have the same specific weight. Friction losses due to flow are generally less at higher pressure. This is primarily due to more laminar, or streamlined, flow and less turbulent flow in the tubes.

When boiling occurs in a tube, bubbles of vapor are formed and liberated from the surface in contact with the liquid. This bubbling action creates voids (Fig. 1.4c) of the on-again-off-again type, because of the rapidness of the action. This creates a turbulence near the heat-transfer surfaces, which generally increases the heat-transfer rate. But the loss of wetness as the bubbles are formed may diminish heat transfer.

Pressure has a marked effect on the boiling and heat-transfer rate. With higher pressures (Fig. 1.4d) bubbles tend to give way to what is called *film boiling,* in which a film of steam covers the heated surface. This phenomenon is very critical in boiler operation, often causing watertube failures due to starvation, even though a gauge glass may show water. It is further compounded by the formation of scale and other impurities along the boiling area of a tube.

Radiation is a continuous form of interchange of energy by means of electromagnetic waves without a change in the temperature of the medium between the two bodies involved. Radiation is present in all

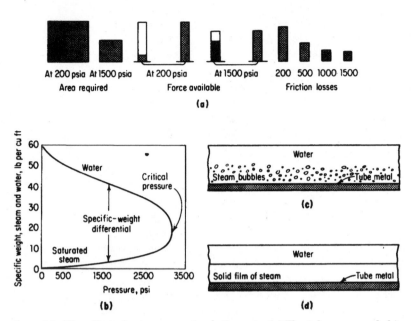

Figure 1.4   The effect of pressure on circulation rate. (a) The tube area needed is higher at low pressure; the force to produce the circulation is less at high pressure; friction loss is greater at low pressure. (b) At critical pressure, water and steam have the same specific weight (3206.2 psia). (c) At low pressure, steam bubbles form near the tube metal. (d) At high pressure, a solid film or layer of steam is formed at the tube metal surface.

boilers. In fact, all boilers utilize all three means of heat transfer: *conductance, convection,* and *radiation.*

Properties of Steam and Boiler Systems

A brief review of some properties of steam will also assist in differentiating boiler systems. A book of steam tables is necessary for computing boiler efficiency. The standard in the United States is *Thermodynamic Properties of Steam* by Keenan and Keyes, published by John Wiley & Sons Inc., New York. For data based on temperature, use Table 1 in Fig. 1.5. Use Table 2 if you know the pressure. All pressures in these tables are absolute. To get absolute pressure, just add 14.7 psi to the gauge pressure (15 psi is close enough).

For properties of superheated steam, use Table 3 in Fig. 1.5. This table of superheated steam must be used with the absolute pressure (gauge pressure plus 15) and with the *total* steam temperature, not the degrees of superheat. This total temperature is the saturation temperature (also given in the table) plus the degrees of superheat.

Table 1. Saturation, Temperatures

Temp, °F	Abs press, psi	Specific vol Sat liquid	Specific vol Sat vapor	Enthalphy (heat) Sat liquid	Enthalphy (heat) Evap	Enthalphy (heat) Sat vapor
32	0.08859	0.01602	3304.7	0.01	1075.5	1075.5
40	0.12170	0.01602	2444	8.05	1071.3	1079.3
50	0.17811	0.01603	1703.2	18.07	1065.6	1083.7
60	0.2563	0.01604	1206.7	28.06	1059.9	1088.0
70	0.3631	0.01606	867.9	38.04	1054.3	1092.3
80	0.5069	0.01608	633.1	48.02	1048.6	1096.6
90	0.6982	0.01610	468.0	57.99	1042.9	1100.9
100	0.9492	0.01613	350.4	67.97	1037.2	1105.2
110	1.2748	0.01617	265.4	77.94	1031.6	1109.5
120	1.6924	0.01620	203.27	87.92	1025.8	1113.7
130	2.2225	0.01625	157.34	97.90	1020.0	1117.9
140	2.8886	0.01629	123.01	107.9	1014.1	1122.0
150	3.718	0.01634	97.07	117.9	1008.2	1126.1
160	4.741	0.01639	77.29	127.9	1002.3	1130.2
170	5.992	0.01645	62.06	137.9	996.3	1134.2
180	7.510	0.01651	50.23	147.9	990.2	1138.1
190	9.339	0.01657	40.96	157.9	984.1	1142.0
200	11.526	0.01663	33.64	168.0	977.9	1145.9
212	14.696	0.01672	26.80	180.0	970.4	1150.4
220	17.186	0.01677	23.15	188.1	965.2	1153.4
240	24.969	0.01692	16.323	208.3	952.2	1160.5
280	49.203	0.01726	8.645	249.1	924.7	1173.8
300	67.013	0.01745	6.466	269.6	910.1	1179.7
340	118.01	0.01787	3.788	311.1	879.0	1190.1
380	195.77	0.01836	2.335	353.5	844.6	1198.1
400	247.31	0.01864	1.8633	375.0	826.0	1201.0

Table 2. Saturation, Pressures

Abs press, psi	Temp, °F	Specific vol Sat liquid	Specific vol Sat vapor	Enthalpy (heat) Sat liquid	Enthalpy (heat) Evap	Enthalpy (heat) Sat vapor
0.50	79.58	0.01608	641.4	47.6	1048.8	1096.4
1.0	101.74	0.01614	333.6	69.7	1036.3	1106.0
5.0	162.24	0.01640	73.52	130.1	1001.0	1131.1
10	193.21	0.01659	38.42	161.2	982.1	1143.3
14.7	212.00	0.01672	26.80	180.0	970.4	1150.4
15	213.03	0.01672	26.29	181.1	969.7	1150.8
20	227.96	0.01683	20.089	196.2	960.1	1156.3
25	240.07	0.01692	16.303	208.5	952.1	1160.6
30	250.33	0.01701	13.746	218.8	945.3	1164.1
40	267.25	0.01715	10.498	236.0	933.7	1169.7
50	281.01	0.01727	8.515	250.1	924.0	1174.1
60	292.71	0.01738	7.175	262.1	915.5	1177.6
70	302.92	0.01748	6.206	272.6	907.9	1180.6
80	312.03	0.01757	5.472	282.0	901.1	1183.1
90	320.27	0.01766	4.896	290.6	894.7	1185.3
100	327.81	0.01774	4.432	298.4	888.8	1187.2
110	334.77	0.01782	4.049	305.7	883.2	1188.9
120	341.25	0.01789	3.728	312.4	877.9	1198.4
130	347.32	0.01796	3.455	318.8	872.9	1191.7
140	353.02	0.01802	3.220	324.8	868.2	1193.0
150	358.42	0.01809	3.015	330.5	863.6	1194.1
200	381.79	0.01839	2.288	355.4	843.0	1198.4
250	400.95	0.01865	1.8438	376.0	825.1	1201.1
300	417.33	0.01890	1.5433	393.8	809.0	1202.8
350	431.72	0.01913	1.3260	409.7	794.2	1203.9
400	444.59	0.0193	1.1613	424.0	780.5	1204.5

Table 3. Superheated Steam

Abs pressure, psi (sat temp)	•	Sat liquid	Sat vapor	300	400	500	600	700	800	900	1000
15 (213.03)	v	0.016	26.29	29.91	33.97	37.99	41.99	45.98	49.97	53.95	57.93
	h	181.1	1150.8	1192.8	1239.9	1287.1	1334.8	1383.1	1432.3	1482.3	1533.1
20 (227.96)	v	0.016	20.09	22.36	25.43	28.46	31.47	34.47	37.46	40.45	43.44
	h	196.2	1156.3	1191.6	1239.2	1286.6	1334.4	1382.9	1432.1	1482.1	1533.0
40 (267.25)	v	0.017	10.498	11.040	12.628	14.168	15.688	17.198	18.702	20.20	21.70
	h	236.0	1169.7	1186.8	1236.5	1284.8	1333.1	1381.9	1431.3	1481.4	1532.4
60 (292.71)	v	0.017	7.175	7.259	8.357	9.403	10.427	11.441	12.449	13.452	14.454
	h	262.1	1177.6	1181.6	1233.6	1283.0	1331.8	1380.9	1430.5	1480.8	1531.9
80 (312.03)	v	0.018	5.472	· · · · ·	6.220	7.020	7.797	8.562	9.322	10.077	10.830
	h	282.0	1183.1	1230.7	1281.1	1330.5	1379.9	1429.7	1480.1	1531.3	
100 (327.81)	v	0.018	4.432		4.937	5.589	6.218	6.835	7.446	8.052	8.656
	h	298.4	1187.2		1227.6	1279.1	1329.1	1378.9	1428.9	1479.5	1530.8
150 (358.42)	v	0.018	3.015		3.223	3.681	4.113	4.532	4.944	5.352	5.758
	h	330.5	1194.1		1219.4	1274.1	1325.7	1376.3	1426.9	1477.8	1529.4
200 (381.79)	v	0.018	2.288		2.361	2.726	3.060	3.380	3.693	4.002	4.309
	h	355.4	1198.4		1210.3	1268.9	1322.1	1373.6	1424.8	1476.2	1528.0
300 (417.33)	v	0.0189	1.5433			1.7675	2.005	2.227	2.442	2.652	2.859
	h	393.8	1202.8			1257.6	1314.7	1368.3	1420.6	1472.8	1525.2
400 (444.59)	v	0.0193	1.1613			1.2851	1.4770	1.6508	1.8161	1.9767	2.134
	h	424.0	1204.5			1245.1	1306.9	1362.7	1416.4	1469.4	1522.4
500 (467.01)	v	0.0197	0.9278			0.9927	1.1591	1.3044	1.4405	1.5715	1.6996
	h	449.4	1204.4			1231.3	1298.6	1357.0	1412.1	1466.0	1519.6
600 (486.21)	v	0.0201	0.7698			0.7947	0.9463	1.0732	1.1899	1.3013	1.4096
	h	471.6	1203.2			1215.7	1289.9	1351.1	1407.7	1462.5	1516.7
800 (518.23)	v	0.0209	0.5687				0.6779	0.7833	0.8763	0.9633	1.0470
	h	509.7	1198.6				1270.7	1338.6	1398.6	1455.4	1511.0
1000 (544.61)	v	0.0216	0.4456				0.5140	0.6084	0.6878	0.7604	0.8294
	h	542.4	1191.8				1248.8	1325.3	1389.2	1448.2	1505.1
1200 (567.22)	v	0.0223	0.3619				0.4016	0.4909	0.5617	0.6250	0.6843
	h	571.7	1183.4				1223.5	1311.0	1379.3	1440.7	1499.2
1400 (587.10)	v	0.0231	0.3012				0.3174	0.4062	0.4714	0.5281	0.5805
	h	598.7	1173.4				1193.0	1295.5	1369.1	1433.1	1493.2

Figure 1.5 Pressure and temperature relationships of water and steam. Use Table 1 for data based on temperature, Table 2 for data based on pressure, and Table 3 for data based on superheated steam.

Enthalpy means the heat content of the fluid. In dealing with water and steam, three enthalpies are to be noted:

1. Enthalpy of saturated liquid [in British thermal units (Btu)], which is the heat content of the water at a certain pressure and temperature under consideration
2. Enthalpy of evaporation (Btu), which is the heat required to evaporate 1 lb of water to steam at that pressure and temperature
3. Enthalpy of saturated vapor (Btu), which is the heat content of the saturated steam at the pressure and temperature being considered

The enthalpy of saturated steam is thus a sum of the enthalpy of saturated liquid and the enthalpy of evaporation, or the *total* heat content of the saturated steam in Btu per pound.

Tables 1 and 2 in Fig. 1.5 give the properties of water and of saturated steam. The only difference is that in Table 1 we enter with the boiler temperature, while in Table 2 we enter with the boiler pressure (psia). For example, Table 1 shows that for water to boil at 100°F, the absolute pressure must be 0.95 psi. Table 2 shows that at 40 psia, water boils at 267°F. It is not necessary to use all the digits given in the table. Most practical work does not require it. Engineers rarely need to figure water temperatures to closer than the nearest degree, or heats or enthalpies to closer than the nearest Btu.

Sat liquid means liquid water at the saturation or boiling temperature; *sat vapor* means steam at the boiling temperature. When water is boiling in a closed container, both the water and the steam over it are in a saturated condition. Steam is saturated when generated by a boiler without a superheater. For steam, *saturated* means steam that contains no liquid water yet is *not* superheated (still at boiling temperature). Note that the absolute pressure is gauge pressure plus about 15 lb. Now, in Table 2, try reading across the line for 50 psia (35 psig).

Boiling temperature is 281°F. At this temperature 1 lb of water fills 0.0713 ft³ and 1 lb of saturated steam fills 8.51 ft³. Specific volume is in cubic feet per pound of water or steam. Thus it takes 250 Btu to heat the pound of water from 32°F to the boiling point and another 924 Btu to evaporate it, making a total of 1174 Btu. As mentioned, enthalpy used to be called *heat* in the old steam tables, and it is given in Btu per pound. The last three columns of the old tables were labeled *heat of the liquid, heat of vaporization,* and *total heat.*

Example A boiler generates saturated steam at 135 psig (150 psia). The enthalpy, or heat of the final steam, is 1194 Btu/lb. The amount of heat required to produce this steam in an actual boiler will depend on the temperature of the feedwater. Suppose the feedwater temperature is 180°F.

Table 1 in Fig. 1.5 shows that the heat in the water is 148 Btu. Then the heat supplied to turn this water into steam is merely the difference, or 1194 − 148 = 1046 Btu.

It is easy from this to figure the boiler efficiency. Let us say the boiler generates 10 lb steam per pound of coal burned and the coal contains 13,000 Btu/lb. Then, for every 13,000 Btu put in as fuel, there is delivered in steam 10 × 1046 = 10,460 Btu.

The efficiency of any power unit is its output divided by its input, so here 10,460/13,000 = 0.805, or 80.5 percent efficiency.

For most purposes, Table 1 in Fig. 1.5 is not needed to get a close value of the heat of the liquid. Just subtract 32 from the water temperature. For example, the enthalpy of water at 180°F is the heat required to raise it from 32 to 180°F, or a difference of 148°F. This takes about 148 Btu. But it will not work out so closely for very high temperatures. Take water at 300°F. Table 1 in Fig. 1.5 gives 269.7 Btu, while our simple method gives 300 − 32 = 268 Btu, close enough for most purposes.

To use the steam tables for superheated steam, the first column of Table 3 in Fig. 1.5 gives the absolute pressure and (directly below it in parentheses) the corresponding saturation temperature, or boiling temperature. In the next column, v and h stand for volume of 1 lb and its heat content. For example, at 150 psia the volume of 1 lb is 0.018 ft³ for liquid water and 3.015 ft³ for saturated steam. The corresponding heat contents of 1 lb are 330.5 and 1194.1 Btu.

The temperature columns give the volume and heat content per pound for superheated steam at the indicated temperature. Take steam at 150 psi, superheated to a total temperature of 600°F. Look in the 600°F column opposite 150 psi. The volume is 4.113 ft³, as against 3.015 ft³ for saturated steam at the same pressure. This is natural because steam expands as a gas when superheated. Also, the heat content is naturally higher, 1325.7 instead of 1194.1 Btu. Note that this table gives the actual temperature of the superheated steam rather than the degrees of superheat, which is a different thing. If the steam has been superheated from a saturation temperature of 358 to 600°F, the superheat is

$$600 - 358 = 242°F$$

These superheat tables are used similarly to the saturation tables. Let us take a problem. How much heat does it take to convert 1 lb of feedwater at 205°F into superheated steam at 150 psia and 600°F? The heat in the steam is 1325.7 (1326) Btu. The heat in the water is 205 − 32 = 173 Btu. Then the heat required to convert 1 lb of steam is 1326 − 173 = 1153 Btu.

To calculate boiler efficiency, the method is the same as that for finding the efficiency of practically any other piece of power equipment; namely, efficiency is the useful energy output divided by the energy input. For example, if we get out three-quarters of what we put in, the efficiency is ¾, or 0.75 percent. In the case of a boiler unit, we feed in Btu in the form of coal, oil, or gas, and we get out useful Btu in the form of steam. Thus, the first method states that boiler efficiency can be figured directly from the total fuel burned in a given period and the total water evaporated into steam in the same period. It is more common to figure first the evaporation per pound of fuel fired and then, from this, the efficiency.

ASME test code.   This is a procedure to determine larger boiler outputs and includes heat balance calculations. This requires calculating output and efficiency by *subtracting* from the fuel energy input all the losses that occur in a steam-generating unit, such as:

Loss due to moisture in the fuel

Loss due to water that may be formed from hydrogen in the fuel

Loss due to moisture in the air used for combustion

Loss due to the heat, or Btus carried up the stack by flue gas

Loss due to incomplete combustion of carbon in the fuel

Loss due to unconsumed combustibles in the solid residue or ash

Losses due to unconsumed hydrogen or hydrocarbons in the fuel

Losses due to radiation, leaks, and other unaccounted for losses

Chapter 15 covers some methods of calculating boiler efficiency and the methods used to improve the efficiency as it applies to smaller boiler plants.

Boiler Definitions

The following definitions of boilers usually are found in state laws and codes on boilers in reference to installation or reinspection requirements as well as engineer-licensing laws for operating this type of equipment.

A *boiler* is a closed pressure vessel in which a fluid is heated for use external to itself by the direct application of heat resulting from the combustion of fuel (solid, liquid, or gaseous) or by the use of electricity or nuclear energy.

A *high-pressure steam boiler* is one which generates steam or vapor at a pressure of more than 15 pounds per square inch gauge (psig). Below this pressure it is classified as a *low-pressure steam boiler.* Small high-pressure boilers are classified as *miniature boilers.*

According to Section I of the Boiler and Pressure Vessel Code of the American Society of Mechanical Engineers (ASME), a *miniature high-pressure boiler* is a high-pressure boiler which does not exceed the following limits: 16-inch (in.) inside diameter of shell, 5-cubic-feet (ft^3) gross volume exclusive of casing and insulation, and 100-psig pressure. If it exceeds any of these limits, it is a *power boiler*. Most states follow this definition. The welding requirements for these small boilers are not as severe as for the larger boilers.

A *power boiler* is a steam or vapor boiler operating above 15 psig and exceeding the miniature boiler size. This also includes hot-water-heating or hot-water-supply boilers operating above 160 psi or 250 degrees Fahrenheit (°F). Power boilers are also called *high-pressure boilers*.

A *low-pressure boiler* is defined as a steam boiler that operates below 15-psig pressure or a hot-water boiler that operates below 160 psig or 250°F.

A *hot-water-heating boiler* is a boiler in which no steam is generated, but from which hot water is circulated for heating purposes and then returned to the boiler, and which operates at a pressure not exceeding 160 psig or a water temperature not over 250°F at or near the boiler outlet. These types of boilers are considered low-pressure heating boilers, built under Section IV of the Heating Boiler Code part of the ASME Boiler Codes. If the pressure or temperature conditions are exceeded, the boilers must be designed as high-pressure boilers under Section I of the Code.

A *hot-water-supply boiler* is completely filled with water and furnishes hot water to be used externally to itself (not returned) at a pressure not exceeding 160 psig or a water temperature not exceeding 250°F. These types of boilers are also considered low-pressure boilers, built to Section IV (Heating Boiler) requirements of the ASME Code. If the pressure or temperature is exceeded, these must be designed as high-pressure boilers.

A *waste-heat boiler* uses by-product heat such as from a blast furnace in a steel mill or exhaust from a gas turbine or by-products from a manufacturing process. The waste heat is passed over heat-exchanger surfaces to produce steam or hot water for conventional use.

The same basic ASME Code construction rules apply to waste-heat boilers as are applied to fired units, and the usual auxiliaries and safety features normally required on a boiler are also required for a waste-heat unit.

Engineers prefer to use the term *steam generator* instead of *steam boiler* because *boiler* refers to the physical change of the contained fluid whereas *steam generator* covers the whole apparatus in which this physical change is taking place. But in ordinary use, both are essentially the same. Most state laws are still written under the old, basic boiler nomenclature.

A *packaged boiler* is a completely factory-assembled boiler, water-tube, firetube, or cast-iron, and it includes boiler firing apparatus, controls, and boiler safety appurtenances. A shop-assembled boiler is less costly than a field-erected unit of equal steaming capacity. While a shop-assembled boiler is not an off-the-shelf item, generally it can be put together and delivered much faster than a field-erected boiler; installation and start-up times are substantially shorter. Shop-assembled work usually can be better supervised and done at lower cost.

A *supercritical boiler* operates above the supercritical pressure of 3206.2 pounds per square inch absolute (psia) and 705.4°F saturation temperature. Steam and water have a critical pressure at 3206.2 psia. At this pressure, steam and water are at the same density, which means that the steam is compressed as tightly as the water. When this mixture is heated above the corresponding saturation temperature of 705.4°F for this pressure, dry, superheated steam is produced to do useful high-pressure work. This dry steam is especially well suited for driving turbine generators.

Supercritical pressure boilers are of two types: once-through and recirculation. Both types operate in the supercritical range above 3206.2 psia and 705.4°F. In this range the properties of the saturated liquid and saturated vapor are identical; there is no change in the liquid-vapor phase, and therefore no water level exists, thus requiring no steam drum as such.

Boilers are also classified by the nature of services intended. The traditional classifications are stationary, portable, locomotive, and marine, defined as follows. A *stationary boiler* is installed permanently on a land installation. A *portable boiler* is mounted on a truck, barge, small riverboat, or any other such mobile-type apparatus. A *locomotive boiler* is a specially designed boiler, specifically meant for self-propelled traction vehicles on rails (it is also used for stationary service). A *marine boiler* is usually a low-head-type special-design boiler meant for ocean cargo and passenger ships with an inherent fast-steaming capacity.

The *type of construction* also distinguishes boilers as follows.

Cast-iron boilers are low-pressure heating units manufactured by casting the pressure components in sections from iron, bronze, or brass. The usual types manufactured are further classified by the manner in which the cast sections are arranged or assembled—by means of push nipples, external headers, and screwed nipples. Three types of cast-iron boilers are:

1. Vertical sectional cast-iron boilers have their sections stacked or assembled vertically one above the other, similar to pancakes, with push nipples interconnecting the sections.

2. Horizontal sectional cast-iron boilers have their sections stacked or assembled horizontally so that the sections stand together like slices in a loaf of bread.

3. Small cast-iron boilers are also built in one-piece, or single casting. These are generally smaller boilers used primarily in the past for hot-water-supply service.

See Chap. 3 on cast-iron boilers for further details on construction.

Steel boilers can be of the high-pressure or low-pressure type and today are usually of welded construction. They are subdivided into two classes:

1. In *firetube boilers,* the products of combustion pass through the inside of tubes with the water surrounding the tubes. Firetube boilers are described in detail in later chapters.

2. In *watertube boilers,* the water passes through the tubes, and the products of combustion pass around the tubes.

Firetube boilers generally are used for capacity up to 50,000 pounds per hour (lb/hr) and 300-psi pressure; above this capacity and pressure, watertube boilers are used. Firetube boilers are classified as shell boilers. Water and steam are confined to a shell. This arrangement limits the volume of steam that can be generated without making the shells prohibitively large, and with respect to pressure, the thickness required would become too expensive to fabricate.

Boiler-output rating terminology. Boiler output can be expressed in horsepower, pounds per hour, Btu per hour, and, for utility boilers, the capability of generating so many megawatts of electricity. Heating boilers can also be rated in horsepower, pounds per hour, and Btu per hour, but their output is also described in terms related to heat-transfer area needed for a space. For example, *equivalent square feet of steam radiation surface* is a measure of the heat-transfer area needed in a room that will use steam as a heat source.

A *boiler horsepower* (boiler hp) is defined as the evaporation into dry saturated steam of 34.5 lb/hr of water at a temperature of 212°F. Thus 1 boiler hp by this method is equivalent to an output of 33,475 Btu/hr, and was commonly taken as 10 square feet (ft^2) of boiler heating surface. But 10 ft^2 of boiler heating surface in a modern boiler will generate anywhere from 50 to 500 lb/hr of steam. Today the capacity of larger boilers is stated as so many pounds per hour of steam, or Btu per hour, or megawatts of power produced.

The term *heating surface* is also used to define or relate to the output of a boiler. The heating surface of a boiler is the area, expressed in square feet, that is exposed to the products of combustion. The following surface parts of boilers must be considered in determining the amount of heating surface that may be available for producing steam or hot water: tubes, fireboxes, shell surfaces, tube sheets, headers, and furnaces.

A comparison of output ratings based on horsepower, heating surface, and pounds per hour can be made by assuming a boiler has a nominal horsepower rating of 500 hp.

1. The heating-surface rating would be 5000 ft^2 under the old rule of 10 ft^2/hp.

2. The steam output in pounds per hour would be

$$500 \times 34.5 = 17{,}250 \text{ lb/hr}$$

3. For a hot-water-heating boiler, the output would be

$$500 \times 33{,}475 = 16{,}737{,}500 \text{ Btu/hr}$$

The pounds-per-hour rating often guaranteed by the manufacturer system of rating boilers is a measure of the capacity at which a boiler can be operated continuously. The peak output of a boiler for a 2-hr period is usually set 10 to 20 percent above the maximum continuous output. The pounds-per-hour rating usually is expressed in pounds of steam at the design temperature and pressure for the boiler. Low-pressure boilers are also rated by heating contractor code requirements as well as pounds per hour or Btu per hour.

In *heating-load calculations,* the terms IBR-rated, SBI-rated, and EDR are often used. These terms affect the output rating of a boiler. Thus they are important in sizing a boiler for heating a certain size space. They also affect the safety valve required on a boiler. They are defined as follows.

The acronym IBR stands for the Institute of Boiler and Radiator Manufacturers, which rates cast-iron boilers. Usually IBR-rated boilers have a nameplate indicating net and gross output in Btu per hour. Gross output is further defined as the net output plus an allowance for starting, or pickup load, and a piping heat loss. The net output will show the actual useful heating effect produced. The ASME Code states that it is the gross heat output of the equipment that should be matched in specifying relief-valve capacity.

The acronym SBI stands for the Steel Boiler Institute. The nameplate data shown on SBI-rated boilers are not uniform, but the style or product number may be shown. The manufacturer's catalog will often show an SBI rating and an SBI net rating. The SBI rating tends to show the sum of SBI net ratings and 20 percent extra for piping loss, not including the pickup allowances noted under IBR ratings. Thus, it is difficult to obtain the true gross output to determine safety relief capacity from these data. But the SBI does require the number of square feet of heating surface to be stamped on the boiler. With this, the ASME rule of minimum steam safety-valve capacity in pounds per hour per square foot of heating surface is used.

EDR stands for equivalent direct radiation. Specifically it refers to equivalent square feet of steam radiation surface. It is further defined as a surface which emits 240 Btu/hr with a steam temperature of 215°F at a room temperature of 70°F. With hot-water heating, the value of 150 Btu/hr is used with a 20°F drop between inlet and outlet water. This term is used by architects and heating engineers in determining the area of heat-transfer equipment required to heat a space. Thus boiler capacity is obtained indirectly from a summation of the EDRs.

The following ratings are also often noted on heating boiler specifications.

American Gas Association rating. This rating method is used by the American Gas Association (AGA) and is applied to boilers designed for gas firing. The rating is expressed as maximum boiler output in Btu per hour, and it reflects 80 percent of the AGA-approved input rating as determined by performance tests described in the "American Standard Approval Requirements for Central Heating Appliances." For all practical purposes, AGA output ratings are equivalent to gross SBI and gross IBR ratings.

Mechanical Contractors Association rating. The Mechanical Contractors Association (MCA) of America (formerly the Heating, Piping, and Air Conditioning Contractors National Association) has adopted methods for rating boilers that are expressed on a net-load basis in square feet of EDR of steam.

The MCA has also adopted a Testing and Rating Code for Boiler-Burner Units which they apply to the rating of commercial sizes of steel heating boiler units fired with oil or gas fuel. This code allows a higher rating than is permissible under the SBI Code. A gross output is established with certain limiting factors applying to flue-gas temperature, carbon dioxide, efficiency, and quality of steam. This output is divided by 1.5 to determine the net MCA rating.

American Boiler Manufacturers Association rating. This rating method, developed by the Packaged Firetube Branch of the American Boiler Manufacturers Association (ABMA), is generally subscribed to by manufacturers of packaged boilers and by a few manufacturers of steel firebox and cast-iron boilers. The ratings are established by performance tests in accordance with the ASME Power Test Code for

Steam Generating Units and are usually expressed as maximum guaranteed Btu output at the outlet nozzle or similar output rating.

Classification by System Application

Boiler system designations will usually provide an immediate idea of capacity, pressures, and temperatures that will be required. Fuel to be used is another important designation, as is the value of the plant.

Systems can be grouped by the following applications:

1. Steam-heating system.

2. Hot-water heating system.

3. High-pressure steam process system.

4. Steam-electric power generation, using fossil fuels.

5. Steam-electric power generation, using nuclear fuel.

6. Systems using a different working fluid than water, such as dowtherm for high temperature, but low-pressure, process use. These fired systems are referred to by the ASME Code as organic fluid heater systems. (See Chap. 4.)

Steam-heating boiler systems (Fig. 1.6). Steam-heating boilers are usually low-pressure units of cast-iron or steel construction, although high-pressure steel boilers may also be used for large buildings or for large, complex areas. Usually if this is done, pressure-reducing valves in the steam lines lower the pressure to the radiators, convectors, or steam coils. The term *steam heating* also generally implies that all condensate is returned to the boiler in a closed-loop system. The maximum pressure allowed on a low-pressure steam-heating boiler is 15 psig.

Cast-iron boilers for steam use are limited to a maximum working pressure (MWP) of 15 psig by the ASME Heating and Boiler Code. Cast-iron boilers are specifically restricted by the ASME Code, Section IV, to be used exclusively for low-pressure steam heating. If they were used for process work, this usually would mean heavy-duty service of continuous steaming and heavy makeup of fresh cold water. This will cause rapid temperature changes in a cast-iron boiler, resulting in cracking of the cast-iron parts. Thus the Code restricts their use to steam-heating service only.

Steam-heating systems use *gravity* or *mechanical condensate-return systems*. Their differences are as follows. When all the heating elements (such as radiators, convectors, and steam coils) are located above the boiler and no pumps are used, it is called a *gravity return,* for all the condensate returns to the boiler by gravity. If traps or

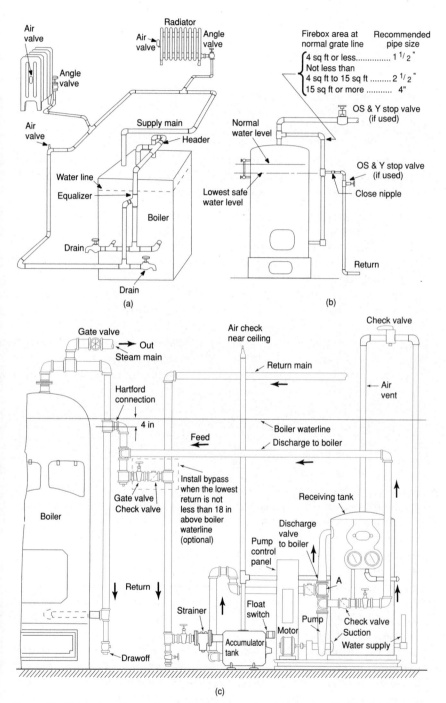

Figure 1.6 Steam-heating systems. (*a*) One-pipe air vent system. (*b*) Hartford return-pipe loop. (*c*) Vacuum-return pipe system.

pumps are installed to aid the return of condensate, the system is called a *mechanical return system*. In addition to traps, this system usually includes a condensate tank, a condensate pump, or a vacuum tank or vacuum pump (Fig. 1.6c).

ASME Section 4 protective devices required. As a result of several serious low-pressure steam-heating boiler explosions in the past, the ASME now requires redundancy controls for boilers with input ratings over 200,000 Btu/hr. These boilers are operated automatically with practically no operator attendance and only spot-checks made by the owner or a maintenance person. This is the reason the ASME requires the following for steam-heating boilers:

1. Each steam-heating boiler must have a steam pressure gauge with a scale in the dial graduated to not less than 30 psi nor more than 60 psi. Connections to the boiler must be not less than ¼-in. standard pipe size; but if steel or wrought-iron pipe is used, it should be not less than ½ in.
2. Each steam-heating boiler must have a water gauge glass attached to the boiler by valve fittings not less than ½ in. and with a drain on the gauge glass not less than ¼ in. The lowest visible part of the gauge glass must be at least 1 in. above the lowest permissible water level as stipulated by the boiler manufacturer.
3. Two pressure controls are required on automatically fired steam-heating boilers:
   a. An operating-pressure cutout control that cuts off the fuel supply when the desired operating pressure is reached.
   b. An upper-limit control set no greater than 15 psi which backs the operating-pressure limit control so that the fuel is shut off when the operating-pressure control does not function.
4. An automatically fired steam-heating boiler must have a low-water fuel cutoff located so that the device will cut off the fuel supply when the water level drops to the lowest visible part of the water gauge glass. Low-water fuel cutoffs must be connected to the boiler with nonferrous tees on Y's not less than ½-in. pipe size and must also have ¾-in. drains if embodying a chamber for the low-water fuel-cutoff device, so that the chamber and connected piping can be flushed of sludge periodically. This drain also permits testing of the low-water fuel cutoff as the level in the chamber drops during blowdown.
5. Each steam-heating boiler must have at least one safety valve of the spring-loaded pop type, adjusted and sealed to discharge at a pressure not greater than the maximum allowable pressure of the boiler. No safety valve can be smaller than ½ in. or greater than 4½ in. The capacity of the safety valves must exceed the output rating

in pounds per hour of the boiler, but in no case should the capacity be less, so that with the fuel-burning equipment firing at maximum capacity, the pressure cannot rise 5 psi above the stamped maximum allowable pressure of the boiler.

6. All electric control circuitry on automatically fired steam-heating boilers must be positively grounded and operate at 150 volts (V) or less. The wiring system must include a grounded neutral as well as equipment grounding.

7. Automatically fired steam-heating boilers must be equipped with flame safeguard safety controls as mentioned in the controls for hot-water-heating boilers.

Stop valves on the steam supply line are not required for a single-boiler installation that is used for low-pressure heating, if there are no other restrictions in the steam and condensate line and all condensate is returned to the boiler. But if a stop valve (or trap) is placed in the condensate-return line, a valve is required on the steam supply line. A stop valve is required on the steam supply line where more than one heating boiler is used on the same steam supply system and also on the condensate-return line to each boiler.

Hot-water systems.   There are three general classes of hot-water systems: *hot-water supply systems* for washing and similar uses, *space-heating systems* of the low-pressure type, often referred to as building heating systems (see Fig. 1.7), and *high-temperature high-pressure water systems,* also referred to as *supertherm systems,* operating at temperatures of over 250°F and pressures of over 160 psi. (See Chap. 4.)

Both the hot-water-heating system and the high-temperature hot-water systems require some form of *expansion tank* in order to permit the water to expand as heat is supplied, without a corresponding increase in pressure. A common problem of hot-water-heating systems is that expansion tanks lose their air cushion, so that the water system can no longer expand without raising the pressure of the system. If this problem is neglected, pressure can build up to the point where the relief valve may open and dump water in the property. Thus periodic checking of the pressure and possibly draining of the expansion tank is necessary to re-establish the air cushion.

Protective devices for hot-water-heating systems.   The ASME Heating Boiler Code requires some minimum protective devices on hot-water-heating boiler systems. Among these are the following:

1. A pressure or altitude gauge is required on the hot-water boiler with a scale on the dial graduated to not less than 1½ times nor more than 3 times the pressure at which the relief valve is set.

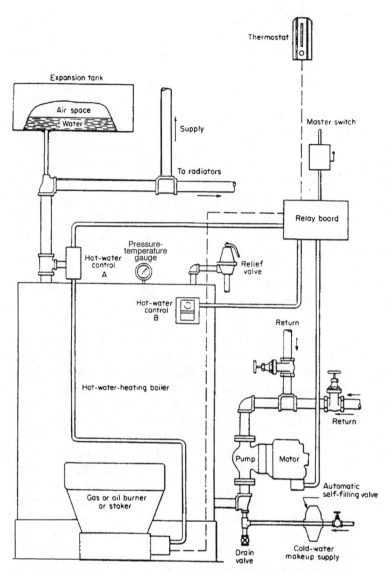

Figure 1.7 Components needed for a hot-water-heating boiler system.

2. A thermometer gauge is needed on the hot-water boiler that is located and connected so that it can be read when the pressure or altitude on the boiler is noted. Graduation of the thermometer must be in degrees Fahrenheit, and the thermometer must be located so that the water temperature in the boiler is measured at or near the outlet of the heated hot water.

3. Two temperature controls are required in automatically fired hot-water boilers:

   a. An operating limit control that cuts off the fuel supply when the water temperature reaches the desired operating limit.

   b. An upper-limit control that backs up the operating-limit control and cuts off the fuel supply. This upper-limit control is set at a temperature above the desired operating temperature, but must be set so that the water temperature cannot exceed 250°F at the boiler outlet.

4. A low-water fuel cutoff is required on automatically fired hot-water boilers with heat inputs greater than 400,000 Btu per hour (Btu/hr). It must be installed so that it cuts off the fuel when the water level drops below the safe permissible water level established by the boiler manufacturer.

5. All electric control circuitry on automatically fired hot-water boilers as well as on steam-heating boilers must be positively grounded and operated at 150 V or less. The wiring system must include a grounded neutral as well as equipment grounding.

6. A hot-water-heating boiler must be equipped with spring-loaded ASME-approved relief valves set at or below the maximum stamped allowable pressure of the boiler. The minimum size of valve is ¾ in., and the maximum permitted size is 4½ in. Capacity must be greater than the stamped output of the boiler, but in no case should the pressure rise more than 10 percent above the maximum allowable pressure if the fuel-burning equipment operates at maximum capacity.

7. Automatically fired hot-water-heating boilers and steam-heating boilers must also be equipped with flame safeguard safety controls that cut off the fuel when an improper flame (or combustion) exists by the burner. The ASME Code makes reference to other nationally recognized standards for further requirements. These usually include pilot and main-flame proving, as well as prefiring and postfiring purging cycles.

Officially rated ASME pressure-relief valves must be used in hot-water boilers. An officially rated ASME pressure-relief valve is stamped for its pressure setting and its Btu-per-hour relieving capacity. Also, it must be equipped with a manual test lever, must be spring-loaded, and must not be of the adjustable screw-down type.

Low water can occur in a hot-water-heating type of boiler for numerous reasons, such as the following: (1) Loss of water due to carelessness in (a) draining the boiler for repair or summer lay-up without eliminating the possibility of firing, (b) drawing hot water from the boiler; (2) loss of water in the distribution system because of (a) leaks in the piping caused by expansion breakage or corrosion, (b) leaks in the boiler, (c) leaks through the pump or other operating equipment; (3) relief-valve discharge caused by overfiring; (4) closed or stuck city makeup line.

In addition to an ASME pressure safety relief valve, a low-water fuel cutoff for an automatic-fired boiler should be installed.

Steam systems for electric power generation. Most utility boilers used for electric power generation are of the supercritical or subcritical type. The steam generator is an important element in power generation. Figure 1.8 is a simplified flow diagram of a basic power plant. Its three most important components are the steam generator (boiler), shown at the left; the turbine generator set, shown coupled together at the right; and the condenser, located beneath the turbine. The principal element that ties together the three pieces of equipment is steam, often called the working medium produced by a high-pressure boiler. The steam travels in succession from the steam generator to the turbine to the condenser. The feedwater cycle, also shown in the diagram, completes this path by making the flow continuous from the condenser back to the boiler. Thus, at the high-temperature end of the cycle, the steam generator transfers heat energy from the fuel to heat energy in the form of superheated steam. The turbine then transfers the heat in the steam to do mechanical work and then to drive the generator which is coupled to the turbine. The generator, in turn, transforms this mechanical energy to electric energy.

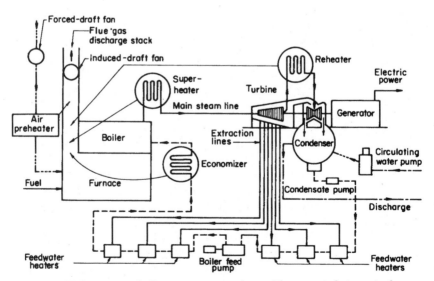

Figure 1.8 High-pressure boiler-steam power plant with connected steam turbogenerator, condenser, feedwater heaters, and boiler feedwater pump.

By adding auxiliaries and other components such as the heaters, superheaters, reheaters, and preheaters shown in Fig. 1.8 greater efficiencies for modern utility plants can be attained.

Nuclear-steam-power generation*. Steam for electric generation is also produced by heat from a *nuclear-powered reactor.* In the boiling-water reactor (BWR) system shown in Fig. 1.9*a,* the reactor vessel supports and contains the reactor core and supplies the necessary flow paths for fluid entering the core and steam leaving it. Water passing over the hot core generates steam, which travels through steam-water separators inside the reactor vessel and then through dryers, where the steam's moisture content is reduced. The steam then passes through the steam line directly into the turbine generator, as shown.

The pressurized water reactor system shown in Fig. 1.9*b* has a reactor vessel and core somewhat similar to the BWR type, but the fluid passes through the reactor (primary loop) and does not mix with that passing through the steam line on the turbine side. The heat is transferred from the reactor system to the turbine system in the steam generator. Actually, the water in the reactor and primary loop does not boil, even at 600°F, for example, because it is kept under very high pressure. In the steam generator, however, this water passes through tubes that are surrounded by water from the turbine loop,

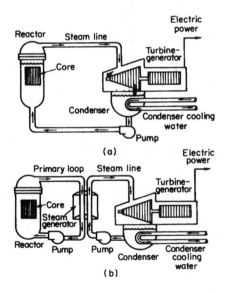

Figure 1.9 Two types of nuclear-power reactor systems. (*a*) Boiling-water reactor; (*b*) pressurized-water reactor.

---

*See also Chap. 5.

which is at a much lower pressure. To use the same example, a transfer of 500°F to the lower-pressure turbine loop is adequate to boil the water and produce the steam necessary to operate the turbine generator. Having given up most of the heat, the water in the primary loop is pumped back to the reactor to be reheated for use again.

High-pressure process systems. High-pressure process systems may use firetube or watertube boilers, depending on the pressure or capacity needed. The steam is used for mechanical drive turbine power to drive compressors, pumps, and similar equipment, or for process use to provide high pressure or temperature for the manufacturing cycle's needs.

Organizations Concerned with Standards

ASME Boiler Code. The ASME Boiler Code and the National Board of Boiler and Pressure Vessel Inspectors Inspection Codes are important source documents for legal requirements in the various states and municipalities that have adopted boiler safety laws. In addition to maintaining active boiler and pressure-vessel committees in order to keep the published Codes up to date with developing technology, the ASME issues to qualified manufacturers, assemblers, material suppliers, and nuclear power plant owners Code symbol stamps indicating that the manufacturer has received authorization from the ASME to build boilers and pressure vessels to the ASME Code.

A fundamental principle of the ASME Boiler and Pressure Vessel Code is that a boiler or pressure vessel, to be stamped ASME Code-designed, must receive third-party authorized inspection during construction for compliance with the prevailing Code requirements. Most third-party inspections are performed by authorized boiler and pressure-vessel inspectors who have appropriate experience and have passed a written examination in a jurisdiction. They must be employed either by the state or by an insurance company licensed to write boiler and pressure-vessel insurance in the jurisdiction where the boiler or pressure vessel is to be built, and in some cases the installation's location also must be considered. With uniform requirements for inspectors that have been prompted and implemented by the National Board of Boiler and Pressure Vessel Inspectors, a boiler or pressure vessel inspected by a properly credited National Board inspector will generally be accepted in all jurisdictions.

The manufacturer or contractor who wishes to build or assemble boilers or pressure vessels under an ASME certificate of authorization must first agree with an authorized inspection agency that Code inspections will be performed by the agency. This is usually arranged by both parties signing a contract with the inspection work done on a fee basis.

Figure 1.10*a* lists the stamps issued by the ASME for boilers, unfired pressure vessels, storage water heaters, power piping and safety valves. The ASME can provide details to manufacturers on what stamp may be required when considering manufacturing components as represented by these stamps.

Figure 1.10*b* lists the stamps required for power, nuclear components, heating and electric boilers with the ASME Code books, or Sections, considered necessary to comply with the ASME Code requirements for that stamp.

Nuclear Components

(a)

CODE SYMBOL STAMPS
A - Assembly of Power Boilers
M - Miniature Boilers
PP - Pressure Piping
S - Power Boilers

CODE BOOKS REQUIRED
SECTION I - Power Boilers
SECTION II - Material Specifications
Part A - Ferrous Materials Part B - Nonferrous Materials
SECTION V - Nondestructive Examination
SECTION IX - Welding and Brazing Qualifications
B31.1- Power Piping

CODE SYMBOL STAMPS
H - Heating Boilers
HLW - Lined Potable
Water Heaters
CODE BOOKS REQUIRED
SECTION II - Material Specifications
Part A - Ferrous Materials Part B - Nonferrous Materials
Part C - Welding Rods, Electrodes and Filler Metals
SECTION IV -Heating Boilers
SECTION IX - Welding and Brazing Qualifications

CODE SYMBOL STAMPS
E - Electric Boilers

CODE BOOKS REQUIRED
SECTION I - Power Boilers
SECTION II - Material Specifications
Part A - Ferrous Materials Part B - Nonferrous Materials
Part C - Welding Rods, Electrodes and Filler Metals

N - TYPE CERTIFICATES OF
AND AUTHORIZATION
CERTIFICATES OF ACCREDITATION
N, NA, NPT
CODE BOOKS REQUIRED
SECTION III - Rules for Construction of
Nuclear Power Plants
Components Subsection NCA-General Requirements for
Division 1 and Division 2 Appendices

SECTION V - Nondestructive Examination
SECTION IX - Welding and Brazing Qualifications

In addition, applicatants must have other subsections
of Section III depending upon scope of certificates as follow:

Subsection NB - Class 1
Components
Subsection NC - Class 3
Components
Subsection ND - Class 3
Components
Subsection NE - Class MC
Components
Subsection NF - Component
Supports
Subsection NG - Core Support
Structures
Division 2: Codes for Concrete
Reactor Vessels and Containers

Figure 1.10 (*a*) Applicable boiler and nuclear component ASME Code stamps. (*b*) Sections of Boiler Code books required per ASME Code stamp.

National Board of Boiler and Pressure Vessel Inspectors. The National Board of Boiler and Pressure Vessel Inspectors is composed of chief inspectors of states and municipalities in the United States and Canadian provinces. This organization has established criteria for boiler inspectors' experience requirements, the promotion and conductance of uniform examinations, and testing that are used by the jurisdictions. The National Board issues commissions to inspectors passing an NB examination, which are accepted on a reciprocal basis by most jurisdictions, thus providing a "portability" feature to a credential.

The NB organization also issues a stamp, called the "R" stamp, for organizations wishing to be certified as Code repairers. This also applies to safety valve repairs and nuclear pressure vessel repairs, which also merit a separate stamp from the NB. NB inspectors must obtain a basic NB Certificate of Competency, which qualifies the inspector to perform in-place field inspections for a jurisdiction upon obtaining a jurisdictional "commission" to make these inspections of boilers and pressure vessels.

NB inspectors who perform third-party inspections at manufacturer's or fabricator's facilities, termed "shop inspection," must pass another examination above the Certificate of Competency in order to become designated by an Authorized Inspection Agency as an *Authorized Inspector* for Code inspection work during a boiler or pressure vessel's fabrication. This is termed an NB commission with an "A" endorsement. A similar program exists for NB inspectors doing nuclear pressure vessels work. Special NDT and Quality Assurance tests must be passed before the NB commission can receive an "N" endorsement.

The NB has an extensive list of publications and forms in relation to boiler and pressure vessel safety. Their address is

National Board of Boiler and Pressure Vessel Inspectors
1055 Crupper Ave.
Columbus, OH 43229

Employers of NB-commissioned inspectors must be qualified as Authorized Inspection Agencies. These are jurisdictional bodies or licensed insurance companies. The inspector must be employed by such an agency for financial responsibility reasons under the present Code rules. Thus, there are three types of commissioned inspectors. These inspectors make the legal inspections and reports to a jurisdiction that a boiler or pressure vessel is safe or unsafe to operate or that it requires repairs before it can be operated:

1. State, province, or city inspectors see that all provisions of the boiler and pressure-vessel law, and all the rules and regulations of the

jurisdiction, are observed. Any order of these inspectors must be complied with, unless the owner or operator petitions (and is granted) relief or exception.

2. Insurance company inspectors who are qualified to make jurisdictional Code inspections, and if commissioned under the law of the jurisdiction where the unit is located, can also make the required periodic reinspection. As commissioned inspectors, they require compliance with all the provisions of the law and rules and regulations of the authorities. In addition, they may recommend changes that will prolong the life of the boiler or pressure vessel.

3. Owner-user inspectors are employed by a company to inspect unfired pressure vessels on their premises only and not for resale by such a company. They also must be qualified under the rules of any state or municipality which has adopted the Code. Most states do not permit this group of inspectors to serve in lieu of state or insurance company inspectors.

Most areas of the United States and all jurisdictions in Canada require that high-pressure boilers be subjected to periodic inspection by a jurisdictionally recognized inspector. In most jurisdictions, this consists of annual internal inspection of power boilers and biennial inspection of heating boilers and usually of pressure vessels for those states that have adopted laws on low-pressure boilers or unfired pressure vessels. If the results prove satisfactory, the jurisdiction issues an inspection certificate, authorizing use of the vessel for a specific period.

Figures 1.11a and b and 1.12 list the states, cities, and counties in the United States and Canadian provinces that have some form of installation and periodic reinspection requirements on boilers and some unfired pressure vessels. These laws vary a great deal. For example, on low-pressure boilers, reinspection requirements may be limited to installations located in places of public assembly. Others include all heating boilers, except those located in private residences or in apartment houses with six families or less. Therefore, local or state laws should be checked for more specific requirements.

Other approval organizations. These organizations are concerned with all phases of potential fire hazards or electrical safety. Thus their labels will appear on fuel trains employed in boiler operation and on electrical controls and wiring. Many jurisdictional fire Codes make reference to these approval bodies' labels; therefore, these are important when installing boilers.

Underwriters Laboratory, UL, is active in approving electrical equipment for different applications to established standards of safety, and if satisfactory, applies the UL label.

Figure 1.11 Table (a) United States; (b) Canada

State or province	Accept insurance company reports (X = yes)	Require inspection for		
		High-pressure boilers	Low-pressure boilers	Unfired pressure vessels
(a) United States				
Alaska	X	X	X	X
Alabama	—	X	X	X
Arizona	X	X	X	—
Arkansas	X	X	X	X
California	X	X	—	X
Colorado	X	X	X	X
Connecticut	X	X	X	—
Delaware	X	X	X	X
District of Columbia	X	X	X	X
Florida	X	X	X	—
Georgia	X	X	X	X
Hawaii	X	X	X	X
Idaho	X	X	X	X
Illinois	X	X	X	X
Indiana	X	X	X	X
Iowa	X	X	X	X
Kansas	X	X	X	—
Kentucky	X	X	X	X
Louisiana	X	X	X	—
Maine	X	X	—	X
Maryland	X	X	X	—
Massachusetts	X	X	X	X
Michigan	X	X	X	—
Minnesota	X	X	X	X
Mississippi	X	X	X	X
Missouri	X	X	X	X
Montana	X	X	X	—
Nebraska	X	X	X	X
New Mexico	No law	—	—	—
Nevada	X	X	X	X
New Hampshire	X	X	X	X
New Jersey	X	X	X	X
New York	X	X	X	X
North Carolina	X	X	X	—
North Dakota	X	X	X	—
Ohio	X	X	X	X
Oklahoma	X	X	X	X
Oregon	X	X	X	X
Pennsylvania	X	X	X	X
Rhode Island	X	X	X	X
South Carolina	X	X	—	—
South Dakota	No law	—	X	X
Tennessee	X	X	X	X
Utah	X	X	X	X
(a) United States (Continued)				
Vermont	X	X	X	X
Virginia	X	X	X	X
Washington	X	X	X	X
West Virginia	X	X	—	—
Wisconsin	X	—	X	X
Wyoming	No law	—	—	—
(b) Canada				
Alberta	No	X	X	X
British Columbia	No	X	X	X
Manitoba	No	X	X	X
New Brunswick	No	X	X	X
Newfoundland	No	X	X	X
Northwest Territories	X	X	X	X
Nova Scotia	X	X	No	X
Ontario	X	X	X	X
Quebec	X	X	X	X
Saskatchewan	No	X	X	X
Yukon Territory	No	X	X	X
Prince Edward Island	X	X	X	X

Figure 1.11 (a) States and Canadian Provinces having boiler and pressure-vessel reinspec-

City or county	Accept insurance company reports (X = yes)	High-pressure boilers	Low-pressure boilers	Unfired pressure vessels (UPV)
Albuquerque, N.Mex.	X	X	X	—
Buffalo, N.Y.	X	X	X	—
Chicago, Ill.	No	X	X	—
Dearborn, Mich.	X	X	X	X
Denver, Colo.	No	X	X	X
Des Moines, Iowa	X	X	X	—
Detroit, Mich.	UPV only	X	X	X
E. St. Louis, Mich.	No	X	X	X
Greensboro, N.C.	X	X	X	X
Kansas City, Mo.	X	X	X	X
Los Angeles, Calif.	X	X	X	X
Memphis, Tenn.	X	X	X	X
Miami, Fla.	X	X	X	X
Milwaukee, Wisc.	X	X	X	X
New Orleans, La.	X	X	X	X
New York City, N.Y.	X	X	X	—
Oklahoma City, Okla.	X	X	X	—
Omaha, Neb.	X	X	X	—
Phoenix, Ariz.	X	X	X	X
St. Louis, Mo.	X	X	X	X
San Francisco, Calif.	X	X	X	X
San Jose, Calif.	X	X	X	—
Seattle, Wash.	X	X	X	X
Spokane, Wash.	X	X	X	X
Tacoma, Wash.	X	X	X	X
Tampa, Fla.	X	X	X	X
Tucson, Ariz.	X	X	X	X
Tulsa, Okla.	No	X	X	X
University City, Mo.	No	X	X	—
White Plains, N.Y.	X	X	X	—
Arlington County, Va.	X	X	X	—
Dade County, Fla.	X	X	X	X
Fairfax County, Va.	X	X	X	X
Jefferson Parish, La.	X	X	X	X
St. Louis County, Mo.	X	X	X	X

Figure 1.12 Cities and counties having boiler and pressure-vessel reinspection laws.

Factory Mutual Laboratories, FM, approves equipment submitted by manufacturers and also approves final installations, such as a fuel-burning train, if the installation satisfies their required standards.

Industrial Risk Insurers, IRI, is a stock company organization that has testing laboratories and also inspects each insured location for their approval for the IRI label. This includes satisfying their standards for burner fuel trains on boilers, ovens, dryers, furnaces, and flame-safeguard burner controls. The organization is also involved with labeling fire protection equipment, as is FM, for sprinklers and fire alarm systems.

American Gas Association, Inc., AGA, labels gas-burning equipment that satisfies their standards. For example, on fuel-burning equipment with over 400,000 Btu/hr input, the standard requires electronic flame safeguard controls, including trial for ignition and trial for main-flame ignition.

As can be noted, these approval bodies stress safety before they apply their label to equipment or a system that has a combustion explosion or fire hazard.

ISO 9000 certification.   This is an international quality control management series of standards published in 1987 by the International Organization for Standardization, or ISO. European companies have been the leaders in adopting this quality control management system that establishes a quality control program, a system manual, and the means or checkpoints for implementing the requirements. ISO 9000 parallels the ASME Code requirements in many instances for boilers and pressure vessels and in most instances for nuclear component documentation.

The European Common Market was the impetus for promoting a standardized quality control management system, but U.S. companies involved in international operations are also beginning to consider ISO registration of their quality control system, because purchasers of their equipment or services are specifying ISO 9000 series registration as part of contracts. A company that wants to be certified to these standards first selects the system model it requires, installs or implements the model, and prepares a quality control manual, which is then reviewed by an independent auditor. This review notes whether the quality control manual follows the guidelines of the ISO 9000 series of standards. See Table 1.1. The auditing team then checks the system on-site for the implementation of the quality control manual and for management's commitment. The auditing team may then recommend certification.

As can be noted from Table 1.1, an organization must select the series it wishes to be certified to. This could include its total operation, or it may select particular areas for certification. Recognized registrars or auditors prepare a report of their findings to a balanced committee made up of similar industry representatives. The committee decides if an organization's application for accreditation is approved, and the registrar then issues a certificate of registration to the applicant. This details the scope of activity of the applicant's program, and to which 9000 series it applies.

Periodic reaudits are made by the outside registrar in order to confirm that the ISO 9000 series of requirements are being maintained.

In the United States, auditing groups are listed by the federal government's standards office, the National Institute of Standards and

TABLE 1.1   The ISO 9000 Series of a Quality Control Management System

ISO 9000 checklist	Management responsibility, quality system principles, material control and traceability, inspection and test procedures, measuring and test equipment adequacy, handling, storage and delivery, document control, quality control, training, statistical methods used, internal audit procedures, marketing quality, purchasing control, process control, production control, corrective action procedures, research and development control, after-sales service, product safety and liability.
ISO 9001 activity	Design, production, installation, and servicing a product.
ISO 9002 activity	Applies only to production and installation.
ISO 9003 activity	Applies only to final inspection and testing.
ISO 9004 activity	Applies to the quality management and system elements needed to develop and implement a quality system for the activity. This includes determining the extent to which each system element is applicable to the activity.

Testing (NIST), in Washington, D.C. ISO also has a list of certification bodies, which provides the qualification and expertise area of the registrar.

Environmental regulations. Federal, state, and city regulations affect boiler plant operators. Fuel-burning systems for boilers and nuclear energy systems are required to be designed and operated so that air, water, and waste disposal from these plants will have minimal effects on the environment. Federal regulations that may merit review by boiler plant operators include the Clean Air Act, the Clean Water Act, and regulations concerning hazardous waste disposal, spills and releases, PCBs, underground storage tanks with harmful liquids or gases, and asbestos.

Boilers using fossil fuels must be operated to control the amount of sulfur dioxide and nitrous oxide emitted into the air. Continuous emission monitoring of these pollutants is now required on large boilers. Monitoring of radiation and thermal discharge into rivers or streams is required in nuclear facilities because nuclear plants produce more thermal discharge per unit of output than do fossil-fuel-burning plants. As a result, during hot weather, some nuclear plants must limit their load in order to avoid violating temperature limits on thermal discharge imposed by regulatory agencies.

Asbestos pollution and disposal may be a problem in boiler plants during any repair activities. OSHA has established a permissible limit of 0.1 fiber per cubic centimeter for an 8-hour time-weighted average.

There is a long list of OSHA safety rules that can affect boiler plant operation; these are detailed in OSHA 29 CFR 1910 regulations.

Supervisors in boiler plants should be familiar with these rules and regulations, because they will assist them in maintaining a safe working environment. For example, confined-space work rules of OSHA include:

1. First evaluating temperature and oxygen levels in a confined space before entry is permitted.
2. Providing emergency help procedures for a person within a confined space.
3. Posting precautions near a confined-space entry point.

Legal requirements on boilers and nuclear power plant equipment no longer are limited to establishing safe construction codes. They have been expanded into requirements on controls, on devices to prevent furnace explosions, and on measurements to limit air pollution and radioactive contamination. Owners and operators must periodically review their operation and maintenance practices in order to make sure they comply with these additional legal requirements of the jurisdiction in which the equipment is located.

Questions and Answers

1 How would you define a boiler?

ANSWER: A boiler is a closed pressure vessel in which a fluid is heated for use external to itself by the direct application of heat resulting from the combustion of fuel (solid, liquid, or gaseous) or by the use of electricity or nuclear energy.

2 What is a steam boiler?

ANSWER: A steam boiler is a closed vessel in which steam or other vapor is generated for use external to itself by the direct application of heat resulting from the combustion of fuel (solid, liquid, or gaseous) or by the use of electricity or nuclear energy.

3 What is a high-pressure steam boiler?

ANSWER: A high-pressure steam boiler generates steam vapor at a pressure of more than 15 psig. Below this pressure it is classified as a low-pressure steam boiler.

4 Define a miniature high-pressure boiler.

ANSWER: According to Section I of the ASME Boiler and Pressure Vessel Code, a miniature boiler is a high-pressure boiler which does not exceed the following limits: 16-in. inside diameter of shell, 5-ft^3 gross volume exclusive of casing and insulation, and 100-psig pressure. If it exceeds any of these limits, it is a power boiler. Most states follow this definition.

5   What is a power boiler?

ANSWER: A power boiler is a steam or vapor boiler operating above 15 psig and exceeding the miniature-boiler size. This also includes hot-water-heating or hot-water-supply boilers operating above 160 psi or 250°F.

6   Define a hot-water-heating boiler.

ANSWER: A hot-water-heating boiler is a boiler used for space hot-water heating, with the water returned to the boiler. It is further classified as low-pressure if it does not exceed 160 psi or 250°F. But if it exceeds any of these, it becomes a high-pressure boiler.

7   What is a hot-water-supply boiler?

ANSWER: A hot-water-supply boiler furnishes hot water to be used externally to itself for washing, cleaning, etc. If it exceeds 160 psi or 250°F, it becomes a high-pressure power boiler.

8   What is meant by a boiler horsepower?

ANSWER: A boiler horsepower (boiler hp) is defined as the evaporation into dry saturated steam of 34.5 lb/hr of water at a temperature of 212°F. Thus one boiler hp by this method is equivalent to an output of 33,475 Btu/hr. In the past it was commonly taken as 10 ft^2 of boiler heating surface.

9   The symbol NB is often noted on boilers, with a number following it. What does this stand for?

ANSWER: The acronym NB stands for National Board of Boiler and Pressure Vessel Inspectors. It means that the boiler's design and fabrication were followed in the shop by an NB-commissioned inspector, including the witnessing of the hydrostatic test and signing of data sheets required by the ASME.

10   What is meant by heating surface in a boiler?

ANSWER: This is the (fireside) area in a boiler exposed to the products of combustion. This area is usually calculated on the basis of areas on the following boiler-element surfaces: tubes, fireboxes, shells, tube sheets, and projected area of headers. See later chapters on safety-valve calculations.

11   Define the terms IBR rate, SBI-rated, and EDR.

ANSWER: The acronym IBR stands for the Institute of Boiler and Radiator Manufacturers, which rates the output of cast-iron boilers in net and gross output in Btu per hour. Gross output is further defined as the net output plus an allowance for starting, or pickup load, and a piping heat loss.

The acronym SBI stands for the Steel Boiler Institute. The SBI boiler-output rating tends to show the sum of SBI net ratings in Btu per hour or pounds per hour, plus 20 percent extra for piping loss, not including the pickup allowances noted under IBR ratings. The SBI requires the number of square feet of heating surface to be stamped on the boiler.

The acronym EDR stands for equivalent direct radiation. It specifically refers to equivalent square feet of steam radiation surface. It is further defined as a surface which emits 240 Btu/hr with a steam temperature of 215°F at a room temperature of 70°F. With hot-water heating, the value of 150 Btu/hr is used with a 20°F drop between inlet and outlet water. This term is used by architects and heating engineers in determining the area of heat-transfer equipment required to heat a space.

12 Name three terms used to indicate boiler output.

ANSWER: These three terms are often used with pressure and temperature listings:

1. For steam boilers, the actual evaporation is in *pounds per hour.* For hot-water boilers, the *Btu-per-hour* outputs for the given pressures and temperatures are stamped on the boiler. Today this is the preferred method.
2. Square feet of heating surface.
3. Boiler horsepower.

13 What is a supercritical once-through boiler?

ANSWER: This is a boiler which operates above the supercritical pressure of 3206.2 psia and 705.4°F saturation temperature and which has no fluid recirculation when operating at full pressure and temperature. The fluid is brought up to pressure and temperature in series-connected fluid passes; thus the term *once-through* is applied.

14 Above what pressure and temperature does a hot-water boiler system become a high-temperature hot-water boiler system?

ANSWER: A hot-water boiler becomes a high-temperature hot-water (HTHW) system when the water temperature exceeds 250°F and the pressure is over 160 psi.

15 What is the main reason for using thermal fluids such as dowtherm and glycol in processes requiring heat?

ANSWER: The thermal fluids are used to get high temperatures at low pressures which may be difficult to obtain with ordinary steam boiler equipment. Note that a pressure of 3206.2 psia is needed to obtain a saturation steam temperature of 705.4°F. This temperature can be obtained with some thermal fluids at a pressure below 50 psi.

16 Name three pressure-limiting devices needed on a steam-heating boiler per ASME Code requirements.

ANSWER: The three pressure-limiting devices are:

1. An operating-pressure cutoff switch that automatically cuts off the fuel supply when the desired pressure is reached

2. An upper-limit pressure control switch, set no greater than 15 psi, that automatically cuts off the fuel supply when the upper pressure is reached

3. At least one spring-loaded pop-type safety valve, set and sealed to discharge at a pressure not greater than the maximum allowable working pressure of the boiler and with a capacity sufficient that the pressure on the boiler cannot rise 5 psi above the stamped maximum allowable pressure on the boiler

17 What is the purpose of the water-level control on a steam boiler?

ANSWER: To maintain the proper water level in the boiler by the control starting the boiler feed pump when there is a demand for water by the lower limit level switch, and by shutting the pump off when the proper water level is reached (upper limit).

18 What should be done if on starting a boiler it is noted that the water level is above the gauge glass?

ANSWER: The boiler should be shut down, and the boiler drained to the proper level. If the level does not drop, it is an indication of (1) leaking tubes or (2) malfunctioning level controller, and the boiler should be secured so that necessary repairs can be made.

19 What is the purpose of the low-water fuel cutoff?

ANSWER: To shut the fuel off to the burner before the water in the boiler drops below the safe permissible level, and thus prevent overheating damage to the boiler.

20 If the water drops below the bottom of the sight glass, and the burner does not shut off, what steps should be taken?

ANSWER: Shut off the fuel valve to the burner, and shut off the electricity to the boiler, assuming this is a small automatic-type steam boiler. Let the boiler cool off, and then check the controls in order to determine why the level controller, and the low-water fuel cutoff did not function as designed.

21 How often should the low-water fuel cutoff be tested on automatic boilers?

ANSWER: At least once per day, and each shift if more than one shift of operation is in effect.

22 Why should the water column and the low-water fuel cutoff chamber be flushed daily by draining the respective devices?

ANSWER: This will keep the water column and the low-water fuel cutoff chambers free of mud and sediments, thus allowing the gauge glass to accurately display the water level, and will permit the low-water fuel cutoff to operate when needed.

23   Why should air openings to a boiler room be kept clean and clear of any obstructions that could hinder airflow into the room?

ANSWER: Both conditions could cause a lack of air to flow into the boiler room, and this will result in possible incomplete combustion to occur in the boiler. This can create air-fuel ratio problems with the burner control and also create a possible health hazard with the formation of carbon monoxide from incomplete combustion.

24   What is the output in pounds per hour and Btu per hour of a boiler rated at 750 hp?

ANSWER: Output = 750(34.5) = 25,875 lb/hr
Output = 750(33,475) = 25,116,250 Btu/hr

25   Calculate boiler efficiency, using the steam generated versus the fuel consumed. You are given that for one calendar month of regular operation, the coal consumed is 682,000 lb and the steam generated is 6.4 million lb at 179 psig and superheated to a total temperature of 520°F. Assume that the heat content of coal as fired is 13,260 Btu/lb, and that the feedwater temperature is 208°F.

ANSWER: First, the actual evaporation per pound of coal fired is

$$\frac{6,400,000}{682,000} = 9.40 \text{ lb steam/lb coal}$$

The absolute steam pressure is

$$179 + 15 = 194 \text{ lb abs}$$

Then, the steam tables show that the total heat of 1 lb of steam at 194-lb absolute pressure and 520°F is 1280.4 Btu. With the feedwater temperature at 208°F, its heat content above water at 32°F is merely 208 − 32 = 176 Btu. Thus the heat put into each pound of steam produced by the boiler is 1280.4 − 176.0 = 1104.4 Btu. The heat put into 9.4 lb of steam will be 10,381 Btu. Then, boiler efficiency equals heat put into 9.4-lb steam divided by the heat in 1 lb of coal, or

$$\frac{10,381}{13,260} \times 100\% = 78.3\% \text{ efficiency}$$

26   (a) Feedwater at 300°F is converted into superheated steam at 800 psia and 900°F. How many Btus are needed to do this?
(b) If this was a boiler burning coal with 13,600 Btu/lb heat value and with evaporation being 9.4 lb of water per pound of coal, determine boiler efficiency using output-input method.

ANSWER:

(a) From the superheat steam table (Table 3 in Fig. 1.5), enthalpy at 800 psia and 900°F is 1455.4 Btu/lb. From Table 1, saturated liquid at

300°F has 269.6 Btu/lb enthalpy. Heat needed to convert the liquid into superheated steam is 1455.4 − 269.6 = 1185.8 Btu/lb.

(*b*) Output is thus 1185.8 × 9.4 = 11,140 Btus. Input is given as 13,600 Btus. Efficiency is 11,140/13,600 = 0.82 or 82%.

27 Why should a boiler not be drained while the fire side is hot?

ANSWER: This is necessary to avoid baking sludge on the tubes and other heating surfaces, and also to avoid high thermal stresses developing from too rapid cooling. This will prevent leakages at gasketed joints and tube rolls and similar seams affected by expansion and contraction.

28 What is a compound gauge?

ANSWER: A compound gauge is a pressure gauge that shows pressure on one-half of its dial, while the other half shows vacuum. It is used where the heating system may vary from a pressure to vacuum conditions.

29 What are the advantages in converting boiler controls and instrumentation to digital-based technology?

ANSWER: Digital-based instrumentation and controls permit finer tuning of input and output on a boiler system. Closer control of thermal efficiency is achieved by approaching set points to designed operating limits. Improved emission is also achieved by more precise control of the combustion process, such as air-fuel ratios, and also of any downstream cleanup process equipment, if so installed, as in coal-burning plants. Fewer personnel are required to operate the plant, because a central control room provides video displays of operating conditions for each loop of the boiler system. Appropriate warnings, alarms, and trips can be incorporated into the control system, usually employing appropriate computer technology.

30 How are diagnostic capabilities incorporated in modern boiler controls?

ANSWER: One manufacturer of moderate capacity boilers provides many diagnostic displays, such as: Boiler functions of input and output displayed on video screens, with the operator controlling any variable from set points by touch screen buttons or key pad interface; start-up or other operating problems displayed on the video screen with explanation as to cause. Typical situations displayed: water level too low, fan not operating, gas pressure too low, air curves to show the air required for the fuel being burned. The diagnostic display on the cause of a problem permits the operator to correct the problem at once instead of initiating a search for cause.

31 Differentiate between ISO 9000, 9001, 9002, 9003, and 9004.

ANSWER: The combined standards are designed to establish a quality control management system. ISO 9000 and 9004 define the terms and provide the system elements required to meet the applicable standard, while 9001, 9002, and 9003 present the various activity system models. ISO 9001 specifies quality assurance in design, production, installation, and servicing of a designated product. ISO 9002 applies only to production and installation. ISO 9003 applies to final inspection and testing. See Table 1.1 for more details on the ISO 9000 series of standards.

32 Briefly describe the ASME Code requirements in Section 1 (Power Boilers) for a quality control program for fabricating a Code boiler.

ANSWER: The quality control program required from the fabricator includes:

1. Management support by naming a quality control manager to implement the program
2. An organizational chart showing where quality control is to be applied in design, material selection, inspection, record documentation, and the check-off sheets to be applied at each step of the manufacturing cycle
3. Provisions for correcting defects per Code rules
4. Periodic checking of gauges, NDT equipment, and similar instruments
5. Maintaining forms and data of the recorded quality control performed for future reference
6. Maintaining records of welder and welding procedure qualification tests as well as records of NDT and heat treatment performed

33 What stamp is required on a field-assembled power boiler when it is assembled by a contractor and not the boiler manufacturer?

ANSWER: The contractor must be qualified by the ASME to stamp the assembled boiler with an "A" stamp.

34 Why are flat surfaces above a certain area stayed or braced in a boiler?

ANSWER: Pressure acting on flat surfaces that are not thick enough will bulge the plate into a spherical shape and possibly cause a rupture. The braces or stays strengthen the flat surfaces in order to avoid this.

35 Why should steam leaks in a boiler system be repaired as soon as possible?

ANSWER: All leaks should be immediately repaired to avoid further damage and also to prevent personal injury. Loss of steam from leaks is also energy wasted, which increases fuel costs to the plant.

36 What hazard can be created by opening a steam valve quickly?

ANSWER: The sudden change in pressure and temperature on connected piping may cause water hammer, vibration, and even pipe rupture at abrupt pipe run changes, such as elbows.

37 What are recognized Authorized Inspection Agencies per the ASME Boiler Code?

ANSWER: The Code designates Authorized Inspection Agencies as those inspection organizations employing inspectors qualified by written examination under the rules of a jurisdiction, consisting of a state or municipality of the United States or Canadian province that has adopted one or more sections of the Boiler Code, or of an insurance company authorized to write boiler and pressure vessel insurance in the jurisdictions.

38 What additional training must an NB-commissioned inspector obtain to merit the "A" endorsement to make manufacturer's shop inspection per NB rules?

ANSWER: The "A" endorsement requires successful completion of 30 hours of formal training beyond the NB Commission and an additional 80 hours of supervised on-the-job training in a manufacturer's shop.

39 What experience requirements must an individual employed regularly by an Authorized Inspection Agency have before being permitted to take the 2-day examination for an NB Certificate of Competency as an NB-commissioned inspector of boilers and pressure vessels?

ANSWER: The present requirements are practical experience in design, construction, maintenance, operation or inspection of high-pressure boilers and pressure vessels that must consist of the following:

1. ME degree plus 1 year of practical experience
2. Other engineering degree plus 2 years of practical experience
3. At least a high school education plus 3 years of practical experience

The candidate must also study the following ASME Code Sections: Section 1—Power Boiler; Section IV—Heating Boilers; Section IX—Welding Qualifications; Section V—Nondestructive Examination, and the National Board Inspection Code.

40 What power boiler is classified as a miniature boiler?

ANSWER: A high-pressure boiler that does not exceed any of the following: inside diameter of shell, 16 in.; 20 ft^2 of heating surface; 5 ft^3 gross pressure part volume; 100 psig working or allowable pressure.

41 Is gauge or absolute pressure used in the ASME Boiler Code for maximum allowable pressure for a boiler?

ANSWER: The ASME Boiler Code refers to gauge pressure, or the pressure above atmospheric pressure in psi, because this is what is read by operating people or users of boilers.

42 How does a section of the ASME Boiler Code become a legal requirement in a jurisdiction?

ANSWER: Usually a legislative body of a jurisdiction establishes the mandatory applicability of the Code rules on a section basis, such as Sections I, IV, and VIII, or even parts thereof, and then this becomes a legal requirement for a location within that jurisdiction. Jurisdictional enforcement machinery and additions or deletions to the ASME Codes may be adopted by the jurisdiction, such as state certificate of operation fees, and state inspection fees, which are not covered by the ASME Code.

2

Firetube Boilers

The firetube boiler is the most prevalent boiler used for heating and commercial and industrial process applications. The early boilers required extensive bracing. Boiler configurations are influenced by heat-transfer requirements so that as much of the heat released by a fuel may be extracted as material and economic considerations permit. The effect of shape is of immense importance to the strength and bracing requirements of a boiler. It is a well-known law of science that pressure of a fluid exerts itself in an equal amount in every direction. Following this law, an irregularly shaped vessel subjected to internal pressure always *tends* to be forced into a perfect spherical shape. The first tendency of an oval-shaped drum or shell would be to change its cross-sectional shape to a true circle.

With regard to these facts, one may look at a flat head of a vessel operated at above very low internal pressure and see that bracing is used. However, if a semispherical or dished head is used, the bracing is dispensed with. Heads for high-pressure boilers are practically always dished or elliptical in form. No braces are required for these heads.

Most firetube boilers have flat surfaces that require stay-bolting or staying with through stays or diagonal stays, because the plate would have to be too thick to resist the pressure imposed on it. Internal furnaces such as those in SM boilers are under compression and bending stress because the pressure inside the boiler is on the outside circumference and length of the furnace. This force tends to collapse the furnace, and if the furnace plate is not thick enough, stiffening of the furnace is also required. This is in the form of corrugated plate or stiffening rings placed along the length of the furnace.

Calculations of firetube boiler components are detailed in the chapter on Code Strength and Stress Calculations.

Figure 2.1 Four packaged 400-hp (13,800-lb/hr) SM boilers equipped to fire oil or gas supply steam for heating and process use in military installation. (*Courtesy Cleaver-Brooks Co.*)

Types and arrangement. Firetube boilers are classified into horizontal-return-tubular (HRT), economic or firebox-type, locomotive firebox-type, scotch-marine-type, vertical tubular, and vertical tubeless boilers. The HRT boiler now represents only about 5 percent of the boilers in service of the total firetube boilers still being operated. The scotch marine (SM) design (see Fig. 2.1) is the dominant firetube type for both heating and industrial process use, up to about 50,000-lb/hr capacity. Above this capacity watertube boilers are generally used.

General construction details. The firetube boiler has tube ends exposed to the products of combustion and has other flat surfaces that require staying with structural steel in order to avoid excessively thick plates. Tubes in all firetube boilers must be rolled and beaded (Fig. 2.2), or rolled and welded. If they are rolled and welded in high-pressure boilers, see Power Boilers, Section I, ASME Boiler Code for details on welds. Tubes are beaded to prevent the ends from being burned off by the hot gases in this area. Beading also increases heat transfer near the tube sheet and tube juncture. The edges of tube holes are chamfered about $\frac{1}{16}$ in. after the holes are drilled so that

Bead   Tube sheet   Seal weld

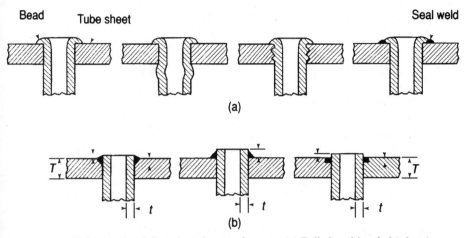

(a)

(b)

Figure 2.2   Code-permitted firetube tube attachments. (*a*) Rolled and beaded tubes in a firetube boiler. (*b*) Rolled and welded tubes in a firetube boiler. See Code for permissible tube extensions and weld dimensions.

there will be no sharp edges to cut into the tube when it is expanded. Firetube holes are finished $\frac{1}{16}$ in. larger in diameter than the outside diameter of the boiler tube so as to permit a tube with scale to be drawn from the tube sheet without damage to the tube-sheet hole.

Fire tubes are normally under external pressure; thus they may collapse but not burst. The biggest problems are loosening of tubes in the tube sheet; cracking, burning, and corrosion of tube ends; waterside pitting and corrosion leading to leakage; and fireside corrosion and pitting leading to leakage or pulling out of the tube sheet (as a result of poor rolling). Also scale buildup on the waterside leads to overheating and possible sagging and loosening in the tube sheet.

Cylinders or barrels of a drum or shell must be circular within a limit of 1 percent of the mean diameter. This also applies to drums or shells of watertube boilers. During inspection for overheating damage of drums or shells, this Code requirement can be used as a guide to whether the boiler still meets this requirement of out-of-roundness limits.

Firetube shell and drum thicknesses must be not less than the following per Code rules:

I.D. of shell or drum	Minimum thickness required
36 in. or under	0.25 in.
Over 36 to 54 in.	0.3125 in.
Over 54 to 72 in.	0.50 in.
Over 72 in.	0.56125 in.

Tube-sheet thicknesses for firetube boilers must be not less than the following:

I.D. of shell or drum	Minimum thickness required
42 in. or under	0.375 in.
Over 42 to 54 in.	0.4375 in.
Over 54 to 72 in.	0.50 in.
Over 72 in.	0.56125 in.

Flat surfaces are stayed with either stay bolts or stays. Stay bolts can be threaded, welded, and hollow. They are generally used to strengthen short spans, such as the inner and outer plates of water legs.

See Fig. 2.3a and b. The ends of the stay bolts, or stays screwed through the plate, must extend beyond the plate not less than two threads when installed, after which the ends must be riveted over without excessive scoring of the plate. They may also be fitted with threaded nuts, provided the stay bolts, or stays, extend through the nut.

If stay bolts are solid, 8 in. in length or less, and threaded, they must be drilled with telltale holes (Fig. 2.3a) at least 3⁄16 in. in diameter to a depth extending at least 1⁄2 in. beyond the inside of the plate. If the stay bolt is reduced in diameter between the ends, the telltale hole must extend 1⁄2 in. beyond the point where the reduction in diameter begins. For bolts over 8 in. in length, telltale holes are not required, nor if the stay bolt is attached in by welding and Code rules have been followed. See Fig. 2.3b. Since stay bolts usually break near

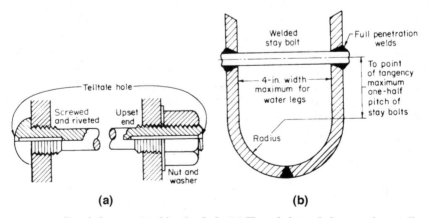

(a)   **(b)**

Figure 2.3 Stay bolts permitted by the Code. (a) Threaded stay bolts must have telltale holes to show if cracks exist on the inside of the stay bolt. (b) Welded stay bolt details. These types of stay bolts do not require telltale holes.

the plate supported, warning is given by water flashing from the tell-tale hole.

The stay bolts and stays must be inserted in countersunk holes through the plate, except for diagonal stays, which can be fillet-welded to the shell but not the head, provided the weld is not less than ⅜ in. and the fillet weld continues the full length of the stay. Stay bolts inserted by welding cannot project more than ⅜ in. beyond the plate exposed to products of combustion. The welding must be stress-relieved after it is completed. Radiographing (x-ray photograph) is not required.

Staying of tube-sheet segments. In firetube boilers the segment of the tube sheets above the top row of tubes requires staying. For this, there are three common methods. In a boiler that does not exceed 36-in. diameter or 100-psi working pressure, structural shapes such as angle irons or channel irons may be welded to the segment. The outstanding legs are proportioned to have sufficient strength in bending to resist the pressure load.

For boilers exceeding 36-in. diameter or 100-psi working pressure, the flat segment of the tube sheets requires staying either by diagonal stays between the tube sheet and shell or by through-stays running the entire length of the boiler. The former are usually preferable, for they leave more room inside the boiler for cleaning and inspection. The through-stays may make it quite difficult for a worker or an inspector to move around over the tubes in the boiler.

Older boilers had diagonal stays riveted to the shell and tube sheet as shown in Fig. 2.4a. Modern boilers use fillet welds as shown in Fig.

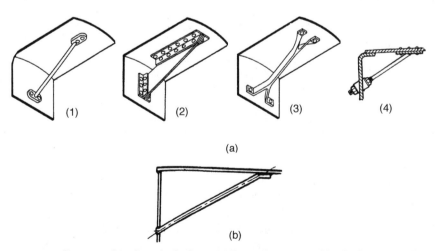

Figure 2.4 Stays used in firetube boilers. (*a*) Riveted stays on older boilers: 1, scully; 2, riveted gusset; 3, crowfoot diagonal; 4, palm stay. (*b*) Welded palm stay on modern boilers requires full penetration welds on tube sheet, and full fillet weld on palm.

2.4*b*. Older stays had designations such as Huston, MacGregor, and Scully as shown in Fig. 2.4*a*(1). Diagonal stays are placed in tension by fastening one end and then heating the stay before attaching the other end. On cooling, the stay is placed in tension.

Welding of diagonal stays must follow Code rules. Among the requirements:

1. Fillet welds shall have a size of not less than ⅛ in., and must be made all around the stay in contact with the shell.
2. Fillet welds are not allowed on the tube sheet for diagonal stays. As Fig. 2.4*b* shows, a full penetration weld is required.
3. All welding of stays requires postweld heat treatment per Code rules.

Overheating of tube sheets or shells will usually show some effect on diagonal stays. They will be found to have sagged, or lost their tension. The diagonal stays are stretched into position before the tubes are installed in the boiler. In order to prevent distortion of the head through the tension of the stays, one or more heavy steel bars are clamped across the tube sheet as a beam, the tube sheet thus being held against the tension of the stays until some of the tubes are installed. This beam is known as a *strongback*.

The staying of the section of the tube sheet below the tubes of HRT and similar boilers is usually effected by through-to-head stays. These stays are "spooled off" at the rear tube sheet. The front ends of the through-to-head stays—usually two in number—pass through the front tube sheet with inside and outside nuts and washers. They are usually made tight against leakage by grommets or soft metallic packing of various types beneath the outside nuts and washers.

The reason why the rear ends of the stays do not pass through the rear tube sheet but are spooled off inside is that the heat of the fire would damage the nuts and threaded ends.

Manholes and handholes. ASME Code specifies that "all boilers or parts thereof must be provided with suitable manhole, handhole, or other inspection openings for examination or cleaning, except special types of boiler where such openings are manifestly not needed or used...."

A manhole is required in the upper part of the shell or head of fire-tube boilers over 40 in. in diameter. A handhold above the tubes will suffice for very small boilers.

If the boiler is 48in. in diameter or larger, a manhole is required in the front head below the tubes. A manhole or handhole is required at this location in smaller boilers. When a handhole is used in the front head below the tubes, it is preferable to provide also a handhole in the rear head below the tubes—whether required by a local code or not. Elimination of this rear handhole may cause considerable difficulty in cleaning the lower internal surfaces of the shell.

Water level. Horizontal firetube-type boilers must be set so that the lowest reading in the water gauge glass is at least 3 in. above the highest point of the tubes. Locomotive-type boilers 36 in. or less in diameter require at least 2 in. above the tubes, but if over 36 in. in diameter, also require 3 in. water level above the tubes.

Heating surfaces. On all boilers heating surfaces are measured on the side receiving heat. For firetube boilers, this includes projected tube area (diameter times length) and any extended surface on the furnace side. In computing the heating surfaces, add tube areas, fireboxes, tube sheets, and the projected areas of headers exposed to the products of combustion. There is an exception for vertical firetube steam boilers, which, by Code rules, require only tube surfaces up to the middle of the gauge class be used in calculating their contribution to the heating surface.

The Code required, prior to the 1995 Code changes, a manufacturer of a completed boiler to stamp the boiler's calculated heating surface, except for electric boilers, where kilowatt power input is required. The stamping now required must show the maximum designed steaming capacity in pounds per hour for the fuel fired.

Heat release rates. High furnace heat release rates on SM boilers have caused boiler furnace and tube-sheet problems in the past. This was due to competition between manufacturers in providing a higher-capacity boiler at lower cost. Two heat release rates are used:

1. *Volumetric* is the heat release rate ratio of

$$\frac{\text{Maximum fuel input at boiler rating} \times \text{fuel's high heat value}}{\text{ft}^3 \text{ of furnace volume}}$$

Recommended guidelines:

Oil-fired 35 MBtuh/ft^3; maximum 45 MBtuh/ft^3

Gas-fired 40 MBtuh/ft^3; maximum 50 MBtuh/ft^3

2. *Effective Projected Radiant Surface* is the ratio of

$$\frac{\text{Fuel release rate}}{\text{ft}^2 \text{ of furnace radiant surface}}$$

Recommended guidelines (note some jurisdictions have limits on this ratio):

Oil-fired 180,000 Btuh/ft^2

Gas-fired 200,000 Btuh/ft^2

The calculation for this guideline is made by taking the fuel heat release rate and dividing it by the furnace area normal to the flame axis. Flame impingement on boiler heating surfaces must be avoided for all firing rates.

Firetube boilers, because of their compact design, automatic operation, and resultant reduced maintenance, have a lower life expectancy in general than do watertube boilers. Life estimates are affected by overfiring, rapid starting and cooling, and poor water treatment programs.

Flue gas temperature tracking, especially on the compact SM boiler, is an excellent method of determining if a boiler of the packaged design is being overfired or is in need of cleaning due to heat transfer surfaces becoming obstructed by scale or deposits on the fire side. The flue gas temperature should be recorded when the boiler is new, or after cleaning, to establish benchmark readings. One SM boiler manufacturer recommends limiting stack temperatures, and thus firing, to a maximum of 150°F above the design steam or water temperature.

The following table can be used as reference for tracking flue gas temperatures.

Boiler steam pressure, psig	SM boiler maximum allowable stack temperature, °F
15	400
25	417
50	448
75	470
100	488
125	503
150	516
175	528
200	538
250	547
275	564
300	577

For hot water heating boilers with 210°F maximum water temperature, stack temperature should not exceed 360°F.

Boiler turndown ratios.   This term is used as a guide in order to note the range of outputs over which a boiler can be operated automatically but still maintain peak and near-peak efficiency. On packaged firetube boilers, a 5:1 turndown is common, or a load from 20 to 100 percent rated is the guaranteed turndown efficiency.

Older Firetube Boilers

A brief description will be provided of the HRT, economic, and locomotive boilers. Although these boilers have generally been replaced by SM boilers, license examinations for operators and inspectors may have questions on these older boilers, because some are still in use.

Horizontal-return-tubular boilers

The HRT boiler (Fig. 2.5a and b) consists of a cylindrical shell, today usually fusion-welded, with tubes of identical diameter running the length of the shell throughout the water space. The space above the water level serves for steam separation and storage. A baffle plate (or dry pipe) is ordinarily provided near the steam outlet to obtain greater steam dryness.

The HRT boiler is simple in construction, has a fairly low first cost, and is a good steamer. It is more economical than the vertical tubular

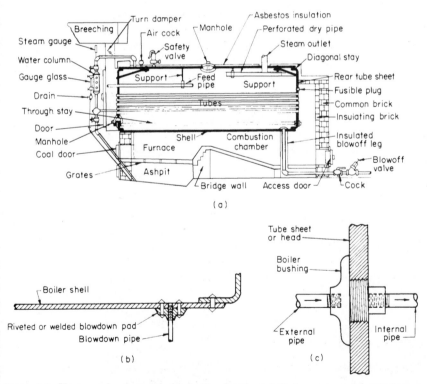

Figure 2.5 Horizontal-return tubular boiler details. (a) Major components; (b) blowdown connection requiring reinforcement; (c) bushing detail as inserted through tube sheet.

or locomotive types, but the scotch marine boiler has replaced it. One disadvantage is that hard deposits of scale are difficult to remove from water surfaces of inner rows of tubes. Another disadvantage is the danger of burning the shell plates above the fire if thick scale or deposits of mud form on the waterside on these plates. But difficulty of cleaning scale from the tubes holds true for all other types of fire-tube boilers.

Horizontal-return-tubular boilers are not very practical in shell sizes over 96 in. in diameter or for pressures exceeding 200 psi. Thick plates for higher pressures exposed directly to the flames would deteriorate very rapidly from overheating, because of poor heat transfer to the water. This could lead to bags (bulges) in the shell.

Bags also result when mud or other sediment from water settles on the bottom of the shell. For this reason, an HRT boiler should be pitched 1 to 2 in. toward the blowdown connection. Also, proper blowdown and water chemical treatment must be used to avoid scale or sediment formation and to keep solids to a minimum.

On older HRT boilers, the blowdown line is in the path of the hot gases and thus must be protected from overheating. A V-shaped pier of firebrick, usually in front of the blowdown pipe, deflects the hot gases. If not properly protected, the blowdown pipe may melt under certain conditions and drain the boiler, possibly leading to a serious explosion.

Blowdown connection.   The blowdown pipe cannot be merely screwed into the shell, for there would not be a sufficient number of threads in contact to give proper support. Instead, a pad is riveted or welded to the bottom of the shell, and the blowdown pipe is screwed into this pad (Fig. 2.5b).

The minimum size of blowdown pipe for boilers over 10 horsepower (hp) (100 ft^2 of heating surface) should be 1 in. A minimum size of ¾ in. is permitted by most states for boilers of less than this size. The maximum-size blowdown pipe in any case should be 2½ in.

See Fig. 2.6. This type of steel structural support is required on all HRT boilers 72 in. in diameter and over, since brickwork cannot support this heavier weight. On older boilers, the pad on the shell was riveted to the shell, with rivets designed with a safety factor of 12.5. But if the pads are welded, full peripheral fillet welds are required, and these must be stress-relieved on high-pressure boilers. Radiographing is not required.

Feedwater inlet.   The feedwater should enter through the upper part of the shell or front head, and it should discharge clear of any riveted or welded seam or part of the boiler exposed to radiant heat or high temperatures.

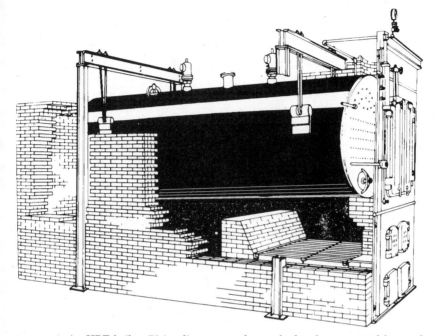

Figure 2.6 An HRT boiler, 72-in. diameter and over, had to be supported by steel structure shown that was independent of the brick walls, due to the heavy weight. (*Courtesy Power magazine.*)

If the boiler is over 40-in. in diameter, a manhole is provided above the tubes. Then the feed piping should enter through a boiler bushing (Fig. 2.5c). An internal feed pipe should discharge the water approximately three-fifths the length of the boiler from the hottest end—usually from the front—so that solids in the feedwater will not be precipitated onto the hot shell plate at the front, where overheating and damage might result.

Restricted circulation in the HRT boiler is uncommon, though it is possible if extremely heavy scale deposits build up and block the space between tubes. In such cases, it is probable that damage would occur to the boiler by overheating resulting from restricted circulation in the affected parts.

Rating. The conventional means of rating an HRT boiler is according to its heating surface. Most authorities accept the standard of 10 ft^2 of water-heating surface per boiler horsepower. The water-heating surface of this type of boiler is taken as one-half the area of the shell plus the total area of all tubes, based on their inside diameter, plus two-thirds the area of the rear tube sheet minus the aggregate area of the tube holes.

$$HS = \frac{(L \times \frac{1}{2}\pi D) + (N \times d \times \pi \times l) + (0.5236D^2 - N \times 0.7854d^2)}{144}$$

where L = length of boiler shell exposed to products of combustion, in.
D = shell diameter, in.
N = number of tubes
l = length of tubes, in.
d = inside diameter of the tube, in.
HS = water heating surface, ft^2

The area of the front tube sheet may be neglected, for the gas temperature is usually too low to cause much evaporation at this point. One-tenth of this result will be the boiler horsepower at 100 percent rating. The evaporation equivalent of a boiler horsepower, 34.5 lb/hr of steam with feedwater at 212°F, may be exceeded by forcing the boiler at a higher rating than that found by the heating-surface standard, but about 125 to 150 percent is the usual practical limitation for this type of boiler. Forcing the output above this point may result in difficulties.

Inspection.   Because the shell is exposed to fire, an HRT boiler requires careful internal inspection for scale, bulging, and blisters.

During an inspection, some of the areas to check carefully on an HRT boiler are the following: Internally, on the section above the tubes, check for corrosion and pitting. Look for grooving on the knuckles of heads, shells, welds, rivets, and tubes. Check the seams for cracks, broken rivet heads, porosity, and any thinning near the waterline of the shell plate. Check all stays for soundness and proper tension. Examine the internal feed pipe for soundness and support, and see that it is not partially plugged. Check the openings to the water-column connections, safety valve, and pressure gauge for scale obstruction. Also check shell and tube surfaces for scale buildup. Follow the same procedure internally below the tubes. Then check the opening to the blowdown connections and make sure that the bottom of the shell is pitched toward blowdown and that it has no blisters or bulges.

Externally, remove the plugs from the crosses of water-column connections and make sure they are free of scale. Examine the blowoff piping pad and the blowoff pipe to make sure it is protected from the fire, that the pipe is sound, and that the blowoff valves are in good order. Examine tube ends and rivets or welds for cracks and weakening of the tube to the tube-sheet connection. Check for fire cracks around the circumferential seam and for leakage at the caulked edge. Then examine the setting and supports for soundness.

Economic boiler

The economic-type boiler (Fig. 2.7) was an adaptation of the HRT boiler, giving somewhat greater heating surface per square foot of floor space. An added advantage is that the required amount of brick-work is much less since the boiler is self-supporting in its special casing. The rear end is supported by an iron cradle under the shell. The boilers of this type may be shipped as a unit with furnace walls held in position by a permanent steel casing. The construction principles, piping connections, and circulation are essentially the same as those of the HRT boiler. Also, the practical limitations for pressure and capacity are similar.

Because the rear course of the shell plate is oval, the flat surface on each side requires bracing as a stayed surface. This requirement is met by through-braces passing between horizontal rows of tubes from side to side. The ends of these braces are riveted over. The pitch and size of these braces are calculated by the same method as for diagonal stays in an HRT boiler.

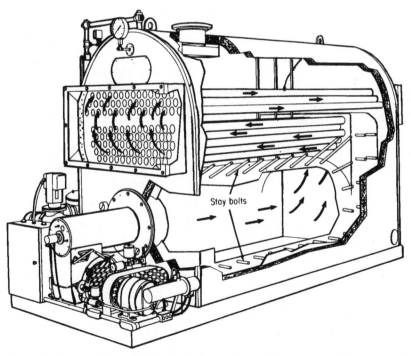

Figure 2.7 Economic-type firetube boiler consisting of three passes with an arched crown sheet and stayed surfaces.

In the economic boiler, the fusible plug, if used, is located in the rear head. If the distance between the top row of tubes and the top of the shell is over 13 in., the fusible plug should be at least 2 in. above the top row of tubes. Otherwise, it may be located not lower than the upper part of the top row of tubes.

Locomotive firebox boilers

See Fig. 2.8. Like the vertical tubular and scotch marine types, the locomotive firebox boiler is an internally fired firetube unit. But its shell is horizontal, and the firebox is not contained within the cylindrical portion of the boiler. The firebox is rectangular with a curved top known as a *crown sheet*. This crown sheet is supported by radial stays screwed into the crown sheet and the outer wrapper sheet. Ends of radial stays are riveted over. The inner sheets of the firebox are connected to the outer side sheets by stay bolts. The space between these sheets is called the *water legs*. The fire tubes are within the barrel and run from the firebox tube sheet to the smokebox head of the barrel. This head in the smokebox is formed by extending the barrel beyond the tube sheet.

The firebox front sheet above the crown sheet and the smokebox tube sheet above the tubes are supported by longitudinal stays. In some cases diagonal stays also are used for this purpose. The steam dome provides additional steam storage space and allows the main steam outlet to be taken off at a considerable height above the waterline, thus reducing the possibility of water carrying over with steam. The steam space extends over both the furnace and the barrel, which usually has 3-in. tubes. All tubes are of one diameter and length.

As is true of most internally fired firetube boilers, some water spaces are very difficult to clean, either mechanically or manually. Also, the locomotive firebox boiler is limited as to pressure and capacity, just as is the HRT boiler.

Water leg. The water space between the outside sheets and the firebox sheets is known as the water leg, as used on all horizontal or vertical firebox boilers which have this type of furnace construction. The inner and the outer sheets are fastened together at the bottom by either a mud ring or an "ogee" flange (Fig. 2.9). The former is more expensive but is usually more satisfactory, for it is easier to keep clean. An inherent disadvantage of the ogee flange is that the breathing action due to expansion and contraction of the furnace may cause fatigue stress and cracking. Modern construction would flange the inner and outer sheet together and butt-weld them as shown in Fig. 2.3.

Most locomotive boilers are equipped with a steam dome because the steam space and the water surface are limited and more space for

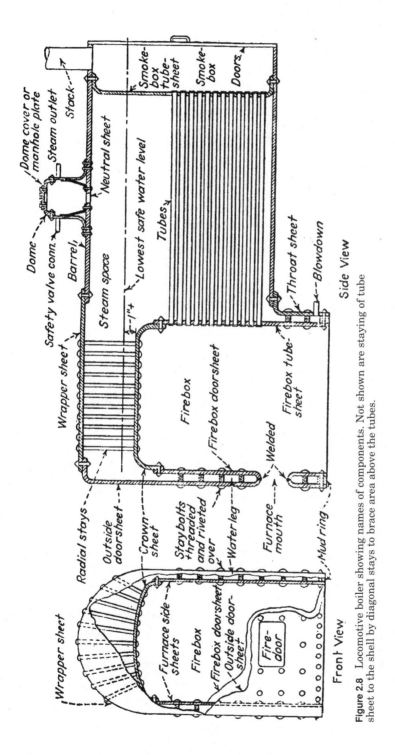

Figure 2.8 Locomotive boiler showing names of components. Not shown are staying of tube sheet to the shell by diagonal stays to brace area above the tubes.

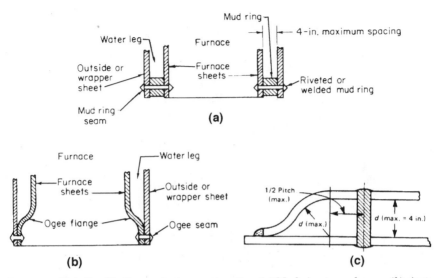

Figure 2.9 Details of bottom waterleg construction. (*a*) Mud-ring type closure; (*b*) riveted ogee-ring closure; (*c*) welded ogee-ring closure.

the steam flow to disengage drops of water is desirable. An additional service for the steam dome in railroad service is to house the throttle valve. The steam dome is usually installed with its vertical seam on one of the two long sides of the dome—this, of course, on one side of the boiler shell. Thus, the longitudinal seam of the dome will not have to be flanged so sharply.

It is important to protect the furnace sheet from overheating at the mud ring. Also, the bottom few inches of the water leg may become filled with sediment, and that area, too, should be protected.

In hand-fired installations, the grate is well above this zone. But with mechanical firing, a firebrick wall should be built up about 6 in. above the mud ring on all sides of the firebox.

Inspections. For internal corrosion, check: (1) waterline and top row of the tubes because of oxygen and other impurities released when boiling; (2) top of the crown sheet, on and around the ends of the radial stays and stay bolts; (3) water legs because of oxygen release and the presence of corrosive sediment; and (4) bottom of the barrel where pitting occurs if the boiler is improperly laid up when out of service.

External corrosion on a locomotive firebox boiler should be especially looked for around handhole plates and at the bottom of the first head. Rainwater wetting the sooty smokebox will corrode its bottom and also the handholes in this area. The bottom of the barrel should be watched for corrosion because of possible dampness or leakage.

Corrosion due to sediment deposits forming at the bottom of water
legs and their handholes may eat through the plate and gaskets of
handholes. Leaking tubes or stay bolts cause corrosion at the furnace
end of the tube sheets.

Scotch marine boilers

The largest number of boilers in use today for commercial and small
industrial plants is the scotch marine (SM) boilers. This boiler was
originally used for marine service because the furnace forms an integral part of the boiler assembly, permitting very compact construction
that requires a small space for the capacity produced. The SM boiler
is sold as a package consisting of the pressure vessel, burner, controls,
draft fan, draft controls, and other components assembled into a fully
factory-fire-tested unit. Most manufacturers test their models as a
unit before it is shipped to the site, basically delivering a product that
is pre-engineered and ready for quick installation and connection to
services such as electricity, water, and fuel. Some manufacturers provide starting service as part of the purchase price. A factory-trained
specialist starts up the unit, readjusts controls, checks out the unit in
operation, makes necessary adjustments, and trains an operator to
troubleshoot on controls.

The SM boiler is built either as a wet-back furnace, as shown in
Fig. 2.10b, or as a dry-back, shown in Fig. 2.10a. This boiler is an
adaptation to stationary practice of the well-known SM wet-back boiler. It consists of an outer cylinder shell, a furnace, front and rear tube
sheets, and a crown sheet. The hot gases from the furnaces pass into
a refractory-lined combustion chamber at the back (sometimes built
into a hinged or removable plate) and are returned through the fire
tubes to the front of the boiler and then to the uptake. This boiler is
suitable for coal, gas, and oil firing.

In the wet-back design (Fig. 2.10b), the shell, tube, and furnace
construction are similar to the dry-back type, but the combustion
chamber, being inside the shell, is surrounded by water. Thus no outside setting or combustion-chamber refractory is needed. The dry-
back type is a quick steamer because of its large heating surface. It is
also compact and easily set up and shows fairly good economy.

The internal furnace is subject to compressive forces and so must
be designed to resist them. Furnaces of relatively small diameter and
short length may be self-supporting if the wall thickness is adequate.
For larger furnaces, one of these four methods of support may be
used: (1) Corrugating the furnace walls; (2) dividing the furnace
length into sections with a stiffening flange (Adamson ring) between
sections; (3) using welded stiffening rings; and (4) installing stay bolts
between the furnace and the outer shell. If solid fuel (coal, wood, etc.)

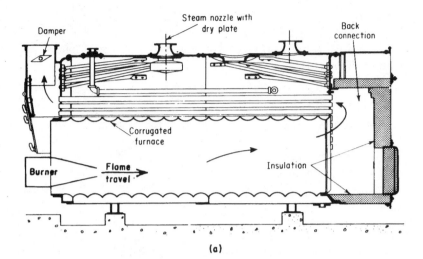

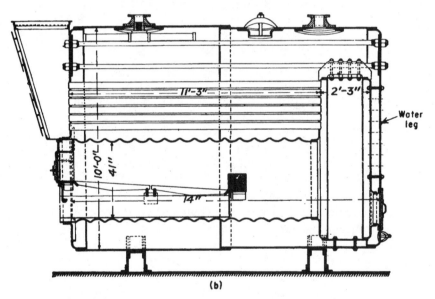

Figure 2.10 Scotch marine boiler with corrugated furnace. (*a*) Dry-back type; (*b*) wet-back type.

is to be fired, a bridge wall may be built into the furnace at the end of the grate section.

Welded furnaces on SM boilers are rolled to a cylindrical shape, and must be formed to a true circle with Code-permitted plus or minus deviations: (1) For furnaces equal to or less than 24 in. OD, the deviation permitted is 1 percent of the OD; (2) for furnaces greater than 24 in. OD, the chart shown in Fig. 2.11 must be used. For example, the

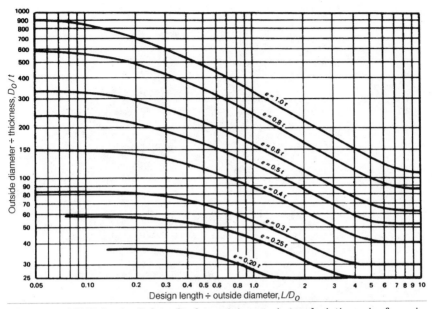

Figure 2.11 ASME Section I chart for determining maximum deviation, e in. for a circular furnace, where t = thickness of furnace wall (in.), D_o = outside diameter of furnace (in.), and L = furnace length. (*Courtesy American Society of Mechanical Engineers.*)

furnace has ⅜-in. wall thickness, 36-in. ID, and is 12 ft long. The chart from Section 1 of the ASME Boiler Code requires all dimensions to be in inches.

$$\frac{D_o}{t} = \frac{36.75}{0.375} = 98 \qquad \frac{L}{D_o} = \frac{144}{36.75} = 3.92$$

Using the chart, e = permissible deviation from circle = $0.72t$, or $e = 0.72 \times 0.375 = 0.27$ in.

This Code guide on out-of-roundness of furnaces in SM boilers can be used when inspecting the furnace for overheating damage, especially after a low water incident.

The scotch marine boiler may be by far the largest in diameter of any firetube boiler, being built up to about 15-ft diameter. Since the area of the segment of heads above the tubes is large, diagonal stays are usually precluded because of the great number that would be required. Instead, it is customary to use a smaller number of head-to-head through-stays of 2- to 3-in. diameter.

In boilers of large diameter, it is the practice to use more than one furnace. Two, three, or even four furnaces are used in the large boilers of this type. Figure 2.12 is a cutaway view of a four-pass model.

Figure 2.12   Four-pass SM boiler. (*Courtesy Cleaver-Brooks Co.*)

This unit maintains a continuously high gas velocity. As the hot gases travel through the four passes, as shown in Fig. 2.13 they transfer heat to the boiler water and thus cool and occupy less volume as the gases pass through the different tube passes. The number of tubes is reduced proportionately to maintain the high gas velocity and thus keep producing as much as possible a constant heat transfer.

While more passes can be used, today four is the practical limit.

The furnace of an SM boiler may provide up to 65 percent of the boiler output even though it may have only about 7 to 8 percent of the total heating surface. In the furnace, most of the heat is transferred by radiation. The furnace should have sufficient volume to permit complete combustion of the fuel-air mixture before the flue gases reach the tube passes. Most designers try to limit the heat-release rate in the furnace to below 150,000 Btu/(hr/ft^3) of furnace volume; otherwise, the ratio of air to fuel becomes critical. Rates above 150,000 Btu/(hr/ft^3) of furnace volume can cause the fuel to be still burning at entry to the first pass of tubes, and this can cause cracking of tube ends or cracking of welds between the furnace and the rear

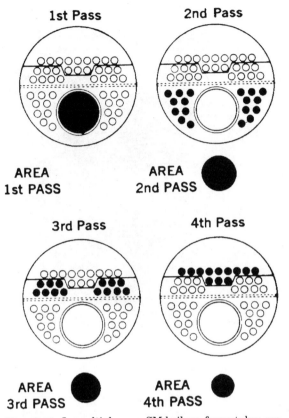

Figure 2.13 In multiple-pass SM boilers, fewer tubes are used as the gas cools and thus occupies less volume. (*Courtesy Cleaver-Brooks Co.*)

tube sheets. Any scale or other deposits can aggravate this cracking with high furnace heat-release rates. Good feedwater treatment is thus essential for SM boilers with high heat-release rates in the furnace.

There is also the problem of corrosion on the fire side when sulfur-burning fuels are used. Corrosion can occur when the plate or tube temperatures drop below the acid dew point, such as may occur from on-off firing, and reaches a maximum rate when the temperatures fall to less than the water dew point. On-off firing usually requires purging of the furnace, and this can also produce thermal gradients in the boiler which might produce cracking from expansion and contraction effects.

Compact designs also tend to make surfaces less accessible for inspection and cleaning. Thus today's higher heat-transfer rates can

Figure 2.14 Collapsed furnace of SM boiler caused by low water. (*Courtesy Royal Insurance Co.*)

easily cause overheating, especially if forced. See Fig. 2.14. This results in loose tubes in tube sheets, cracks between ligaments on tube sheets, weld cracks in high-heat zones, bulged furnaces, and low water. Even more dangerous is the complete reliance on automatic controls to safely cycle a boiler, without periodically checking the controls for (1) conditions of electric contacts; (2) electric connections; (3) water-column connections; (4) waterside plugging of pressure switches; (5) low-water fuel cutoffs; (6) soot accumulation in tubes; (7) operation of solenoid valves in fuel-cutoff lines; (8) firing-equipment timing and operation of flame-failure devices; and (9) operation of safety valves.

Vertical tubular boiler

The vertical firetube boiler is used where floor space is at a premium and the pressure and capacity requirements come within the scope of this type of boiler. The vertical tubular (VT) boiler is an internally fired firetube unit. It is a self-contained unit requiring little or no brickwork. Requiring little floor space, it is popular for portable service, such as on cranes, pile drivers, hoisting engines, and similar construction equipment. Vertical tubular boilers are used for stationary service where moderate pressures and capacity are required for process work, such as pressing, drying-roll applications in various small laundries, and in the plastics industry.

The coil-type watertube boiler is a competitor of the VT boiler for small capacities and lower pressures to 150 psi. But the VT boiler is limited in capacity and pressure even more than the horizontal firetube boiler. For this reason, most VT boilers of the firetube type seldom exceed 300 hp, or about 10,000-lb/hr capacity with a maximum pressure of 200 psi. There are five general classifications:

1. Standard straight shell type with dry top (Fig. 2.15a)
2. Straight shell type with wet top (Fig. 2.15b)
3. Manning boiler with enlarged firebox
4. Tapered-course bottom with enlarged firebox
5. For even smaller capacity, the vertical tubeless unit

The Manning and the tapered-course types provide a greater grate area and furnace volume, which permit somewhat better combustion efficiency.

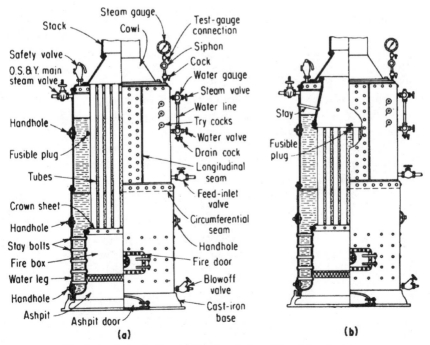

Figure 2.15 Vertical tubular boiler. (a) Dry-top design; (b) wet-top design.

The submerged-head type was developed to prevent overheating of the upper ends of the tubes, which are in the steam space in the standard vertical firetube boiler. Overheating and damage result sometimes when the latter boiler is forced or too hot a fire is maintained when starting up. A slow fire is essential until steam generation starts. Then the upper ends of the tubes may be "cooled" by steam.

The advantages of VT firetube boilers are (1) compactness and portability; (2) low first cost; (3) very little floor space required per boiler horsepower; (4) no special setting required; and (5) quick and simpler installation.

Disadvantages are that (1) the interior is not easily accessible for cleaning, inspection, or repair; (2) the water capacity is small, making it difficult to keep a steady steam pressure under varying load; (3) the boiler is liable to prime (carry over with steam) when under heavy load because of the small steam space; and (4) the efficiency is low in smaller sizes because hot gases have a short, direct path to the stack, so much of the heat goes to waste.

Deposits settle in the 4-in.-wide water legs (maximum per Code) because they have restricted circulation. Cleanout openings should be periodically opened around the circumference in the water legs and bottom tube sheet so all areas are accessible for cleaning. Some units have a continuous drain in the bottom of the water legs. When this boiler is opened for cleaning, a length of chain can be pulled around inside to get the sludge and scale to a cleanout opening for removal.

The Code indicates no specific water level for the dry-top boiler, except to state: "It shall be at a level at which there shall be no danger of overheating any part of the boiler when in operation at that level." But the level is generally taken as a minimum at a point two-thirds the height of the shell, above the bottom head, or crown sheet or tube sheet. This is the same minimum requirement as for miniature vertical firetube boilers.

For the submerged wet-top type, the minimum water level must be at least 2 in. above the top tube ends, except for miniature boilers, where it can be 1 in.

Internal and external corrosion usually take place in a VT boiler in the following areas:

1. At the waterline, usually on the tubes. This is due to oxygen and organic materials being released there during the process of boiling.

2. In the vicinity of the feedwater discharge, as a result of oxygen release.

3. On top of the lower tube sheet, because of scale formation.

4. On and around the ends of the stay bolts, as a result of the stresses imposed and the subsequent expansion and contraction (this leads to stress corrosion).

5. In the water legs, especially at the bottom, where grooving may occur in addition to the usual pitting under the scale.

External corrosion occurs:

1. On top of the top tube sheet, bottom tube sheet, and tube ends, because of acid formed by the damp soot in contact with the products of combustion.

2. Around all manhole, handhole, and washout openings. Here corrosion is from leakage due to poor gaskets, improper handhole-cover installation, and thermal expansion and contraction which loosen opening closures.

3. At the bottom of shell, water legs, and furnace sheet, as a result of soot attack.

4. Around all openings including gauge connections, safety-valve connections, steam-outlet connection, feedwater connection, blowdown connection, and water-column connection, because of leakage.

Multiple Boiler Systems

Multiple boiler systems are used in many smaller boiler plants in order to obtain progressively larger output by sequencing the firing of each boiler based on demand. (See Fig. 2.1.) The advantage of this arrangement in comparison to operating one boiler from 25 to 100 percent of rating, is that by sequencing per load, each boiler can be operated at its rated capacity, and thus achieve design efficiency. This reduces thermal and standby-duty losses, especially where a variable-temperature heating load exists. Individual boilers can also be isolated for repairs and maintenance.

The system can be equipped with automatic controls to match boiler sequencing with load demand. Factory-assembled prewired control panels expedite installation and servicing of the controls if needed. Prewiring extends to providing multiple-zone heating systems, either through automatic valves, or circulators.

Questions and Answers*

1 What would cause a fire tube to burst or explode?

---

*Use ASME Section 1, Boiler Code for Code questions.

ANSWER: Fire tubes are normally under external pressure. They may collapse, but do not burst.

2 What boiler has a crown sheet that is braced similarly to the locomotive crown sheet?

ANSWER: The scotch marine boiler.

3 What internally fired boiler frequently has more than one furnace?

ANSWER: The scotch marine or the scotch dry-back often has two to four circular furnaces when the boiler is of large diameter.

4 How are internal furnaces braced?

ANSWER: If flat, they are stay-bolted. If circular, they may be thick enough to be self-supporting for moderate pressures. If they are not thick enough, they may be stay-bolted, corrugated, or braced by Adamson rings.

5 If used, where should the fusible plug be located in any boiler?

ANSWER: Not lower than the lowest safe water level.

6 (*a*) What is the minimum thickness required for a steel boiler tube in order that a fusible plug may be installed?
    (*b*) What are the maximum pressure and temperature allowed on copper tubes or nipples in firetube boilers?

ANSWER: (*a*) 0.22 in. (*b*) 250 psi and 406°F.

7 What fusible metal is used in fire-actuated fusible plugs? What is its required melting point?

ANSWER: Tin; 445 to 450°F.

8 (*a*) State the reason for beading the ends of tubes of firetube boilers.
    (*b*) State the reason for flaring the ends of tubes of watertube boilers.

ANSWER: Both beading and flaring increase the holding power of tubes. Tests have indicated that a flared tube has more holding power than a beaded tube.
    (*a*) Beading increases the holding power of the tubes and eliminates to a large degree the burning and erosion of tube ends. Therefore, beaded tube ends are used in firetube boilers.
    (*b*) Flaring of tube ends increases the holding power of the tubes. Since fire is not in contact with a watertube end, tubes in watertube boilers are usually flared.

9 Explain the meaning of externally and internally fired boilers.

ANSWER: Externally fired boilers have a separate furnace built outside the boiler shell. The HRT boiler is probably the most widely known example of the externally fired boiler. In internally fired boilers, the furnace forms an integral part of the boiler structure. The VT, locomotive firebox, and the scotch marine are well-known examples of internally fired boilers.

10 Where are stay bolts most likely to break in the locomotive firebox boiler?

ANSWER: Usually the top row of stay bolts and the first row of radial stays, with fracture occurring close to the inner surface of the outer sheet (wrapper sheet). This area is a high-heat zone, causing large expansion and contraction movements.

11 What are the most common staying methods used on an HRT boiler?

ANSWER: Four methods are used:

1. Through-stays are used below the tubes because there is insufficient room for other stays, which would accumulate deposits more readily.
2. Diagonal stays are used above the tubes to support the flat, unstayed tube sheet above the tubes.
3. Gusset stays are used above the tubes.
4. If not over 36 in. in diameter and not over 100 psi, structural shapes can be used if they are arranged according to ASME Code requirements.

12 Name three causes of bagging (bulging) of the shell of an HRT boiler.

ANSWER:

1. Oil (from lubricated pump rods, etc.) getting into the boiler feedwater and being carried to the lower part of the shell, which is exposed to fire, and thus causing overheating
2. Scale or mud (from sediment in the water) deposits on the lower portion of the shell, restricting heat transfer
3. Excessively localized flame on a portion of the shell, causing overheating

13 How are tube holes made per Code requirements?

ANSWER: They may be punched to not over ½ in. less than finished diameter and finished by machining. They are often cut with a pilot drill and rotating cutter.

14 (a) How much larger than a fire tube may the tube hole be?
    (b) How much larger for a water tube?

ANSWER:

    (*a*) $\frac{1}{16}$ in. larger diameter.

    (*b*) $\frac{1}{32}$ in. larger diameter.

15 How are the ends of shell plates and butt straps formed for the longitudinal seam?

ANSWER: They should be formed by rolling or pressing and not by blows.

16 How far through the sheets should the ends of stay bolts extend before they are riveted over?

ANSWER: Not less than two threads.

17 Why should the feedwater not discharge into the water leg?

ANSWER: The cooling action of the water against the hot furnace sheets would cause serious stresses and probable damage.

18 What purposes do smokebox doors serve in a locomotive boiler?

ANSWER: Inspection, cleaning, and tube replacement.

19 Name four types of firing-door construction in a VT boiler.

ANSWER: Flanged together and riveted, flanged together and butt-welded, both sheets flanged out and riveted (sucker-mouth), riveted doorframe ring.

20 What are the four types of VT boiler?

ANSWER: Standard straight-shell, Manning, tapered-shell, and submerged-head.

21 What are the reasons for or advantages in the deviations from the standard type of the other types named in the answer to question 20?

ANSWER: The submerged head protects the upper ends of the tubes from overheating. The Manning and tapered-shell types allow a larger grate area and furnace volume.

22 On what point of construction is the maximum allowable pressure based in the standard VT boiler?

ANSWER: On the ability of the shell or furnace, whichever is the weaker, to resist rupture from internal pressure or collapse of the furnace from pressure.

23 What additional method is used in calculating the maximum allowable pressure of a Manning boiler?

ANSWER: It is necessary to figure the strength of the shell at its point of greatest diameter, between the reverse flange and the top row of stay bolts.

24   Is the economic boiler internally fired?

ANSWER: No. Its combustion chamber is steel-encased, but the casing is not a pressure part of the boiler.

25   Name four internally fired boilers.

ANSWER: Vertical tubular, locomotive, scotch marine, and scotch dry-back.

26   When breaking a gasketed joint on a power boiler, should the gaskets be reused?

ANSWER: Handhole and manhole gaskets should not be reused but replaced by new gaskets to prevent leakage when the boiler is returned to service.

27   What precautions should be taken before closing drum manholes and setting doors on a steam generator?

ANSWER: Make sure all tools, rags, and similar foreign objects are removed and that all personnel that worked in the steam generator have exited from the unit.

28   A Code boiler is installed in a plant. Why is it necessary for a jurisdictional or authorized inspector to make an inspection again before the boiler is placed in operation?

ANSWER: A Code boiler is inspected during construction by an authorized inspector; however, the installation must also be inspected for jurisdictional requirements on boiler support, piping, safety relief valves, blowdown, stop and feed valves and connections, instrumentation and controls, and combustion safeguards that must meet jurisdictional installation requirements.

29   How are stays and stay bolts checked for soundness during inspections?

ANSWER: All stays should be checked to see if they are still in even tension, the welded or riveted connections should be checked for signs of cracks, weld washout or similar signs of excessive wear and tear. Stay bolts are usually "sounded" by tapping the end of the bolt to note if the ring or sound is the same on all stay bolts. A cracked or broken stay bolt will give a different sound.

30   How is circulation obstruction checked on a firetube boiler?

ANSWER: To check for scale and deposits beyond the top of the tubes, a small light can be lowered between the tubes to observe if scale and deposits are plugging the space between the tubes or in water legs.

31   Why should a boiler operator be familiar with manual operation on a boiler equipped with automatic combustion controls?

ANSWER: Manual operations should be periodically practiced in the event that emergency conditions develop with the automatic controls.

32   Should fuel solenoid valves on automatic boilers be of the manual reset type?

ANSWER: On power boilers, it is important to determine why the fuel solenoid valve shut the fuel off. For example, it could be because of poor combustion at the burner activating the flame safeguard system. Therefore, the fuel solenoid valve should be of the manual reset type to prevent a dangerous condition developing. An operator can first check out why the fuel was shut off.

33   Two high-pressure boilers are connected to a common steam pipe header. What are the Code requirements on the connections to the header from each boiler?

ANSWER: The Code and ANSI B31.1 (Power Piping) require boilers connected to a common steam pipe header to be fitted with two stop valves having a free-blow drain between them to check on the tightness of valve closure. The first stop valve near the boiler must be an automatic nonreturn valve to prevent back flow. The second valve should be of the outside-screw-and-yoke type, and if steam outlet is over 2 in. it must be of the rising spindle type.

34   Explain the difference between a bag and blister if found on the shell of a firetube boiler.

ANSWER: A bag is a deformation or bulge caused by local overheating of the shell and extends through the thickness of the shell. A blister is a partial separation on the shell that does not extend through the thickness, but is of a scablike nature, usually caused by impurities getting into the metal during the steel rolling operation.

35   Why are manholes and handholes usually found oval in shape?

ANSWER: It facilitates removing the covers from the boiler during maintenance or inspection periods. Oval manholes provide extra length for climbing into the boiler without making the opening so large that extra material would have to be removed, requiring more reinforcement around the opening. The long axis of manholes is always placed girthwise on the shell.

36   When inspecting an old HRT boiler, it is found that the boiler has three stays on one side of the top manhole and four stays on the other side. Must the pressure be reduced?

ANSWER: This condition should be called to the attention of a jurisdictional inspector. He will determine the permissible pressure by checking the loading and pitch between stays that are necessary to carry the present boiler allowable pressure. If three stays satisfy Code requirements on load carrying and permissible pitch, nothing further may need to be done. If four stays are required on each side, then an additional stay will have to be installed or pressure reduced.

37   A solid stay bolt under 8 in. requires what size telltale hole, and for what distance?

ANSWER: Per Code, the hole must be at least ³⁄₁₆-in. diameter and must extend ½ in. beyond the inside surface of plate being stayed.

38   What effect does the length of a furnace have on its strength?

ANSWER: Length increases the bending moment the furnace must resist against collapsing. This is the reason for stiffening the furnace by installing, for example, an Adamson ring along its length.

39   What is the maximum distortion permissible by the Code in the diameter of a welded boiler shell or drum?

ANSWER: 1 percent of mean diameter.

40   Is it permissible to locate a tube hole in a welded joint?

ANSWER: Per Code rules, it is permissible; however, the weld after the hole is made must be checked for cracks and tears by the magnetic particle inspection to make sure it was not affected.

41   Under what conditions may stay bolts be welded per Code rules?

ANSWER: The joint must be prepared per Code rules for welding, and the welding procedure and welder must meet Code qualifications.

42   What minimum thickness must a firetube shell and tube sheet be on a 60-in.-ID-diameter boiler?

ANSWER: Per Code, ⅜ in. for shell and ½ in. for tube sheet.

43   On what type of boiler would you find an Adamson furnace, and per Code, what is the minimum section length and thickness of an Adamson furnace?

ANSWER: SM boiler, 18-in. length, ⁵⁄₁₆-in. thick per Section 1, PFT part.

44   A wet-bottom boiler is to be installed in the boiler house. What minimum distance is required between the bottom of the shell and the floor?

ANSWER: A distance of 12 in. is required so that the bottom of the shell is accessible for inspection or repair.

45   A gauge glass on a horizontal firetube boiler shows the lowest possible water level in the glass. How many inches above the tubes should this level be in a Code boiler?

ANSWER: The Code requires water to be 3 in. above the tubes when the water in the glass is at its lowest visible level.

46   The Code does not permit welds in shells of a firetube boiler to be in contact with primary furnace gases. Define primary furnace gases.

ANSWER: Primary furnace gases are those that are in the fire-side zone area where the temperatures exceed 850°F.

47   For a firetube boiler, per Code, what is the minimum size of washout plugs and access doors into setting?

ANSWER: Washout plugs must be at least 1½ in., and access doors must have a minimum area presented by a 12- by 16-in. rectangle with 11 in. being the least permissible dimension.

48   Why were steam domes used in locomotive boilers?

ANSWER: To provide drier steam and also to provide additional steam storage space for use during steep train climbs.

49   (*a*) Why do the upper ends of tubes in VT boilers which are not covered with water not burn out?
(*b*) Why do the rear ends of tubes on a high-temperature boiler leak in case of low water?

ANSWER:

(*a*) Because the products of combustion at this point are at such a temperature that cooling of the tube ends is provided by the steam and prevents the upper ends of the tubes from burning.

(*b*) The products of combustion at the front tube sheet are at a much lower temperature than at the rear tube sheet, so that when low water occurs, the tubes overheat and expand; at the same time, the rear tube sheet overheats and the tube hole expands and gets larger, so that the seating of the tube produced by the expander is destroyed and, therefore, leakage occurs at the rear ends.

50   Which type of boiler, fire tube or water tube, has the most heating surface in the tubes?

ANSWER: The Code states heating surface shall be computed for that side of boiler surface exposed to products of combustion. Therefore, a water tube contains more heating surface than a fire tube of the same diameter.

51   When is a manhole required in the front tube sheet below the tubes in HRT boilers?

ANSWER: When the boiler is 48-in. diameter or larger.

52   When is a manhole required above the tubes of a horizontal firetube boiler?

ANSWER: If externally fired, for boilers of 40-in. diameter or over; if internally fired, for boilers of 48-in. diameter or over.

53   (*a*) What is the minimum face width of a manhole flange for a gasket bearing surface? (*b*) What could be done if the flange were only ⁷⁄₁₆-in. thick?

ANSWER: (*a*) ¹¹⁄₁₆ in. (*b*) A steel ring could be shrunk onto the flange, and the double face could be machined for a bearing surface.

54   Name four ways of supporting a circular furnace subjected to external pressure.

ANSWER: It may be: (1) self-supporting; (2) stay-bolted; (3) corrugated; or (4) equipped with Adamson rings.

55   What was the maximum permitted length of a course in an HRT boiler of riveted construction?

ANSWER: 12 ft.

56   What are the advantages of submerged top tube sheets in a VT boiler?

ANSWER: A submerged top tube sheet of a VT boiler protects the upper tube sheet and tube ends from overheating and possible fire cracks.

57   What size limitations must not be exceeded for a firetube boiler to be classified as a miniature boiler?

ANSWER: The following cannot be exceeded: inside diameter, 16 in.; gross volume, 5 ft³; water-heating surface, 20 ft²; allowable working pressure, 100 psi.

58   Name three advantages in using a corrugated furnace in a scotch marine boiler.

ANSWER: The corrugated SM type of furnace offers the following advantages over a plain furnace: (1) The corrugations stiffen the furnace so it does not collapse as easily from external pressure as does a similar thick, plain furnace; (2) the corrugated furnace permits more expansion and contraction to take place as a result of the bellowslike action of the corrugations; (3) there is a slight increase in the heat-absorbing area over a plain furnace of the same diameter and length.

59   How can the water level in a steam boiler be checked if the water gauge glass is temporarily out of service?

ANSWER: By using the water gauge test cocks that are located on the water column.

60   What space consideration is required when installing horizontal firetube boilers?

ANSWER: Sufficient room must be left in the front or back of the boiler so that tubes can be replaced.

61   Why did the Code require HRT boilers over 72-in. diameter to be supported by outside suspension type of settings?

ANSWER: Brickwork settings used on smaller HRT boilers could not support the heavier load that a larger boiler full of water would impose on the setting, and crushing or crumbling of the support could result.

62 What is a strongback?

ANSWER: It is a bar bolted to the tube sheet to prevent the sheet from buckling when the stays are put in tension before the tubes are installed.

63 In checking an HRT boiler, what is the difference between a through-brace and a through-to-head brace? Where is each used? Why?

ANSWER: The through-brace has washers and nuts on each end. The through-to-head brace has nuts and washers on one end; the other end is forged into an eye that is held clear of the rear head by a pin or other construction. The through-stays may be used above the tubes, for the rear outside nuts are protected from burning off by the rear arch.

The through-to-head braces are used below the tubes where the rear ends have to be protected from the high-temperature gases.

64 A locomotive boiler is to be exhibited at a state fair and requires approval for operation.

(*a*) What are the sheets of a locomotive boiler?

ANSWER: Barrel or shell, smokebox tube sheet, dome shell, dome head, wrapper sheet, throat sheet, firebox side sheets, firebox tube sheet, crown sheet, inside-door sheet, and outside-door sheet.

(*b*) Where is the most dangerous part of the boiler in case of low water?

ANSWER: The crown sheet.

(*c*) From which end are tubes removed and replaced?

ANSWER: The smokebox end.

(*d*) How are the firebox sheets supported?

ANSWER: The side, door, and throat sheets are supported by stay bolts. The crown sheet is supported by radial or girder stays.

(*e*) Where is the usual location for the manhole?

ANSWER: In the dome head, or in the shell if no dome is used.

(*f*) How many manholes are used in a locomotive boiler? Explain your answer.

ANSWER: One. The part of the barrel over the tubes is the only part of the interior that is accessible.

(*g*) What should be the lowest level for the bottom nut of the gauge glass?

ANSWER: The lowest visible part of the glass should be at least 3 in. above the highest part of the crown sheet.

65   For what size firetube boiler does ASME CSD-1 and addendas titled, "Controls and Safety Devices for Automatically Fired Boilers" apply?

ANSWER: This is also an ANSI standard and may be required by jurisdictional adoption as an installation requirement for automatically fired boilers with inputs of 400,000 to 12,500,000 Btu/hr. See later chapters on fuels, firing, and controls.

Chapter

3

Watertube Boilers

The chief differences between the watertube boiler and the firetube boiler is that in the former the water circulates *through* the tubes instead of around them. The hot gases pass around the tubes.

Firetube boilers are designed with the tubes contained in the shell. The tubes of most watertube boilers are located outside the shell or drum. There are two advantages to this feature of the watertube boiler: (1) Higher capacity may be obtained by increasing the number of tubes independent of shell or drum diameter; (2) the shell or drum is not exposed to the radiant heat of the fire.

The biggest advantage over firetube boilers is the freedom to increase the capacities and pressures. That is impossible with firetube boilers because the thick shells and other structural requirements become prohibitive over 50,000-lb/hr capacity and over 300 psi. The large capacities and pressures of the watertube boiler have made possible the modern, large utility-type steam generators.

Modular construction of large watertube boilers permits factory assembly of waterwall, superheater, and reheater panels that are lifted into place at the site and then connected to drums or headers. This expedites erection, provides factory quality control in fabrication, and is more economical. However, small packaged watertube boilers also have been developed to compete with the firetube boiler, and to tap the large market that exists for smaller boilers. Advantages stressed are the following:

1. By installing multiple packaged units instead of a single large boiler, firing is based on load per boiler, thus providing flexibility during periods of low steam demand. The boilers are quick steamers; therefore, weekend shutdown is possible without experiencing the high refiring costs of a large boiler.

2. Small steam generators can be placed near the steam load, thus avoiding steam and condensate line heat losses and reducing the size of boiler room space needed.

3. Reserve or peaking capacity can be provided with a multiple boiler system.

Early Watertube Boilers—Straight Tubes

The early watertube boilers followed some firetube designs, being straight tubes suspended over a furnace with the tubes rolled into two types of headers: (1) a box header or (2) a sinuous header.

Sinuous- and box-header boilers

Figure 3.1a shows an early straight-tube watertube boiler. The tubes are placed in the furnace, and the shell above is used primarily as a storage tank for water and steam. Circulation from the drum is down the back headers, through the water tubes, and through the front header. With this arrangement, the tubes on boilers began to be separated from the internal shell, in contrast to firetube boilers. The design shown has one drum; larger boilers had two or three drums. The drum runs from the front to the rear of the boiler. The inclined straight steel tubes, usually of 4-in. OD, are connected with the drum by pressed-steel headers of the sectional type, the tubes being staggered in pairs. A mud drum below the rear headers was used to collect sediment and was blown out from time to time. Tube headers are in one piece for each vertical row of tubes. Header handholes (for tube cleaning) are closed by bolted covers with machined joints.

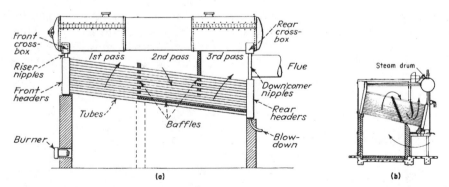

Figure 3.1 Early straight-tube watertube boilers. (a) Longitudinal-drum type; (b) cross-drum type.

The sectional header, which is also referred to as a *sinuous,* or *serpentine, header* (Fig. 3.2*a*), is either a casting or a forging. In older low-pressure boilers, the headers were also constructed of cast iron. But for high pressure, they were limited by the ASME Code to a maximum design pressure of 160 psi and 350°F. In contrast, headers constructed of forged steel have been used for pressures up to 1200 psi.

A *box header* (Fig. 3.2*b*) is constructed of flat plates, referred to as a tube sheet and tube cap sheet. But these surfaces must be stayed to prevent deformation. The sides, top, and bottom are flanged and riveted (welded on new boilers) to the tube sheets and cap sheets. The staying of sheets limits pressure for a box header to about 600 psi.

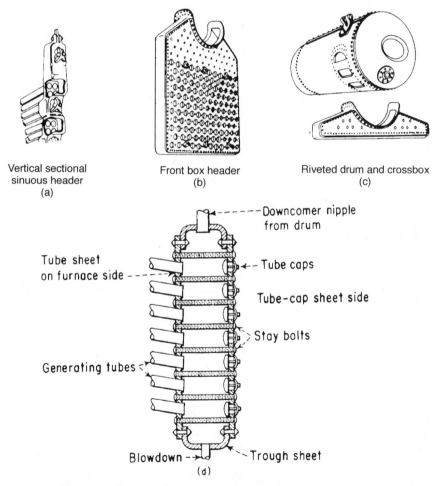

Vertical sectional sinuous header (a)

Front box header (b)

Riveted drum and crossbox (c)

Figure 3.2 Sinuous and box headers were used on straight-tube watertube boilers. (*a*) Vertical sectional sinuous header; (*b*) front box header; (*c*) cross-box connecting shell to box header; (*d*) box-header detail.

Cross-drum boilers. Cross-drum watertube boilers were constructed with either sinuous or box headers, as with the longitudinal-drum types. (See Fig. 3.1*b*.) The rear headers are supplied with water by a row of downcomer nipples connected to the lower part of the cross drum. These are often the same diameter as the generating tubes. The upper ends of the front headers are connected to the front side of the cross drum by one or more rows of horizontal circulating tubes which carry the mixture of steam and water (delivered to the front headers by the generating tubes) into the drum.

Both the sinuous-header and box-header types of boilers have tube caps, which is a favorite source of leakage if gaskets and the caps are not kept tight. On sinuous-header boilers, corrosion can take place on the tube-entrance outer plate as a result of leakage from handholes or tube caps. Leakage and corrosion also take place between the sinuous headers, which are usually packed with asbestos. On box-header boilers, leakage at handhole plates causes corrosion on the wrapper or external sheet where the closing caps are located. Leakage of these handholes finds its way down to the bottom of the headers (Fig. 3.2*d*), which is usually concealed in brickwork. Thus leakage from above causes undetected corrosion. When inspecting box-header boilers having bricked-in bottom headers, always remove the brickwork and check this surface for corrosion.

Some of the manufacturers who built header-type watertube boilers were Heine, Union Boiler, Murray Boiler, E. Keeler Co., Springfield Boiler Co., Babcock and Wilcox Co., and the Wickes Boiler Co.

Vertical watertube boilers—straight tube type

The vertical watertube boiler is a type requiring moderate headroom and small floor space per unit of capacity. The watertube boiler in Fig. 3.3*a* represents this type. These boilers make use of a Dutch oven or furnace extension to secure proper combustion space. Some installations use the front row of generating tubes to form a roof waterwall in the furnace extension.

The upper drum is known as the steam drum; the lower, as the mud drum. The tube sheets are braced by sling stays of proper size and pitch. The heads opposite the tube sheet in the steam drum and in the mud drum do not need staying, for they are dished to the proper radius to be self-supporting at the design pressure.

Tube replacement may be accomplished by insertion through a circumferential row of handholes provided in the upper head of the steam drum for this purpose. Access to these drums is by a manhole in the dished head of each.

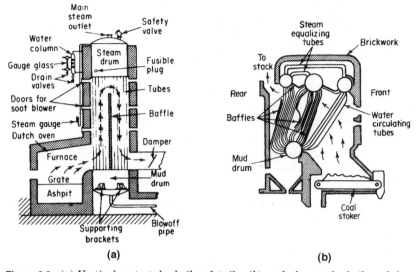

Figure 3.3 (*a*) Vertical watertube boiler details; (*b*) early bent-tube boiler of the Sterling type.

Bent-Tube Watertube Boilers

Bent tubes are more flexible than straight tubes. Boilers can be made wide and low, where headroom is limited, or narrow and high, where floor space is at a premium. Also, bent-tube boilers allow more heating surface to be exposed to the radiant heat of the flame. Drums serve as convenient collecting points in the steam-water circuit and for separation of steam and water. Thus boilers with two, three, and sometimes four drums have been used. As boilers grew in size (made possible by bent-tube design), the demand for more active furnace cooling increased. It was then that waterwalls and other improvements in design were made. A better knowledge of fluid dynamics resulted in simpler and much safer methods for the circulation of waterwall fluids, on both the gas and the steam sides.

Two-drum boilers, even boilers with but one drum on top and one or two large headers at the bottom, became commonplace. Thousands of boilers of the vertical-header type and of multidrum bent-tube design are still operating today. The tubes are bent for several reasons:

1. Heat-transfer reasons make it impossible to use straight tubes.

2. The bent tube allows for free expansion and contraction of the assembly, usually on the lower mud-drum end, since the upper drum (or drums) is separated or suspended by steel structures.

3. The bent tubes enter the drum radially to allow many banks of tubes to enter the drum.

4. Bent tubes allow greater flexibility in boiler-tube arrangement than is possible in straight tube boilers.

The drilling of tube holes into the shell or drum of watertube bent-tube-type boilers weakens the shell or drum in resisting longitudinal and circumferential stresses. This is treated in the ASME Boiler Code by assigning a ligament efficiency to the tube hole arrangement in calculating the maximum allowable pressure for the shell or drum. The portion of the shell or drum that has the tube holes is also referred to as the tube sheet, following the firetube designation. Calculation of ligament efficiency and allowable pressure on shells and drums is covered in later chapters. Generally, the tube sheet section of a drum or shell is made thicker in order to strengthen the shell or drum as a result of the tube hole drilling.

Stirling boilers

The Stirling boiler was one of the first types of bent-tube boiler to come into common use. Figure 3.3*b* shows a unit with front and middle boiler banks connected to both the front and the middle upper drums. This arrangement equalizes the discharge of the steam-and-water mixture to improve circulation and reduce carryover with the steam. Boilers of this general type were usually designed for pressures from 160 to 1000 psi and capacity range from 7500 to 350,000 lb/hr of steam. Both the top and the bottom drums have brickwork built partly around them. Because the Boiler Code required the longitudinal joint to be away from furnace heat, the joint was usually under this brickwork. Even if it means removing some brickwork, always check the condition of longitudinal and circumferential riveted or welded joints, including the caulked edges, and look for caustic embrittlement affecting the drums.

Feedwater entrance and circulation. The feedwater pipe enters through the top of the rear steam drum or through the upper rear part of its manhole head. Feedwater discharges into a long trough riveted or welded along the rear part of this drum. It then spills over the front of the trough to enter the boiler-water circulation.

The water circulates downward through the rear bank of tubes to the mud drum and supplies the middle and front banks of tubes through which it rises to the respective steam drums. It circulates from the front to the middle steam drum through the short circulating tubes, thus tending to equalize the water level in these two drums. As the rear rows of tubes from the middle drum pass back to join the rear tube bank where lower gas temperatures permit a downward circulation, the water level in the middle and rear steam drums tends to equalize through the mud drum.

The mud drum is large in diameter, and all the riser and downcomer generating tubes are remote from its bottom. Sediment is deposited here, for this section is not disturbed by rapid circulation. The blowdown connects into a forged-steel fitting riveted or welded to the bottom of the mud drum. In the case of large boilers with long mud drums, more than one blowdown may be provided.

Baffles are extensively used in bent-tube boilers as shown in Fig. 3.3*b*. These require inspection to make sure they are still functional in order to avoid high temperatures in other sections of the gas passages. This can cause circulation problems and overheating of tube sections.

Modern bent-tube boilers

Modern packaged boilers have grown in popularity and size for industrial occupancies. Today, most packaged watertube boilers follow one of the three designs shown in Fig. 3.4. These are known as A, D, and O types.

A typical factory-assembled D-type watertube boiler consists of two drums and comes equipped with a low-pressure air-atomizing burner. The furnace walls are water-cooled at the front and rear walls, and the outside walls, floor, and roof are cooled with tangent tubes. The boiler can be equipped for either top or side flue-gas outlets. Both the inner and outer casings are constructed of 10 gauge steel. The inner casing is completely seal-welded and pressure-tested at the factory. The heavy 10 gauge outer seal-welded casing is suitable for outdoor installation. The inner furnace walls are made of tangent tubes. See Fig. 3.5.

Packaged watertube boilers usually are equipped with safety combustion-ignition and flame-failure equipment. Both the pilot and the

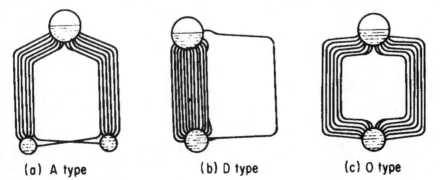

(a) A type   (b) D type   (c) O type

Figure 3.4 Modern factory packaged boilers are arranged in A, D, and O-type configuration.

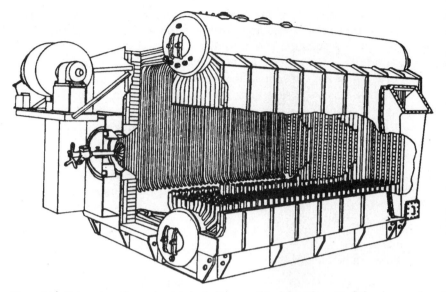

Figure 3.5 D-type two-drum, bent-tube boiler, equipped for oil or gas firing in a water-cooled furnace. (*Courtesy Combustion Engineering, Inc.*)

main burner are monitored by a flame-sensitive scanner (see later chapters). Safety shutdown, safety interlocked starting, and main-flame operation features are provided by most manufacturers. Limit switches are included with interlocks for low water, high steam pressure, fan failure, low oil pressure and temperature for units firing oil, high and low fuel gas pressure for those firing gas, and flame-failure devices suitable for either fuel.

Coal-fired packaged boilers. Stoker-fired, watertube boilers generally are built up to about the 250,000 lb/hr rating. Above this rating, pulverized-coal and cyclone-fired units are generally used. Modern stoker-fired boilers are usually of the two-drum type. Long-drum and cross-drum designs are used with bent tubes. On a long-drum boiler, the flue gases flow lengthwise to the drums, while on a cross drum they flow across or perpendicular to the drum. Figure 3.6 is an artist's drawing of a 500-hp steam generator fired by a single-retort underfeed stoker with the design patented by the E. Keeler Co.

Combination Fuel Burning

Boilers burning combination fuels, such as shown in Fig. 3.7, require special considerations. Among these are the following. When oil and/or gas capability is added to a solid-fuel boiler, consider (1) superheater and attemperator to be used; (2) excess air requirements; (3)

Figure 3.6 Long drum on top and short drum on bottom is a distinguishing feature of a 500-hp CP steam generator fired by underfeed stoker. (*Courtesy E. Keeler Co.*)

NO_x emission standards to be observed; (4) tube spacing needs to avoid fouling and draft restrictions; (5) air-heater metal temperatures to be expected; (6) final flue-gas temperatures to be obtained.

When solid fuel is burned on a grate, air cooling must be provided for oil or gas burner parts to withstand the heat released by radiation from the furnace interior. This may affect efficiency by burning with excess air. When oil or gas from wall burners is burned, a flow of cooling air is required through the grates in order to prevent overheating. For this reason, it is wise to consult with the original boiler manufacturer for design recommendations when considering converting a boiler to a multifuel burning boiler.

Changes in fuel

Since most boiler furnaces are designed for a certain fuel, it is necessary to check with the boiler manufacturer when operation with some other fuel is being considered. Some of the factors that must be considered are quantity of combustion air needed, fan sizes, furnace volume in order to obtain complete combustion in the furnaces, and not

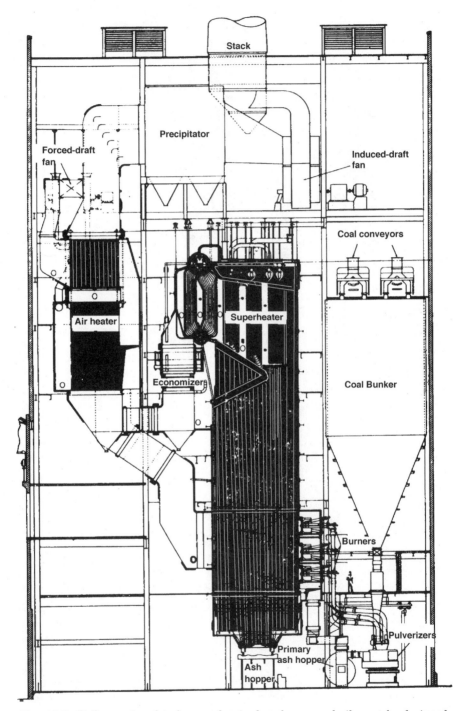

Figure 3.7 Boiler equipped to burn pulverized coal, gas, and oil must be designed accordingly. (*Courtesy Babcock & Wilcox Co.*)

in flue passage areas, the size of the waste-gas-handling equipment, and whether this equipment can handle the different fuels' products of combustion without corroding, eroding, or not meeting environmental standards, such as smoke, SO_2, NO_x, and toxic discharges.

Utility Boilers

Large-capacity utility boilers are classified by whether they operate at subcritical or supercritical pressure. Further classification is also made to indicate whether natural or forced circulation is used.

Supercritical boilers

Supercritical boilers (see Fig. 3.8) are those that operate above the critical pressure of the water-steam loop of 3206.2 psia, and an oper-

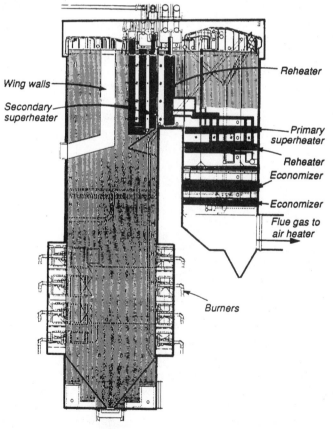

Figure 3.8 A supercritical steam generator rated 9,775,000 lb/hr, 4200 psi outlet pressure, 1010°F superheat, 1000°F reheater outlet, burning pulverized coal, and supplying steam to a 1300-MW turbogenerator. (*Courtesy American Electric Power System*)

ating temperature above the critical temperature of water of 705.6°F. Critical pressure is the pressure at which water and steam have the same density, while the critical temperature is the temperature above which water cannot exist as a liquid, no matter how high the pressure. This means water at a temperature of 705.6°F will also be at the critical pressure of 3206.2 psia. There is no latent heat of vaporization above the critical pressure, because water and steam have the same density. There is no drum to separate water and steam.

Supercritical boilers operate "once-through" the water tubes in contrast to conventional units where the liquid may circulate through the tubes more than once before it is converted to steam. This is done by forced circulation in a once-through unit by the boiler feedpump, or auxiliary pump, in contrast to the natural circulation in subcritical units. Feedwater in a supercritical unit is pumped into the boiler inlet and passes in a continuous path to the boiler outlet. Feedwater is introduced at a pressure above the critical pressure, while the water temperature is below the critical temperature but picks up temperature in going through the boiler.

There is a transition zone where water becomes steam. It is necessary for the transition zone to be in a location hot enough to carry the water beyond the critical temperature, and cool enough to prevent tubes overheating. Therefore, supercritical boilers must have stringent feedwater chemistry control in order to avoid tube failures because of impurities in the water. It is necessary to purify the feedwater to a much greater degree than with conventional drum boilers, from which deposits can be eliminated by periodic blowdown.

Figure 3.8 shows a supercritical boiler. The maximum continuous rating is 9,775,000 lb/hr, at 3845 psi and 1010°F at the secondary superheater outlet. Reheat is at 1000°F. It supplies a 1300-MW turbogenerator.

Fluid flow through the unit starts at the economizer, through economizer banks and stringers to two outlet headers. Two downcomers bring the fluid to the furnace and convection pass circuitry and to the primary and secondary superheaters. Four crossover lines between the primary and secondary superheaters contain the superheater attemperators and the bypass piping connections. There are two secondary superheater outlets and two reheater outlets.

Figure 3.9 is a supercritical unit (3500 psi and 1000°F supplying a single 600-MW turbogenerator) and uses superheater bypass valves to pass steam to a flash tank on startup. The flash tank acts like a steam drum by scrubbing moisture particles from the flashed fluid. This scrubbed steam is returned to the steam generator via the steam return valve and the primary superheater supply downcomer. This mode of bypass startup operation reduces the silica carryover to the

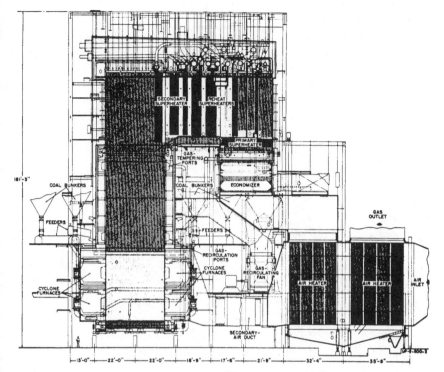

Figure 3.9 Cyclone coal-fired boiler 600-MW, 3500 psi, 1000°F, two-reheater steam generator. (*Courtesy Indiana & Michigan Electric Co.*)

superheaters. The bypass steam is directed through feedwater heaters before going to the condenser, thus reducing heat losses.

Controlled circulation boilers. The LaMont-type boiler shown in Fig. 3.10*a* is called a *controlled-circulation boiler* because the quantity of water passing through the boiler is from 3 to 20 times the amount evaporated. Thus two pumps are required, one for circulating the high rate of flow through the tubes (no natural circulation), the other as a conventional boiler feed pump. The feed pump operates on the same principle as most boiler feed pumps by maintaining a constant level of water in the drum.

The function of controlled circulation is to establish a flow through the first section of inlet to the boiler to prevent the water in the tubes from evaporating to complete dryness. Instead, evaporation is only to the extent that dissolved solids and salts will remain in solution. This solution (a mixture of water and steam) passes to the steam-and-water drum, where the steam is separated while the excess water is removed. The separated water, along with feedwater, is returned to the pumps through downcomers.

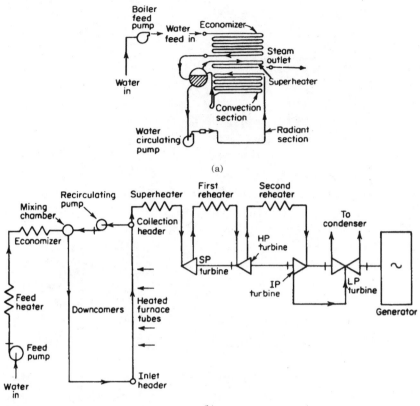

Figure 3.10 Forced circulation boilers require a separate pump in addition to the boiler feed pump. (LaMont design) (*a*) LaMont principle requires separate pump to overcome frictional resistance to flow. (*b*) Boiler-turbine arrangement using forced recirculation in a supercritical unit.

By means of continuous or intermittent blowdown, some water and solids in solution are removed. The separated steam is passed through the superheater for final usage.

Controlled circulation in large watertube boilers permits water to be carefully apportioned to furnace walls, boiler-tube sections, parallel tubes or tube, in accurate, predetermined amounts. Water can even be changed in total flow or distribution within a certain range at any subsequent time in operation. This is usually done by installing circulating pumps between the boiler drum and water inlet to the heat-absorbing surfaces. The result is positive flow in one direction at all times, regardless of heat application.

Figure 3.10*b* is a schematic diagram of a supercritical boiler-turbine arrangement using forced recirculation and equipped with two reheaters.

Supercritical boiler code exceptions. Special ASME Code exceptions or rules have been made in reference to once-through forced-flow steam generators which have no fixed waterlines and steam lines. For example,

1. It is permissible to design the pressure parts for different pressure levels along the path of water-steam flow.

2. No bottom blowoff pipe is required.

3. If stop valves are installed in the water-steam flow path between any two sections of line, certain safety valves or power-actuated pressure-relief valves, with control impulse interlocks, are required. (This is different from the typical boiler with safety valves protecting the entire boiler.)

4. Pressure gauges required are more numerous at various sections. No water-gauge glass or gauge cocks are required.

Coil Watertube Boilers

Coil boilers were developed to satisfy industry's need for a compact, fast-steaming, factory-assembled packaged boiler. They find special application where a process requires high-pressure steam in one part of the process flow and the capacities required are moderate. A packaged unit is placed where the load need exists, and this makes it unnecessary to operate large, centralized boilers at reduced capacity during periods of operation when other parts of the plant may have low demand. Several packaged boilers can be placed close to the steam loads of a plant at widely separate locations, thus avoiding long steam-line losses that may exist with a centralized steam plant. Coil-type boilers are used over packaged firetube types when high pressures and capacities may be required. Pressures up to 900 psi are possible with coil-type watertube boilers. Capacities generally are below 10,000 lb/hr, but units of greater size are available.

The generating tubes of coil-type boilers consist of small-diameter helical, spiral, or horizontal coils of tubing. Some large units consist of a series of bundles connected to make a continuous coil. In case of a coil failure, the affected coil section need only be removed and connected to the remaining coil sections. Figure 3.11 shows the steam-and-water flow of a coil-type unit. Note the boiler uses forced circulation and forced recirculation, which permits smaller tubes to be used with high steam velocities and high heat-transfer rates.

The flow of water and steam in the system shown in Fig. 3.11 starts with water entering the feedwater section of the water pump with the water pumped directly to the steam accumulator. The feedwater rate is controlled by the water-level control, which in turn responds to the

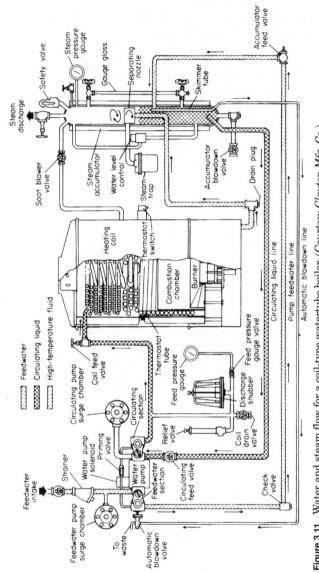

Figure 3.11 Water and steam flow for a coil-type watertube boiler. (*Courtesy Clayton Mfg. Co.*)

Feedwater

Circulating liquid

High-temperature fluid

Steam discharge

Safety valve

Steam pressure gauge

Gauge glass

Separating nozzle

Soot blower valve

Skimmer tube

Steam accumulator

Water level control

Steam trap

Accumulator blowdown valve

Drain plug

Accumulator feed valve

Heating coil

Thermostat switch

Combustion chamber

Burner

Circulating pump surge chamber

Coil feed valve

Circulating section

Thermostat tube

Feed pressure gauge

Feed pressure gauge valve

Discharge snubber

Circulating liquid line

Pump feedwater line

Automatic blowdown line

Water pump solenoid

Priming valve

Relief valve

Water pump

Feedwater section

Coil drain valve

Circulating feed valve

Check valve

Feedwater intake

Strainer

To waste

Feedwater pump surge chamber

Automatic blowdown valve

liquid-level control in the accumulator. The circulating liquid from the accumulator is then pumped to the heating coils of the boiler. Flow of liquid in the coils is downward through the coils spiral-wound in a counterflow direction to the combustion gases. When leaving the spiral generating section, the fluids pass through the ring thermostat and the helically wound waterwall section and then to the separating nozzle in the accumulator. The centrifugal action of the nozzle separates dry steam from the liquid and allows excess liquid to return to the lower section. The dry steam from the accumulator is discharged on top through the discharge valve.

An unusual feature is the ring thermostat tube shown in Fig. 3.12*a*. This tube is an integral part of the heating coil, actuated directly by combustion heat. In case of water shortage or any excessive-heat condition, the thermostat control will expand beyond the set point and thus shut off the fuel supply.

Another design has three nests of coils and inlet and outlet headers. It is widely used in small sizes up to 300 hp and pressures to 250 psig.

Malfunctions in the loop may cause a coil failure from overheating. This could come about from pump failure, partial blockage of inlet and outlet lines, blocked tubes (scale), fire-side soot accumulation in concentrated heat zones, malfunction of controls, etc. Thus it is essential to keep both the water side and fire side of the coils clean; proper feedwater treatment is vital. A pressure-differential chart is often used to show whether the difference between suction and discharge pressure on the recirculation pump exceeds a certain pressure differential. See Fig. 3.12*b* and *c*. If so, it is a sign of tube or flow obstruction. In a conventional boiler of multitube design, a leaky tube can be plugged, then the boiler operated until it is convenient to replace the affected tube. With a coil-type boiler, the entire coil must usually be replaced. All coil-type packaged boiler manufacturers supply excellent instruction manuals with their units which include maintenance and feedwater treatment sections as well as a description of the controls used. Prominent coil-type boiler manufacturers are Clayton, Besler, Francis, Mund-Parker, Vapor-Clarkson, and Sid Parker.

Large Watertube Boiler Components

The term *steam generator* is used to indicate a large boiler with many heating surface components, as shown in Fig. 3.13. This includes waterwalls, economizers, superheaters, reheaters, and air heaters. As shown in Fig. 3.13, it includes fuel-burning equipment, draft systems, discharge gas or ash removal equipment as well as feedwater-treating loops on the water supply. Large watertube boilers have far more

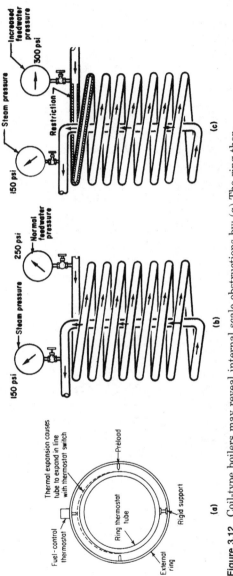

Figure 3.12 Coil-type boilers may reveal internal scale obstructions by: (*a*) The ring thermostat shuts the burner off from overheating temperatures; (*b*) abnormal feedwater pressure from that established when coils were new, or when benchmarks were established; (*c*) increased feedwater pressure above normal (may be a sign of flow restrictions).

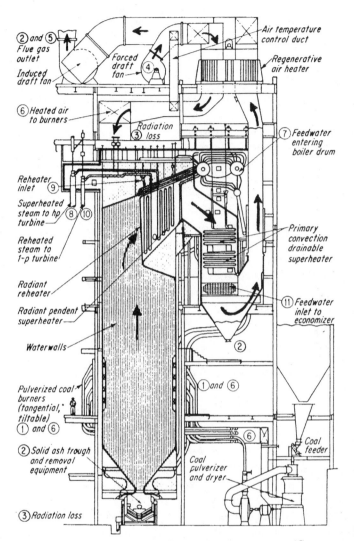

② and ⑤
Flue gas
outlet

Induced
draft fan

Forced
draft
fan ④

Air temperature
control duct

Regenerative
air heater

⑥ Heated air
to burners

Radiation
③ loss

⑦ Feedwater
entering
boiler drum

Reheater
inlet ⑨

Superheated
steam to hp ⑧ ⑩
turbine

Reheated
steam to
t-p turbine

Primary
convection
drainable
superheater

Radiant
reheater

Radiant pendent
superheater

⑪ Feedwater
inlet to
economizer

Waterwalls

②

Pulverized coal
burners
(tangential,
tiltable)
① and ⑥

① and ⑥

⑥

Coal
feeder

② Solid ash trough
and removal
equipment

Coal
pulverizer
and dryer

③ Radiation loss

Figure 3.13 Components of a large steam generator. (*Courtesy Power magazine.*)

extensive heat-absorbing components than do firetube-type boilers, as well as other construction details that need to be reviewed.

Older boilers, especially the bent-tube type, used to be set up in pairs with a common parting wall between them which covered the two heads next to each wall. This could be a steam drum or mud drum. Because the external surfaces of the heads were covered, the term *blind head* was applied to these drum heads.

There is a tendency for soot, ashes, and moisture to collect in the space between the brickwork and external surfaces of the blind heads. And this condition is not directly observable. In older inspection days, it was common for an inspector to insist that the brickwork be removed from around these blind heads. Then the external surface was available for inspection and drill-tested for thickness. Today, ultrasonic instruments can be used to get a profile of the thickness of the head. These checks on deterioration are made without disturbing the brickwork unless the ultrasonic check indicates it is necessary.

Firing doors on watertube boilers are required to be of the *inward-opening* type or a type provided with self-locking door latches of a style omitting springs or friction contact, so that the door will not be blown open from pressure in the furnace in case of a tube rupture or furnace explosion and thus possibly burn or scald an operator standing by the boiler. Explosion doors, if used and if located in the setting walls within 7 ft of the firing floor or operating platform, must be provided with substantial deflectors to divert any blast.

Baffles deflect the hot gases back and forth between the tubes a number of times to enable greater heat absorption by the boiler tubes. They also permit designing for better temperature differences between tubes and gases throughout the boiler. Baffles help maintain gas velocity, eliminate dead pockets, deposit fly ash and soot for proper removal, and prevent high draft losses. When a furnace baffle breaks, the gases short-circuit one or more passes, causing excessive flue-gas temperatures and a loss in efficiency and capacity. Overheating and damage might result in those parts of the boiler designed for low gas temperatures. Thus on any outage inspection, the baffles should always be carefully checked for erosion, breaks, leakage (around tubes), or dislocation, for tube failure may result.

Waterwalls consist of relatively close-spaced vertical tubes forming the four walls of the furnace. They were originally developed to cool and protect the furnace lining. One design of large power-generating boilers has 144-ft-high, 0.340-in.-thick tubes at the hottest furnace zone (below 85-ft elevation) but only 0.320-in.-thick tubes above.

Depending on the type of boiler, the waterwall heating surface may account for only 10 percent of the boiler's total heating surface, yet represent as much as 50 percent of the total heat absorption. Waterwalls perform three basic functions: (1) protect the insulated walls of the furnace; (2) absorb heat from the furnace to increase the unit's generating capacity; and (3) make the furnace airtight (on pressurized furnaces with tangent-welded tubes).

Heat is transferred to the waterwall tubes as radiant heat from the zone of highest temperature in the furnace. Because of the great amount of heat absorbed by that part of the boiler, feedwater must be of the best quality. Also, the circulation of water must be rapid and plentiful to ensure positive flow through each tube at all times.

Blowdown valves are required for each header at the bottom of a series of waterwall tubes, for the same reasons that the boiler itself needs them. Sediment accumulation in a header supplying wall tubes might cause interruption of circulation, with consequent overheating and failure of the tubes.

Warning. Under no circumstance should waterwall headers be blown down while the boiler is operating. If they are blown, the boiler's normal circulation will be upset, and the overheated tubes will bulge or rupture.

Figure 3.14 shows some typical arrangements. The waterwall in Fig. 3.14a is designed for moderate cooling. This design has the tubes spaced apart and the wall surface composed of part firebrick. The brick is backed with several layers of insulation and strong steel casing. Reinforcing metal lath is often used in wall construction. Figure 3.14b shows how tangent tubes are placed in the furnaces of many large and small boilers. The staggered-tube arrangement offers a high heat-absorbing surface that is backed by solid block, or plastic, insulation and a strong steel external casing. Figure 3.14c shows steel lugs or longitudinal fins welded to nontangent wall tubes. In some designs the lugs protrude from the tube into the furnace and are covered with a chrome-base refractory or slag. To ensure furnace tightness, adjacent fins are often welded.

Figure 3.14d shows newer gastight casings, known as membrane walls. Here tightness is obtained by welding a flat strip of metal between the tubes. This eliminates the casing and many of its problems. Insulation is applied directly to the outside of the tubes, while metal lagging is attached to give the outer surface durability and good appearance. Figure 3.14e shows how the outer casing, insulation, and steel-skin casing are often constructed.

Waterwall strengthening. The firing of fuel in suspension in large furnaces can create a delayed ignition fire-side explosion. In some furnaces, the resultant pressure rise can be quite large; however, most steam generator designers use waterwall reinforcements to protect the waterwalls against common puffs or minor explosions that seem to assume about a 5 psi rise in pressure. It is something to review with the boiler manufacturer, as some furnace explosions have reached much higher pressures. This caused tubes to bulge outward, and in some cases, pull out of header attachments.

To strengthen furnace walls, boiler designers use bars and channel irons welded to tubes to form a band around the waterwalls at different levels, based on the forces that must be resisted from a furnace explosion. See Fig. 3.15. *Buckstays* are used to transmit the force

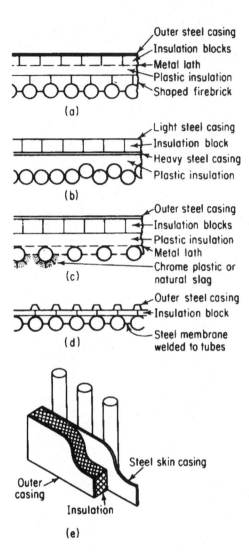

(a)

Outer steel casing
Insulation blocks
Metal lath
Plastic insulation
Shaped firebrick

(b)

Light steel casing
Insulation block
Heavy steel casing
Plastic insulation

(c)

Outer steel casing
Insulation blocks
Plastic insulation
Metal lath
Chrome plastic or natural slag

(d)

Outer steel casing
Insulation block
Steel membrane welded to tubes

(e)

Steel skin casing
Outer casing
Insulation

Figure 3.14 Possible waterwall tube arrangements in a steam generator.

from the bands surrounding the waterwalls to steam generator support beams, which are in turn connected to steel columns.

Superheaters. Each pressure of saturated steam has a corresponding temperature. Heat added to the dry steam at this pressure is known as superheat and results in a higher temperature than that indicated on the curve for the corresponding pressure.

The advantage of superheated steam in prime movers is twofold: (1) Work may be done down through the superheat range before condensation starts to take place. This represents an increase in steam-utilization efficiency. (2) This period of work performed with dry

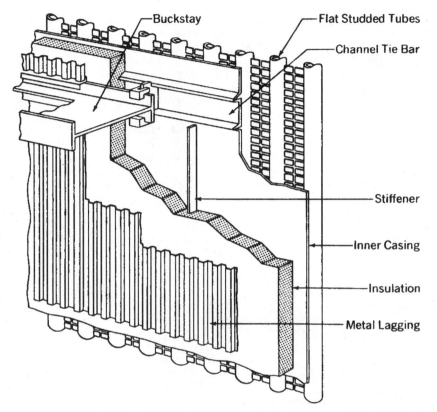

Figure 3.15 Waterwalls are reinforced with structural steel supports to withstand abnormal furnace pressure rises. (*Courtesy Babcock & Wilcox Co.*)

steam eliminates corrosive and erosive effects of condensate. The deterioration of high-speed turbine blades caused by impingement of drops of condensate may be considerable.

Superheat is produced by passing the flow of saturated steam from the boiler through a superheater of one or both of two types, radiant and convection.

Figure 3.16 shows some typical superheaters and their locations in the boiler. The general classification includes *radiant* and *convection* types, depending on whether they absorb radiant or convection heat. The *interdeck* type has tubes arranged between banks of primary boiler tubes. The *pendant type* is a suspended series of coils, usually shielded against radiant heat by a screen of boiler tubes. It is often arranged as the first steam heater before the steam goes to the superheater outlet header. The *platen type* is similar to the pendant type, but the tubes are in one plane. Usually the steam goes through the platen superheater before it enters the pendant superheater.

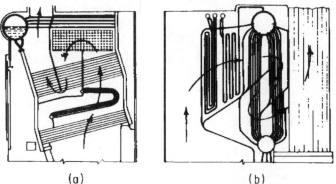

(a)   (b)

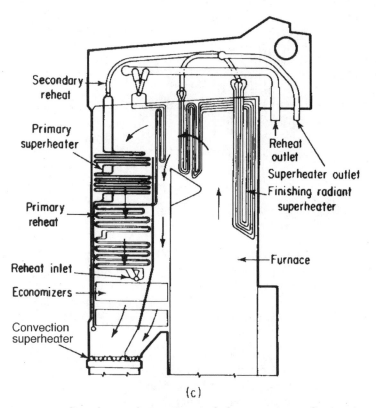

(c)

Figure 3.16 Superheater designations include convection, radiant, primary, secondary and finishing, depending on steam generator arrangement.

Constant temperature of superheated steam is the desire of most designers, for a steam turbine is designed for the particular steam temperature at which it will operate most efficiently. Characteristics of the convection-type superheater may produce a drooping temperature curve with increasing combustion rates, whereas the reverse may be true with radiant types. Thus a combination of the two types in certain installations, in order to obtain a practically constant superheated steam temperature with varying loads, is employed. Damper control of gas through the passes is another development for control of superheat temperature.

Economizers serve as traps for removing heat from the flue gases at moderately low temperature, after they have left the steam-generating and superheating sections of the boiler. The general classification is:

1. Horizontal- or vertical-tube, according to the direction of gas flow with respect to the tubes in the bank

2. Parallel-flow or counterflow, with regard to the relative direction of gas and water flow

3. Steaming or nonsteaming, according to thermal performance

4. Return-bend or continuous-tube types

5. Plain-tube or extended-surface types, according to the details of design and the form of the heating surface

The tube bank may be further designated as of the staggered or in-line arrangement, with regard to the pattern and spacing of tubes which affect the path of gas flow through the bank, its draft loss, heat-transfer characteristics, and ease of cleaning.

Theoretically, economizer heating surface could be added to a boiler until the exit temperature neared the outside air temperature. But an abnormally high heat-surface area would be needed. Further, each fuel burned has a *dew-point temperature* which can cause moisture accumulation on the economizer and corrode the surface in a short time. The amount of heating surface which should be used in the economizer is limited by the final gas temperature at the exit. If the gas temperature is cooled below the dew point, condensation (sweating) may result. Sulfur in the soot then unites with moisture to produce a sulfurous acid, which is extremely corrosive to all steel construction contacted between the economizer and the stack.

Feedwater of low oxygen content is recommended in using the steel-tube type of economizers. Relief valves (water safety valves) are required on economizers to protect them against excessive pressure that might be built up by the feed pump if the regulator or feed valve to the boiler were closed.

A *reheater* is essentially another superheater used in modern utility boilers for boosting plant efficiency. While the superheater takes steam from the boiler drum, the reheater obtains used steam from the high-pressure turbine at a pressure below boiler pressure. This lower-pressure steam passing through the reheater is heated to 1000°F and then is introduced into the intermediate, or low-pressure, turbine. Reheaters, like superheaters, are also classified according to their location in the boiler as convection or radiant. Convection super-heaters and reheaters may be of the horizontal or pendant type.

Air heaters make the final heat recovery from boiler flue gases with which they preheat the incoming furnace air for its combustion with fuel. Thus some fuel is saved which would otherwise be used in heating the air-fuel mixture to its ignition point. But the temperature of the flue gas must not be reduced below its dew point since moisture would condense out of the flue gas. That would cause water to combine with sulfur and possibly carbon dioxide, also carbon monoxide, to form highly corrosive sulfurous and carbonic acids.

A *regenerative air heater* is shown in Fig. 3.17. This type of air heater offers a large surface of contact for heat transfer. It usually consists of a rotor which turns at about 2 to 3 revolutions per minute (r/min) and is filled with thin, corrugated metal elements. Hot gases pass through one half of the heater; air passes through the other half.

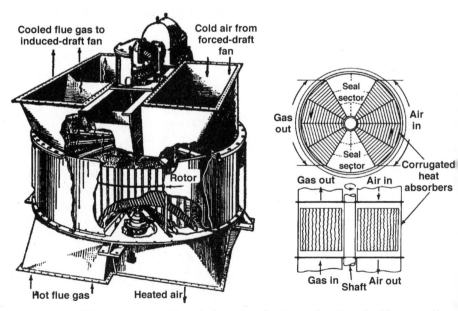

Figure 3.17 Rotating regenerative air heater preheats combustion air. (*Courtesy Air Preheater Corp.*)

As the rotor turns, the heat-storage elements transfer the heat picked up from the hot zone to the incoming-air zone.

Figure 3.18 illustrates the *internals* of a typical drum, which performs two essential functions: separates steam from water to provide the downcomer system with steam-free water necessary for proper and safe circulation, and it separates moisture from steam to provide high steam quality. The drum internals shown provide both functions by means of two stages of separation. The normal water level is 1½ in.

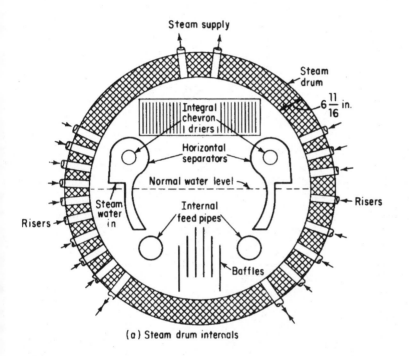

(a) Steam drum internals

(b) Horizontal steam separator

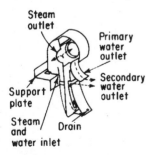

(c) Unit chevron drier

Figure 3.18 Internals of a high-pressure steam drum showing steam separator and chevron drier.

below the horizontal centerline of the drum. Vortex eliminators separate the steam and water passages in the drum.

The total circulating steam-and-water mixture from the steam-generating tubes is directed to the horizontal turboseparators (Fig. 3.18). The steam-and-water mixture enters the separators and is centrifuged by following the curved contour of the separator dome. Most of the separated water is discharged horizontally at the water level of the drum. The separated steam flows to the chevron driers at the top of the drum. The steam flows from the chevron through the dry box, then out the top of the drum via steam tubes across the roof to the partition-wall superheater.

The opening to the main steam line will usually be adequate to handle the design flow capacity. The ASME Boiler Code has definite rules on the opening to a safety valve. For example, internal collecting steam pipes, splash plates, or pans are permitted to be used near safety valve openings, provided "the total area for inlet of steam thereto is not less than twice the aggregate areas of the inlet connections of the attached safety valves. The holes in such collecting pipes must be at least ¼ in. in diameter."

In the case of steam scrubbers or driers, the ¼-in.-diameter opening does not apply, provided "the net free steam inlet area of the scrubber or drier is at least 10 times the total area of the boiler outlets for the safety valves." Therefore, when inspecting the internals of drums, check the condition of the ¼-in.-diameter holes in the collecting pipes. They must be free and clear. The same applies to the openings on driers, because plugged driers on collecting pipes could lead to restrictions of the safety-valve openings. That would reduce relieving capacity flow, which could be dangerous.

Steam generator drum supports.   Drums of smaller watertube boilers are generally bottom-supported as is shown in Fig. 3.19b. Larger boilers, or steam generators, are top-supported as is shown in Fig. 3.19a. The drums must be able to move in a longitudinal direction with temperature changes on the unit. Bottom-supported units will have rollers on one end of the drum so that it can expand or contract with temperature changes. Enclosures usually consist of waterwalls, and with top-supported drums, portions of the enclosure are supported by pressure parts, but at higher elevations, the enclosure may be supported by hanger rods attached to structural steel. Bottom-supported drums may have the enclosure also supported by the common foundation on which the drum rests.

Soot blowers are used extensively in pulverized-fuel-burning plants in order to keep heat-absorbing surfaces free from soot and ash buildup. Changes in flue gas temperatures at constant load and draft loss are indications of the need for soot blowing of tubes in the flue

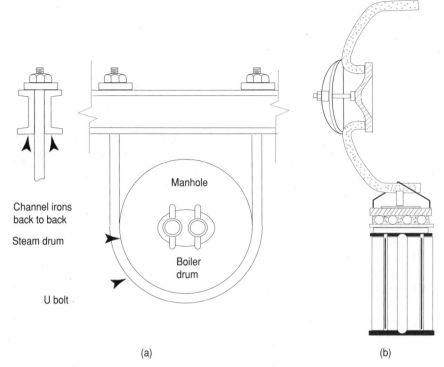

Channel irons
back to back

Steam drum

U bolt

Manhole

Boiler
drum

(a)

(b)

Figure 3.19 Top and bottom supports for boiler drum. (*a*) U-bolt top-supported drum; (*b*) bottom-supported drum rests on saddles with rollers permitting longitudinal movement.

gas passages. It is important to check soot blower alignment and condensate leakage from soot blowers that may affect the tubes by erosion, or corrosion. Frequency of blowing soot or ashes will depend on the fuel burned and conditions found during inspections. Usually, the boiler manufacturer will provide an initial schedule, which can then be adjusted if required by conditions found during inspections or operations.

Watertube Boiler, General or Code Requirements

Watertube attachments. For high-pressure service, the ASME Code stipulates the following attachment of water tubes to tube sheets, headers, and drains: (1) Expanded or rolled and flared (Fig. 3.20); (2) flared not less than ⅛ in., rolled, and beaded; (3) flared, rolled, and welded; (4) rolled and seal-welded, provided the throat of the seal weld is not more than ⅜ in. and the tubes are rerolled after welding; (5)

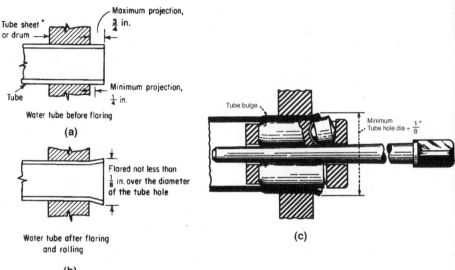

Figure 3.20 Watertube boiler tubes are flared. (*a*) and (*b*) Code-required flaring details; (*c*) expander rolls the tube and bells, or flares, at the same time.

superheater, reheater, waterwall, or economizer tubes may be welded without rolling or flaring, provided that the welds are heat-treated after welding and the welding is done according to Code requirements.

Inspection openings. The ASME Code requires suitable manholes, handholes, or other inspection openings so that examination and cleaning of the drum or shell is possible. Coil-type boilers do not have the typical shell or drum, but do have accumulators and separators outside the firing area of the boiler. Most inspectors want to see if any sediment is accumulating in this vessel.

Manhole openings must be a minimum of 11 by 15 in. or 10 by 16 in. if elliptical, and 15 in. in diameter if circular.

Handhole openings cannot be less than 2¾- by 3½-in. size.

Threaded openings used for inspection or washout must be not less than 1-in. pipe size.

On shell- and drum-type boilers, it is common to find a standard-size manhole in at least one head of each drum. Handholes are provided in external mud drums to permit cleaning and inspection.

Water-column connections. The water column is connected to the upper and the lower part of the front head of the main steam drum. See Fig. 3.21. If more than one drum is installed at the same level, it is not necessary usually to provide a separate water column for each drum, for the water level should equalize in all the drums.

Figure 3.21 Manhole, water column and gauge glass are located in top drum of this D-type package boiler. (*Courtesy Cleaver Brooks Co.*)

Water columns must be so mounted and be visible to the operator so that they will show at all times if the normal water line for the boiler is being maintained. The pipes connecting the column to the boiler must be a minimum diameter of 1 in. For boilers operating over 400 psi, sleeves or shields are required on the lower pipe connection to reduce the temperature differential between the column and the boiler so that accurate water levels exist in the column.

Inspection and cleaning openings are required on the pipes connecting the column to the boiler. No shutoff valves are allowed between the column and the boiler unless the valves or cocks are of a through-flow construction that prevents sediments from accumulating in the pipe connections. The valves or cocks must have indicators to show if they are open or closed, and must be locked or sealed open in operation.

Gauge glasses showing water level in the boiler must be visible to the operator stationed in the boiler operating area. Since large water-tube boilers can be quite high, it may be difficult to see the gauge glass from the operating area. Under these conditions, the image of the gauge glass and water level is usually transmitted to the operating area by remote level indicators, such as mirrors, closed-circuit television, and similar devices. However, the Code requires that if the gauge glass cannot be seen from the operating area, there must be *two independent remote level* indicators installed.

A similar rule exists on boiler *pressure gauges,* or steam gauges as used by operating people. Pressure gauges must be installed in enclosed control rooms remote from the operating area; however, the remote pressure indicators of the pneumatic, electrical, or other remote method of transmission must also be of the two independent types. The Code also requires that under all conditions of remote pressure reading, there must be at least one direct-reading boiler pressure gauge within sight of the boiler operating area.

Many jurisdictions require *permanent platforms and ladders* for boilers with manholes above a certain height. These should also be equipped with railings and toe rails so that proper and safe inspection and maintenance can be performed on the boiler.

Rating of Watertube Boilers

Large watertube boilers do have complex circulation loops consisting of waterwalls, screen walls, radiant platen superheaters which are designed for mass-flow passes through tubes at different loads with the aim of keeping the tube metal within a safe temperature limit. *Mass flow* is defined as the weight of fluid mixtures in pounds per hour passing through the tube. As boilers become larger, it is obvious that extensive calculations must be made in design to consider the variables involved with mass flows, such as velocity of flow, pressures, temperature zones, specific heat of fluid, conductivity, viscosity, tube diameter, and internal surfaces that could affect the friction of flow through a tube. In comparison, the early watertube boilers were relatively simple in design.

The rating of all watertube boilers, as of firetube boilers, is based on the square feet of heating surface. This is the sum of the areas of drum surface exposed to the products of combustion, the area of all watertubes so exposed based on their outside diameter, and the aggregate projected area of all headers so exposed as well as other areas specified by the ASME Code. Minimum safety valve capacities for watertube boilers is reviewed in the chapter on safety valve requirements.

The ASME Code requires manufacturers to install or stamp their boilers with the *maximum* designed steaming capacity in pounds per hour on the basis their designers have calculated outputs per mass flow calculations. It is thus essential to become familiar with this stamped capacity rating when operating a watertube boiler in order to prevent overloading the boiler above its stamped rating.

Watertube Heating Boilers

Cast-iron boilers

The cast-iron boiler is basically a watertube type because the water is inside the cast sections (no tubes) and the products of combustion are on the outside. But because of the limitations of cast iron, the Boiler Code treats cast-iron boilers as a special type, without considering the heat-transfer method. Many cast-iron boilers are stamped by the manufacturer as cast-iron watertube boilers, which should not be misinterpreted as a Code classification.

Cast iron is a term applied to many iron-carbon alloys which can be cast in a mold to make a particular shape. But for cast-iron boilers, gray cast iron is generally used. When the casting is cooled slowly in the molds, part of the carbon separates out as graphite. This makes the gray cast iron less brittle and easier to machine. Also, when it is alloyed with nickel, chromium, molybdenum, vanadium, or copper, considerable tensile-strength properties can be achieved. The general practice is to classify cast iron by class:

Class no.	Ultimate tensile strength, lb/in.^{2}
20	20,000
25	25,000
30	30,000
35	35,000
40	40,000

Cast-iron boilers are built to various shapes and sizes, but can be grouped into the three following broad classifications:

1. Round cast-iron boilers (Fig. 3.22a) consist of a firepot (furnace) section with base, a crown-sheet section, one or two intermediate sections, and a top or dome section. The sections are held together by tie rods or bolts with push nipples interconnecting the water side of the section. Thus water circulates freely through the nipples from section to section. Fuel is burned in the center furnace, with the flue gases rising and flowing through the various gas passages of the water-filled sections, then out to the stack. The round cast-iron boiler used to be popular for hot-water-supply service and was often stamped "hy-test" to indicate an allowable pressure of 100 psi for domestic hot-water service. But it is rapidly being replaced by fired steel-welded water heaters or by electrically heated water heaters.

2. Sectional boilers (vertical) consist of sections assembled front to back, with sections standing vertically and assembled by means of push nipples or screwed nipples.

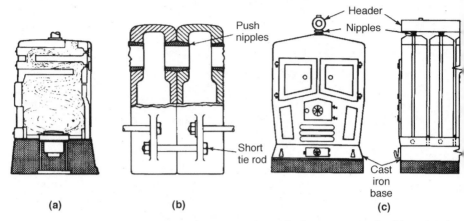

(a)   **(b)**   **(c)**

Figure 3.22 Types of cast-iron boilers. (*a*) Round unit for hot-water supply; (*b*) sectional boiler with sections held together by tie rods and push nipples; (*c*) screwed-nipple type of sectional header boiler.

3. Sectional boilers (horizontal) consist of assembled sections stacked like pancakes. Here each section is laid flat in relation to the base. This type of vertical stacking may be supplemented by having three vertically stacked boilers side by side and interconnected to gain additional capacity. In this arrangement a common supply and return header is used with no intervening valves between the vertically stacked boilers. These units are usually gas-fired, with a burner for each vertical stacking.

Two methods used to aid in assembly of vertical cast-iron boilers are the following:

1. Internal, tapered push nipples are inserted into holes of the vertical section. Then by means of through tie rods, or short tie rods (Fig. 3.22*b*), the sections are pulled together by tightening nuts against washers on the tie rod. That interconnects the sections on the water side, enabling them to withstand pressure as assembled sections.

2. In the external header type (Fig. 3.22*c*), the sections are individually assembled to headers (supply drum and return drums) by means of threaded nipples, locknuts, and gaskets. This type of assembly allows replacing an intermediate section, because only the locknuts, gaskets, and threaded nipple have to be removed on each header to slide a section out. In contrast, with through tie-rod construction, all the sections in front of the intermediate section to be replaced have to be removed first to get at the affected section.

When checking new boilers, it is necessary to check the type of nuts securing the tie rods. If solid-steel or brass nuts are used, they should be only hand-tight or backed off a few threads. The first choice on new boilers is collapsible washers, with shallow split brass nuts. The second choice is split shallow nuts backed off hand-tight (or cloverleaf-type nuts that fail when a slight expansion takes place). On an older unit idle during summer months and located in a damp basement, make sure the holding nuts are not tight. Also determine whether the tie rods are rusted into their holes, which may have the same binding effect as tight securing nuts. Obviously, if the rods are free and nuts slacked off, there should be no problem of expansion cracking. If rust growth and tight tie rods are the cause of cracking, the boiler must be dismantled and the rust buildup removed by chipping.

Header-type boilers have no tie rods or tapered nipples, so nothing can be adjusted to allow for abnormal expansion. In addition to rust depositing between the sections, rapid start-up can cause serious damage to these units. Make sure older boilers without good blow-down facilities do not develop scale buildup, because it causes cracking. Restriction of water supply and circulation can also be caused by scale buildup in these units, resulting in overheating.

When controls are not operating properly, or if the safety relief valve is stuck or of the wrong size, a cast-iron boiler will explode. Many have. The big difference between a shell-type steel boiler explosion and a cast-iron boiler explosion is that a cast-iron boiler will usually fragment into smaller pieces of the affected sections. A steel boiler, on the other hand, rips and tears along the sheet of a drum or shell. Then it flies apart in the form of curved panels of steel. But in each explosion, the danger to life and property is very great.

In cast-iron boilers of different poured shapes, unpredictable stress-concentration areas, geometry, service factors, waterside fouling, soot and carbon accumulation on the fire side, and rapid temperature changes are all causes for cracking. Thus they may never fail twice in an identical manner. Careful investigation will usually point to one or more of the following causes for unexpected cracking failures:

1. Rapid introduction of cold water into a hot boiler, as may occur with poorly operated manual makeup or an automatic-feed device

2. Makeup line connected to a section instead of to the return line to temper the cold water with the returning condensate or hot water (the Code requires makeup to *enter return lines*)

3. Controls not functioning and thus overstressing the boiler by either pressure or temperature

4. Insufficient water in the boiler

5. Internal concentrated deposits, blanking off proper heat transfer or obstructing circulation from section to section

6. Defective material or casting which does not become noticeable until after several years of service

7. Poor assembly or work or improper installation of the boiler as to pitch, alignment, tie-rod tightening, screwed-nipple fitting, etc.

Flexible-tube boiler

Figure 3.23 shows a Bryan patented "flexible-tube" watertube boiler built to capacities of about 6600 lb/hr and pressures to 350 psi for hot-water heating. Note the individual tubes are bent into several S-type

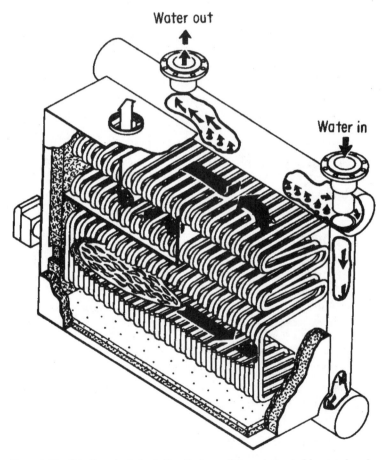

Figure 3.23 Small watertube boiler features S-type patented bent tubes for avoiding thermal shock in hot-water service. (*Courtesy Bryan Steam Corp.*)

shapes, and this provides expansion and contraction in a vertical direction with temperature changes. Tubes are replaceable. They are inserted into the top and bottom shells' tube holes, and secured into the hole by tapered bushings. The end of each tube has a welded steel tapered bushing, which is drive-fitted into the steel or cast-iron header. Studs and clamps are also used on some models to prevent the loosening of the steel bushing. This small watertube boiler comes completely factory assembled for easy installation, primarily for heating.

Questions and Answers

1 Why are tubes flared in a watertube boiler?

ANSWER: To add to the holding power of the tubes after rolling and to prevent the tubes from pulling out of the tube holes if the holes should become enlarged from overheating caused by low water or other reasons.

2 What is the difference between a watertube and a firetube boiler?

ANSWER: In the firetube boiler, the products of combustion pass through the tubes and the water surrounds them. The reverse is the case in watertube boilers.

3 How is a new tube placed in position for installation in a vertical watertube boiler?

ANSWER: Through one of the handholes provided for this purpose in the top head of the steam drum.

4 What type of firing door is required on watertube boilers? Explain your answer.

ANSWER: The inward-opening type or a type provided with self-locking door latches of a style omitting springs or friction contact, so that the door will not be blown open in case of tube rupture or furnace explosion and possibly injure a person standing nearby.

5 Are tubes beaded or flared in the straight-tube watertube boiler? Explain your answer.

ANSWER: They are flared, for the ends are not contacted by hot gases and nothing would be gained by beading.

6 To what points are water-column connections made in most watertube boilers?

ANSWER: To the upper and lower parts of one head of the main steam drum.

7 How many gas passes are there in most standard-type watertube boilers?

ANSWER: Three.

8 Explain the method to be used in checking the operating condition of a water column and gauge glass when there are shutoff valves in the column connection to the drum. What causes water to return slowly into the gauge glass?

ANSWER: First close upper valve and blow column and gauge glass through drain valve with lower column connection valve open. The water should slowly rise in gauge glass from open bottom connection. Then reverse the procedure by closing the bottom connection with the top water-column connection open. Watch water rising in gauge glass. If water rises in gauge glass during both procedures, it indicates connections are not obstructed. A slow return of water to the gauge glass indicates obstruction in the column connection or the bottom of the gauge glass is obstructed. This can be checked by using the gauge glass cocks.

9 What is the maximum allowable working pressure and temperature on a water column made of cast iron and malleable iron per Code?

ANSWER: Cast iron: 250 psi and 450°F. Malleable iron: 350 psi and 450°F.

10 What is the minimum and maximum tube projection from a drum that a tube may have if it is to be flared?

ANSWER: Per Code: ⅛ in. minimum and ¾ in. maximum.

11 How would you define a steam drum in a watertube boiler used for electric power generation?

ANSWER: This is a pressure vessel in which steam is separated from the steam-water mixture in order to be directed to the superheater section of the steam generator for an increase in steam temperature.

12 What is a reheater as used in utility boilers?

ANSWER: A reheater receives steam from a steam turbine after it has given up some of its original heat in doing work in the turbine's high-pressure stages and is reheated to a higher temperature in the steam generator, then returned to the lower pressure stages of the turbine to do more work in driving (usually) an electric generator. The steam line carrying the steam from the turbine to the reheater is called the *cold reheat line,* while the steam line carrying the reheated steam from the reheater to the turbine is called the *hot reheat steam line.*

13 Define the term "economizer."

ANSWER: This is a heat-absorbing section of a large watertube boiler that preheats the incoming feedwater to the boiler, or steam generator, by absorbing heat from the outgoing flue gases.

14 What are the advantages of using tangent inner furnace walls in a watertube boiler?

ANSWER: By using tangent tubes, no tile, refractory, fins, or strip are used with the tangent tubes enclosing the furnace walls to ensure maximum setting cooling and uniform heat absorption in the furnace area.

15 What important safety device must be evaluated when changing fuels in a boiler?

ANSWER: Fuels have different Btus; therefore, the capacity of the safety valves must be checked to make sure they exceed the potential heat release rate of the new fuel.

16 Why is it necessary to defer draining a steam generator until it has cooled sufficiently?

ANSWER: Cooling to a safe temperature before draining will prevent sludge baking on the internal surfaces, and also will make it safer for personnel to enter parts of the steam generator for inspection and maintenance.

17 Why should positive furnace pressure operation be avoided on those units not designed for positive pressure operation?

ANSWER: Positive furnace pressure will cause hot gases to escape from crevices and openings in the setting, thus causing boiler-room pollution, unsafe conditions for personnel, and deterioration of the boiler setting or enclosure.

18 What safety practice should be followed when blowing down a boiler when the gauge glass is not visible?

ANSWER: A person should be stationed by the gauge glass who is able to signal to the person blowing down in the event the water level reaches an unsafe level, so that blowdown procedures can be immediately stopped.

19 Name two precautions in using steam sootblowers.

ANSWER: To avoid tube erosion, the steam should be dry, and condensate should be drained from the sootblowers before they are used in blowing deposits off tubes.

20 What would happen if a baffle broke down?

ANSWER: Gases would short-circuit one or more passes, excessive flue-gas temperatures and a loss in efficiency and capacity thus resulting. Overheating and damage might result in parts of the boiler designed for low gas temperatures.

21 Is the volumetric capacity of each gas pass in a watertube boiler the same? Explain your answer.

ANSWER: No, it decreases in each succeeding pass. The gases contract as they cool, and in order to maintain the high gas velocity necessary to sweep off stagnant gas films to effect good heat transfer, the cross-sectional area of the passes must decrease as the gases require less space.

22   In what type of watertube boiler are stay bolts used?

ANSWER: The box-header types.

23   What is a downcomer nipple?

ANSWER: It is a short length of boiler tube between the steam drum and header carrying downward circulation of boiler water.

24   Where are interdeck nipples used?

ANSWER: In double-deck boilers where two sets of headers are installed one above the other. Interdeck nipples connect them vertically.

25   Name the two main sheets of a box header.

ANSWER: The tube sheet and the handhole (or tube-cap) sheet.

26   What is the narrow plate sometimes used to form the bottom of a box header called?

ANSWER: The header trough.

27   How many courses are there in a cross-drum boiler? Explain your answer.

ANSWER: Usually one, for a girth seam might unnecessarily interfere with the tube ligament.

28   Is it more difficult to withdraw a tube from a watertube boiler or from a firetube boiler? Explain your answer.

ANSWER: It is usually more difficult in a firetube boiler, for any scale will be on the outside of the tube.

29   How many drums are there in a Sterling bent-tube boiler?

ANSWER: Usually two to four.

30   What is an advantage of the multidrum type? What may be a disadvantage?

ANSWER: The greater steam- and water-storage capacities enable the multidrum type to meet peaks in load fluctuations with less pressure drop. The added cost of the additional drums is the disadvantage.

31   Which is the shortest drum in a Stirling boiler? Explain your answer.

ANSWER: The lower (mud) drum is the shortest. It is suspended by the tubes, and the heads of the drum are inside the boiler sidewalls.

32   Is the water level equal in the steam drums of a bent-tube boiler of a type having the drums at the same level?

ANSWER: It is when the boiler is idle or operating at low ratings. At high ratings, the water level in the drum supplying the downcomer tubes is often lower than that in the other drums.

33 For what reasons are bent tubes used in watertube boilers?

ANSWER: To allow for expansion and contraction; to permit tube replacement; to allow tubes to enter drums perpendicular to surface tangent; to allow flexibility in design in regard to pressure and capacity.

34 What is the purpose of a bridge wall in a Stirling boiler?

ANSWER: It protects the mud drum from exposure to the direct heat of the fire.

35 What might happen if the Stirling bridge wall collapsed?

ANSWER: The longitudinal seam might be overheated and damaged.

36 Where is the manhole in bent-tube boiler drums?

ANSWER: In at least one head of each drum.

37 What is the purpose of circulating and equalizing tubes in bent-tube boilers?

ANSWER: Circulating tubes connect the water space of adjacent steam drums (usually at the same level) and aid in equalization of water level. Equalizing tubes connect the steam space of such drums to equalize the steam pressure.

38 What is the difference between a mud drum in a Stirling and in a straight-tube type of boiler?

ANSWER: A Stirling mud drum is one of the main drums in the circulatory system. In straight-tube boilers, the mud drum may be a trough in the steam drum (Heine boiler) or an external box at the bottom of the headers (as in the Babock and Wilcox boiler).

39 Where are the blowdown connections on bent-tube boilers, and how many are there?

ANSWER: In the bottom of the lowest, or mud, drum; one or more, depending on the length of the drums.

40 What is the difference between the tube sheet in a bent-tube boiler and in an HRT boiler?

ANSWER: In the bent-tube boiler, the tube sheet is that portion of the drum into which the tubes are rolled. In the HRT boiler, the tube sheet is a flat head.

41 (*a*) Define a superheater as used with a steam boiler.

(*b*) Name two separate ways of producing superheated steam.

ANSWER:

(a) A superheater is a form of heater used to raise the temperature of saturated steam above the temperature due to its pressure. It usually consists of an inlet and an outlet header with interconnecting tubes.

(b) One separately fired and the other ones located in the boiler setting or enclosures and sometimes in the walls of the furnace.

42

(a) Define an air preheater and mention some advantages gained in using it as part of the steam-generator unit.

(b) Define air-cooled walls used in boiler installations and what is gained by their use.

ANSWER:

(a) An air preheater is a heat exchanger usually installed in the smoke stack or in the boiler breeching. The cool incoming air is heated by heat in the gases that would otherwise be wasted, and the heated air is used in the furnace again, saving fuel or heat needed to heat the air.

(b) Air-cooled walls are in the furnace setting. Cool air keeps the furnace brickwork from burning, and the air acts as an insulating blanket to prevent heat loss through the brickwork of the setting.

43 What is a windbox on a watertube boiler?

ANSWER: This is a chamber surrounding the fuel nozzles where preheated air under slight pressure is supplied for combustion of the fuel.

44 How do water and steam circulate in a straight-tube watertube boiler?

ANSWER: This type of boiler has natural circulation with the water and steam rising along the inclined tubes to the higher front header, then through the headers to the drum. The water then circulates through the downcomers to the rear header to the inclined tubes to complete the cycle.

45 Name two classifications used to describe furnace enclosures for watertube boilers.

ANSWER: Two classifications are furnace enclosures consisting primarily of refractory to keep the heat within the furnace and furnaces with water-cooled walls.

46 What is usually the weakest part (from the viewpoint of construction) of a bent-tube boiler?

ANSWER: The tube ligament. See Chap. 9 on strength calculations.

47 What are the types of tubes that are used in watertube boilers?

ANSWER: Operating conditions have encouraged manufacturers to seek new solutions to boiler operating problems, such as when burning abrasive or corrosive fuels in the furnace. See Fig. 3.24. Tubes made for these special operating conditions include:

(a) Seamless tubes for ordinary service made of carbon steel, hot or cold drawn. Thickness range varies from 0.049-in. minimum and up.

(b) Welded carbon steel tubes per Code requirements with thickness also starting at 0.049-in. minimum.

(c) Alloy and stainless steels, hot or cold drawn, starting at 0.049-in. minimum thickness.

(d) Bimetallic tubing as shown in Fig. 3.24a, which has a stainless steel outer tube bonded by rolling to a carbon steel tube, which is sized to resist design pressure, while the stainless steel outer tube provides corrosion and erosion resistance. These tubes were developed to be used in recovery boilers in the papermaking industry, but are also being used in trash-burning facilities, and similar corrosive firing services.

(e) Ribbed tubes, as shown in Fig. 3.24b, are used to promote circulation and thus avoid DNB, or departure from nucleate boiling. This tubing has internal, spiral-formed ribbing.

(f) External finned tubes are extensively used for membrane walls in watertube boilers. See Fig. 3.24c. This construction eliminates welding bars to tubes to make a membrane wall; the tubes are fabricated as shown. Only the fins are welded together, to form an usually airtight furnace wall.

48 Explain the advantage of drainable superheaters and reheaters.

ANSWER: The drainable-type superheater and reheater permits establishing flow through the tubes on start-up of large boilers, and thus avoids possible

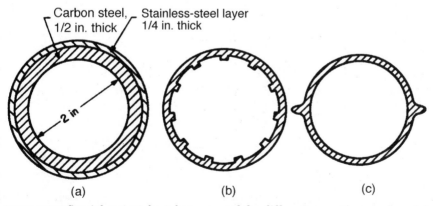

Carbon steel, 1/2 in. thick   Stainless-steel layer 1/4 in. thick

2 in.

(a)   (b)   (c)

Figure 3.24 Special watertube tubes are used for different operating services. (a) Bimetallic tubes to resist corrosion and erosion on the fire-side of some boilers; (b) ribbed tubes for promoting natural circulation; (c) external finned tubes for connecting tubes in membrane walls, and also in waste-heat boilers, to increase heat-transfer surfaces.

tube overheating. The drainable type also permits better chemical cleaning of superheater and reheater tubes by ensuring that the chemical solution is circulated uniformly through all the tube circuits. This also permits better neutralization and improves the flushing out of chemical and scale residue when the chemical cleaning is finished.

49   What does the term "ramp rate" on boilers mean?

ANSWER: This is the manufacturer's recommended rate of permissible pressure and temperature increase per unit load increase. This is especially applicable to utility boilers. Excessive changes in loading with corresponding pressure and temperature changes will place additional stresses on boiler components. Metal temperature increases due to rapid ramp rates may increase superheater and reheater metal temperatures above design, and this can cause tube failures.

50   What is the difference between a "once-through" and supercritical steam generator?

ANSWER: The two terms apply to supercritical steam generators where the water goes through the water tubes only once in contrast to natural circulation where the liquid may circulate through the tubes more than once before it is converted to steam. The fluid in a once-through steam generator is under forced circulation by means of a circulating pump in contrast to the natural circulation in a subcritical steam generator. The feedwater in a supercritical steam generator is above the critical pressure of water, but below the critical temperature.

51   What are the pressures and temperatures that define a supercritical steam generator?

ANSWER: An operating pressure above 3206.2 psia and an operating temperature above 705.6°F.

52   What is the purpose of downcomers in large watertube boilers?

ANSWER: Downcomers are pipes that carry water from the top of the boiler to the bottom of the boiler, external of any steam-generating zones, in order to ensure the delivery of steam-free water to the lower headers or drum to promote natural circulation in the boiler.

Coil-boilers

53   What limitation does the Code place on hot-water coil boilers that do not have a steam space?

ANSWER: The Code does not cover this type of boiler if the following restrictions are complied with: (1) tubing of ¾-in. diameter with no drums or headers attached; (2) maximum 6-gal water-containing capacity; and (3) no steam generated in the coil.

54 Name two reasons why a coil-type boiler may develop excessive coil temperatures.

ANSWER: Low water and insufficient water being circulated through the coil because of partial obstructions or pump failure. For this reason, coil-type boilers should have a high-temperature cutout and possibly a flow switch that will shut off the firing mechanism when high coil temperature is developed. This is usually done by a thermostatic switch.

55 How is the feedwater rate controlled in a Clayton-type coil boiler?

ANSWER: The feedwater is controlled by the water-level control which in turn responds to the liquid level existing in the accumulator. When the accumulator water level drops, the water-pump solenoid de-energizes to allow the feedwater pump to operate.

56 What procedure is recommended in checking a coil type for leaks?

ANSWER: Most manufacturers recommend the boiler be hydrostatically tested; if failure is small, repair the leak by an approved welding process. The coil should be replaced if difficulties are encountered in the repair.

Cast-iron boilers

57 Does a cast-iron boiler require a bottom blowoff pipe and valve?

ANSWER: Yes. The ASME Code requires each boiler to have a blowoff pipe connection fitted with a valve or cock, of not less than ¾-in. pipe size. It must be connected with the lowest water space practicable.

58 What is the minimum size of pipe required for connecting a water column to a steam-heating boiler?

ANSWER: The minimum size of ferrous or nonferrous pipe must be of 1-in. diameter.

59 What *hydrostatic test* is required on hot-water or hot-water-supply boilers built of cast iron and operating over a pressure of 30 psi?

ANSWER: Each section of a cast-iron boiler must be subjected to a hydrostatic test of 2½ times the maximum allowable pressure *at the shop* where it is built. Cast-iron boilers marked for working pressures over 40 psi must be subjected to a hydrostatic test of 1½ times the maximum allowable pressure in the field (when erected and ready for service). After the boiler is in service and a hydrostatic test is required, the test shall be at 1½ times the maximum allowable pressure.

60 Name the two pressure controls required on a steam-heating boiler and the temperature controls required on a hot-water-heating boiler.

ANSWER: Automatically fired steam-heating boilers require an operating-pressure cutout switch set lower than the allowable pressure and an upper-limit cutoff switch set at a pressure no higher than 15 psi. For hot-water-heating boilers, an operating-temperature cutout switch set less than the maximum allowable temperature is required as well as an upper-limit temperature cutout switch set no higher than 250°F.

61 When does the Heating Boiler Code require a low-water fuel cutout for hot-water boilers?

ANSWER: When the boiler is automatically fired with heat input greater than 400,000 Btu/hr.

62 What hydrostatic tests are required on sections of a steam-heating boiler made of cast iron?

ANSWER: The individual sections must be tested with a hydrostatic pressure not less than 60 psi. The assembled boiler is tested at a hydrostatic pressure not less than 45 psi.

63 What hydrostatic tests are required for a cast-iron boiler to be used for hot-water heating?

ANSWER: For boilers with working pressure not over 30 psi, the hydrostatic test pressure must be at least 60 psi for each individual section. Those with working pressures over 30 psi require a test pressure 2½ times the maximum allowable pressure on each section. In both of the above, the assembled boiler requires another hydrostatic test not less than 1½ times the maximum allowable working pressure. These requirements apply to the manufacturer of the cast-iron boilers. The test pressures required must be controlled within a 10-psi range.

64 Name the three methods or types of cast-iron boilers.

ANSWER: Horizontal sectional, vertical sectional, and one-piece types.

65 What *stamping* does the ASME require on cast-iron boilers?

ANSWER: The marking must consist of the following: manufacturer's name, maximum allowable pressure in pounds per square inch, and capacity in pounds per hour for steam or Btu per hour for water service.

66 What is the minimum size of safety valve allowed on a cast-iron boiler to be used for steam-heating or hot-water-heating service?

ANSWER: ASME rules require a minimum ¾-in. safety valve for both steam-heating and hot-water-heating cast-iron boilers.

67 What pressure rise is permitted on a boiler before the safety-valve capacity may be considered inadequate for heating boilers?

ANSWER: For steam-heating boilers, the safety-valve capacity shall be sufficient at maximum firing rate that the pressure does not rise more than 5 psi. For hot-water-heating boilers, the relief-valve capacity must be sufficient to prevent the pressure from rising more than 10 psi with the burners operating at the maximum firing rate.

Electric and Special Application Boilers

Electric Boilers

Two basic types of electric boilers are available. First, units for low capacity and voltage generally consist of the resistance type. In the resistance type, current generates heat by flowing through resistance elements. This is wire encased in an insulated metal sheath, and these are submerged in water to generate usually moderate pressure steam at low capacities. These types of units do not depend on the conductivity or resistance of the water for generating heat. Second, in electrode boilers, the current flows through the water, as shown in Fig. 4.1a, b, and c, and not through wires. The liquid in the boiler converts electric energy to heat energy.

The energy crisis and air pollution regulations have created a demand for higher-output electric boilers above the 10,000-lb/hr ratings. Where electric power is economical to buy, high-voltage units are now available for increased capacities. A high-voltage unit is classified as boilers with energy input between 2300 and 15,000 volts (V). The high-voltage steam electrode boiler shown in Fig. 4.1a has three generating compartments, one for each phase or electrode. Steam is generated as current flows from the electrodes to the neutral walls of the cylindrical compartments that contain them.

High-voltage sprayed electrode boilers, as shown in Fig. 4.1b and c, control the output by regulating the amount of water sprayed on electrodes that are suspended in the steam space. The current path is created when the water flows through ports in a storage compartment to the electrodes. Water that is not converted to steam runs down the electrodes and falls to a counterelectrode. This creates a second cur-

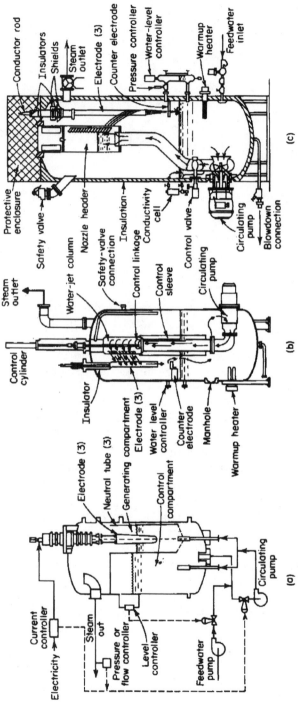

Figure 4.1 Types of electrode boilers. (a) Current flows through the fluid to the wall of the generating compartment; (b) high-voltage sprayed electrode boiler with control sleeve to regulate water flow; (c) high-voltage sprayed electrode boiler with water flow regulated by a rotating distributor.

rent path for steam production—from the electrodes to the counter-electrode. The remaining water returns to the reservoir.

Electrode boilers are packaged; they come fully equipped with controls and safety devices. Most boilers have controllers to maintain conductivity within the manufacturer's limits by monitoring the water and adding prescribed chemicals as needed, as well as blowing down the boiler when necessary. With electrode boilers it is necessary to control the conductivity within specified ranges of the manufacturer's specifications. If the conductivity is too low, the boiler output may not be attained; if it is too high, short circuits may occur. Water treatment is also important with these boilers because faulty treatment can affect conductivity as well as steam problems such as foaming.

Most states include electric-type boilers in their rulings. The ASME Boiler Code states that electric boilers of a design employing a removable cover, which will permit access for inspection and cleaning of the shell, and having a normal water content not exceeding 100 gallons (gal) need not be fitted with washout or inspection openings.

The usual access for inspection and cleaning is the electrode cover connection to the inside of the boiler. This must be pulled out to check the internal conditions of the boiler. At the same time, check the condition of the electric elements.

The capacity of the safety valve is determined by the kilowatt input. The minimum safety-valve capacity must be at least 3.5 lb/(hr/kW) input. This is true whether it is a high- or low-pressure boiler. On a Btu-per-hour basis, the requirement is 3500 Btu/hr for each kilowatt input.

While low-water-level controls are required for the electric-resistance boilers, they are not needed for the electrode type. The reason is that they cannot generate steam when the water is low or when no steam is required. To ensure that the boiler carries the proper salinity, a conductivity control is generally supplied. This means that salts are added or the unit is blown down, depending on the condition of the water. All electric boilers should be built in accordance with the ASME Boiler Code and also approved by Underwriters Laboratories.

High-Temperature Hot-Water Boilers

A high-temperature hot-water (HTHW) system shown in Fig. 4.2 is a heat-utilization boiler that uses water at an elevated temperature, usually over 300°F, with no specific pressure limitations. However, the ASME Boiler and Pressure Vessel Code, Section I, defines a HTHW boiler system as one using water at a temperature over 250°F and at a pressure over 160 psi. The HTHW systems usually are not designed for temperatures exceeding 500°F. Operating temperatures

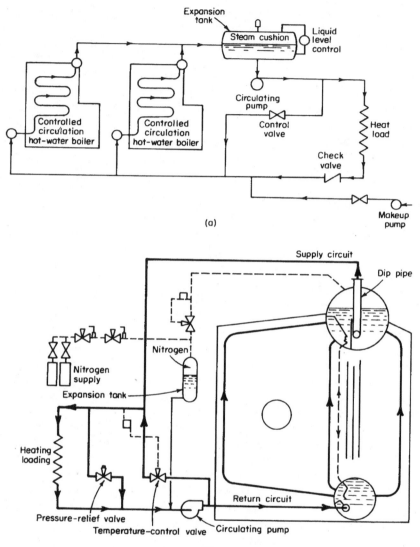

Figure 4.2 High-temperature hot-water boilers use (*a*) expansion tank, (*b*) drum of boiler to permit expansion and contraction of water with temperature changes.

within the range of 350 to 450°F are the most frequently encountered. Above 500°F, other fluids are used to obtain high temperature, such as Dowtherm and similar inorganic fluids.

At these temperatures, water in its liquid form can exist only at pressures above the corresponding saturation pressures. The HTHW systems have widespread use for supplying the heating needs of large airports, military bases, office buildings, hospitals, colleges, and other large multibuilding complexes. The application of HTHW systems has been extended to many industrial processes in the fields of chemistry, plastics, rubber, metal plating, paper, textiles, etc.

Advantages of HTHW systems. For many applications, HTHW systems are claimed to have distinct advantages over steam systems. The major advantages are:

1. Because of its large heat-storage capacity, an HTHW system permits very close control of temperature, which is important with many process applications.

2. The large quantity of heat in the system forms a heat reserve so that fluctuating loads have a minimal effect on the boiler, permitting use of a lower-capacity boiler unit.

3. The absence of return-line corrosion in a HTHW system, which is frequently a problem in a steam system, is advantageous.

4. No traps, pumps, receivers, vents, or other condensate-return equipment are needed on HTHW systems, thus producing lower first cost, less maintenance, and no steam losses.

5. A small amount of makeup is required; thus there is no need for an expensive feedwater-treatment system.

Disadvantages of HTHW systems. There are several disadvantages, one of the most important being the large water content of the system. This produces the following disadvantages:

1. The system takes longer to heat up initially.

2. More time is needed for cooldown when a repair or alteration has to be made and much water has to be discharged.

3. If a pipe or pressure-vessel break occurs, the water content, being above the atmospheric boiling point, will partly flash into steam with a powerful disruptive effect. Even a relatively small leak can result in considerable water damage before the system can be depressurized.

Both firetube and watertube boilers are used for HTHW systems. The main advantage of firetube boilers for this application is that they are generally less expensive in the smaller sizes below 600 hp. Above this size, watertube boilers are used. Surprisingly, the box-header boiler is prominently used for lower capacities; however, the forced-circulation unit (LaMont type) is used extensively.

Watertube hot-water generators are available in packaged ratings up to 150 million Btu/hr for water temperatures up to 650°F. The combustion equipment used is the same as that used for steam boilers. All types of fuels can be used.

Pressurization and flow control. Three considerations in the design of an HTHW system, which are not factors in steam-system design, are:

1. How to maintain the circulating-water pressure above the saturation pressure for the water temperature required

2. How to provide for expansion of heating-system and generator water

3. How to ensure uniform flow rates under all operating conditions

There are three pressurization methods used—steam cushion, gas, and mechanical—and two ways to accommodate water expansion—in a separate accumulator or in the upper drum of the boiler for drum-type units serving small heating systems.

The steam-cushion arrangement generally is used for large-capacity, continuously operated hot-water systems, such as those used at many airports, because it maintains tighter control of water temperature than the other two. See Fig. 4.2a and b.

High-temperature hot water from the boiler discharges into a separate expansion drum, where a small amount of the fluid flashes to steam to maintain system pressure. The boiler's firing rate is controlled by drum pressure. The drum is located high enough above the boiler to permit free vapor release and to ensure a reasonable suction head for the circulating pump. It is necessary to avoid the introduction of large slugs of cold return water into the drum to prevent the cushion from collapsing. Such a condition causes flashing and water hammer in the piping system.

In the *inert-gas pressurization system,* a nitrogen blanket is maintained in an expansion vessel, which is partially filled with water from the circulating system. The pressure of the nitrogen is kept higher than that of the saturated liquid.

High-temperature hot-water boilers should have a low-water fuel cutout in case the water level drops below a safe level. On forced-circulation units, a pressure-differential switch or flow switch is advisable, so that the fuel to the burner is cut off in the event no water circulation is occurring. High temperature hot-water boilers may not require a water gauge glass or gauge cocks if the boiler is completely filled with water and has an external expansion tank. However, if the boiler is a natural-circulation boiler with a drum, then the Code requires a gauge glass and cocks, because the drum may have fluctuating water levels that must be displayed to the operator who notes if it is proper. Figure 4.2b shows a D-type, natural-circulation boiler in which the drum acts as an expansion tank. This boiler would require a gauge glass per Code rules.

High-temperature heat-transfer liquid and vapor organic fluids. Thermal fluids such as oils, silicates, glycols, and similar liquids with high boiling points are used where higher temperatures are demanded by process requirements but at low operating pressures. Uses include drying of fabrics, clay, wood, paint, etc., which previously may have been done by direct firing of natural gas. Temperatures up to 750°F are now practical with thermal fluids; a 1000°F limit reportedly is not far off.

Where HTHW systems can be used, they are preferred to thermal liquids, because they have about twice the heat capacity, do not deteriorate in use, and are cheap.

Thermal-liquid systems also can be used in conjunction with steam systems. Plants that require high-pressure steam for specific applications and that can apply thermal liquids economically for other uses as well can install a thermal-liquid-to-steam heat exchanger and use part of its output to generate steam. This eliminates installation of a separate gas- or oil-fired boiler.

When an organic substance is in the liquid state, the heating unit is referred to as a heater, and if it is vaporized, it is called a vaporizer. The heating may be performed in electric-fired units, up to about 1 million Btu/hr, with units above this size generally being fired by gas, oil, and even coal. Units are generally packaged for automatic operation to a maximum capacity of 300 million Btu/hr. Multiple units are used for greater capacities as a rule.

The systems used can be classified as follows:

1. Vapor systems are generally of the firetube type using either gravity returns or pumped condensate returns. See Fig. 4.3b.

2. Liquid systems use watertube boilers with either natural circulation or the more pronounced forced circulation, as shown in Fig. 4.3a. Figure 4.4a shows a horizontal, helical watertube type. The once-through forced-circulation La Mont design is used for larger capacities.

Among the organic fluids used are Dowtherm, Aroclor, Therminol FR, and Tetralin. Each of these substances has similar properties, but complete details should be obtained from the manufacturer of the fluids. Since many of these fluids are hydrocarbons, a tube failure can create a fire hazard in the unit. Some fire insurance inspection departments require a steam smothering device or an inert-gas smothering system to be installed to smother a fire in the unit from a leaking pressure part. Safety controls should include a high-tempera-

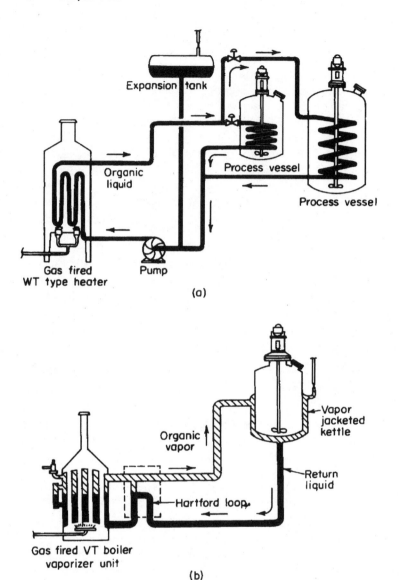

Figure 4.3 Organic heaters. (*a*) Liquid heater with forced circulation; (*b*) vapor heater with gravity condensate return. (*Courtesy Dow Chemical Co.*)

ture cutout, pressure cutout, and safety relief valves set to the maximum allowable pressure of the unit.

Dowtherm is one of the better known organic fluids, and a review of some of its characteristics is somewhat a review of other organic fluids. In Fig. 4.4*b* it will be noted that, at a temperature of 650°F, a Dowtherm pressure of about 53 psig is required. To attain this temperature with

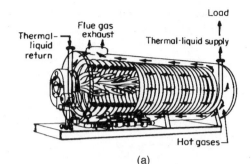

(a)

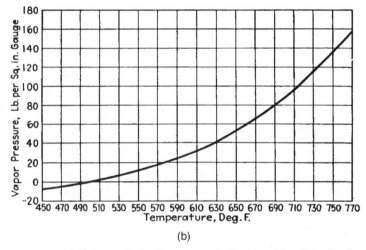

(b)

Figure 4.4 (a) Coil-type boiler heats organic liquid to 20 million Btu/hr.
(*Courtesy Power magazine*) (b) Dowtherm A vapor-pressure curve.
(*Courtesy Dow Chemical Co.*)

saturated steam, a pressure of 2196 psi would be required. The comparative constructural costs of boilers for these pressures are obvious.

Dowtherm is an organic material. Dowtherm A, which boils at 500°F and is recommended for temperatures up to 750°F, is composed of 26½ percent diphenyl and 73½ percent diphenyloxide. Dowtherm C is used in only the liquid phase and, for temperatures up to about 800°F, Dowtherm is nontoxic.

It is more difficult to maintain pressure-tight joints with Dowtherm than with steam or water, and so joints of the boiler proper are welded instead of riveted and tubes are welded into the drums. Whereas a hydrostatic test under cold-water pressure may show leakage at defects in a steam boiler, no indication would be shown for leaks that hot Dowtherm might find. The most satisfactory test for Dowtherm boilers is to introduce ammonia gas into the system at a pressure of up to about 1 psi or to fill the boiler half full of aqua ammonia. Air pressure at about 50 percent of the rated boiler pressure is then applied. A lighted sulfur candle or a dilute hydrochloric acid swab is passed in front of seams, joints, or points where leakage is considered likely. If leakage is occurring, a white smoke will result from the indicator.

Some decomposition may occur if the recommended temperatures are exceeded greatly, owing to faulty operation or localized overheating. Such decomposition may produce a fixed gas, which may be vented during operation, and soluble triphenyls. If the triphenyl content becomes excessive, purification of the Dowtherm may be effected by a distillation or fractional crystallization method.

Blowdown connections are provided on Dowtherm boilers, but are used only when the boiler is off the line and cool and for the purpose of emptying the boiler or to withdraw liquid for purification. Since Dowtherm is a valuable chemical, no waste is permissible. All vents, the discharge of safety valves, and so on are conducted back through a condenser to the storage system. Welding or burning on equipment with organic fluids generally requires complete draining of the fluid and then flushing with steam until traces of the fluid are no longer noted before welding can be performed. Tube failures occur from coking of the organic fluid due to overheating; contamination of the fluid, causing it to break down chemically to carbon; flame impingement, leading to hot spots; and poor circulation for the heat input.

Thermal cracking and *oxidative breakdown* of organic fluids is a major concern in maintaining organic fluid heaters. Oxidation can produce organic acids, which can damage equipment and also increase fluid velocity. To prevent oxidation, nitrogen blanketing of storage and expansion tanks has been recommended by vendors. Antioxidants may affect the temperature limits of the organic fluid. Thermal cracking of the organic fluid may change its viscosity, which can affect pumps by causing cavitation. Coke deposits from thermal cracking can lead to tube overheating. Some petroleum-based fluids are reported to avoid coke deposition by forming crystalline carbon granules that cannot adhere to heated surfaces. Operators should consult with their organic fluid supplier for the grade and manufacturer of such a product if the organic fluid has to operate above 700°F.

Waste-Heat, Combined-Cycle, and Cogeneration Systems

Waste-heat boilers.   There are a number of manufacturing processes that give off considerable quantities of high-temperature gases. The most prominent among these are the exhaust from gas turbines as used in combined-cycle power plants. The value of heat recovery depends primarily on three considerations:

1. The cost of producing an equivalent amount of heat by other means

2. The cost of heat-recovery equipment

3. The operating and maintenance cost of the waste-heat-recovery equipment

Steam boilers may be designed to use waste heat as all or part of the steam-generating medium. Since the gas temperature is usually 500 to 1100°F, whereas combustion products in the conventionally fired installation may enter generating passes at about 2000°F, some means of compensating for the lower gas temperature must be employed. Otherwise, to have an appreciable steam-generating capacity, the boiler would have to be beyond all reason in size. Other factors to consider besides pressures and temperatures of available waste gases are the physical and chemical properties of the gas, their effect on boiler parts, the effect on the heat-recovery system by plant-process disturbances, and similar considerations involving continuity of service. Recovery of exhausted heat from industrial processes and combustion equipment can often reduce overall plant fuel consumption with minimal capital investment.

Supplementary firing is used when the waste-heat gases do not have sufficient heat to produce the desired final pressure or temperature of the steam. Figure 4.5 shows some typical waste-heat gases and temperatures usually available from processes.

Depending on the properties of the waste gases and the pressure and capacity needed, the following waste-heat boilers are used:

1. Firetube boilers, both the vertical and horizontal types, if waste gas is relatively clean

2. Straight-tube watertube boilers, for clean or moderately dust-laden waste gas

3. Watertube of the bent-tube (Stirling) type, for very heavy dust loadings

4. Positive circulation boilers, for clean, low-temperature gases

5. Pressurized or supercharged boilers, for gas turbine exhaust (Velox type). Used mostly in Europe.

Because waste gases quite often have inert gases and solid entrapped particles in the mixture, special material and other design

Source of Gas	Temp, F
Ammonia oxidation process	1350-1475
Annealing furnace	1100-2000
Cement kiln (dry process)	1150-1500
Cement kiln (wet process)	800-1100
Copper reverberatory furnace	2000-2500
Diesel engine exhaust	1000-1200
Forge and billet-heating furnaces	1700-2200
Gas turbine exhaust	850-1100
Garbage incinerator	1550-2000
Open-hearth steel furnace, air blown	1000-1300
Open-hearth steel furnace, oxygen blown	1300-2100
Basic-oxygen furnace	3000-3500
Petroleum refinery	1000-1100
Sulfur ore processing	1600-1900
Zinc-fuming furnace	1800-2000

Figure 4.5 Waste-heat gases with the temperatures available from industrial processes for steam generation.

factors must be considered in the application of a waste-heat boiler. These factors and operation and maintenance practices will prevent the unexpected type of failures that can occur on these units also. Among the possible failures are the following:

1. Tube failures from scale and mud buildup, erosion of tubes from particles in the gases, corrosion of tube material from chemical attack from the gas side

2. High-heat-content mass gas input, causing the equivalent of overfiring damage

3. Fireside plugging of gas passages, resulting in localized overheating

4. Malfunction of controls or equipment, causing a low-water condition, with tube and other damage

A factor or condition often neglected on waste-heat steam boilers is the possibility of developing a dangerous low-water condition which, if undetected, has caused tubes in the convection passes to melt. This is an extremely important design consideration. It is also a very important operating check that should be made periodically to ensure that the unit is in working condition. This test should be the same as when low-water fuel cutoffs are tested on a suspended fuel-fired standard boiler. Basically the design should include a mechanism, either a heavy damper or some other quick-closing mechanism. In this way in case of low water, the waste gases can be cut off from the boiler and bypassed to the stack. Then the basic process will not be interrupted, and the boiler will also be saved from serious damage due to overheating.

A preferred method is a device that automatically diverts the gases as soon as the water level drops to a predetermined dangerous level. *Time* is of extreme importance in a low-water condition, so the heat input must be quickly removed.

Combined-cycle power plants and cogeneration are sometimes grouped together, but need definition. In a *combined-cycle plant,* a prime mover such as a gas turbine drives a generator, and the exhaust gas from the gas turbine is directed to a heat recovery boiler, called a heat recovery steam generator, or HRSG, which generates steam, and this steam is directed to the second prime mover, a steam turbine that also drives an electric generator. This system thus combines the Brayton cycle on the gas turbine with the Rankine cycle on the steam turbine. *Cogeneration* is the simultaneous generation of electric power with a by-product being process heat, or steam for process use.

The combined-cycle power plant using gas turbines has shown remarkable growth since the 1970 energy crisis, because the system has improved overall power plant thermal efficiencies, thus saving fuel, and also because of material and design improvements on the gas turbine, which permits higher temperature operation. Utilities are using combined-cycle technology to retrofit older plants with the gas turbogenerator supplying the exhaust heat to make steam that is then directed to rebuilt older steam turbogenerators.

Cogeneration of electric power and steam for process use has been stimulated by the federal government's PURPA legislation (Public Utility Regulatory Policies Act), which encourages extracting the maximum energy out of fuels for energy conservation reasons. Independent power producers have arisen to sell electricity to the local utility per PURPA legislation, and to sell steam to a nearby industrial plant or to produce steam for in-house use, and selling the surplus electric power to the local utility.

Figure 4.6 is a combined-cycle cogeneration plant. The key units are two heavy-duty industrial gas turbines, each rated at 40.5 MW at 65°F ambient temperature; two waste-heat boilers each rated to produce 142,000 lb/hr, 900 psi, 900°F steam; a medium-pressure section at 60 psi to produce a total of 50,000 lb/hr steam; and a 10-psi section

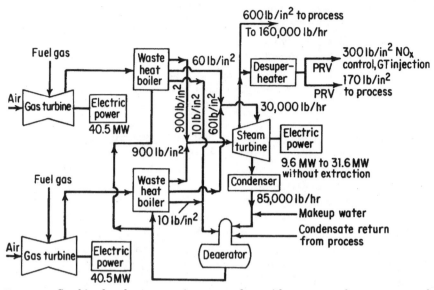

Figure 4.6   Combined-cycle cogeneration power plant with two gas turbogenerators and two waste-heat boilers, or HRSGs. Plant is capable of generating up to 112.6 MW of electric power, and also provides 100,000 lb/hr of 600-psi steam.

for the deaerator. The 900 psi steam is directed to a single steam turbogenerator, rated from a base of 9.6 to 31.6 MW without 600 psi extraction. Gas turbine exhaust enters the waste-heat boiler at 990°F and exits the stack at 276°F.

A distributed control system will automatically regulate all plant systems, except for the gas and steam turbines, which have their own controls. However, the integrated control scheme does provide for monitoring the performance of the turbines for integrating this performance into the main automatic network.

Boilers of the watertube type (Fig. 4.7) are used in combined-cycle plants, as shown in Fig. 4.6. They depend on large convection heat-transfer surfaces since there is no furnace to provide radiant heat. Supplementary firing is used on larger cogeneration plants during peak steam demands, with duct burners being used between the gas turbines and the boilers.

Supplementary fired HRSGs usually employ a duct burner to raise the temperature of the gas turbine exhaust, and thus increase the steam production, such as may be needed during peak periods of load demand. Casing design considerations limit the maximum gas temperature after the duct burner to about 1700°F in order to prevent duct liner material warping which would cause leakage problems.

Vertical tubes are used, and this promotes natural circulation as the steam bubbles form and then rise to the steam drum. Finned

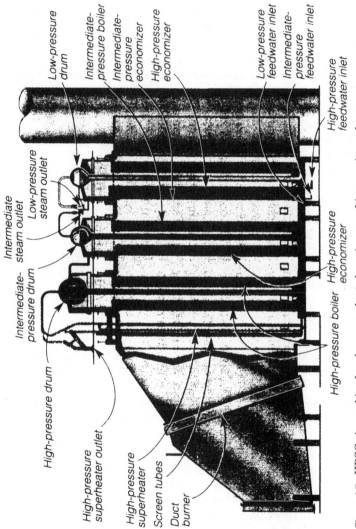

Low-pressure
drum

Intermediate-
pressure boiler

Intermediate-
pressure
economizer

High-pressure
economizer

Low-pressure
feedwater inlet

Intermediate-
pressure
feedwater inlet

High-pressure
feedwater inlet

Intermediate
steam outlet

Low-pressure
steam outlet

Intermediate-
pressure drum

High-pressure drum

High-pressure
superheater outlet

High-pressure
superheater

Screen tubes

Duct
burner

High-pressure boiler

High-pressure
economizer

Figure 4.7 HRSGs in combined-cycle power plants have increased in capacity and generate multiple pressure steam for the plant's steam turbogenerator, gas turbine steam injection, and other plant steam users. (*Courtesy Power magazine.*)

tubes are also used in order to provide extended heat-transfer surfaces in the HRSG. As Fig. 4.7 shows, as the combined-cycle power plant has grown in size, multiple pressure HRSGs have been developed to provide high-pressure steam for the steam turbine prime mover, but also to provide lower-pressure steam requirements in the power plant or for process use.

Combined-cycle power plants must meet jurisdictional emission standards on CO and NO_x discharge to the atmosphere. Various methods are used, one of which is steam injection into the gas turbine stream to lower the gas temperatures. This steam also comes from the HRSG. Much development work is going on in further improving the combined-cycle plant. One is the conversion of coal to a gas under pressure for burning in the gas turbine.

Another is the *Cheng cycle*. This cycle injects steam into the gas turbine after the steam is superheated for increasing the electrical output. This requires steam purity in the parts-per-billion range, Cheng cycle units use a combination of internal and external steam separators to remove noncondensibles and other impurities before the steam is injected into the gas turbine.

Most combined-cycle plants are computer controlled. However, some basic boiler-operating practices must still be observed in order to avoid possible trouble. This includes maintaining proper water level and periodically checking safety controls, including low-water cutouts. Operators must be familiar with the strategies to be employed in such emergency conditions. Tubes have been melted in HRSGs due to faulty operating and safety controls and perhaps to operator confusion as the event took place.

Waste-fuel boilers. Fuel characteristics must be considered when the many available waste fuels are burned such as wood and trimmings from lumber, paper, furniture, and similar industries using lumber as a basic raw material. Liquid and gas wastes offer another burning problem. The furnace design must be ample so that these waste fuels are burned efficiently in the furnace without causing obnoxious fumes to be emitted from the stack. Regulations on permissible emissions now govern many designs, including whether the boiler will be of the firetube or watertube type. Escalating fuel costs and shortages that may develop in the future have caused industry and governments to re-examine the potentials of waste fuels of all types as a combustive alternate to the once-plentiful supply of fuels such as oil and gas.

Solid by-product fuels are many. Among them are wood chips, sawdust, hulls from coffee and nuts, corn cobs, bagasse (waste product from sugar cane), coal char (residue from low-temperature carbonization of coal), and petroleum coke (final solid residue from a refinery). Each product must be handled in a special manner because of differences in moisture content, consistency, specific weight, and heat content. The furnace rather than the steam generator is affected when these special fuels are used. Products like bagasse, which has about 50 percent moisture, require a Dutch oven. The Ward furnace is a popular design for bagasse, both in the United States and abroad. Here bagasse fuel is partly dried and burned in refractory cells below a radiant arch. The combustion of gases is completed above the arch. Spreader stokers can also be used as shown in Fig. 4.8a.

In refining sugar from cane, the juice squeezed out of the cane eventually is processed into sugar. The remaining fibrous, tenacious, and bulky crushed cane is called bagasse. It is also moist. Depending on where it is grown and the efficiency of the juice extractor, bagasse contains 30 to 50 percent wood fiber and 40 to 60 percent water. Heating value is 8000 to 8700 Btu/lb as a dry solid, with a yield of about 4500 Btu/lb at around 45 percent moisture content.

Figure 4.8b shows a combination scrap wood and oil burning boiler.

Refuse-to-energy boilers. Municipal solid waste disposal in landfills is under attack because of the long-term environmental effects such landfills may have on the land area. In addition, municipal solid waste has long been recognized as having fuel value, as shown by Fig. 4.9. This fuel can be used to generate steam and electric power, and thus save the earth's limited oil and gas supplies. Designs have now reached a level where pollution from burning refuse can be minimized to acceptable state and city levels. Economic stimulus has been provided under the PURPA legislation, which encourages the use of alternate fuels to save oil and gas, and which requires local utilities to buy the power from independent electric power producers. Cogeneration plants using refuse-derived fuels are thus increasing very rapidly in the United States. They are being built with sophisticated controls that incorporate rate of firing with load swings in addition to pollution controls. See Fig. 4.10a for a typical solid-waste-burning plant, and Fig. 4.10b for the inclined grate details. Burning of the solid waste is by inclined grates or by rotary-type kiln combustors. Some units burn the solid waste outside the furnace, after presorting, and only the hot gases burn in the furnace. These are called *RDF boilers* (refuse-derived fuel). *Mass burning* boilers have very little presorting of the refuse, but do employ shredders, with burning employed in the furnace of the boiler.

Practically all refuse boilers are of the watertube type. Fireside corrosion is a problem on these types of boilers, because of the contents in the refuse fuel. Refractories are used to protect susceptible furnace areas, but these require above-normal maintenance and have slagging problems. Earlier designs used carbon steel tubes, which deterio-

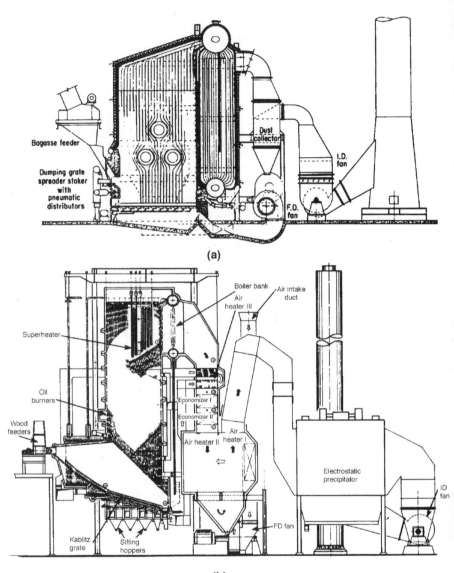

(a)

(b)

Figure 4.8 (a) Bagasse-burning watertube boiler with dump-type spreader stoker and auxiliary oil burners; 100,000-lb/hr capacity at 350 psi. (*Courtesy Riley Corp.*) (b) Wood- and oil-burning combination boiler generates 317,500 lb/hr with wood, and 476,000 lb/hr with oil at 928 psi and 860°F. (*Courtesy Gotaverken Energy Systems Inc.*)

Property	Average value
Higher heating value, Btu/lb	3800 min–5200 max
Moisture, %	20.0
Carbon, %	25.2
Hydrogen, %	3.5
Oxygen, %	22.3
Sulfur, %	0.1
Inerts, %	28.3
Bulk density, lb/yd^3	270–875

Figure 4.9   Typical properties of municipal solid waste. (*Courtesy Power magazine.*)

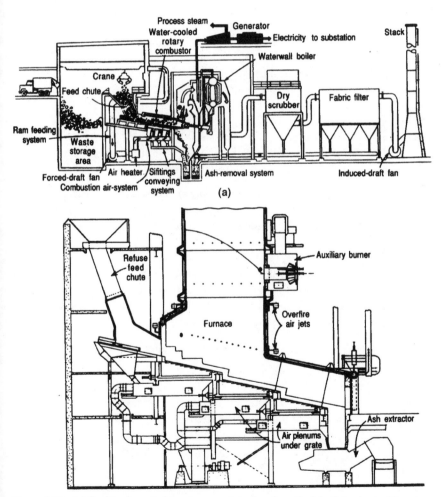

Figure 4.10   (*a*) Inclined-grate burning municipal refuse fuel boiler plant. (*Courtesy Power magazine.*) (*b*) Details of a refuse-burning grate and furnace. (*Mechanical Engineering Magazine, Dec. 1988.*)

rated rapidly. Weld-deposited alloys, such as Inconel 625, have been used to protect the carbon steel tubes, while Inconel 825 is being used in high gas temperature zones.

Refractory maintenance is above normal in refuse-burning boilers, due to the heavy wear or erosion that is also influenced by temperature cycling and chemical attack. Each heatup from a cold condition causes expansion effects between the refractory and housing and this can cause thermal shock effects to the refractory. Chemical attack can come from a variety of chemicals in the trash fuel. For example, sodium in trash fuel tends to replace aluminum in the refractory and forms sodium silicate, or water glass. This creates a brittle surface slag on the refractory, and when it breaks off, it takes some of the refractory surface with it, reducing the refractory thickness. It is essential to inspect the refractory at regular intervals and repair any attritioned refractory by patching or replacement.

Residue from the combustion of municipal waste creates bottom ash and fly ash, with the bottom ash residue on grate-burning boilers representing 80 to 90 percent of the total ash.

Combustion of municipal solid waste with high plastic content results in the formation of acid gases, mostly hydrogen chloride. High excess air of 80% is recommended in burning this type of fuel.

Regular thickness tests of tubes must be followed in order to prevent unexpected tube failures from fire-side attacks.

Modifications being made on plain carbon steel tubes. Practically all boiler manufacturers are correcting the abnormal tube wear by stainless steel overlay on carbon steel tubes after the thickness reaches the minimum required.

Some manufacturers use stainless steel-clad tubes in original design, and this is certainly good engineering practice for the environmental conditions faced by waste-burning boiler tubes. See Fig. 3.24a.

For those plants with carbon steel tubes, operators should immediately determine the following:

1. Is a tube thickness testing program in effect?

2. Have plans been made to stainless steel overlay the tubes when minimum code thickness for the tubes is reached? Size of tubes and required minimum thickness should be included in this evaluation of wear and the correction of same.

3. If tubes were clad originally, a periodic inspection of the cladding should be established to make sure it is still in place as a corrosion-resistant material.

Spare boiler feed pumps. There is an inherent heat source in grate- and refractory-lined boilers that can cause overheating damage if only electric-driven boiler feed pumps are available, and the electric power is interrupted.

The ASME recommends spare *steam-driven feed pumps* or injectors for stoker-fired boilers, or boilers fired by other methods whose furnaces contain large amounts of refractory, or which may have or accumulate large amounts of hot slag in the furnace setting. The steam-driven pump is required even if there are spare electric-driven pumps, *unless* there is on site a fully independent auxiliary source of electric power. The purpose of these recommended practices is to avoid overheating the boiler from a loss of electric power, on those boilers with large amounts of latent heat stored on stokers, refractories, or hot slag in the furnace area.

Black-liquor boilers.   Black liquor is a by-product of wood-pulp processing in the papermaking industry. See Fig. 4.11. Chips of wood are cooked by steam in a solution of sodium sulfide and sodium hydroxide in a large tank known as a "digester." Strong liquor from the digester flows to a storage tank where it is joined by weak liquor from pulp washers. In order that this liquor may sustain combustion, it is then concentrated by evaporation and crushed salt cake is added until it contains over 58 percent solids.

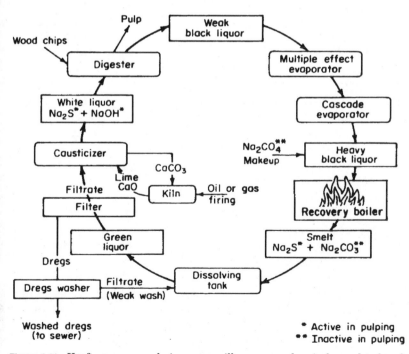

Figure 4.11   Kraft recovery cycle in paper mills recovers chemicals used to break up wood fibers in digesters by burning black liquor in a recovery boiler.

The concentrated liquor is pumped at about 220°F through oscillating burners which spray it onto the furnace walls; deposits of combustible char build up until they are heavy enough to drop to the furnace floor, where combustion is assisted by primary air nozzles. The gases and a small percentage of fuel particles rise to the upper part of the furnace where secondary air is admitted to complete the combustion process.

A considerable percentage of chemical, a form of soda ash, is recoverable from the ash of this process. Thus, the combustion of this black liquor is twofold, namely, steam generation and for soda-ash recovery. The capacity of a black-liquor recovery boiler is also expressed in the amount of tons of dry solids it can burn in 24 hr. As an example, a unit designed for 1600 psi with a capacity of 392,000 lb/hr of steam is rated at the same time as being able to burn 800 tons/day of black liquor. The two major manufacturers in the United States are B&W and Combustion Engineering (CE). The B&W units use a slanting-bottom furnace (Fig. 4.12a). The oscillating burners spray the black liquor on the furnace walls. The deposited liquor dries, forms char, and falls to the hearth where sodium chemicals are smelted, and the organics in the char are turned to gas which burns in the furnace. The smelted chemicals drain down the sloping floor through the water-cooled spout and then into the dissolving tank.

The CE unit shown in Fig. 4.12b sprays the black liquor into the center of the furnace. The bottom of the furnace in the CE unit is flat and is called a decanting furnace. Most other boiler details are similar for the two manufacturers' designs.

Serious *furnace explosions* on recovery boilers have instituted research as to the cause. It is generally believed that the smelt in the furnace can chemically combine with water to produce a chain-reaction type of detonation in the furnace. This has produced rapid pressure buildup in the furnace with shock-wave results to its structural components. The paper industry, insurance companies, and boiler manufacturers have drawn up guidelines, including emergency procedures, in order to prevent smelt-water explosions. Tube integrity is stressed, and NDT inspections of welds in the furnace area are recommended, as is black-liquor-concentration monitoring. Among the novel procedures recommended is rapid draining of the boiler to a level 8 ft above the low point of the furnace floor at any time when water is suspected of entering the furnace area.

Operators of black-liquor boilers have to make sure all black-liquor pumps are simultaneously shut down when the emergency rapid drain procedure is initiated, otherwise severe dry-firing may result. The paper industry, their insurance companies, and boiler manufacturers have formed a very active Black Liquor Recovery Boiler Advisory

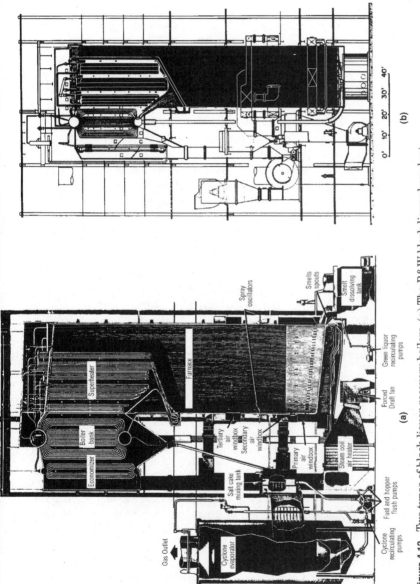

Figure 4.12 Two types of black-liquor recovery boilers. (*a*) The B&W black liquor has slant-ed-bottom furnace for smelt removal to dissolving tank. (*Courtesy Babcock & Wilcox Co.*) (*b*) The CE black-liquor boiler features a flat-bottom furnace with the fuel sprayed into the center of the furnace for burning. (*Courtesy Combustion Engineering, Inc.*)

Labels in figure (a):
- Gas Outlet
- Economizer
- Boiler bank
- Superheater
- Cyclone evaporator
- Cyclone recirculating pumps
- Fuel and hopper flush pumps
- Salt cake mixing tank
- Steam coil air heater
- Primary air windbox
- Secondary air windbox
- Tertiary air windbox
- Furnace
- Spray oscillators
- Smelts spouts
- Smelt dissolving tank
- Forced Draft fan
- Green liquor recirculating pumps

Labels in figure (b):
- 0' 10' 20' 30' 40'

Committee (BLRBAC), in order to develop and implement safe installation, inspection, and maintenance procedures on these types of boilers, and operators should become completely familiar with those guidelines and procedures. Another organization active in developing recovery boiler safety audit programs is the American Paper Institute. Teams from different companies visit site locations in order to note if any unsafe practices are followed in the operation, maintenance, and inspections of black-liquor recovery boilers. Items stressed are personnel safety; pressure part integrity, which includes material, welds, corrosion and erosion checking by visual and NDT methods; water treatment programs; BLRBAC recommendation compliance, including latest reports or recommendations; safety interlock systems and failsafe designs recommended and implemented and the checking of these devices; emergency operating procedures that are in effect for the location; existing training program for the black-liquor boiler operators; maintenance programs being applied to black-liquor recovery boilers; and the operating history and reliability of the location.

The purpose of the BLRBAC and the API programs is to avoid the severe property and human injury losses that can occur from a fireside explosion on black-liquor recovery boilers.

Red-liquor (MgO)–fired boilers. The black-liquor recovery boiler is used in the papermaking industry where hydrogen sulfite is used as the chemical to break up the lignin in wood that is cooked in digesters. The process is considered alkaline. Another process is called the *red-liquor acid process*. Red liquor with a concentration of about 50 percent solids is fired with steam-atomizing burners, as shown in Fig. 4.13.

The red liquor burns in suspension and forms little or no slag in the furnace. Magnesium oxide ash is removed from the furnace floor as well as from the flue gas. Sulfur dioxide is also recovered from the flue gas by passage of the flue gas through absorption towers where a magnesia oxide slurry absorbs the sulfur dioxide. The result is a cooking acid that is reused in the digesters.

Black and red liquors do not burn alike; thus the steam generators used are not alike. Black liquor in the kraft process is very difficult to burn. Large furnaces are needed to keep the temperature relatively low because the liquor has a high content of low-fusion-temperature ash. Smelt collects on the refractory sloping hearth, and a reducing atmosphere must be maintained in the lower part of the furnace for chemical conversion. Also, since superheater and boiler surfaces have a tendency to coat with slag, they operate at low absorption rates. Thus frequent soot blowing and shot cleaning of heating surfaces is necessary.

Red liquor in the MgO (magnesia oxide) process, on the other hand, burns completely in suspension, making little or no slag. Thus a

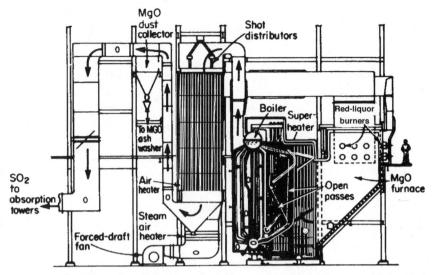

MgO
dust
collector

Shot
distributors

To MGO
ash
washer

Boiler
Super-
heater

Red-liquor
burners

Open
passes

SO2
to
absorption
towers

Air
heater

Steam
air
heater

Forced-draft
fan

MgO
furnace

Figure 4.13 Red-liquor burning watertube boiler used in the acid process of chemical recovery and steam generation in the paper industry. (*Courtesy Power magazine.*)

smaller steam generator can be used for an equivalent amount of steam production.

Fire-side corrosion.   Corrosion and erosion are a continuous problem on black-liquor recovery boilers. The paper industry has established a tube thickness testing program because of the inherent risk that a tube failure would allow water to enter the furnace and thus expose the unit to a smelt-water chemical reaction, and possible furnace explosion. Extruded composite tubes are also being used on new boilers as shown in Fig. 3.24*a*. Finally, when minimum wall thickness is imminent, older tubes are metal-sprayed with stainless steel–type material. Periodic inspections are still required in order to determine the condition of the cladding or the metal spray.

Acoustic emission is used to warn operators of a tube failure, as are strategically located thermocouples, especially on pendant-type superheaters, which require attention in order to assure circulation on startup or during any plant upsets.

Fluidized-bed boilers.   Fluidized-bed boilers are being developed in order to take advantage of the large coal deposits that exist as well as to overcome the pollution problem that can exist in burning coal. The merits in using fluidized-bed combustion are the following: (1) High-sulfur fuels can be burned without resorting to flue-gas treatment. This is accomplished by injecting limestone into the bed which absorbs the sulfur dioxide. (2) Higher combustion efficiencies are

obtainable in fluidized-bed burning. (3) Lower combustion temperatures are possible which minimize nitrogen oxide and furnace slag
formation. (4) The waste product formed at a lower bed temperature
is easier to handle and dispose.

A fluidized bed consists of granular particles lying on a nonsifting
grid through which air is blown at a velocity sufficient to lift and float
the particles. Bubbles are formed to the extent that the mass now
acts like a boiling liquid. See Fig. 4.14. The particles held in suspension in the bed are usually limestone. Crushed coal is injected into
the limestone and burned. To capture the sulfur in the fuel, the bed
temperature is controlled at approximately 1550°F, at which optimum

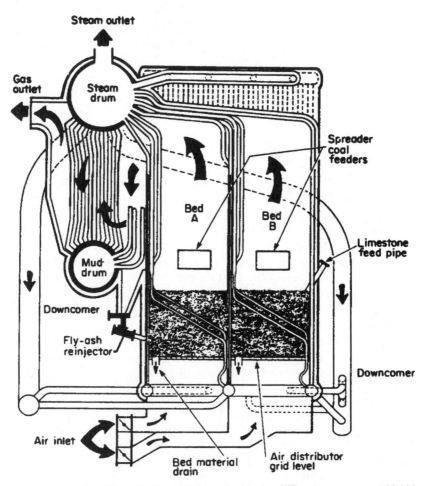

Figure 4.14   Atmospheric fluidized-bubbling-bed boiler, WT type, generates 100,000
lb/hr at 625 psi. (*Courtesy Foster Wheeler Corp.*)

sulfur capture on lime occurs to form calcium sulfate. Raw limestone is continually injected into the bed while a gravity drain system withdraws the spent material including large coal-ash particles. The low operating temperature in the fluidized bed creates low nitrogen oxide formations. Particles being carried by the flue gas can be captured by conventional electrostatic precipitators or bag-house filters.

Atmospheric beds have the combustion in the furnace operating at atmospheric pressure. These are further classified as follows. See Fig. 4.15, which shows the three types.

1. *Bubbling bed:*   Fig. 4.15a. In this bed, the depth during operation is about 4 to 5 ft, and collapses to 1 ft without high-velocity airflow.

2. *Deep bed:*   Fig. 4.15b. Operates the same as the bubbling bed, but with deeper depths of fuel, sand, and limestone.

3. *Circulating fluidized bed:*   Fig. 4.15c and d. In this configuration the bed is deeper than the other two with the fluidization extending from the bottom of the combustion chamber to the top. The burning particles of fuel and the inert material are carried from the combustion chamber into a cyclone-type separator. This permits hot combustion gases and some fine inert particles to pass into the steam generator section of the boiler assembly. The cyclone separator has passages in the flow to produce a spiraling effect that separates out designed particle sizes from the gas stream.

Note that feedwater is heated in the external heat exchanger with steam generation occurring in the convection section of the fluidized-bed boiler.

Pressurized fluidized bed combustion is being developed in demonstration plants; however, the basic research and development has been completed so that commercial use of this technology is progressing. The pressurized fluid-bed system operates at pressures above 150 psi, and the hot gases, after suitable cleanup, are directed to a gas turbogenerator as is shown in Fig. 4.16. The main elements are a boiler where steam is produced and superheated to power a steam turbogenerator and the gas turbogenerator that uses the combustion gases in a combined-cycle arrangement. The power generated is about 50 percent per prime mover. The topping combustor shown in Fig. 4.16 burns a low-Btu gas from a pyrolyzer prior to delivery of the gas to the gas turbine in order to increase gas temperature to about 2500°F. Pressurized fluidized-bed technology under development in demonstration projects has the ultimate aim of burning coal cleanly and efficiently, and using the clean gases to power a gas turbine in a combined cycle arrangement. This is expected to boost the thermal efficiency of coal-fired plants.

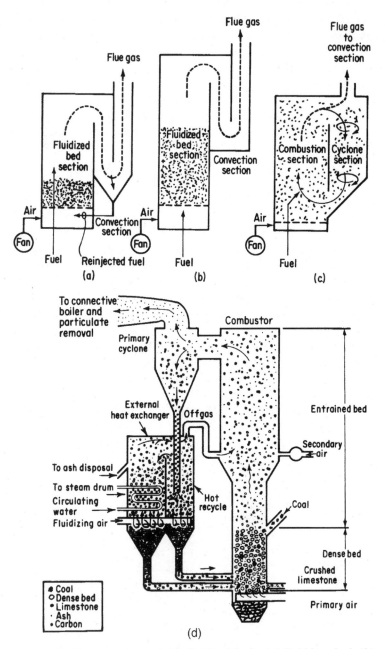

Figure 4.15 Types of atmospheric fluidized beds. (*a*) Bubbling bed; (*b*) deep bed; (*c*) circulating bed. (*d*) Cyclone separates out particles in circulating bed.

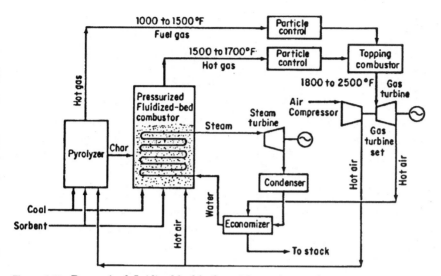

Figure 4.16   Pressurized fluidized-bed boiler with topping combustor in a combined-cycle power plant arrangement.

Tube leaks are of concern in fluidized-bed boilers for various reasons, including fly ash erosion in economizer, convection sections, and waterwall heating surfaces. High gas velocities and abrasive flue gas particles are cited as causes. Erosion control on tubes is practiced by spray metallizing the tubes with wear-resistant alloys. Sacrificial pad welding has also been used.

Tube-wear problems.   Special application boilers pose wear-and-tear problems on many boiler components; however, new materials, such as erosion-resistant tube materials, are continuously being developed to reduce abnormal wear and tear. Metals being used today to resist corrosion and erosion vary by the application: (1) For flame-spraying tubes, nickel aluminide and nickel titanide are used by specialty companies. (2) For plasma spray coating, nickel titanium and chromium–nickel titanium are employed. (3) For hard-face welding, austenitic alloys containing 18 percent chromium or more are presently used.

Questions and Answers

1   What is the minimum thickness of plate permitted for shells or heads subjected to pressure on an electric boiler?

ANSWER: The plate must be at least ³⁄₁₆ in. thick per Section I (ASME Code) rules.

2   When are postweld heat treatment and radiographs not required on a welded joint of an electric boiler?

ANSWER: When the following limitations are not exceeded: 16-in. inside shell diameter, 100-psi pressure.

3   When must an electric boiler have a built-in inspection opening?

ANSWER: Electric boilers over 5 ft^3 in volume and not having a manhole must have an inspection opening in the lower portion of the shell or head so that the unit can be periodically checked for scale or other deposits and cleaned accordingly. The opening cannot be smaller than 3-in. pipe size.

4   If an electric boiler has a power input of 750 kW, what Code requirements on capacity and number of safety valves must be met?

ANSWER: Since the input is over 500 kW, two or more safety valves are required. Their total capacity must be at least equal to $750 \times 3.5 = 2625$ lb/hr.

5   How small can the feedwater connection be on an electric boiler?

ANSWER: It cannot be smaller than ½ in.

6   What type of electric boiler does not require a low-water fuel cutoff?

ANSWER: Electrode-type boilers do not require a low-water cutoff because the water in the boiler completes the electric circuit from electrode to electrode. Low water would interrupt the electrical flow, thus serving the same purpose as a low-water cutoff.

7   When are two gauge glasses required on electric boilers?

ANSWER: Non-electrode-type boilers operating at pressures over 400 psi require two water gauge glasses.

8   How many blowoff valves are required on an electric boiler operating over 15 psi and having a water content of 95 gal?

ANSWER: The Code requires only one blowoff valve if the water content does not exceed 100 gal.

9   How is the minimum required safety relief capacity determined for a high-temperature water boiler?

ANSWER: Use the stamped designed maximum Btus per hour output for the fuel used and divide this by 1000 to obtain pounds per hour relieving capacity.

10   What is a fluid-vaporizer generator?

ANSWER: This is a closed vessel in which a heat-transfer medium, other than water, is vaporized under pressure by the application of heat with the heat-transfer medium used externally to the closed vessel.

11   Do fire insurance companies and local fire regulations have requirements on liquid-phase heaters or on vapor-phase heaters?

ANSWER: Yes, because a distinct fire hazard may exist with some of the organic fluids, including mineral oil.

12   May rupture disks be installed between an organic vaporizer and the required safety relief valve?

ANSWER: The Code permits this arrangement to minimize loss of the organic fluid, but certain Code requirements must be adhered to. For example: (1) The rupture disk must have a pressure and temperature rating for bursting that cannot exceed the allowable pressure of the vaporizer and must be at a pressure below that of the safety valve setting. (2) There must be sufficient cubic volume after the rupture disk so that the opening to the safety relief valve is not restricted if the rupture disk functions. (3) There must be a pressure gauge, try cock, or open vent in the space between the rupture disk and the inlet to the safety relief valve so that rupture disk leakage can be determined.

13   What kind of gauge glass is required on an organic-fluid vaporizer generator?

ANSWER: It must be of the flat gauge-glass type with forged-steel frames. Gauge cocks cannot be used in order to avoid spilling or draining the organic fluid.

14   Can a safety valve have a lifting lever when installed for overpressure protection on an organic-fluid vaporizer?

ANSWER: No lifting lever is permitted in order to avoid discharging accidentally the organic fluid. The safety valve should be removed at least once per year and tested off the unit for pressure setting and capacity. The discharge from an installed safety valve should be directed to a condenser or, as a minimum, to a safe point outside a building.

15   To what pressure must a vaporizer be designed?

ANSWER: At least 40 psi above the operating pressure at which it will be used with the usual safety factor applied to this design per Code rules.

16   What two requirements govern the selection of safety relief valves for liquid- and vapor-phase organic generators?

ANSWER: The pressure setting can be no higher than the maximum allowable pressure of the unit, and the capacity must equal at least the maximum Btu-per-hour output of the generator.

17   What kind of valves should be used for HTHW application?

ANSWER: Valve seats, plugs, and bodies should be made only of cast steel, forged steel, or steel alloys. Valve seats should be stainless chrome-nickel steel, to avoid corrosion and erosion by flow of water. Pressure ratings should follow the Power Boiler Code rating on stop valves, feed valves, and blow-down valves, which is 125 percent of boiler allowable working pressure (AWP).

18 Do HTHW generators require a water gauge glass?

ANSWER: Only if a natural-circulation boiler has a drum which is used as an expansion tank. If the boiler is completely filled with water under pressure and has an external expansion tank, the Power Boiler Code does not require a water gauge glass or gauge cocks.

19 On forced-circulation boilers for HTHW systems, what provision should be made to prevent hot spots, or lack of circulation, in case the pump fails?

ANSWER: A pressure-differential or flow switch should be installed to shut off the fuel-burning equipment in case no water is flowing.

20 Should a low-water fuel cutout be used on an HTHW generator of the suspended fuel-fired type?

ANSWER: Some states now require a low-water fuel cutout on low-pressure hot-water-heating systems. Thus it follows that this safety device is even more necessary on an HTHW boiler. It should be installed on the boiler to shut off the fuel in case the water level drops below a dangerous level.

21 Name three methods for pressurization of an HTHW system.

ANSWER: Three methods are: (1) steam cushion; (2) gas pressurization such as use of a nitrogen gas cushion; and (3) mechanical pressurization which makes use of an expansion tank and a pump.

22 When chemical water treatment is applied to an electrode boiler to mini-mize scale formation and similar water considerations, what other factor in the treatment must be considered?

ANSWER: The effect to conductivity of the treatment. If the conductivity becomes too high (low resistance to current flow), short circuits may occur that may damage the electrodes and the insulation. If the conductivity is too low (high resistance to current flow), the unit may not be able to reach its rated output.

23 Name some typical advantages in burning waste fuels in a steam generator.

ANSWER: Advantages to a plant that has waste fuel as a by-product of a manu-facturing process are: (1) An inexpensive or economical means is available to produce steam in comparison to fossil-fuel-fired boilers which require the pur-chase of fuel; (2) A fuel is available that may not be affected by shortages of oil or gas; and (3) The by-product can be disposed of in the steam generator; thus waste-disposal regulations may be avoided when the waste has to be dumped in a landfill or similar disposal means that could affect the environment.

24   Name three burning requirements in the burning of wood refuse in a steam generator.

ANSWER: In the burning of wood refuse, one must consider the design of the furnace, the type of burning equipment to be used, and similar factors to accomplish the following: (1) evaporation of the moisture in the wood; (2) distillation and combustion of the volatile components in the refuse; and (3) burning of the remaining carbon material in the wood.

25   What is the Code rule for establishing the allowable pressure on a bimetallic tube as used in waste-fuel-burning boilers?

ANSWER: Per Code, only the inside, or core tube thickness and tube diameter, is used in calculating the allowable pressure on bimetallic tubes. The stainless steel cladding is considered as wear-resistant material, and thus as not contributing to the strength of the tube in resisting pressure.

26   In a combined-cycle plant, how is the capacity of the required safety relief valve determined for the HRSG that uses supplementary firing?

ANSWER: The manufacturer of the boiler is required to calculate and stamp the maximum output in pounds per hour for the combination of waste-heat-steam generation plus that added by supplementary firing. The safety relief valve capacity must be at least equal to this capacity. This applies to the different pressures on a multiple pressure waste-heat boiler.

27   What is cogeneration?

ANSWER: Cogeneration is the coincident generation of electricity and process-steam needs for a manufacturing complex. Usually high-pressure steam is produced in on-site steam generators. This high-pressure steam is passed through a turbogenerator to obtain electric power and process steam, by extracting the steam from the turbine at a pressure needed for the process. Several extraction pressures can be used. The overall heat balance for a cogeneration plant can thus be attractive in comparison to just buying power and generating only process steam.

28   Name three reasons why waste-heat boilers are usually of the watertube type.

ANSWER: The watertube boiler type is favored for the following reasons: (1) Large mass flows of waste heat are possible; (2) solid entrapped particles are more readily recovered in hoppers and similar equipment; and (3) a better tube arrangement is possible to avoid slagging and erosion problems.

29   Name three possible sources of a furnace explosion in a recovery boiler of the black-liquor type.

ANSWER: (1) Reaction between water and smelt from a tube failure and a similar source of water entering the furnace; (2) reaction between a weak or low solid concentration black liquor that is sprayed into the furnace and then because of its high water content, reacts with the smelt in the furnace; and (3) conventional ignition of unburned fuel from auxiliary burners using gas or oil as a fuel, with the ignition leading to a fire-side explosion when the accumulation of unburned fuel is ignited.

30 Name the three types of atmospheric-bed fluidized combustions.

ANSWER: Bubbling bed, deep bed, and circulating bed.

31 What is the principle fire-side corrosion cause in black-liquor recovery boilers?

ANSWER: Sulfidation corrosion, caused by sulfur-bearing gases under reducing conditions, appears to be the main cause of fire-side tube metal wastage, although the corrosion mechanism is not fully understood. It is, thus, essential to establish a tube thickness testing program on furnace or other tubes exposed to the furnace smelt bed in order to prevent any corroded-through tube from spraying water into the smelt bed, and thus cause a violent smelt-water reaction-type furnace explosion in the recovery boiler.

32 What is the main difference between the circulating fluidized-bed and the bubbling-bed type of fluidized-bed burning?

ANSWER: In the circulating fluidized-bed burning, there are higher air velocities through the combustion zone compared to the atmospheric bubbling bed. This increased air velocity propels small particles of burning fuel, limestone, and ash upward through the furnace. This increases the thermal capacity of the circulating particles and hot gases, which produces a more uniform temperature throughout the furnace. The longer combustion and sulfur capture time in the furnace compared to a noncirculating fluidized-bed design results in: (1) increased combustion efficiency; (2) reduced CO formation and NO_x emissions; (3) better utilization of limestone in capturing SO_2. Also, a mechanical cyclone in a circulating fluidized-bed burning after the furnace section assists in removing entrained particles in the flue gas, and then recirculates these hot particles back to the fluidized bed, thus helping to convert any unburnt fuel particles into heat energy.

5

Nuclear Power Plant Steam Generators

The generation of electric power by nuclear energy is a technology that differs considerably from fossil-fuel burning plants, and requires special types of operators as well as inspection and maintenance staffs. Nuclear power systems depend on a controlled fission of nuclear fuel for a heat source instead of a boiler that burns fossil fuel. However, both use a medium to pick up the heat, and this medium transfers the heat either directly in a boiling water reactor, or indirectly in a pressurized-water reactor, to make steam that is then used by a steam turbogenerator.

The fission process involves atom splitting, and this is the commonly used atomic energy term employed in describing the nuclear power generation process.

Atomic Structure and Atomic Energy

A brief review of the source of atomic energy and the special risks involved in generating power is being offered as introductory material for further study of advanced texts on this subject.

A review of the structure of an atom will reveal that the nucleus is made of protons and neutrons defined as follows:

A *proton* is a *positive-charged particle* of matter. There is usually one proton in the nucleus for each negative-charged particle, or electron, circling the nucleus. The positive-charged proton attracts the negative-charged electron whirling around the nucleus, and this prevents the electron from flying off by centrifugal force just as the earth is held in orbit around the sun by gravitational forces.

A *neutron* is matter in the nucleus with no electric charge, and it weighs about the same as a proton. There is a strong binding force holding together protons and neutrons. The number of neutrons is not fixed, but varies with atoms, although this number does affect the atomic weight. Neutrons do not affect the chemical properties of the elements, as protons and electrons do. Since neutrons have no charge, they are very useful in nuclear-bombardment applications. They travel in straight lines until they come in contact with other matter. This makes them very useful in splitting the nucleus of an atom.

Electrons are negatively charged particles, quite a bit lighter than protons. In fact, they weigh about $\frac{1}{1800}$ the weight of a proton. Electrons are arranged in rings around the nucleus, with definite numbers of electrons per ring. That is, the first ring holds 2 electrons; the second, 8; the third, 18; until a maximum of 32 is reached. The total number of electrons in an atom is equal to the number of protons and neutrons in the nucleus, which is also the atomic weight of the atom. Uranium has the highest atomic weight at 92 and also has the greatest number of electrons surrounding the nucleus.

Material that is capable of capturing neutrons, and therefore splitting an atom into two or more particles, is called a *fissionable material*. Uranium 235, uranium 233, and plutonium 239 are fissionable materials. When an atom is split by a neutron traveling at high speed so that two or more fragments are split from the atom, it is called *fission*. During fission, two to three neutrons are released from the split atom. If a chain reaction is to result, one of these free neutrons must be captured by another atom and cause fission of that atom; then in turn it must produce another free neutron to establish the chain reaction of fission. The ratio of the number of neutrons which cause fission to the number of neutrons initially produced is called the *effective multiplication*.

The fundamental law of physics on which a nuclear reaction is based is Einstein's law of the interrelationship of mass and energy, namely, energy equals mass times velocity squared ($E = MV^2$). One must realize, though, that the velocity term is near the speed of light. It has been established, for example, that 1 lb of uranium 235, when hit by a neutron bullet, will *explode* to form lighter atoms and spare neutrons. Some of the mass, about one-thousandth of a pound, is lost in the form of energy which is equivalent to about 11.4 million kilowatthours (kWhr) of energy. Many other principles of nuclear physics are involved in a nuclear reaction, but they are beyond the scope of this book.

The present method of nuclear power generation is by fission, or the splitting of the nuclei of certain heavy atoms by bombarding them with neutrons. Hitting the uranium-isotope ^{235}U nucleus, these neutrons split the atom to form new elements, such as krypton and barium, release energy in the form of heat, and liberate additional neutrons from the nucleus that can bombard another nucleus to keep the reaction going. Other atoms that fission are ^{233}U, thorium, plutonium, and ^{238}U.

Figure 5.1 is a schematic of the indirect cycle, where the coolant from the reactor transfers heat to a separate working fluid in a steam generator. This is a pressurized-water reactor system (PWR). In a direct cycle, the heated coolant is water steam, and this is used as the working medium in a steam turbogenerator.

There are many differences between a nuclear power plant and a fossil-fuel burning plant, such as:

1. The fuel inserted into the core of a reactor lasts a long time compared to a fossil-fueled plant where fuel is fed continuously into a boiler to maintain combustion, or heat generation.

2. Special equipment and handling is required in a nuclear plant because of the radioactivity that exists in the core and the equipment that is in contact with the core fuel or reaction. This affects inspection techniques as well as operation and maintenance.

3. To prevent the spread of radioactivity, special containment designs are required in a nuclear power plant, that are not required in fossil-fuel burning plants.

4. The nuclear fuel requires precise preparation and special processing in the form of fuel rods before they are inserted into the reactor core.

5. Elaborate safety procedures in the form of specially designed instruments and controls are necessary to avoid the radiation hazard.

6. Since no combustion air is used to burn the fuel in the reactor, there is no stack discharge.

Some nuclear terms

These are presented because they do not exist in a fossil-fuel burning plant.

Criticality is the point at which the nuclear fuel can sustain the chain reaction.

Fission products are the atoms formed when the nuclear fuel, such as uranium, fissions in the reactor.

Decay heat is the heat produced by the nuclear fuel after the reactor has been shut down.

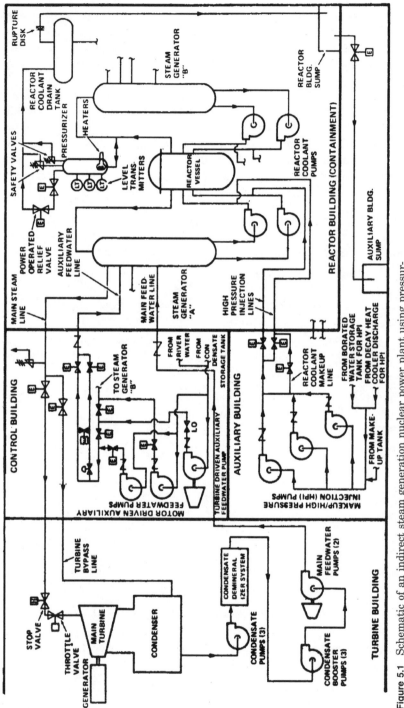

Figure 5.1 Schematic of an indirect steam generation nuclear power plant using pressurized water as coolant and two heat exchangers as steam generators. (*Source: Federal Register.*)

Core is the central part of a nuclear reactor that contains the fuel assemblies.

Control rods are made of a material that can absorb neutrons and, when inserted into the nuclear fuel, shut down the reactor by causing the fissioning process to stop.

Fuel rods are cylindrical shaped rods of special material that contain within them the nuclear fuel pellets.

Cladding in uranium nuclear fuel plants is a zirconium alloy cover surrounding the uranium fuel, which acts as a barrier between the fuel and the reactor system's coolant water.

Maximum permissible dose (MPD) is the legal limit of radiation exposure that a person can be subjected to from a nuclear plant as established by the Nuclear Regulatory Commission (NRC). For the general public, the limit is 500 mrems per year. For nuclear plant workers, it is 5000 millirems per year.

Millirem is the unit used to measure radiation dosage, and is defined as 1/1000th of a rem, where rem, or roentgen equivalent man, is a measure of radiation that indicates potential impact on human cells.

Nuclear Regulatory Commission is the federal government agency that establishes safety regulations, and oversees inspections of nuclear power plants, because of the inherent interstate risk if a nuclear accident should occur.

Residual heat removal system is a series of pumps and heat exchangers that removes the heat from the reactor after the fissioning process has stopped and the nuclear reactor shut down.

Emergency core cooling system is a backup system of cooling water designed to cool the reactor in case the primary cooling system fails.

Primary coolant is the water that transfers the heat from the nuclear reaction either to steam, as in a boiling-water reactor, or to a secondary coolant loop heat exchanger to make steam, as in a pressurized-water reactor.

Dosimeter is an instrument used by personnel to measure the amount of radiation received in a given time.

Radioactivity is the ability of some elements to spontaneously give off energy in the form of waves or charged particles. Radiation may have alpha, beta, or gamma rays.

Alpha radiation is the least penetrating type and can be stopped by a sheet of paper.

Beta radiation is emitted from the nucleus of an atom during fission, but its penetration can be stopped by thick cardboard.

Gamma radiation is also emitted from the nucleus, has deep penetration capability, and can only be stopped by heavy shielding, such as lead or concrete.

Background radiation is naturally occurring radioactivity for the particular area under consideration.

Types of Reactor Systems

Reactor vessels are used to house the fuel elements in a nuclear plant where the chain reaction of fission takes place. Reactor vessels, fuel elements, coolants, moderators, and control rods are hardware common to most types of reactors. See Fig. 5.2.

Reactors are generally classified by the type of coolant used to extract the heat from the fission reaction. Most common are the *pressurized-water reactor, boiling-water reactor, heavy-water reactor, gas-cooled reactor,* and *metal-cooled reactor (sodium).*

Figure 5.3*a* shows a *boiling-water reactor (BWR) schematic* where the nuclear-reaction-heated water in the reactor evaporates in the same pressure vessel and then is directed to a steam turbogenerator. The steam is condensed, passed through feedwater heaters, and returned to the reactor by a feed pump.

Figure 5.3*b* illustrates a *pressurized-water reactor (PWR) schematic.* The coolant is high-pressure water which is pumped through the core of the reactor to remove heat by primary coolant pumps. The pressurized hot water is passed through water-to-steam heat exchangers, with the tubes carrying the radioactive water within the shells containing the steam going to the turbogenerator. The water must be at a higher temperature in the heat exchangers than the steam, with the water pressure in the reactor being about 2200 psi to prevent boiling at high water velocities. A pressurizer is used to keep the water under constant pressure; it is similar to an expansion tank on an HTHW system.

Figure 5.4*a* illustrates a *heavy-water reactor flow schematic,* which resembles the PWR type except that in this system, heat transfer is from fuel rods to heavy-water coolant and then to light water in steam generators or water-to-steam heat exchangers. This design is popular in Canada.

Figure 5.4*b* illustrates a *gas-cooled reactor.* This system uses a large compressor for circulating the helium through the reactor core and the steam generators. Helium is at 700-psia pressure and a temperature of 1430°F, with the steam produced at higher temperatures (1000°F) and pressure than is possible with the BWR or PWR types.

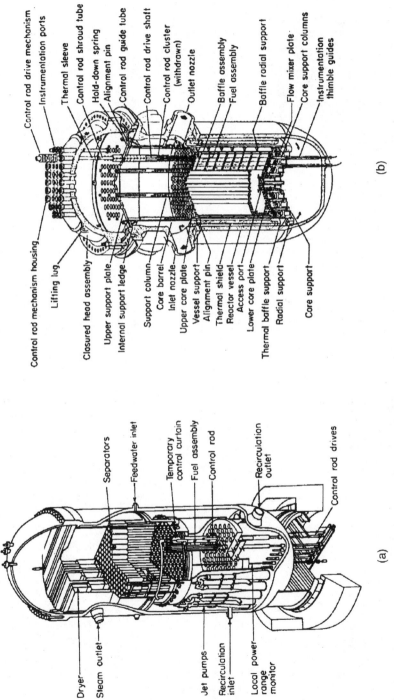

Control rod drive mechanism
Instrumentation ports
Thermal sleeve
Control rod shroud tube
Hold-down spring
Alignment pin
Control rod guide tube
Control rod drive shaft
Control rod cluster (withdrawn)
Outlet nozzle
Baffle assembly
Fuel assembly
Baffle radial support
Flow mixer plate
Core support columns
Instrumentation thimble guides

Control rod mechanism housing
Lifting lug
Closured head assembly
Upper support plate
Internal support ledge
Support column
Core barrel
Inlet nozzle
Upper core plate
Vessel support
Alignment pin
Thermal shield
Reactor vessel
Access port
Lower core plate
Thermal baffle support
Radial support
Core support

(b)

Separators
Feedwater inlet
Temporary control curtain
Fuel assembly
Control rod
Recirculation outlet
Control rod drives

Dryer
Steam outlet
Jet pumps
Recirculation inlet
Local power range monitor

(a)

Figure 5.2 Internals of reactors. (*a*) Boiling-water (BWR) type; (*b*) Pressurized-water (PWR) type.

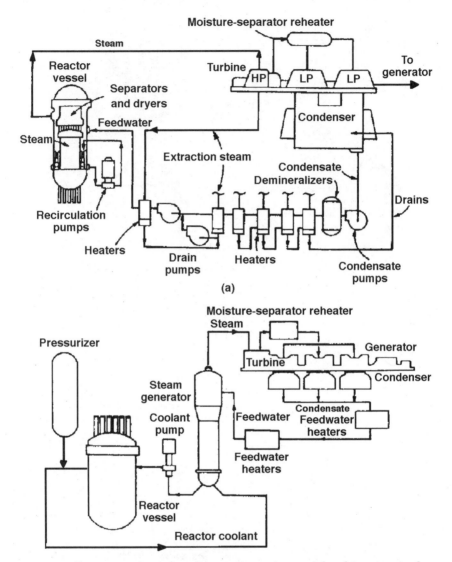

Figure 5.3 Two commonly used nuclear reactor systems employed to generate electric power. (*a*) Boiling-water reactor (BWR); (*b*) pressurized-water reactor (PWR).

Under development are breeder reactors for power generation. The
breeder reactor is so named because it produces more fissionable fuel
material than it consumes. The now wasted uranium 238 and low-
grade thorium ores would be converted in a breeder reactor, through
neutron bombardment, to a fissionable fuel from the dwindling supply of the present fuel used, uranium 235.

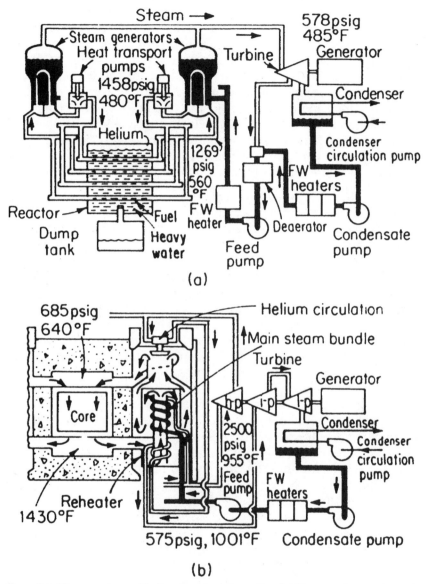

Figure 5.4 Heavy-water and helium reactor schematics. (*a*) Heavy-water reactor uses water flowing through the fuel-rod tubes, and heavy water as the moderator in the reactor. Helium is used as an expansion cushion. (*b*) This helium gas-cooled reactor system uses helium as a coolant and helium to steam heat exchangers.

Fusion power plants are still in the conceptual and development stages. Fusion requires temperatures to 100 million degrees Fahrenheit which no known material can withstand. It is necessary to develop intense magnetic fields to confine and also to obtain the hydrogen-gas plasma that is the fusion fuel. In the fusion process, hydrogen or its isotopes, deuterium and tritium, are fused to form helium, the next higher element in the atomic scale. The joining or fusion takes place in nuclear ovens which generate intense heat that must be extracted for the production of useful energy. Intense development is going on all over the world in an effort to find an economical method of extracting usable energy from a fusion nuclear reaction.

Generated heat. The heat produced in a nuclear reactor is usually expressed in so many megawatts of power. The heat produced depends on the thermal or neutron flux developed in the nuclear chain reaction per unit of volume, the energy produced per fission, and similar nuclear physics criteria. One concern in nuclear reactors that is similar to the case of fired steam generators is the possibility for overheating damage. The limiting factors are the temperature limits imposed on fuel assemblies in order to avoid burnout of the fuel cladding and/or melting of the oxide fuel. These factors also influence coolant flow rates in order to avoid overheating damage of the fuel rods. This also requires maintaining proper coolant levels in the reactor system, just as one has to maintain proper water levels in a steam generator.

Core meltdown from a loss of coolant has received great attention by designers and regulatory agencies. To prevent core meltdowns and consequently possible release of radioactive material to the biosphere, reactor systems are equipped with numerous safety devices to forewarn of a developing incident and also to initiate backup emergency core-cooling systems if needed. This system is intended to replenish cooling water that might be lost through a rupture of the primary cooling system. All emergency core-cooling systems are designed to inject water into the pressure vessel rapidly enough to keep the core cooled. Boiling-water reactors employ a core spray and an independent low-pressure coolant injection system.

Reactor pressure vessels must be constructed to the strict standards of the ASME Boiler and Pressure Vessel Code, Section III. The reactor vessel becomes highly radioactive during operation because of neutron capture. Massive biological shielding is used to surround the reactor pressure vessel.

Thermal shields are layers of steel, or other forms of shielding, that are placed between the core and the reactor vessel wall. This has the effect of absorbing some gamma rays from the core as well as reducing the intensity of fast neutrons reaching the reactor vessel wall.

Fast-neutron bombardment of the vessel wall can produce long-term radiation effects on the vessel's material, such as embrittlement over a long time period. In the design of the reactor, this effect is considered by assigning an expected life for the vessel. Also of great importance in the design is to use material that has a nil-ductility transition temperature, or ndt, well below the temperature normally encountered in operation. The ndt is the temperature at which a material changes from ductile to brittle behavior.

In-Depth Safety Defense Systems

Nuclear safety begins with designing a plant to have numerous defenses and redundancy systems in order to avoid an unforeseen accident. See Fig. 5.5. The design incorporates strict quality-control requirements that are also generally, as a minimum, legal requirements. Protective systems include instruments and measuring devices that are constantly scanning vital areas of the plant and which are capable of securing the plant from abnormalities of operation. Multiple safeguard systems are also used to protect the plant from postulated accidents. See Fig. 5.5a for a sketch of the system in a boiling-water reactor that provides a defense against the release of radioactive material from a reactor to the environment.

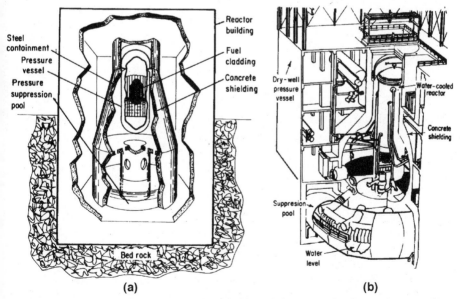

(a)   (b)

Figure 5.5 (a) The reactor vessel is surrounded by a defensive in-depth containment. (b) Suppression pools are used to condense steam from a reactor leak into the containment vessel, thus limiting pressures and volume required in the containment vessel.

The reactor containment shell is a thin cylindrical or spherical, steel pressure vessel surrounding the shielding. Its purpose is to accommodate energy released by sudden, uncontrolled fission in the event of a reactor accident. Since personnel work inside, the shell must be provided with access locks. The pressure within is maintained below atmospheric pressure so that any leakage of radioactive gas from the reactor is retained in the shell and passes through specially filtered and monitored ducts.

The pressure-suppression containment system eliminates the need for a large containment shell, at least in the case of water-cooled reactors. Instead, the reactor pressure vessel and its associated pumps and pipework are enclosed in a second pressure vessel sized to accommodate the steam formed in the event of vessel or pipe rupture. Ducts are connected to this outer vessel. They lead to an annular suppression pool in which the steam is condensed. Entrained fission products are then removed by absorption.

Thermal and biological shielding—radiation hazard. The radiation hazard in nuclear plants has received wide publicity, but the extent of exposure becomes significant only when the pressure-containing boundary is pierced. Waste disposal problems are under control, but the long-term effect of storage is a matter of concern to many environmentalists. The main fear of radiation is its ability to cause cancer and genetic defects; however, the threshold limit required for this to occur is a medical matter.

Three distinct kinds of radiation are emitted from radioactive material. Alpha and beta rays have little penetrating power, but gamma rays can penetrate great thicknesses. Neutrons, too, have great penetrating power and are the primary radiation hazard in an operating reactor.

To safeguard personnel against neutrons, gamma rays, and heat, the reactor, and much of its auxiliary equipment, must be enclosed within thermal and biological shielding. Gamma rays can be absorbed by a number of materials, particularly those with the greatest density, such as lead or steel. The same effect is obtained by using a much greater thickness of water or concrete. For land-based reactors, space and weight limitations are not major considerations. The cost of material and construction is usually more important. The cheapest and most widely used shielding material is concrete. When several feet thick, a concrete biological shield is an excellent neutron absorber. The addition of some percentage of denser material to the mix, such as iron or barytes, will improve local gamma shielding. Alternatively, the use of more expensive magnetic concrete affords protection against both gamma rays and neutrons.

The intensity of radiation follows an inverse-square law as the distance from the source increases. Therefore, the amount of shielding required can be reduced by building the shield at a greater distance from the core. Since concrete is not able to withstand high heat, a thermal shield is constructed between the main shield and the reactor. This may be steel or a separate, thin concrete vessel. Coolant channels may be provided in the concrete vessel to carry away heat. The major problem is preventing radiation-leaking paths where coolant inlet and outlet pipes, control rods, and other hardware leave the reactor. Any such annular paths must be stepped to avoid a direct line-of-sight path.

Control and instrumentation. The energy potential of the fuel mass in a nuclear rector and the lethal radioactivity of many fission products demand more stringent safety measures than a conventional power plant. They may be considered under two heads: control and instrumentation for safe reactor operation and means to prevent escape of radioactivity during normal operation, possible reactor runaway, or other failure.

Reactor instrumentation can be grouped into three classes: *control, safety,* and *monitoring.* The major requirement in control instrumentation is measurement and display of the heat rate produced in the core over a complete range from subcritical to full-power operation. This is done by measuring the neutron flux level, or rate of neutron fission, with neutron detectors. Different types of instrumentation are provided for shutdown, start-up, and low-power and high-power operation.

For shutdown and start-up measurements, pulse counters are used, often taking the form of ionization chambers filled with boron trifluoride gas. Operating at a particular voltage, they give a pulse proportional to the incident radiation. Alternatively, fission chambers, coated with a fissile material, are capable of detecting neutrons by fissions inside the chamber. Each neutron produces an electric pulse. In either case, the output passes through a pulse amplifier to counting rate meters, logarithmically scaled since neutron flux increases exponentially. Start-up and similar low-power detectors retract into the biological shield during high-power operation. For power measurements in the normal operating range, where temperature effects on reactivity become important, instruments are linearly scaled. Ionization chambers act directly on high-impedance potentiometer recorders.

Reactor emergency tripping is based on set limits being executed for such factors as neutron density change, fuel-element temperature, power level, and coolant flow. Then reactor "scram" takes place: control rods are automatically inserted into the core, either by power or by releasing the drive mechanism and allowing them to fall by gravity. "Scram" time is usually 2 to 3 seconds (sec).

Containment of fission products demands several lines of defense. For example, a reactor vessel is surrounded by a thermal and biological shield system. A stainless-steel thermal wall protects the shielding from heat; borated graphite acts as a neutron absorber. Gamma rays emitted by the absorber are captured in the outer concrete shell. A suppression pool (Fig. 5.6b) is a safety measure to condense steam if the water-cooled reactor pressure vessel or the piping ruptures.

Monitoring the reactor power level is the purpose of the neutron detector (fission counter or ionization chamber) placed within the core (Fig. 5.6a). Detectors may be alternatively placed outside the reactor, treating the core as a point source of neutrons, as shown in Fig. 5.6c. The extreme range of power level possible requires multiple detectors with overlapping ranges. To gain data on fuel, heat transfer, and metallurgical characteristics, extensive in-core sensing instrumentation may be applied. An instrumented fuel assembly (Fig. 5.6b) is used in a boiling heavy-water reactor to determine fuel power limits.

Regulations and Administration of Safety

Nuclear Regulatory Commission (NRC). The federal government's Nuclear Regulatory Commission, previously called the Atomic Energy Commission, has broad powers over nuclear-using plants, and has established extensive inspection and administrative controls that involve all major areas affecting the utilization of nuclear energy. Included are:

1. Ownership of nuclear fuel material requires a license from the NRC, and this applies to organizations that receive, possess, or fabricate nuclear fuel.

2. Nuclear facilities must be licensed by the NRC starting with construction permits, operating licenses, and nuclear plant operators.

3. The NRC establishes rules on the release of radioactive wastes.

4. The NRC also establishes limits and rules for permissible radiation doses for nuclear facility workers and the general public.

5. Minimum insurance requirements are dictated by the NRC for nuclear facilities.

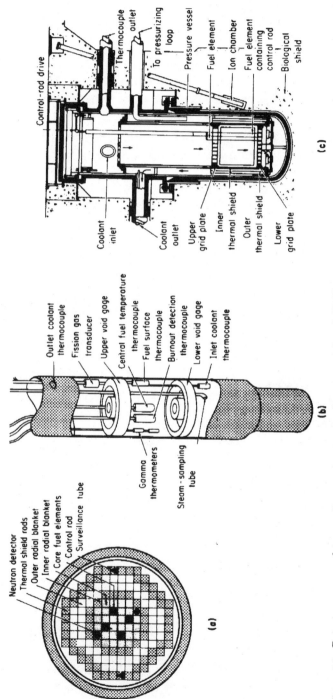

Figure 5.6 Detectors are used to monitor reactor-core performance. (*a*) and (*b*) Neutron detectors monitor core performance. (*c*) Detectors are placed outside the reactor for core monitoring.

175

6. Broad operating, maintenance, and inspection guidelines are established by the NRC for nuclear power plants that involve safety issues.

Some responsibilities on the use of nuclear fuel and power are divided between the NRC and various state agencies, as well as with other federal agencies, such as the Interstate Commerce Commission, Coast Guard, Navy, and Department of Labor.

ASME Boiler and Pressure Vessel Code, Nuclear Sections. By adoption as legal requirements by federal and state jurisdictions, ASME Code Sections involving nuclear construction and in-service inspections have materially affected nuclear power generation as well as periodic in-service inspections of these plants. The National Board of Boiler and Pressure Vessel Inspectors now require inspectors involved with nuclear pressure vessel construction and in-service inspections to receive special instructions, especially on NDT, and pass examinations in order to be considered nuclear qualified on their commissions.

ASME Nuclear Code Sections. There are a large number of ASME Code Sections that are applicable to nuclear power plants as respects construction and installation, and quality control, documentation, data reports, inspection, and nondestructive testing (NDT). For further study, here is a list:

Section III, Subsection NCA—General Requirements for Division 1 and Division 2

Section III, Division 1

   Subsection NB—Class 1 Components

   Subsection NC—Class 2 Components

   Subsection ND—Class 3 Components

   Subsection NE—Class MC Components

   Subsection NF—Supports

   Subsection NG—Core Support Structures

Section III, Division 2—Code for Concrete Reactor Vessels and Containment

Section V—Nondestructive Examination

Section XI—Rules for Inservice Inspection of Nuclear Power Plant Components

National Board inspectors that seek a nuclear endorsement to their commission must be familiar with these sections that deal with nuclear power plant components, as well as personnel involved with operation, maintenance, and inspection of these plants.

A separate Nuclear Code on pressure vessels had to be adopted for the following reasons:

1. Nuclear vessels have many unusual design considerations which require more stringent rules than are needed for boilers and other pressure vessels.

2. There are various types of nuclear vessels, defined as follows: *primary vessels* are those subjected to the primary coolant, which may be radioactive; *secondary vessels* are not subject to radioactivity; *pressure vessels* are basically containment vessels to guard against radioactive contamination in the event of failure of the primary vessels.

3. Since a nuclear-reactor vessel may be radioactive for years, periodic internal inspections in the usual sense as applied to boilers and pressure vessels are impossible. Sophisticated testing by nondestructive means has to be employed.

4. The hazard in a nuclear reactor vessel is not only a pressure explosion, but also a radioactive contamination hazard far more serious to life than a steam pressure explosion only. A whole area may be affected by radioactive fallout. Thus more stringent design and fabrication rules are required than for boilers or pressure vessels.

5. Nuclear reactor vessels can be subjected to very sudden heating or cooling temperature changes. This creates abnormal thermal stresses and cycling fatigue stresses. The Power Boiler Code does not have enough provisions for calculating these stresses, but the Nuclear Vessel Code requires these to be calculated.

6. The material to be used on a nuclear vessel may have to withstand radiation effects which may affect its properties. Materials for nuclear vessels thus require this consideration, whereas materials for boilers and pressure vessels may not.

7. Inspection during fabrication of a nuclear-reactor pressure vessel has to be far more thorough than of a boiler because of the hazards involved. Rigid quality control must also be exercised far above the usual practice in boiler and pressure-vessel construction.

Class 1, 2, and 3 components. The ASME Construction Code provides for three levels of quality for design and construction, depending on the element of radioactive risk that may be involved with the component under consideration. The choice of quality level for a given component requires a knowledge of how the component functions in the process or system. Class 1 components are those that are a part of the reactor-coolant pressure boundary, i.e., those components containing coolant for the core and piping and similar components that cannot be isolated from the core and its radioactivity. Class 2 components are those involved with reactor auxiliary systems which are not part of the reactor-coolant pressure boundary, but which are in direct communication with it, such as a residual heat-removal system. Class 3 systems are those components that support Class 2 systems without being part of them.

The design of Class 1 vessels is far more exacting in terms of Code requirements than for conventional boilers. Complete stress analysis is required on all major components of Class 1 vessels, and this must be done by a registered professional engineer. All forms of stress must be considered and combined to obtain the actual stress on each critical section of the vessel. These include normal, shear, discontinuity, bending, thermal, cycling, and fatigue stresses.

Corrosion allowance must be provided if analysis shows that it is required. All loading conditions must be considered, including: (1) internal and external pressure or combinations thereof; (2) the weight of the vessel and its contents; (3) superimposed loads, such as other vessels, insulation, piping, and cladding; (4) wind loads, snow loads, and earthquake loads; (5) reactions of supporting lugs, rings, saddles, and other supports; (6) temperature effects; and (7) environment effects due to radiation.

The allowable design stress is based on the yield strength, not on the ultimate strength, as is the case for boilers and unfired pressure vessels.

Welding requirements are *stiffer*. No backing strips are allowed to remain. Full radiographing is required on all welds subject to stress caused by pressure. Some welds require ultrasonic examination also. The material must be certified by the material manufacturer to comply with all sections of the Nuclear Vessel Code. Heat-treatment test coupons are required. Ultrasonic examination is required for steel plates on reactor vessels and all other plates 4 in. and over in thickness. Standards are established for the ultrasonic technique employed. The ultrasonic test is by means of a pulse-echo instrument.

Forgings must be inspected ultrasonically and their surface by the magnetic-particle method. Forgings of nonmagnetic material must be examined on the surface by the liquid-penetrant method. Castings must be examined by radiographing, ultrasonic, magnetic-particle, and liquid-penetrant methods.

Pipes, tubes, and fittings must be examined over their full length by radiographic, ultrasonic, magnetic-particle, liquid-penetrant, or eddy-current methods, based on the material to be examined. Bolts and bolting material must be examined by the wet magnetic-particle method or by the liquid-penetrant method. Carbon steel, alloy steel, and chromium-alloy steels must be subjected to various impact tests to determine nil-ductility temperature and brittleness.

During the fabrication phase, it is important to keep complete and detailed records of all the inspections, to be sure that the inspections were made at the proper time, in the proper sequence and not after the fact, and on all material released in fabrication.

Testing.   Vessel test plates must be made. The welding procedure and the nondestructive examination methods used shall be the same as those used in the fabrication of the vessel. Important test specimens must be made. A final hydrostatic test of $1\frac{1}{4}$ times the design pressure is stipulated. Pneumatic tests may be used in lieu of the hydrostatic test.

Pumps and valves.   The ASME Code also has requirements on valves and pumps involved with reactor cooling. The primary purpose of code-required inspections and tests is to provide assurance that the valves and associated pumps as used with reactor systems will have the integrity to operate per design during normal and emergency conditions so that coolant is furnished to the reactor for a safe shutdown or operation. Among the requirements for valves and pumps:

1. All valves which have devices at remote locations to indicate if they are open or closed must be tested at defined intervals in order to make sure that the indicating devices are operating correctly.

2. Regular exercising of valves is required in order to make sure that the valves move up and down properly and are not frozen.

3. Pumps must be test run periodically to make sure they are functional when needed.

4. There must be blackout power available on-site, such as a diesel generator, and this must also be test run at stipulated intervals in order to note if it is functional for emergency use.

Marketing, stamping, and reports.   Design calculations and specifications must be filed with the state enforcement authority responsible, at the point of the installation, for nuclear and other pressure vessels. The design specifications must be certified as to Code compliance and suitability for use per vessel classification and per detailed report on expected operating conditions to be applied so that a complete evaluation of the design by a Code required registered professional engineer experienced in pressure-vessel design can be made. The pressure vessels must then be inspected during construction and installation by a qualified Nuclear Code inspector according to ASME Nuclear Code Rules. All necessary data sheets must be included. The vessels to be marked with the "N" symbol must be stamped with the following information: (1) class of vessel; (2) manufacturer; (3) design pressure at coincident temperature; (4) manufacturer's serial number; and (5) year built.

Certificates of authorization. The ASME Code for Nuclear Power Plant Components requires certificates of authorization to be obtained and retained by the following organizations that may be active in nuclear power plant work: manufacturers, engineering organizations, fabricators, installers, material manufacturers and suppliers, owners and agents of a proposed plant. The aim and purpose of certification is to have a quality assurance program from all accredited organizations that have met certain minimum Code standards. The ultimate aim is that all work performed will be conducted in rigid compliance with approved and controlled specifications and procedures, that critical activities will be verified by qualified personnel from an organization independent of work performance (third-party inspection program), that measurement and test results will be completely documented and analyzed for conformance to established specifications and Code requirements, and that appropriate actions will be taken to preclude the recurrence of discrepancies and deficiencies. Most certificate holders are required to have a quality assurance manual. This document must be kept current and must explain in detail the controlled manufacturing system that is being followed in order to achieve Code compliance with specifications.

Inspection agencies. Another document that affects the construction of nuclear power plants is American National Standard Institute (ANSI) N626, titled "Qualification and Duties for Authorized Nuclear Inspection," which the ASME has also adopted as a requirement. This document defines authorized inspection agencies and authorized nuclear inspectors and supervisors. Great emphasis is placed on knowledge of welding and how it can affect metals in the examinations given to potential nuclear inspectors. Knowledge of NDT methods and interpretation of results are also highlighted in the requirements. Quality control procedures in monitoring a construction job are also stressed in the duties of a nuclear inspector, as are traceability of documentation on materials, welding procedures, qualification of welders, and similar areas where variables in construction may exist.

Baseline inspections. The Code requires on-site initial examinations and inspections on nuclear power plant components prior to any operation. These inspections must be thoroughly documented for future reference, and consist of visual, NDT, and any other examination methods which can be duplicated in the future of plant operation. Periodic examination and testing are stipulated in the Code, or by the

NRC. Components in the primary radioactive area are required, as a minimum, to be inspected at 10-year intervals.

In-service inspection—Section XI. The primary objective of Section XI of the ASME Code is to provide a means of ensuring that the mechanical integrity of the primary coolant system is maintained throughout the operating life of the facility. This objective is accomplished through the requirement for the conductance of minimum periodic inspections of critical nuclear components, such as Class 1 vessels and their weld zones and other highly stressed areas. A preservice inspection is required. Usually ultrasonic inspection techniques are used that can be duplicated later during required reinspections. This permits comparisons to be made to note changes. Automatic inspection devices are utilized as much as possible. The use of remotely operated inspection equipment under properly planned procedures has reduced the radiation hazard to examining and inspection staffs. The frequency of inspections and acceptable criteria are detailed in the ASME Code for nuclear power plant components. Fracture mechanics is used to analyze the seriousness of flaws where no Code standards exist.

Inspection agencies and jurisdictional authorities are notified if any conditions are found that require approval for repairs or replacement. For example, steam generator *tube problems* in the form of operating degradations such as stress corrosion and erosion have been detected by the use of eddy-current inspection methods on the U-tubes in the steam generator in pressurized-water plants. This ongoing testing program has resulted in changing tube material from the original stainless steel to Inconel 600, a Ni-15Cr-9Fe alloy, and to the present U.S. practice of using Inconel 690TT, a Ni-30Cr-10Fe alloy with thermal treatment. These types of problems are being solved as a result of in-service inspection programs.

Operator qualifications. Operator qualification and training are governed and determined by federal requirements. The criteria for the selection and training of nuclear power plant personnel are contained in American National Standard 3.1-1978, entitled "Selection and Training of Nuclear Power Plant Personnel." Most nuclear plants require a minimum presence during operation of at least one senior reactor operator, two reactor operators, and two auxiliary operators who may not be licensed as yet. Operator training is receiving increased attention, as is the need for supporting staff such as a safety engineer to supplement the operating staff in case an emergency situation develops in operation. The many valves and subsystems in a nuclear power plant as well as electrically activated controls require a broad knowledge of the interplay between systems. See Fig. 5.1. The many pumps and valves shown can lead to errors of valve opening and closing and also increase the possibility of electrical and mechanical failures. Complete reliance on automatic redundancy systems cannot anticipate all the possibilities of malfunction or sense that a malfunction is taking place; therefore, skilled operators well versed and trained in the design and operation requirements of nuclear power plants are needed.

Nondestructive testing personnel. Nondestructive testing personnel must be qualified, and these qualifications are graded as follows:

A Level 1 person must have experience or training in the performance of the inspections and tests that he or she is required to perform. This person should be familiar with the tools and equipment to be employed and should have demonstrated proficiency in their use. She or he must be familiar with inspection and measuring equipment calibration and control methods and be capable of verifying that the equipment is in proper condition for use.

A Level 2 person must have experience and training in the performance of required inspections and tests and in the organization and evaluation of the results of the inspections and tests. This person must be capable of supervising or maintaining surveillance over the inspections and tests performed by others and of calibrating or establishing the validity of calibration of inspections and measuring equipment. She or he must have demonstrated proficiency in planning and setting up tests and must be capable of determining the validity of test results.

A Level 3 person must have broad experience and formal training in the performance of inspections and tests and should be educated through formal courses of study in the principles and techniques of the inspections and tests that are to be performed. This person should be capable of planning and supervising inspections and tests, reviewing and approving procedures, and evaluating the adequacy of activities to accomplish objectives. He or she must be capable of organizing and reporting results and of certifying the validity of results.

Personnel involved in the performance, evaluation, or supervision of nondestructive examinations, including radiography, ultrasonic, penetrant, magnetic-particle, or eddy-current methods, must meet the Level 3 qualification specified in SNT-TC-1A of the American Society for NDT and supplements. Those personnel involved in the performance, evaluation, and supervision of gas-leak test methods must meet the qualification requirements specified for a Level 2 person.

Personnel who are assigned the responsibility and authority to perform project functions must have as a minimum the level of capability shown in Table 5.1. When inspections and tests are implemented by

TABLE 5.1   Minimum Levels of Capability for Project
Functions

Project function	Level		
	1	2	3
Approve inspection and test procedures			X
Implement inspection and test procedures	X		
Evaluate inspection and test results		X	
Report of inspection and test results		X	

teams or groups of individuals, the one responsible must participate and must meet the minimum qualifications indicated. A file of records of personnel qualifications must be established and maintained by the owner. This file should contain records of past performance, training, initial and periodic evaluations, and certification of the qualifications of each person.

Fire protection.   Extra measures are required in a nuclear power plant to prevent any fire that could affect the plant's safety system due to the radiation hazard. Property and liability insurance pools that provide coverage for nuclear facilities have been active in making inspections and submitting recommendations for any risk improvement, and so have the inspectors from the NRC.

Great emphasis has been placed on providing fire-resistive barriers of not less than 3-hr rating to separate the following critical areas in a nuclear facility: administration building, battery rooms, boilers used for starting, cable penetrations into reactor building and control rooms or other similar area that jeopardizes the facility, cable shafts, cable tunnels, computer rooms, control building or room, decontamination areas, fire pump rooms, switch gear areas, and all other areas, including turbogenerator building, if there is a fire risk to the facility. The same philosophy is applied to the use of sprinklers or halon fire suppression.

High-level nuclear waste is a severe disposal problem as an environmental concern by many federal and state jurisdictional authorities. Still under development is a system to convert liquid wastes to a solid and immobilize this solid in glass for burial in deep, underground mines in remote areas away from any human environment.

Questions and Answers

1   How is radioactivity detected?

ANSWER: The instruments commonly used are Geiger-Müller counters, scintillation counters, gamma survey meters, and proportional counters. The instrument to be used depends on the type and density of radiation to be measured. Geiger-Müller counters are used to detect beta and gamma radiation and are not effective for measuring alpha radiation.

2   What is radiation as applied to nuclear plants?

ANSWER: Unstable isotopes of certain chemical elements undergo spontaneous change in the atomic structure of the element. This change is called radioactive decay. While this phenomenon is going on, rays of energy are emitted; thus the term *radiation of particles* describes the radiation that is experienced in a nuclear power plant, primarily under controlled conditions in the reactor pressure vessel.

3   How is an isotope defined?

ANSWER: Isotopes are elements that have the same number of protons, but differ in the number of neutrons in their nucleus. Isotopes of an element have the same chemical properties and somewhat the same physical properties, but have different atomic weights as a result of the difference in the number of neutrons in the nucleus.

4   What is half-life of a radioactive substance?

ANSWER: *Half-life* is defined as the time required for a radioactive substance to reach one-half of its radioactive emitting strength.

5   What units are used to measure radioactivity?

ANSWER: The roentgen (R) is the unit most frequently used to measure the quantity of radiation emanating from a body. It is primarily used to measure x-rays and gamma rays. One roentgen is the production of 2.58×10^{-4} coulombs per kilogram (C/kg) of air. Exposure tolerances are usually expressed in milliroentgen (mR), or one-thousandth of a roentgen. The radioactivity of material undergoing radioactive decay is expressed in curies (Ci), which is equal to 3.7×10^{10} atomic disintegrations per second.

6   How are alpha, beta, and gamma radiation defined?

ANSWER: Alpha-radiation particles consist of two protons and two neutrons. This makes them identical to the positively charged nucleus of the helium atom. They cannot penetrate human skin, but are dangerous to human health if inhaled or ingested into the body. This could occur from inhaling alpha-bearing dust or eating alpha-contaminated food or water.

Beta particles are high-speed electrons with penetrating power sufficient to go through aluminum up to 1-in. thick. Thus they are a health hazard to the entire human body.

Gamma radiation is an electromagnetic-type ray similar to light, radio waves, and x-rays with a speed approaching that of light. Its penetrating power can reach up to 3 ft of concrete. This radiation is the most dangerous to human health in the operation of nuclear power plants.

7   Name some typical methods employed to protect workers from excessive radiation doses.

ANSWER: Some typical factors considered in the protective scheme are generally the following: (1) Control the length of an exposure; (2) control the distance between the human body and the radiation source; (3) provide shielding between the body and the source of radiation; (4) establish a strict and precise radiation-monitoring program in the work area; and (5) have medical facilities available at all times to handle any accidental exposures above normal stipulated levels.

8   What procedures are generally followed in decontaminating radioactive material?

ANSWER: Radioactive material cannot be destroyed. Therefore the process of decontamination involves one of the following methods of lessening the hazard of radioactivity: (1) Isolate the area of radioactive contamination until such time as the radioactivity is decreased to a safe level as a result of radioactive decay. The half-life of the contaminant will influence this procedure. (2) Treat the surface so that the radioactive material is absorbed, cleaned off, swept, etc., and then taken to a site where the radioactive substance will not harm people. The most difficult decontamination is where the contaminant is absorbed into porous material such as concrete. Complete removal of walls, floors, and similar contaminated areas may be necessary under this type of contamination.

9   What is a thermal shield in a nuclear reactor?

ANSWER: The thermal shield usually consists of iron or steel plates surrounding the core inside the reactor vessel, so as to conserve heat and reduce the temperature and thermal stresses in the wall of the reactor vessel. The inner layer of the shield next to the core is subject to intense neutron and gamma radiation, which is converted to heat. The thermal shield often is cooled by circulating water.

10   What is a biological shield?

ANSWER: The biological shield consists of high-density concrete or lead plates around the reactor. It prevents the escape of neutrons and radiation out of the vessel wall so as to protect personnel. Radiation is present in operation and shutdown periods because of the radioactivity of the fuel elements.

11   What is a containment vessel?

ANSWER: To prevent the release of fission products in case of a fuel meltdown or explosion of the primary reactor, most power reactors are housed in steel airtight containers. They are not normally pressurized, but are designed to withstand the maximum pressure or shock waves that could develop as a result of an accident. The primary radioactive components are usually inside the containment vessel.

12   Name the types of reactors possible, based on coolant used.

ANSWER: From the many possible arrangements of fissionable and fertile fuel, moderator, and coolant that can constitute a chain-reacting system, six types have emerged as principal contenders for full-scale electric power generation:

(1) pressurized water; (2) the closely related boiling water; (3) sodium-cooled, graphite-moderated; (4) gas-cooled; (5) heavy-water-cooled; and (6) organic-cooled, heavy-water-moderated reactors.

13   What are a shim rod and scram rod as applied to nuclear plants?

ANSWER: A shim rod is a control rod used for making coarse adjustments in the reactivity of a reactor's chain reaction, whereas a regulating rod makes fine adjustments in the reactivity. Reactor control is also achieved by varying the liquid level for those reactors that use a liquid as a moderator of reactivity.

A scram rod is a safety rod that is capable of shutting down a reactor very quickly in the event the shim or control rods fail to control the reactivity within prescribed limits.

14   What is meant by the term *poison*?

ANSWER: The term *poison* applies to fission products in the fuel elements of a reactor that absorb neutrons and thus affect the reactivity of the reactor. The two most prominent fission products considered poisonous are xenon 135 and iodine 135. These are produced when a reactor uses uranium. The poison formed as fission products eventually reduces the output of the reactor; as a result, the fuel elements are spent, which eventually requires the reactor to be refueled. Poison can also be injected into a reactor to scram it or shut it down under critical emergency conditions.

15   What is fertile material?

ANSWER: This is a material that is capable of capturing a neutron and then becoming fissionable. Another term for converting fertile material to fissionable material is *breeding*. As an example, thorium 232 can be converted to uranium 233 by nuclear bombardment.

16   How is spent fuel processed?

ANSWER: The spent fuel is shipped to a reprocessing plant. The fuel cladding is chopped open to leach out uranium, plutonium, and other fission products. Uranium and plutonium are separated and put back into the fuel cycle for reuse. The remaining components are strontium 90 and cesium 137, with half-lives of approximately 30 yr. Some plutonium with a half-life of 24,000 yr is also present. The liquid waste is converted to solids and then canned in leakproof containers for eventual shipping to a federal deposit site under perpetual government control.

17   What is a digital audible dosimeter?

ANSWER: It is an instrument that provides an immediate measurement of the radiation exposure in the area as well as audibly alarming personnel that a radiation exposure exists. Set points for alarming are possible.

18   What are the chief functions of an authorized inspection agency in nuclear power plant work?

ANSWER:

1. To participate in the ASME surveys of any organization that has a nuclear code stamp and for which the authorized inspection agency will provide the Component Code inspection service.
2. To maintain a qualified staff of nuclear inspection supervisors who will monitor the shops with which an inspection agreement has been made to provide Code-required inspections.
3. Provide documented instructions to the authorized nuclear inspectors of the inspection agency on procedures to follow during routine performance and when assistance or guidance may be needed from their supervisors in order to comply with nuclear requirements.

19 Name the chief functions of a nuclear inspector doing work per ASME Nuclear Code requirements.

ANSWER:

1. Verify that work is being done by manufacturers or installers who have the ASME certificate of authorization for the work to be done.
2. Monitor the contractor or manufacturer's quality assurance program in order to note if it is being followed.
3. Verify that all material being used meets Code requirements for data, stamping, and traceability.
4. Verify that all welding is performed to qualified procedures and that the welders employed are Code-qualified.
5. Maintain a written record of inspection activity.
6. Verify that Code design calculations have been made and are available for review.
7. Verify that required heat treatments have been performed properly.
8. Verify and interpret as necessary the required nondestructive examinations required by the Code.
9. Witness required hydrostatic or pneumatic tests.
10. Witness and verify nameplate stamping as complying with specifications and Code requirements.
11. Sign the necessary documents as required by the Code to show that the component meets Code requirements.

20 To what part of pumps does the ASME Code on design of nuclear pumps address itself?

ANSWER: The ASME rules for nuclear pumps address pump casings, inlet and outlets of pumps, covers, clamping rings, seal housing, related bolting, internal pump coolers, nozzles attached to pumps, related piping, and supports.

21 How does the ASME define an owner of a nuclear power plant?

ANSWER: The Code recognizes that in the utility field there are many operating companies of facilities. The word *owner* is stipulated as being an organization that is responsible for the operation, maintenance, safety, and power generation of the nuclear plant.

22   To what standard must personnel performing NDT examinations be qualified?

ANSWER: They must be qualified for the technique and method to be used in accordance with standard SNT-TC-1A entitled "Recommended Practice for NDT Personnel Qualification and Certification" of the American Society for NDT. If this standard does not cover the proposed examination, then the NDT personnel must be qualified for the particular examination by the manufacturer or contractor.

23   Who audits the operators of NDT examinations?

ANSWER: The authorized inspector has to verify that NDT operators have been certified in accordance with SNT-TC-1A. The inspector has the prerogative to audit the NDT examination program in effect at the site. When the authorized inspector has reason to question the performance of an NDT operator, she or he can request requalification of the operator in the written procedure required for the particular examination.

24   How long must welded joints of a nuclear component of the Class 1 type remain exposed?

ANSWER: They must be left uninsulated and exposed for examination by the authorized inspector for leaks during the required hydrostatic test.

25   What hydrostatic test pressure is required on Section III, Class 1 components?

ANSWER: Completed components and appurtenances except brazed joints, pumps, and valves must be subjected to a hydrostatic test pressure not less than 1.25 times the system design pressure. Brazed joints, pumps, and valves must be hydrostatically tested at a pressure 1.5 times the system design pressure.

26   How is the total relieving capacity of Class 1 components' relief devices specified?

ANSWER: The total relieving capacity of the pressure-relief devices must be adequate to prevent a rise in pressure above 10 percent of the system design pressure and temperature within the pressure-retaining boundary of the nuclear system.

27   Besides spring-loaded valves, what other relief valves may be used on Class 1 systems?

ANSWER: Pilot-operated relief valves and power-actuated pressure-relief valves, provided they meet installation requirements of the Code.

28   What are considered flaw indications when one is performing NDT examination during in-service inspections of a nuclear power plant?

ANSWER: Evidence or signals from instruments that reveal the following: cracks, slag inclusions, segregates in the material, aligned or clustered porosity, lack of weld penetration, and plate laminations.

29   Differentiate between surface examination and volumetric examination.

ANSWER: Surface examination includes the liquid-penetrant method and magnetic-particle examination. It is used to verify surface or near surface cracks or discontinuities. Volumetric examination is for the purpose of detecting subsurface discontinuities to the extent that the entire volume of metal below the metal surface may be examined. The two principal examination methods used are radiographic examination and ultrasonic testing.

30   Name the five main components of a PWR nuclear power plant.

ANSWER: (1) Nuclear reactor; (2) steam generators or heat exchangers; (3) piping to circulate the coolant between the reactor and the steam generators; (4) pumps to circulate the coolant; and (5) a pressurizer to prevent the coolant from boiling and to permit volume changes in the coolant without excessive pressure rise with coolant temperature changes.

31   Whose responsibility is it to establish the suitable NDT examination procedures where a reference Code Section, such as Section III, requires NDT examination?

ANSWER: The responsibility for establishing suitable examination procedures and having the examination performed by qualified NDT personnel in accordance to the Code Section requirements is that of the manufacturer of the vessel or boiler or the fabricator if different from the manufacturer, or the installer or assembler of the Code vessel or boiler.

32   What is pressure-suppression containment?

ANSWER: In a pressure-suppression containment, the reactor is in a relatively small pressure vessel, called the dry well, which is connected by vent pipes to a second vessel that is partly filled with water, called the suppression pool. If a reactor system has a sudden leak, such as a pipe failure in the dry well, the steam released from the reactor system is directed to the suppression pool by the vent pipes, and is there condensed. The advantage of this system is that the containment structure can have less volume, and/or lower design pressure than would be the case for a comparable pressure-containment vessel.

33   On what basis are pressure-containment structures designed?

ANSWER: These are designed to contain the energy and fission products that could be released from a worst case accident to the reactor. This design includes the possible pressure that could be released, and potential volume requirements.

6

Material Structure, Required Code Material, and Specifications

ASME Boiler Code Sections

There are extensive material requirements, tests, and markings detailed in the ASME Boiler Code, Section II, titled "Material Specifications." There are four parts to this section:

Part A: Ferrous materials

Part B: Nonferrous materials

Part C: Welding rods, electrodes, and filler metal

Part D: Allowable stress tables for Boiler Code strength calculations

The specifications in Section II are similar to the specifications published by the American Society for Testing and Materials (ASTM) and for the welding material specifications published by the American Welding Society (AWS).

Section I details the material that must be used for Power Boilers. Other sections involving heating boilers, unfired pressure vessels, and nuclear components have their own listing of permissible material for the part under consideration.

Provisions are made in Section I for use of material not identified to a specification as listed in Section II; however, certain conditions must be satisfied such as recertification by a "neutral" party that the material is equivalent to permissible listed material, including its chemical analysis and mechanical properties and that hardness requirements have been met. Full documentation is required to be furnished to the boiler manufacturer and Code inspector.

This chapter provides introductory metallurgical material, and certain definitions stressed in the Code, which is applicable to boiler construction and even repairs. Its main purpose is to introduce the reader to material variables in construction and welding that includes required quality control so that only Code permitted material is used. Metallurgical texts should be consulted for further study as respects the properties of materials, their crystalline structure, and how they are affected by fabrication methods.

Ferrous alloys.   Most pressure parts of boilers are made from ferrous alloys, which signifies that they are manufactured from iron and iron-base alloys. Except for cast-iron boilers and some special heating boilers, all other boilers usually have steel parts that are joined by fusion welding. There is an increased demand for higher service pressures which requires better material and fabrication technology. This is being achieved by sophisticated alloying, vacuum degassing in steel-making processes, multiple heat treatment, extensive nondestructive testing, and high quality-control standards, to name a few of the methods being used to obtain better material. Quality control of the welding procedure used and control of the welding operation have been materially improved by greater emphasis being placed on more stringent specifications, inspection, and documentation requirements.

Material Structure

If a piece of metal is carefully polished, immersed for a short time in an acid or other appropriate reagent, and then examined under a microscope, it will be found to be composed of small particles or crystals. The metal, instead of being perfectly uniform, is built up of these small units of matter.

Materials which appear to be perfectly homogeneous in reality are composed of an aggregate of grains or crystals of distinctly different materials. A piece of the material will therefore have different properties at different points within the piece. Even pure metals are made up of an aggregate of crystals having different properties in different directions.

One convenient method for explaining some of the similarities and differences in the characteristics of metals is to study the arrangement of atoms in the material.

Space lattice.   A crystal of a given metal is composed of atoms arranged in a definite and regular geometric pattern. This pattern is known as a space lattice and is determined by x-ray studies. The atoms in iron at *room temperature,* for example, are arranged in a body-centered cubic lattice. That is, the atoms are located as at the corners of a cube, with one atom in the center, as indicated in Fig. 6.1a1. This pattern repeats itself throughout a single crystal, as indicated in Fig. 6.1a2. Two adjacent crystals in a bar of iron will have the same lattice formation, but their orientation, or the directions of the axes of the space lattice, will be different. Chromium, vanadium, molybdenum, and tungsten also have the body-centered cubic lattice.

Iron at high temperature crystallizes in the face-centered cubic lattice, or with an atom at each corner of a cube and an atom in the center of each face, as illustrated in Fig. 6.1a3. There are a total of 14 lattice systems.

A pure metal crystallizes in cooling through its freezing point; if a nucleus or seed is present, crystallization consists of growth of crystals around a nucleus at the freezing point of the metal. Once a nucleus is established on a pure liquid metal at or below the freezing point, other atoms from the molten liquid start to attach to the nucleus. The atoms deposit along defined directions of the crystal, or space lattices.

A study of space-lattice or crystal formations that are created when metals are cooled to freezing or solidification temperatures assists in

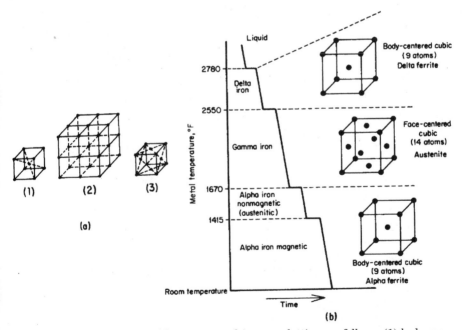

Figure 6.1 (*a*) Atoms in metal are arranged in space lattices as follows: (1) body-centered cubic lattice, (2) unit cubes in crystal, (3) face-centered cubic lattice. (*b*) Crystal forms of iron at different temperatures.

determining the effects of alloying, rate of cooling, and similar metallurgical considerations. Iron, for example, is transformed to various crystal types, and the corresponding space-lattice arrangements for these types are as follows:

Type of iron	Type of crystal structure
Alpha, delta, and ferrite iron	Body-centered cubic (BCC) lattice
Gamma and austenite iron	Face-centered cubic (FCC) lattice

Alpha iron reaches the BCC structure below 1670°F.
Delta and ferrite iron with the BCC structure exist above 2534°F.
Gamma and austenite iron with the FCC structure exist between 1670 and 2534°F.

At room temperature, iron is composed of a body-centered cubic lattice (see Fig. 6.1b). In this form it is known as alpha iron or alpha ferrite, and it is soft, ductile, and magnetic. Upon heating above about 1415°F, alpha iron loses its magnetism, but retains its body-centered crystalline structure. This structure changes to face-centered cubic at about 1670°F, at which temperature alpha iron is transformed to gamma iron and remains nonmagnetic. In continuing upward in temperature, another phase change occurs at 2570°F, when delta iron is formed. The latter is identical in crystal structure (body-centered) to that of the low-temperature alpha iron. It is magnetic and is stable to the melting point. There are no known phase changes in the liquid form (above about 2800°F). On cooling very slowly from the liquid state, the atomic rearrangements described above occur in reverse order.

A difference of 2 or 3 percent in the carbon content, a difference in heat treatment, and a difference in the amount of mechanical working during the shaping process may cause the ultimate tensile strength to vary from 20,000 to 300,000 psi. The elongation at failure may be varied from 50 to 0.1 percent, and the elastic strength may be varied from 10,000 to 270,000 psi.

The variations in strength, ductility, and other properties of an iron-carbon alloy may be explained by referring to a typical equilibrium diagram for a portion of the iron-carbon system, as shown in Fig. 6.2. The temperature range is illustrated from 0 to 1535°C (2795°F), the melting point of pure iron. The range of carbon contents is from 0 to 5 percent, representing the range of materials used. Alloys containing more than 5 percent carbon are too weak and brittle. Wrought iron contains between 0 and 0.12 percent carbon; steel, between 0.12 and 1.7 percent; and cast iron, over 1.7 percent.

Figure 6.2 represents phase changes that occur in iron-carbon mixtures at the indicated temperatures. Up to about 4.3 percent carbon in the solution causes any carbon addition to lower the melting point of the mixture.

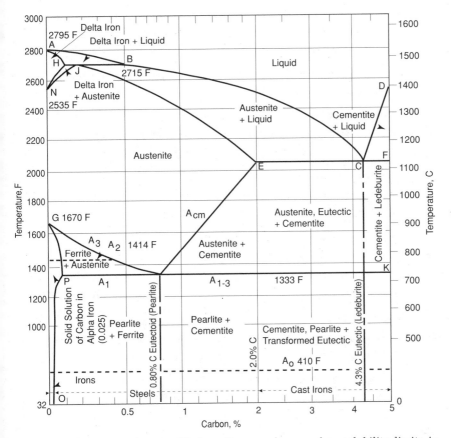

Figure 6.2 This iron-carbon equilibrium diagram shows carbon solubility limits in iron. The percentages of each add up to 100 percent on the bottom of the graph, i.e., 5 percent carbon, 95 percent iron.

Metallurgists use equilibrium diagrams of alloys to predict changes that may occur with temperature as the alloy material precipitates or hardens. In Fig. 6.2, the freezing of iron-carbon alloys occurs at temperatures as shown by the liquidus line ABCD. Carbon content up to 2.0 percent goes into solid solution in iron. If the carbon content is more than 2.0 percent, the excess liquid forms a eutectic with a compound of iron and carbon. Metallurgists can predict from an equilibrium diagram what the alloy's phase will be at different temperatures based on the percentage contents of the alloy.

In carbon steels, only three phases can be present in the solid state under equilibrium conditions:

Austenite is a phase in steels that consist of the gamma form of iron with carbon in solid solution. Austenite is nonmagnetic and tends to work harden when cold worked in those steels that are austenitic at ordinary temperatures.

Ferrite is pure iron.

Cementite is iron carbide, Fe_3C formed when iron and carbon combine chemically at about 1274°F.

In welding, both melting and solidification takes place, and metallurgists study the effect this may have on alloys. Figure 6.2 presents the conditions that can exist with an iron-carbon alloy without any other alloying elements. An increase in the rate of cooling of the alloy material by immersing it in oil or water, or the addition of another alloying element such as nickel, silicon, or chromium, will result in a different equilibrium diagram than that shown in Fig. 6.2.

The size of the crystals and the dispersion of the carbon in the iron also affect the properties of a carbon steel. For example, small crystals and highly dispersed carbon tend to increase the strength and hardness. *Grain refinement* is the reduction of the crystalline structure by heat treating or by a combination of heat treating and mechanical working. Specified grain sizes can range from *coarse* (1 to 5) or *fine* (5 to 8) in steel specifications. The McQuaid–Ehn test is used to specify grain sizes for a steel specification under Metallurgical Structure in Section II, Ferrous Materials.

Transformation temperature. In carbon steels, only three phases can be present in solid state under equilibrium conditions: *austenite, ferrite,* and *cementite.* The temperature interval within which austenite forms on heating, and also the temperature interval within which austenite disappears on cooling, is called the transformation-temperature range. The transformations on heating do not occur at the same temperatures as on cooling, except at almost infinitely slow rates of temperature change. The temperatures of transformation are dependent on carbon content in accordance with the equilibrium diagrams.

Effects of reheating. In general, the effects of reheating a metal are the reverse of those obtained in slow cooling. For example, a steel containing 0.20 percent carbon and consisting of pearlite and ferrite is reconverted to austenite as the alpha iron is transformed to gamma iron in the critical range. However, for this steel, the temperatures of transformation are about 30°C or 86°F higher for heating than for slow cooling. Changes in the carbon content and the presence of other alloying elements affect the temperature differential.

Because they indicate the effects of different cooling rates, isothermal and continuous transformation diagrams are sometimes used by metallurgists in predicting the results of various welding processes. In the arc welding of steel, the structure may range from ferrite to martensite, depending on welding conditions. In spot welding, large quantities of martensite are obtained, and a postweld treatment is essential if the carbon content is higher than approximately 0.25 percent. A transformation diagram will indicate, approximately, the type of microstructure in the zone adjacent to the weld. The type of microstructure desired will indicate also the most suitable welding process to use.

Controlling the Properties of Cast Iron and Steel

Blast furnace. The first step in the production of cast iron and steel is the extraction of the iron from the ore to make pig iron. This is accomplished in the blast furnace. (See Fig. 6.3.)

Cast iron is produced by remelting pig iron and pouring it into molds of the desired shape. The purpose of the melting is to reduce the amount of impurities and to secure a more uniform product than would be obtained by casting the pig iron directly as it came from the blast furnace. There are two types of furnaces in general use for the remelting of pig iron. In the production of most of the ordinary gray cast iron, a cupola is used, while an air furnace, known as a reverberatory furnace, is more commonly used for the better grades of gray cast iron and the cast iron which is to receive heat treatment.

Cast-iron types and properties. Compared with steel, cast iron is decidedly inferior in malleability, strength, toughness, and ductility. The most important types of cast iron are the white and the gray cast irons.

White cast iron. White cast iron is so known because of the silvery luster of its fracture. In this alloy, the carbon is present in combined form as iron carbide (Fe_3C), known metallographically as *cementite*.

Gray cast iron. Gray iron is the most widely used of cast metals. In this iron, the carbon is in the form of graphite flakes that form a multitude of notches and discontinuities in the iron matrix. The appearance of the fracture of this iron is gray because the graphite flakes are exposed. The strength of the iron increases as the graphite-crystal size decreases and the amount of cementite increases. Gray cast iron is easily machinable because the graphite carbon acts as a lubricant for the cutting tool and also provides discontinuities which break the chips as they are cast. Gray iron, having a wide range of tensile strength, from 20,000–30,000 to 90,000 psi, can be made by alloying with nickel, chromium, molybdenum, vanadium, and copper. Permissible stresses per the ASME Code, Section IV, are as follows:

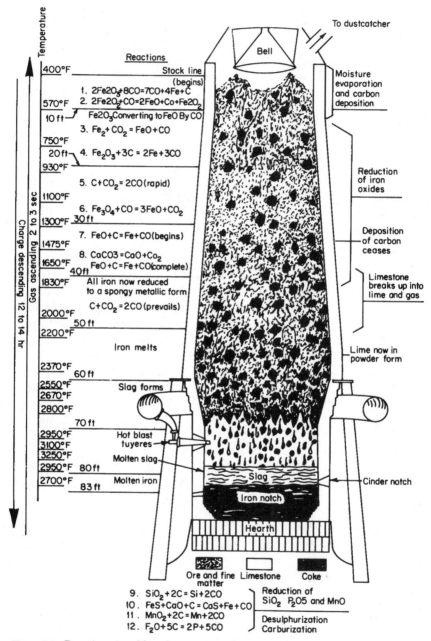

Figure 6.3 Reactions in a blast furnace to make pig iron from iron ore. (*Courtesy CF & I Steel Corp.*)

Class	Ultimate tensile strength, kilopounds per square inch (kips/in.^{2})	Allowable stress in tensile strength, kips/in.^{2}
20	20.0	4.0
25	25.0	5.0
30	30.0	6.0
35	35.0	7.0
40	40.0	8.0

Physical defects. Some of the most common physical defects which may be present in cast iron and which will weaken it are blowholes, cracks, segregation of the impurities, and coarse-grain structure.

Sulfur, which combines with the manganese or the iron to form a sulfide, makes the cast iron brittle (hot-short) at high temperatures. It also increases shrinkage. Hence, its amount is usually limited to less than 0.1 percent in specifications for cast iron.

Phosphorus, in amounts of more than 2 percent, makes the iron brittle and weakens it. However, it also has the effect of increasing the fluidity and decreasing the shrinkage. So it is desirable in making sharp castings for ornamental parts where strength is unimportant.

Manufacture of Steel

The production of steel from pig iron involves the removal of as much of the impurities as is practicable, the adjusting of the carbon content to the desired value, and the addition of such alloying as may be required to alter the properties.

The manufacture of steel starts with pig iron. Pig iron is transformed into steel by the oxidation of the impurities with air, oxygen, or iron oxide, since the impurities combine more readily with oxygen than does iron. Pig iron may be refined by oxidation alone in the acid process or by oxidation in conjunction with a strong basic slag in the basic process. Carbon, silicone, and manganese are removed by both processes, but phosphorus and some of the sulfur in the pig iron can be removed only in the basic process. Phosphorus stays persistently with the iron if the slag is strongly acidic or when it is high in silica and therefore is not removed. The principal methods of manufacturing steel for subsequent rolling or forging are the basic-oxygen process, the basic open-hearth process, and the basic electric-furnace process. Cast steel can be made by the above methods and also by the acid electric and acid open-hearth methods. The furnace charge is steel scrap and pig iron in both the basic-oxygen and basic open-hearth processes and selected steel scrap in the electric-furnace process. Prereduced iron pellets may also be used in these processes as part of the charge.

Figure 6.4   Electric-arc furnace discharging a 150-ton melt. (*Courtesy Lukens Steel Co.*)

Electric-furnace process. High-grade alloy steels are usually manufactured in the electric furnace (see Fig. 6.4) because the electric-furnace process offers the following advantages over the other processes:

1. Desired temperatures can be obtained because of good control of
   temperature within close limits.

2. Oxidizing, reducing, or neutral conditions can be maintained in
   the furnace. This allows the addition of alloying elements to the
   melt with better quality.

3. The melt is free of contamination from burning fuel.

4. Sulfur and solid nonmetallic impurities, as well as occluded gases,
   are significantly eliminated.

The control of quality and composition in the electric furnace is better than in the open-hearth process, but the cost is somewhat higher.
Some steel is melted under vacuum. This procedure holds dissolved
gases such as oxygen, nitrogen, and hydrogen to a minimum and significantly reduces the number of nonmetallic inclusions.

Shaping of steel. As the steel comes from the basic-oxygen furnace,
open-hearth furnace, or electric furnace, it may be cast directly into

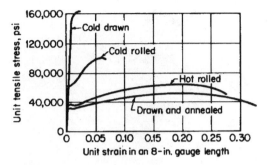

Figure 6.5   The method of shaping operations on carbon steel can effect its ultimate strength.

the desired shape or into ingots weighing from 3 to 10 tons. After solidification, the ingots are given a preliminary shaping by being rolled or forged into billets, which may then be reduced to final dimensions by rolling, forging, drawing, or other operations. Each of the operations will affect the crystalline structure, thereby changing the properties of the final product. Figure 6.5 shows typical stress-strain diagrams for specimens of low-carbon steel shaped by some of the more common manufacturing methods.

The temperature at which steel is mechanically worked has a profound effect on its properties. Cold work increases the hardness, tensile strength, and yield point, but its indices of ductility-elongation and reduction of area are decreased.

Seamless tubes for boilers are made by piercing hot, solid ingot pieces in special piercing machines especially designed for the purpose. Some cold drawing is performed where tubes require finer finish or close tolerance. Some seamless tubes are now made by the hot extrusion press method.

Electric-resistance-welded steel tubing is made by forming flat strips into a tubular shape called a skelp and then welding the edges together by an electric-resistance welding machine. Compared with seamless tubes, electric-resistance-welded boiler tubes have a smoother surface, a more uniform wall thickness, and less eccentricity on the outside diameter of the tubes.

Figure 6.6*a* shows a hot steel plate being formed into a head on a hot spinning machine at a steel company. Heads are used as end closures for boiler drums, unfired pressure vessels, nuclear reactors, and all types of petroleum and chemical process vessels.

Figure 6.6*b* shows a rolling mill making plate steel.

Impurities from steel manufacture.   The strength, ductility, and related properties of the iron-carbon alloys can be affected by the presence of harmful elements such as sulfur, phosphorus, oxygen, hydrogen, and nitrogen.

(a)

(b)

Figure 6.6 Hot shaping of steel. (*a*) Hot spinning of steel plate to form a concave head. (*b*) A rolling mill processes a hot ingot into steel plate. (*Courtesy Lukens Steel Co.*)

Sulfur. Sulfur has the same effect on steel as on cast iron, making the metal hot-short, or brittle at high temperatures. As a result, it may be harmful in steel which is to be used at elevated temperatures or, more particularly, may cause difficulty during hot-rolling or other shaping operations. Most specifications for steel limit its amount to less than 0.05 percent.

Phosphorus. Phosphorus makes steel cold-short, or brittle at low temperatures; so it is undesirable in parts which are subjected to impact loading when cold. However, it has the beneficial effect of increasing both fluidity, which tends to make hot rolling easier, and the sharpness of castings. Since cast iron is brittle anyway, phosphorus is sometimes added to make the castings clean-cut. Most specifications for structural steel limit the phosphorus content to less than 0.05 percent.

Oxygen. When iron is in the liquid state, any free oxygen combines readily with iron to form iron oxide. In the finished steel or iron, the iron oxide usually appears in the form of tiny inclusions distributed throughout the metal. These inclusions introduce points of weakness and increased brittleness, concentration from voids, and stress, which are undesirable, for they promote the formation of cracks that may result in progressive fracture.

Hydrogen. Hydrogen, such as is produced when steel is immersed in sulfuric acid to remove mill scale before cold drawing, makes steel brittle. The hydrogen may be removed by heating the steel for a few hours, or hydrogen will gradually work out of the steel at ordinary temperatures. When present, hydrogen increases the hardenability of the steel.

Nitrogen. Nitrogen has a hardening and embrittling effect on steel. This may be objectionable, or it may be desirable in producing a hard surface on the steel, under controlled manufacturing conditions. For example, in the nitriding process, the steel is exposed to ammonia gas at about 600°C (1112°F) to produce a hard, wear-resisting surface without lowering the ductility of the center of the piece.

Nonmetallic inclusions. Nonmetallic inclusions occur during the plate-rolling operation in steel making. Quite often, these inclusions cause laminations, or planes in the plate where metal separation or voids exist.

The quality of steel plate can be significantly affected by nonmetallic inclusions. The presence of inclusions, such as sulfides and oxides, primarily affects the ductile behavior of the steel. The quality of a particular grade of steel can be improved by eliminating or minimizing inclusions. Nondestructive testing is used extensively for critical-pressure applications, such as nuclear reactors, in order to find inclusions.

Alloying elements in steel. The purpose of adding alloying elements to carbon steel is to impart to the finished product desirable physical or chemical properties which are not available in carbon-steel parts fabricated by standard procedures. These properties could involve desirable electrical, magnetic, or thermal characteristics, as well as engineering considerations such as (1) high tensile strength or hardness without brittleness; (2) resistance to corrosion; (3) high tensile strength or high creep limit at elevated or subzero temperatures; (4) other desirable physical features required to resist special loading.

Carbon. Up to 1.2 percent carbon in iron increases the strength and ductility of steel. When the carbon content is above 2 percent, graphite formation is promoted, which lowers both the strength and the ductility of steel. Carbon content above 5 or 6 percent causes the metal to be brittle with very low strength for any load-resistance application.

Manganese. By combining with sulfur, this element prevents the formation of iron sulfide at the grain boundaries. This minimizes surface ruptures at steel-rolling temperatures (red shortness) which results in significant improvement in surface quality after rolling.

Nickel. This element increases toughness or resistance to impact. In this respect it is the most effective of all the common alloying elements in improving low-temperature toughness.

Chromium. This element contributes to corrosion resistance and heat resistance in alloy steels. A strong carbide-former, chromium is frequently used in carburizing grades and in high-carbon-bearing steels for superior wear resistance. An 18 percent chromium and 8 percent nickel (18-8 nickel-chrome) alloy is widely used as a high-strength stainless steel in pressure equipment requiring strengths and corrosion resistance.

Molybdenum. Like nickel, molybdenum does not oxidize in the steel-making process, a feature which facilitates precise control of hardenability. It markedly improves high-temperature tensile and creep strength and reduces a steel's susceptibility to temper brittleness.

Boron. Since boron does not form a carbide nor strengthen ferrite, a particular level of hardenability can be achieved without an adverse effect on machinability and cold formability that may occur with the other common alloying elements.

Aluminum. In amounts of 0.95 to 1.30 percent, aluminum is used in nitriding steels because of its strong tendency to form aluminum nitride, which contributes to high surface hardness and superior wear resistance.

Silicon. Silicon combines with carbon to form hard carbides which, when properly distributed throughout the alloy, have the effect of increasing the elastic strength without loss of ductility.

Tungsten. Tungsten forms hard stable carbides when it is added to steel. It raises the critical temperature, thus increasing the strength of the alloy at high temperatures.

Vanadium. This element acts as a deoxidizing agent as aluminum does on molten steel. It forms very hard carbides, thus increasing the elastic and tensile strength of low-carbon and medium-carbon steels.

Other elements may be added to improve the rolled strength and toughness of steels in the high-strength low-alloy category.

Heat treatment of steels. When steel is heated to certain temperatures and then rapidly or slowly cooled, its physical properties such as elastic limit, ultimate strength, and hardness can be changed.

Heat treatments fall into two general categories: those which increase strength, hardness, and toughness by quenching and tempering and those which decrease hardness and promote uniformity by slow cooling from above the transformation range or by prolonged heating within or below the transformation range, followed by slow cooling.

Annealing consists of heating steel to a certain temperature and cooling it by a relatively slow process. Annealing may be used to remove stresses such as are produced in forgings and castings, to refine the crystalline structure of steel, or to alter the ductility or toughness of steel.

Stress-relief annealing involves heating to a temperature approaching the transformation range, holding for a sufficient time to achieve temperature uniformity throughout the part, and then cooling to atmospheric temperature. The purpose of this treatment is to relieve residual stresses induced by normalizing, machining, straightening, or cold deformation of any kind. Some softening and improvement of ductility may be experienced, depending on the temperature and time involved.

Normalizing involves heating to a uniform temperature about 100 to 150°F above the transformation range, followed by cooling in still air.

Hardening consists of heating steel to above its transformation temperature range and cooling it suddenly by quenching in water, oil, or some other cooling medium that absorbs heat rapidly.

Quenching is defined as a process of rapid cooling from an elevated temperature, by contact with liquids, gases, or solids. Quenching increases the hardness of steel if its carbon content is 0.20 percent or higher. It also raises the elastic limit and ultimate strength and reduces the ductility; however, it induces internal stresses, and the metal is apt to become brittle. This may be relieved by annealing or reheating the metal or material to an appropriate metallurgically determined temperature for the metal or material involved in the quenching operation.

Physical and Mechanical Properties

Properties of materials for engineering structures are generally described under the following headings:

1. *Physical:* These would describe the material's composition, structure, homogeneity, specific weight, thermal conductivity and ability to expand and contract, and resistance to corrosion.

2. *Formability:* These would relate to the manufacture of the material, such as fusibility (welding), forgeability, malleability, and ability to shape the material by bending or machining.

3. *Mechanical:* Mechanical properties describe the ability of a material to resist applied loads and are usually obtained from tests.

The basic mechanical properties are elastic limits, moduli of elasticity, ultimate strengths, endurance limits, and hardness. Secondary mechanical characteristics determined from the basic ones or simultaneously with them are resilience, toughness, ductility, and brittleness.

Strength depends on the type and nature of loading. The static strength of a material is expressed by the corresponding elastic limit stress. The impact strength is measured by the corresponding modulus of resilience. The endurance strength is expressed by the corresponding endurance limit. See later chapters on stress, pressures, and forces.

Qualities other than strength are also important.

Hardness is a relative characteristic. There are several methods of measuring it, all of an arbitrary nature. The Brinell hardness number (Bhn) is obtained as follows: A hardened steel ball 10 millimeters (mm) in diameter is pressed under a certain load, F kilograms (kg), into the smooth surface of the material to be tested; the diameter D of the indentation is measured in millimeters, and the depth h is calculated from it. The hardness number, Bhn, is then expressed as

$$\text{Bhn} = \frac{F}{10\pi h}$$

In using the Shore scleroscope, a small cylinder of steel with a hardened point is allowed to fall on the smooth surface of the material, and the height of the rebound of the cylinder is taken as the measure of hardness.

The hardness number obtained with the Rockwell instrument is based on the additional depth to which a test point is driven by a heavy load beyond the depth to which the same penetrator has been driven by a definite, lighter load.

A material is *ductile* if it is capable of undergoing a large, permanent deformation and yet offers great resistance to rupture. The measure of ductility is the percentage of elongation or the percentage of reduction of area during a tensile test carried to rupture, and it is used as a relative measure. Ductility helps to relieve localized stress concentration through local yielding. It is a necessary characteristic of a material used to take live loads, especially where concentrated stresses may occur.

Brittleness is a characteristic opposite to ductility and toughness. A material may be considered brittle if its elongation at rupture through tension is less than 5 percent in a specimen 2 in. long.

Toughness is a term used to denote the capacity of a material to resist failure under dynamic loading. The *modulus of toughness* is defined as the amount of energy per unit volume that a material can withstand or absorb without fracture occurring. The modulus of toughness is useful as an index for comparing the resistance of materials to dynamic loads, and it is especially applicable in the design of moving parts of machinery.

ASME Code SA-Number Designations

SA-numbers, such as SA-178, electric-resistance-welded carbon steel boiler tubes, are used in the different sections of the ASME Boiler Code to show the permissible material specification for the different components of a boiler or pressure vessel. The list of permissible materials is quite extensive and is actually growing as steel mills and metallurgists develop new material that finds approval by the Code approval committees. Section II lists the specific requirements for Section I approved material. This now includes allowable stresses per SA listed material per expected temperature of operation. Such a stress table is shown in Figs. 9.2 and 9.3 in Chap. 9 where code strength calculations are reviewed.

A few definitions are provided that are appropriate in reviewing ASME boiler material requirements.

Cladding is material applied over a steel tube for improved corrosion resistance, but does not contribute to the strength of the tube to resist internal pressure per Code rules.

Firebox quality is steel that is suitable for use in pressure vessels that will be exposed to fire or heat and thus able to resist the resulting thermal and mechanical stresses.

Flange quality is steel for use in pressure vessels that are not exposed to fire or radiant heat. Special manufacturing, testing, and marking of the steel as flange quality is required.

Heat of steel is steel produced from one charge of the furnace and consequently practically identical in its properties.

Heat-resisting steels are those steels graded for service at relatively high temperatures because they retain much of their strength and resist oxidation at those high temperatures.

Killed steel is steel to which sufficient deoxidizing agents were added in manufacture to prevent gas evolution during solidification and thus reduce the chance for porosity in the steel.

Hot-shortness is a metal that is, or becomes, brittle at an elevated temperature.

Maraging steels are a group of high-nickel martensitic steels, which have high strength and ductility.

Martensite is a distinctive structure with most steels that is developed by cooling as rapidly as possible from the quenching temperature. In this form, the steel is at its maximum hardness.

McQuaid–Ehn test is used for revealing grain size of steel by heating above the critical range in a carbonaceous medium. This causes the grain to be outlined sharply when polished, etched, and viewed under a microscope. Grain sizes range from No. 8 (fine) to No. 1 (coarse).

Microstructure is the structure of metals revealed by examination of polished and etched samples under a microscope.

Neutron embrittlement results from bombarding a steel with neutrons, such as in a nuclear reactor. In steels, neutron embrittlement causes a rise in the ductile-to-brittle transition temperature.

Pearlite is a relatively hard constituent of steel made up of alternate layers of ferrite (iron) and cementite (iron carbide). When the proportion of carbon is about 0.8 percent, it is called *eutectoid steel.*

Precipitation hardening is the process of hardening an alloy by heating it for the purpose of allowing a structural constituent to precipitate out from a solid solution.

Rimmed steel is a steel that when poured in manufacturing has enough oxygen to evolve appreciable amounts of gas during solidification. The gas evolution results in a finished steel with a very pure surface, but with impurities concentrated in the interior. The pure surface zone is referred to as the rim, hence the name.

Semikilled steel is a steel having properties intermediate between those of killed and rimmed steel, and which is characterized by variable degrees of uniformity and composition when manufactured.

Skelp is steel or iron plate from which pipe or tubing is made.

Solid solution is a condition where one element is dissolved in another element while the dissolving element is in a solid form, and not in a liquid condition.

ASME Code material requirements. Certain procedures must be followed by a fabricator or repairer in order to make sure only Code-specified material is used in boiler construction. It also is part of the responsibility of the authorized inspector to help implement a quality control procedure in order to make sure Code material is used. These controls may include the following:

1. The material to be used for a boiler or pressure vessel must be specified in the section of the Code under which the boiler or pressure vessel is built. For example, if a high-pressure boiler is involved, it must be listed as a permissible material in Section I (Power Boilers), or data must be presented to show that it has the same chemical and physical characteristics as a Code-listed material.
2. The fabricator of the boiler or pressure vessel generally orders permissible Code material from the steel mills. The steel mill is responsible for making the necessary tests to the specifications given in Section II of the ASME Boiler and Pressure Vessel Code.
3. Section II test requirements that the steel manufacturer may have to perform include the following:
   a. Chemical analysis of the steel to determine if it is within Code limits for the specification.
   b. Tests to determine if the metallurgical grain structure is within Code-specified limits.
   c. Inspection of plate or tube to note if defects such as blowholes, slag, laminations, and any other imperfections may be present and whether these are within Code-permissible tolerance.
   d. Tension and bend tests as stipulated in the Code to note if these are within Code specifications.
   e. Notch toughness tests to check on fatigue-failure strength.
   f. Mill test report showing that the material complies to Code specifications; this must be certified by a responsible person of the material testing laboratory of the steel manufacturer.

Plate steel for any part of a boiler subject to pressure and exposed to the fire, or products of combustion, must be of firebox quality. If not exposed to fire or the products of combustion, the plate can be of flange quality. Some firebox-quality steels are specification SA-201 carbon-silicon steel, specification SA-202 chromium-manganese-silicon steel, and specification SA-204 molybdenum steel. Check the Code for other firebox-quality steels and refer to the ASME Material

Specification, Section II, for their physical and chemical characteristics. Seamless steel drum forgings made in accordance with specifications SA-266 and SA-336 for alloy steel can be used for any part of a boiler for which either firebox or flange quality is permitted.

Pipes and tubes may be made of open-hearth, electric-furnace, basic-oxygen, or acid deoxidized bessemer steel pipe or tubing, according to Code specifications. Some tube material listed in Section I include SA-178, electric-resistance-welded carbon steel tubes; SA-192, a seamless carbon steel tube for high-pressure service; and SA-210, a seamless carbon steel boiler and superheater tube material. Others include low-alloy steel tubes in the SA-213 series, a chrome-moly tube material, and high-alloy steel such as the SA-213 TP series, a chrome-nickel tube material. Stay bolts and stays material permitted are threaded stay bolts, SA-36 and SA-675. Stay bolts, stays, and through rods with ends for weld attachment must be SA-36 or SA-675. Seamless steel tubes used for threaded stays must be SA-192 or SA-210.

It is important to check the section under which a boiler, pressure vessel, or nuclear component is built and stamped as a code object, to note if the material for a component is permitted by that section of the ASME Code.

If the pressure does not exceed 250 psi and the temperature does not exceed 450°F, specification SA-278 gray-iron castings may be used for power boiler parts such as pipe fittings, water columns, and valves and their bonnets. Cast iron cannot be used for nozzles or flanges for any pressure or temperature. But this does not apply to low-pressure boilers. The same pressure parts as enumerated for cast iron can be made of malleable iron SA-395, except that the pressure is limited to a maximum of 350 psi and the temperature to 450°F.

The minimum thickness is ¼ in. for plate subjected to pressure. An exception is for miniature boilers of seamless construction, where the minimum plate thickness may be ³⁄₁₆ in. The minimum thickness of tube sheets is ⅜ in., except on miniature boilers where it is ⁵⁄₁₆ in. The plate material must be not more than 0.01 in. thinner than that required for the plate by the formula used to calculate its strength, provided the tolerance in fabrication (or when the plate is ordered) also has this tolerance of not less than 0.01 in.

Code stress tables. After the material has been identified as permitted by the section of the Code under which a component is built or repaired, allowable stresses for different temperatures can be obtained from the stress tables in Section II, Part D. Such a table is shown in Chapter 9 for solving allowable pressure problems by using code equations and allowable stresses for the component under consideration and for the expected mean operating temperature.

Questions and Answers

1 Name some typical mechanical properties that describe the ability of a steel material to resist loads.

ANSWER: Mechanical properties that are important to describe a steel material are ultimate strength, elastic limit, endurance limit, hardness, and modulus of elasticity.

2 What are the three most prominent names given to the method used in measuring the hardness of steel material?

ANSWER: Brinell, Rockwell, and the Shore scleroscope are the most prominent methods used in describing the hardness quality of a steel material.

3 How is a space lattice defined?

ANSWER: A space lattice is the geometric arrangement of atoms in the crystal formation of a given metal, such as face-centered cubic lattice or body-centered cubic lattice, describing the arrangement of atoms in an iron crystal.

4 What is the space lattice of alpha iron?

ANSWER: Alpha iron is composed of body-centered cubic lattices.

5 What is meant by a eutectic?

ANSWER: A eutectic is the alloy of a solution that has the lowest melting point in the solution.

6 What is the transformation temperature in carbon steels?

ANSWER: This is the temperature within which austenite (nonmagnetic steel) forms on heating and also the temperature interval within which austenite disappears in cooling in an iron-carbon solution. The temperatures of transformation are dependent on the carbon content, as noted in iron-carbon equilibrium diagrams for the solution.

7 Name some typical physical defects that may be found in cast iron.

ANSWER: Blowholes, cracks, segregation of impurities, and coarse grain or lack of uniform grain structure are the most common defects.

8 What are the two major chemical steps in the manufacture of steel from ore?

ANSWER: (1) To remove oxygen from the ore in producing pig iron. (2) To remove excess carbon from the pig or cast iron to produce steel.

9 How does steel differ from cast iron?

ANSWER: The carbon content is much lower, the tensile strength is much higher, and the ductility and resistance to shock load are higher.

10   Briefly, how is excess carbon removed from iron to form steel?

ANSWER: By blowing air or oxygen through the molten iron. The oxygen in the air unites with the carbon by combustion.

11   How is the highest-quality steel produced?

ANSWER: By the electric-furnace method, owing somewhat to more accurate control and to a minimum of oxidizing factors.

12   Why is the electric-furnace method not always used?

ANSWER: The expense of manufacture by this method is greater.

13   What are the ASME Code requirements for chemical properties of open-hearth firebox and flange steel?

ANSWER: For open-hearth firebox steel, the maximum carbon content for plates not over ¾ in. thick is 0.25 percent, and for plates over ¾ in. thick, 0.30 percent; the maximum phosphorus (P) content for steel made by the acid method is 0.04 percent, and for steel made by the basic method 0.035 percent; the maximum sulfur (S) content is 0.04 percent.

For open-hearth flange steel, the maximum manganese (Mn) content is 0.80 percent; the maximum phosphorus content if the steel is made by the acid method is 0.05 percent; if the steel is made by the basic method, it is 0.04 percent; and the maximum sulfur content is 0.05 percent.

14   What is the minimum yield point required for steel of either flange or firebox grade?

ANSWER: One-half the tensile strength.

15   How much below specified thickness may a boiler plate be rolled and still be acceptable?

ANSWER: 0.010 in.

16   Where and how shall a boiler plate be stamped?

ANSWER: In at least two places not less than 1 ft from the edges should be the manufacturer's name or label, the test number, the type of steel, and the lowest tensile strength of the 10,000-lb range.

17   What determines the hardness of steel?

ANSWER: The carbon content, the alloy content, and the grain structure often due to heat treatment.

18   How may grain structure and hardness be changed?

ANSWER: By heat treatment at temperatures above the lower critical point, and by mechanical working.

19 What are the major differences between cast iron and malleable iron?

ANSWER: Malleable iron is more ductile. It will withstand safely a considerably greater shock load than cast iron. It possesses a higher tensile strength.

20 What are ductility and malleability?

ANSWER: These terms are practically synonymous. They describe the ability of a material to withstand a comparative degree of deformation without failure or impairment of its strength or other physical properties.

21 What is the common measure of ductility?

ANSWER: The percentage elongation as found in making a test of tensile strength.

22 What is hardness?

ANSWER: It is the property of a material to resist surface deformation on application of an external force.

23 What is brittleness?

ANSWER: It is the tendency of a material to fracture on application of shock load.

24 Is there any definite relation between hardness and brittleness?

ANSWER: Usually not. For example, some of the so-called white metals are quite brittle but soft.

25 Is there any definite relation between hardness and toughness?

ANSWER: Usually not. Toughness is a physical property enabling a material to withstand deformation as well as abrasion, such as is required for the teeth of a power dredge.

26 What are fatigue and fatigue failure?

ANSWER: Fatigue is "tiring" of the metal. It is the breaking through of the crystalline grain or fibrous structure of a material after a number of stress applications or stress reversals. Fatigue endurance is the ability of a material to withstand a great number of stress applications or reversals without failure.

27 In what direction does the fiber run in plates of a boiler, and does it make any difference?

ANSWER: The fiber runs in the direction in which the plate is rolled—usually parallel to its longest dimension. In fabrication the fiber runs circumferentially, because the longest dimension of the plate is usually required in this direction, and to keep the greatest tensile load on the plate in the same direction as the so-called fibrous structure.

28 How much does a piece of boiler plate 12 in. by 12 in. by 1 in. weigh?

ANSWER: The average weight of steel boiler plate is 0.28 lb/in.^{3}; therefore, $12 \times 12 \times 1 \times 0.28 = 40.3$ lb.

29 How would you tell wrought iron from steel?

ANSWER: Wrought iron is of much greater fibrous structure than steel. If a spot is polished on the surface of the metal with a file or emery cloth, the fibrous structure usually can be seen. Sparking the metal on an emery wheel will show a reddish spark from wrought iron and an exploding yellowish spark from steel.

30 What is a laminated boiler plate?

ANSWER: One that contains a stratum of slag, rolled in by accident.

31 What harm does a lamination do?

ANSWER: It is an insulator that prevents free conduction of heat. If the plate is in a high-temperature zone, a blister may result because of overheating of the plate on the fire side of the lamination.

32 What is notch brittleness?

ANSWER: This is the susceptibility of a material to brittle fracture at points of stress concentration, such as a sharp corner. When tensile tests are conducted on materials with a notch, the material is considered notch brittle if the strength of the notch specimen is less than the specimen without a notch.

33 Is the actual tensile strength stamped on steel plate or is there a range for stamping per the ASME Code?

ANSWER: For a carbon steel, the actual tensile strength may be anywhere between 55,000 and 65,000 lb/in.^{2}, but the plate is stamped 55,000 lb, the minimum of the 10,000-lb range. If the plate fell slightly below 55,000-lb/in.^{2} tensile strength, it would be stamped 45,000.

34 What is mill scale, and what harm does it do?

ANSWER: It is an iron oxide scale formed on the surfaces of a plate when, after rolling at high temperatures, it is exposed to the air. It may be a cause of plate deterioration in boiler service, or it may promote and localize corrosion.

35 Where can the exact tensile strength be found?

ANSWER: From the mill test report.

36 What else does the mill test report show?

ANSWER: The thickness; the chemical properties; other required physical properties, namely, elongation and yield point; the manufacturer's name; and the heat and slab numbers.

37 What is the effect of sulfur and phosphorus when they are melted into steel?

ANSWER: Sulfur makes the steel brittle at high temperature, often referred to as hot-short. Phosphorus makes steel brittle at low temperatures, so it is undesirable in parts which are subjected to impact loading when cold. This brittleness when cold is sometimes called cold-short.

38 In making a shop inspection of plate material for ASME-constructed boilers, what should be especially checked by a Code inspector?

ANSWER: The plate should be checked for the following:

1. Examine mill test reports and note if the plate complies with Code specification for that material.
2. Check plate stamping and note if stamping identifies the plate with the mill test report, including tensile strength, thickness, and grade and quality of finish.
3. Make a visual examination of the plate for defects such as cracks, laminations, inclusions, pits, and similar questionable conditions.

39 When may firebox or flange steel be used per the ASME boiler code? What process of manufacture is stipulated?

ANSWER: Firebox-quality steel must be used when parts of the boiler are exposed to fire or products of combustion. For other service, either firebox- or flange-quality steel may be used. Steel should be made by an open-hearth or electric-furnace process by either the acid or basic method.

40 What is the maximum allowable working pressure and temperature on a water column made of cast iron and malleable iron?

ANSWER: Per the ASME Code, for cast iron, 250-psi pressure and 450°F; for malleable iron, 350 psi and 450°F.

41 Who is responsible for stamping the steel plates used in high-pressure boilers?

ANSWER: The steel mill in which the plates are made.

42 Name some typical types of cast iron.

ANSWER: Cast-iron types are white cast iron, gray cast iron, malleable iron, and ductile cast iron. Cast irons are also classified by their ultimate strengths—20,000, 30,000, 40,000, etc.

43 What process is used in fabricating tubes?

ANSWER: Tubes are made by either the seamless method or the butt welding process.

44 Define brittleness and brittle fracture.

ANSWER: Brittleness is the sudden fracture of a material without first undergoing plastic deformation. Brittle fracture is the rapid propagation of a crack with little expenditure of energy as the crack progresses to ultimate metal separation.

45   What is plastic deformation?

ANSWER: When a material is strained (stretched) beyond its elastic limit, permanent set, or distortion, occurs in the metal; however, the distortion occurs in a ductile manner.

46   Does Section II, Part D of the ASME Code provide other information besides allowable stresses of materials?

ANSWER: Section II, Part D lists, besides each material, ultimate and yield strengths and also lists in Sections I, III, or VIII if that particular material is permitted. It also includes external pressure charts for calculating allowable pressure on boiler components under external pressure, such as tubes in firetube boilers.

47   What type of piping requires a Code stamp and certification, even though it is constructed to ASME B31.1 Power Piping Code rules?

ANSWER: The Preamble to Section I defines this as external piping; it consists of piping from the boiler to any Code-required valve(s). This piping requires Code data reports to be prepared by a "S," "A," or "PP" stampholder. The piping involved is steam piping to the stop valve(s), feedwater piping, blowoff and blowdown piping, boiler drains, and miscellaneous external piping such as surface blowoff, steam and water piping for water columns and gauge glasses, and the recirculation line for a high-temperature water boiler.

48   What design pressure is required on the above Code piping?

ANSWER: For steam piping, the pressure setting of the safety valve, but not below 100 psi.

For feedwater and blowoff piping, the design pressure must be at least 125 percent of the maximum allowable working pressure of the boiler, or 225 psi above the maximum allowable working pressure, whichever is less, but it cannot be less than 100 psi.

Most other miscellaneous piping must be designed for a pressure at least equal to the maximum allowable working pressure of the boiler, but not less than 100 psi.

49   May galvanized wrought iron and galvanized steel pipe and fittings be used for blowoff piping on a high-pressure boiler?

ANSWER: Galvanized pipe cannot be used for blowoff piping per ASME B31.1 Power Piping Code rules.

7

Fabrication by Welding and NDT

There are many source documents that provide details including welding requirements, for the construction of boilers, pressure vessels, and nuclear vessels. Each Code Section has a chapter on welding requirements. For example, Section I, Power Boilers has Part PW, titled "Welded Boilers" that provides some requirements for welding joints and the nondestructive testing that must be applied as quality assurance that a sound weld was made for the material under consideration. Other documents that pertain to this chapter's material include:

Section V, ASME Code, Nondestructive Examination that provides details or requirements for the different NDT methods.

Section IX, Welding and Brazing Qualifications for welders, operators, and welding procedures.

ASNT's SNT-TC-1A, Recommended Practice for Qualification and Certification of NDT Personnel. This provides details on qualifying to the three levels of classification recognized by this guide.

The American Society for Nondestructive Testing, ASNT, has now published another new document to supplement the guideline of SNT-TC-1A by an absolute compliance standard, and not a guide, called ANSI/ASNT CP-189-1991 and titled *Standard for Qualification and Certification of Nondestructive Testing Personnel*. This requires certification of NDT personnel to recognized standards.

Threaded and expanded connections. The Code permits threaded connections to be used for joining boiler parts, but these are limited to

3-in. pipe size, and cannot be used for pressure over 100 psi. Expanded connections are used on tubes, but for high-pressure boilers, the expanded connection cannot exceed 6-in. outside diameter.

Joining by Welding

Welding is the chief method employed in the manufacture of boilers and pressure vessels. Materials used in the construction of power boilers that involve pressure parts must be listed in that section as allowable material and with the specification as listed in Section II of the Code. Carbon or alloy steel having a carbon content of more than 0.35 percent cannot be used in welded construction. Welding requires certain procedures, and operators and welders must be qualified in order to perform Code welding. The National Board also requires the welding to be qualified per Code rules when welded repairs on pressure parts are performed. Contractors or repairers can obtain an "R" stamp from the NB if their organizations can demonstrate compliance with the NB repair requirements.

Boiler operating, maintenance, and inspection personnel should have a working knowledge of Code welding, because this knowledge has become a legal requirement in the many jurisdictions that have adopted ASME Code requirements, or NB rules for permissible repairs.

Welding methods. Welding is defined as a localized coalescence (fusing together) or consolidation of metal where joining is produced by heating or fusion temperatures, with or without the application of pressure, and with or without the use of a filler metal. The filler metal, when used, must have properties close to the base metal, including melting point. The weld is that portion which has been melted together during the welding operation. The welded joint is the union of two or more metals produced by the welding process. *Weld reinforcement* is weld metal in excess of that needed to fill a joint, and for butt groove welds, the weld metal deposited beyond the flush base metal surface is considered reinforcement.

Weld reinforcement is permitted by Section I as it helps in having a complete joint penetration of the metals being joined. However, the Code has a limit on reinforcement height, depending on plate thickness. For example, for a 2- to 3-in. plate thickness range, the maximum reinforcement is about ¼ in. The Code also requires the weld surfaces to be free from coarse ripples, grooves, overlaps, abrupt ridges, and valleys to avoid stress concentration. The surface of the weld shall also be suitable so that proper interpretation of radiographic and other NDT can be made. Grinding off of excess weld reinforcement is usually applied when NDT interpretations of the joint may be difficult.

Backing rings, if used on a longitudinal joint of a pipe, shell, or drum, must be removed. Circumferential joint backing rings may not have to be removed if certain Code sizes are not exceeded, and where radiographic examination of the joint is not required.

The most common method of welding pressure parts is by *fusion* (melting) of the metal, the heat being supplied in one of several different ways. In fusion welding, no pressure is applied between the pieces being welded. Arc welding, gas welding, and Thermit welding are classified as fusion welding, but arc welding is the most common.

Arc welding is a localized progressive melting and flowing together of adjacent edges of the base-metal parts, caused by heat produced by an electric arc between a metal electrode, or rod, and the base metal. Both the welding material (welding rod or electrode) and the adjacent base metal are melted. On cooling they solidify, thus joining the two pieces with continuous material.

In *metal arc welding,* welding rods can be of two types: *bare* electrodes and *coated* electrodes. Plain arc welding is used with bare electrodes. Shielded arc welding is used with coated electrodes. The reason is that the flux or coating (Fig. 7.1a) on the electrode protects the deposited metal from oxidation. Then as the welding rod melts, a more reliable weld is obtained than with bare electrodes. Shielded metal arc welding is abbreviated as SMAW.

Coated electrodes provide shielding gas to protect the weld from contamination from the atmosphere, such as oxygen, moisture, and carbon dioxide. These contaminants can cause weld porosity, crack

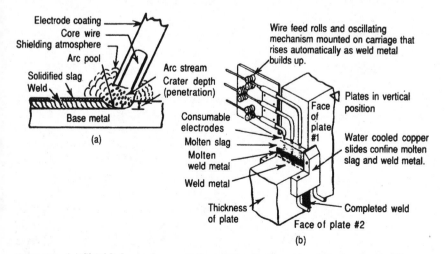

Figure 7.1   (a) Shielded metal-arc welding; (b) basic elements of electroslag welding.

formation, and weld bead oxidation. *Shielding gas* is used in gas tungsten arc welding, gas metal arc welding, and similar welding processes that employ bare electrodes or filler metal. *Purging gas* is used to replace unwanted air and other vapor contaminants from the root pass by using a gas that prevents oxidation during welding. Argon is used for shielding and purging as is helium. Purging is usually recommended when welding stainless steel, nickel alloys, and most nonferrous metals.

In *submerged arc welding,* coalescence (fusing together) is produced by heating with an electric arc, or arcs, between a bare metal electrode, or electrodes, and the work. The welding is shielded by flux, which is a blanket of granular, fusible material on the work. Pressure is not used, and filler metal is obtained from the electrode or sometimes from a supplementary welding rod.

In *gas tungsten arc welding,* coalescence is produced by heating with an electric arc between a single tungsten (nonconsumable) electrode and the work. Shielding is obtained from a gas or gas mixture (which may contain an inert gas). Filler metal is usually added separately from the electrode. This process is also called tungsten inert gas (TIG) welding. If no filler metal is used, the process is also called *autogeneous welds.* Tungsten electrodes are classified on the basis of their chemical composition, such as pure tungsten, zirconiated tungsten electrode, thoriated tungsten electrode, and so on. AWS developed a marking and numbering system for the different types, from EWP to EWG, including color markings from green to gray.

Gas welding is a group of welding processes in which coalescence is produced by heating with a gas flame, or flames, with or without the application of pressure. Filler metal is added (usually) to the heated base metals to be welded.

Oxyacetylene welding is a gas welding process in which coalescence is produced by heating with a gas flame, or flames, of about 6000°F obtained from the combustion of acetylene with oxygen as filler metal is added.

Thermit welding is a group of welding processes in which coalescence is produced by heating with superheated liquid metal and slag resulting from a chemical reaction between a metal oxide and aluminum. Usually no pressure is applied. Filler metal, when used, is obtained from the liquid metal. The Thermit process is useful for very heavy welds, such as parts with thicknesses of 3 in. and over.

In *electroslag welding* (see Fig. 7.1b), parts are joined in one pass with the welding groove in a vertical plane. The slag on top conducts the electric arc that generates the heat required to melt the slag or flux and the weld filler metal. A water-cooled dam is used to confine the molten pool of flux, filler metal, and parent metals within the welding groove until the weld is completed and solidified. Deposit rates are high with this type of welding. The weld is considered coarse-grain-structured and requires further heat treatment in order to refine the grain structure of the weld and the heat-affected zones.

Some other welding processes are *electron beam welding,* where a concentrated beam of high-velocity electrons generates heat of fusion on the surface to be joined; *plasma arc welding,* an inert-gas fusion welding process using a constricted arc, similar to the gas tungsten arc process; and *laser welding and cutting,* which is accomplished by the energy of light concentrating a beam through an optical device called the optical maser or laser. The radiation source in a laser is concentrated with only small beam divergence, which results in high power intensities of up to 10^9 watts per square centimeter (W/cm^2).

The ASME Section I Code permits the following welding processes to be used on power boilers: shielded metal arc, submerged metal arc (SAW), gas metal arc (GMAW), gas tungsten arc (GTAW), plasma arc, atomic hydrogen metal arc, oxyhydrogen, and oxyacetylene. Pressure welding processes allowed are flash, induction, resistance, pressure Thermit, and pressure gas.

Fillet welds are made with certain tolerances to be followed, and it is necessary to define the terms *throat, toe, face, legs,* and *root.* (See Fig. 7.2, showing a full fillet weld.) The terms are defined as follows. The *throat* of a weld is the shortest distance from the root of a fillet weld to the face. The *toe* of a weld is the juncture between the face of the weld and the base metal. The *face* of a weld is the exposed surface on the side from which welding was done. The *root* of a weld is the deepest beginning of the weld which intersects the base-metal surfaces. The *leg* of a weld is the distance from the root of the joint to the toes of a fillet weld. Groove welds are used extensively for butt joints (Fig. 7.2a).

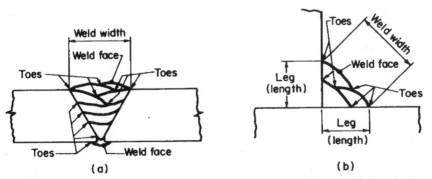

Figure 7.2 Weld terminology for butt and fillet welds. (*a*) Butt-weld description details; (*b*) fillet-weld description details.

Double-butt weld. In double-butt welding, the metal is deposited in the welding groove from both sides of the plate, whereas in single-butt welding the metal is deposited from only one side. Single-butt welding on longitudinal joints of drums is permitted, provided a backing strip is used to ensure full penetration welding of the joint. But single-butt welding can be used only where the inside of the weld is inaccessible for welding. On longitudinal joints, the backing must be removed.

After the first side of a double-welded butt joint has been welded, the second side should be chipped, ground, or melted out to secure a clean surface of proper shape, to ensure fusion of the weld metal on that side without porosity, slag, or voids. The base metals must be clean and free from grease, dirt, rust, paint, or other foreign substances. Sometimes, prior to welding, a linseed oil coating is specified on the base metals.

Welding problems. There are certain welding problems that good welding procedures and welding skill try to avoid. Figure 7.3 lists some common welding defects and the steps to be taken to avoid these problems. A common welding problem is *slag inclusion*. This is a nonmetallic solid material entrapped in a welded joint or between weld metal and the base metal. The slag is objectionable because it prevents metal-joint strength based on metal characteristics. *Porosity*

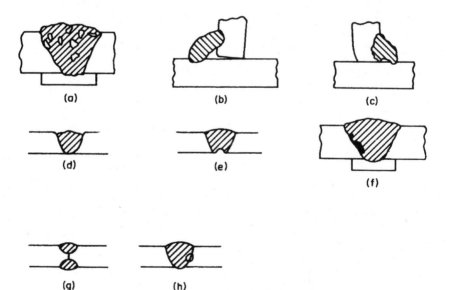

Figure 7.3 Some welding defects. (*a*) Porosity; (*b*) inadequate penetration; (*c*) incomplete fusion; (*d*) incomplete groove weld; (*e*) root cavity; (*f*) slag inclusion; (*g*) incomplete root penetration; (*h*) incomplete side fusion.

of a weld is the voids or gas pockets left in a weld as a result of incorrect welding or welding difficulties. When fusing takes place, the base metals welded are affected in a zone called the *heat-affected zone* (HAZ). The heat-affected zone is that portion of the base metal which has not been melted but where the structural properties of the base metal have been altered by the heat of welding (or cutting).

Locked-up stresses are those internal stresses remaining in the weld metal and adjoining base material when a weld has been made. They are caused by the high concentrated heat in the weld compared to the adjoining cooler metal. This sets up a thermal gradient, leading to nonuniform expansion and contraction, which sets up internal stresses in the weld. Peening and heat treatment will reduce locked-up stresses. This is called *stress relieving,* or postweld heat treatment.

Cracks in welds are illustrated in Fig. 7.4 and are usually caused by some form of expansion and contraction effect during the welding process. However, each type of crack has specific causes, even though cracks are divided into two broad categories: (1) hot cracks that occur prior to solidification of the weld and (2) cold cracks that occur after solidification is completed. Cracks are not permitted by the Code on

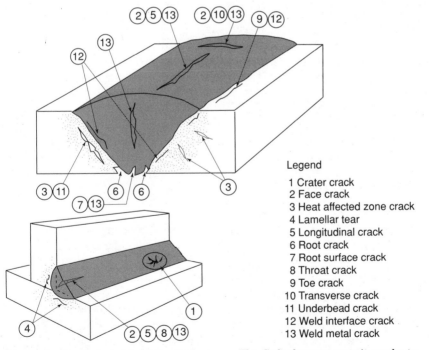

Legend

1 Crater crack
2 Face crack
3 Heat affected zone crack
4 Lamellar tear
5 Longitudinal crack
6 Root crack
7 Root surface crack
8 Throat crack
9 Toe crack
10 Transverse crack
11 Underbead crack
12 Weld interface crack
13 Weld metal crack

Figure 7.4 Cracks in welds have many causes. The Code does not permit cracks to remain uncorrected, because they act as stress concentration for further crack propagation. (*Courtesy American Welding Society.*)

welds, because this type of discontinuity has the potential to propagate and cause a severe failure in pressure parts. Cracks are also caused by metallurgical changes in the welding process that make the welded joint brittle, especially in the heat-affected zone of the weld. The likelihood of forming a brittle structure in the HAZ increases as the carbon and alloy content of the base metal is increased, because the higher carbon content will increase the hardness and decrease the ductility of the HAZ region. This type of cracking can be avoided by using low-hydrogen electrodes, using preheat, and practicing good joint cleanliness preparation free of moisture, grease, oil, and any other hydrocarbon compound. In many cases postweld heat treatment must be used.

Defects in welding usually can be repaired by chipping or grinding out the defective section (voids, heat-affected cracks, lack of penetration, etc.) to sound metal and then rewelding. Generally, nondestructive testing must be used to make sure that the defective section has been completely eliminated; the repair also has to be retested by nondestructive testing. Openings or holes are permitted in boiler-welded joints. However, the weld must have been stress-relieved and radiographed, and the weld at the opening must be examined for cracks on both sides by the magnetic-particle method.

Preheating. Preheating is recommended by welding engineers for alloy steels in order to reduce the level of thermal difference that exists between the base metal area near the weld and the weld pool. This will avoid possible cracks in the weld as the colder base metal resists the contraction of the weld metal as it heats up and cools down. The use of preheat also lowers the cooling rates in the weld metal and in the HAZ, producing a more ductile metallurgical welding structure, and drives off hydrogen or diffuses it, preventing hydrogen cracking. Preheating involves heating the base metals to a temperature of around 150 to 400°F, depending on the materials welded, before welding. This reduces the chances of locked-up stresses when the weld is made.

Postweld heat treatment. This procedure lowers the tensile and yield strength of carbon steel and alloy steel weld metals. The purpose of postweld heat treatment is to relax any residual stresses caused by welding and temper the metal, which will prevent cracking and brittleness.

In addition to stress relieving, postheat is used in low alloy steels to obtain a homogeneous grain structure in the weld and base metal by heat treating at a higher temperature than for stress relief, which may be followed by tempering at a lower temperature, that varies with the type of steel. Depending on the base materials involved, postweld heat treatment may improve, make no change, or degrade *fracture toughness.*

Postweld heat treatment or stress relieving requires heating a welded joint from 1100 to 1200°F after the joint is welded and keeping it at this temperature *one hour for each inch of thickness.* The entire vessel or only the joint may be stress-relieved. It is common practice to complete all welding and then stress-relieve the entire vessel. Stress relieving affects the metal structure in its crystalline form and thus reduces the concentration of locked-up stresses in the weld. After the required temperature is held for the specified time, cooling to 600°F must be gradual, after which the metal can be cooled to room temperature. *Peening* is a mechanical working of metal by means of hammer blows so as to reduce the residual stress in a welded joint caused by the heat of welding.

Metallurgical considerations.   Metallurgical considerations are many when one contemplates a welding process. The following must be considered:

1. Weldability of the parts to be joined. The Code may have severe restrictions.

2. Surface conditions needed and the compatibility of the chemical and mechanical properties of the base metals to the welding process being used. Moisture on the surface of ferrous material as well as on the electrode can result in porosity and hydrogen under bead cracking.

3. The matching of filler metals or electrodes to the base metals in order to avoid possible embrittlement of the weld.

4. Suitability of the welding process for the type, thickness, and alloys of the metals to be joined.

5. Evaluation of the effects of preheat and postwelding heat treatment. For metals with high hardenability rates, slow cooling will prevent heat-affected zone cracking. Face-centered cubic-lattice material may develop coarse-grain crystal structures when cooled slowly.

6. Fluxes. They float when melted on top of the liquid metals, protecting or shielding the metals from the atmosphere. Chemical reactions between the metal and the flux are possible at the interface. If the flux is high in silicon dioxide, a chemical reaction is possible which may produce higher silicon in the weld metal than anticipated.

Code welding tests.   Section IX of the ASME Boiler and Pressure Vessel Code contains details on procedure specifications that must be prepared by a manufacturer and the tests that must be made by a welder in order to be considered a qualified welder.

It might be appropriate to define the terms *reduced-section tension test, free-bend test, root-bend test, face-bend test,* and *side-bend test* since they are used extensively in Code qualifications of welding. See ASME Code, Section IX, for typical weld test specimens. The reduced-section tension test is used for qualifying the procedure that the shop, or contractor, is to use in welding. When broken in tension, it must have an ultimate tensile strength at least that of the minimum range of the plate which is welded (base material), and the elongation of stretch must be a minimum of 20 percent.

The side-bend test is used for qualifying welders. The specimen is subjected to bending against the side of the weld. In the face-bend test (see Fig. 7.5), the specimen is subjected to bending against the surface, or face, of the weld. In the root-bend test, the specimen is subjected to bending against the bottom, or root, of the weld. The free-bend test is a shop- or contractor-qualifying procedure test. The test consists of bending the specimen cold, and the outside fibers of the weld must elongate at least 30 percent before failure occurs.

In order to pass each test, guided-bend specimens must have no cracks or other open defects exceeding ⅛ in., measured in any direction on the convex surface of the specimen *after bending,* except that cracks occurring on the corners of the specimen during testing are not considered unless these occur from slag inclusions or other welding technique defects.

Welding Procedure

Each manufacturer or contractor who is to do boiler Code welding is required to record in detail the procedure to be used. Each procedure

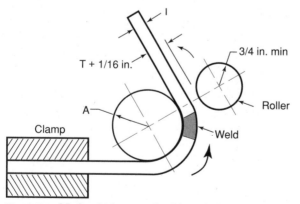

Figure 7.5 Mechanical tests of welds include guided-bend test of groove welds. Illustrated is the face-bend test.

requires testing of the welds to be made by reduced-section test specimens and guided-bend specimens. The variables requiring a new procedure and new test plates are very numerous. Among these are changes in base materials, grouped in ASME Welding Qualifications (Section IX) into "P" numbers. For example, P-1 includes mostly carbon steels, P-2 used to be wrought iron, but is no longer listed, P-3 consists of chrome-moly steels with chromium content below 0.75 percent, and with a total alloy content not exceeding 2 percent. The P numbers range to P-10, so the base-material variable on procedure qualification is large.

The next variable is the electrode and welding rod selection, which ranges from F-1 to F-7. Any change in electrode or welding rod selection requires a new set of test plates or procedure qualification. Weld metal is classified by weld-metal analysis numbers A-1 to A-8. These are related to equivalent P numbers of the base material. Again, changes of weld metal from equivalent base-metal classification require a new set of test plates or procedure qualification.

The thickness of the plate, or pipe, to be welded is another variable. Classification is from $\frac{1}{16}$ to $\frac{3}{8}$ in., from $\frac{3}{8}$ to $\frac{3}{4}$ in., and over $\frac{3}{4}$ in. Each classification requires new test plates as shown in the Code. The ASME Welding Code specifies other variables to consider in requiring a new procedure qualification test, and these should be consulted for specific variables.

On groove welds, one face-bend test and one root-bend test are required for welder qualification for each position to be welded. For fillet welds, a test plate is required according to the Code, but passing the groove-weld test will also qualify a welder for fillet welding. Procedure qualification requires two face-bend tests, two root-bend tests, and two reduced-section tension tests, as illustrated in the Welding Qualification Code of the ASME.

Qualified welder. A qualified welder is one who is capable of performing manual or semiautomatic welding. A welding operator is one who operates a machine or automatic welding equipment. Boiler Code work requires the use of a qualified welder on most of the work where Code welding is to be done.

Qualification tests may be given by responsible manufacturers or contractors. On pressure-vessel work, the welding procedure of the fabricator or contractor must also be qualified *before* the welder can be qualified. Under other codes this is not necessary. To become qualified, the welder must make specified welds, using the required welding process, type of metal, thickness, electrode type, position (see Fig. 7.6), and joint design. Test specimens must be made to standard sizes and under the observation of a qualified person. In most government specifications, a government inspector must witness the making of

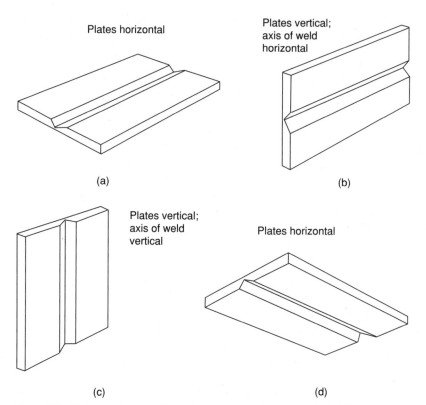

Figure 7.6 Code welding positions for groove welds used to qualify operators. Test positions: (*a*) 1G, (*b*) 2G, (*c*) 3G, and (*d*) 4G.

welding specimens. Specimens must also be properly identified and prepared for testing. The common test is the guided-bend test. However, x-ray examinations, fracture tests, or other tests are also used. Satisfactory completion of test specimens (providing that they meet acceptability standards) will qualify the welder for specific types of welding. Code certification, in general, is based on the range of thicknesses to be welded, the positions to be used, and the materials to be welded.

Remember. Welding boiler parts is under close control of the many variables involved in welding, so check the ASME Welding Code. The following is stressed again:

1. The manufacturer or contractor who is to do Code welding on boilers and pressure vessels (or nuclear vessels) is responsible for conducting procedure qualification tests and welder's qualification tests for work to be done by his or her organization.

2. The welder's qualification remains in effect as long as the welder is employed by the same manufacturer or organization and does welding on a continuous basis. But if the welder changes employment, she or he is no longer considered qualified and thus must take the test again. If the welder has not done any welding for a period of over six months in the position, material, etc., for which she or he is qualified, the welder must be requalified.

3. The manufacturer or contractor is responsible for keeping all records of procedure and welder's qualification tests. These are needed as evidence of the shop's, or welder's, ability to do acceptable Code work. A Code inspector has the right, however, to ask for retests if there is reason to believe the welding is not acceptable by Code requirements.

4. The manufacturer or contractor must assign identifying numbers, letters, or symbols for each qualified welder and welding operator so that the work of the welder or welding operator can be identified and traced if so required.

Unqualified welders may be used on Code-constructed boilers to weld parts only where the strength of the boiler does not depend on the weld in any way and, in addition, the effect of welding heat will in no way affect the boiler components' strength or set up stresses on adjoining parts to the area being welded.

It is not permissible to weld every type of steel for boiler pressure parts. It must be material permitted by the Code and also must be carbon or alloy steel having a carbon content of not more than 0.35 percent. This rule also applies to oxygen cutting or other thermal cutting processes.

Welded boiler drums can have a *limited distortion*. The drum must be circular at any section within a limit of 1 percent of the mean diameter. If necessary to meet this requirement, plates may be reheated, rerolled, or reformed. Furnaces must be rolled (scotch marine boilers), with a maximum permissible deviation from the true circle of not more than ¼ in.

Plates of *unequal thickness* can be welded provided a tapered transition section, having a length not less than 3 times the offset between the adjacent plate surfaces, is provided at joints between plates that differ in thickness by more than one-fourth of the thickness of the thinner plate, or by more than ⅛ in.

There are many Code requirements detailed in Section I as well as requirements in the Heating Boiler Code, Section IV. In addition, general welding requirements are covered quite extensively in Section IX of the Code, and these should be reviewed for details on requirements.

Field Welding Inspections

The following are some areas that plant personnel should cover when monitoring the welding performed in their plant:

1. Conformity of the welding process being used to specifications or written procedure

2. Extent of cleaning of joint prior to welding in order to ensure a good weld

3. Preheat and interpass temperatures maintained and how these compare with the code requirements for the material being welded

4. Joint preparation and conformity to code specifications and print dimensions

5. Filler metal being used and suitability for the welding procedure and the material being welded per code requirements

6. Chipping, grinding, or gouging that is being carried out after each welding pass in order to remove slag and impurities

7. Bracing of the plate during welding in order to control distortion

8. Postweld heat treatment being conducted and compliance with code requirements for temperature and time held

9. Qualification of the welders doing the job and corresponding documentation on file with the manufacturer or contractor

Safety Rules

The AWS has published many guidelines in following good safety rules in performing field welding. Among those cited are:

1. Fumes generated by welding require adequate ventilation around the site.

2. Radiation in the form of ultraviolet light from arc welding requires the use of filters to protect the eyes, and proper clothing to protect the skin.

3. Noise because of adjoining work requires using ear protection plugs or earmuffs.

4. Lifting of heavy parts requires following established safety rules on proper lifting practices.

5. Electric shock can be avoided by making sure that equipment is properly grounded, and no contact is made with live electrical parts.

6. The danger of fire erupting can be avoided by proper purging of any vessel that contained combustible gases or liquids. The adjoin-

ing area of any welding site should be checked for combustible construction, and fire prevention procedures followed, including fire watch and extinguishers being located by the site.

7. It is essential to have a *hot work permit* procedure for the location, involving supervision by both the property's manager and the welding contractor. The property management know the hazards that may exist in their operation from using a potential source of ignition represented by any welding or cutting operation.

8. Confined space entry by OSHA regulations requires purging procedures, and then testing of the confined space to make sure there is at least 19% oxygen present in the vessel's confined space. The regulations also require that all other rules of OSHA for confined space entry are followed so as not to endanger entering personnel.

Nondestructive Testing and Examination

Modern inspection techniques to assure good quality of material, to detect hidden flaws in fabrication, to recheck welds and repairs, and to enhance a plant's loss-prevention effort require the use of nondestructive inspection methods. Nondestructive tests are also used to predict the future life of equipment. For example, periodically checking the thickness of a shell in corrosive service will assist in determining when pressure may have to be reduced as the vessel thins down; this also determines what the wear rate is per year.

The term *nondestructive testing* (NDT) is used to describe a method of testing or inspecting material for soundness (or lack of defects) without affecting the material physically or chemically. Nondestructive testing can involve the following methods: visual examination, hydrostatic or leak testing, radiography, magnetic particle, dye penetrant, ultrasonic, and eddy current. Also being developed are acoustic emission and holographic testing with both requiring computer application to track signals and to note changes in the signals from previous readings.

The purpose of NDT is to detect incipient-type faults such as cracks, inclusions, voids, porosity, lack of fusion in welds, laminations, lack of penetrations, undercuts, shrinkage, and similar defects so that repairs can be made before the defects may cause a serious failure in service. Mechanical properties of welds are also checked. These include bend tests and tensile tests, but these are considered destructive tests because permanent deformation has taken place. Hardness tests are also used to check the metallurgical changes that may have occurred in the heat-affected zone (HAZ) of a weld, and this test can be considered nondestructive.

A brief review of some of the NDT methods employed in boiler fabrication and repair might be appropriate, but exact requirements are detailed in the ASME Codes, especially Section V.

Visual inspections. *Visual inspection* of welding as a NDT method starts with checking if proper Code material is being used. This is usually done by comparing the mill test report with the marking on the material. Electrode selection and storage should also be checked. Written welding procedures must be reviewed as well as the record of the welders who will perform the work. It is important to review fitup and alignment of the parts to be welded in order to make sure they comply to Code specifications. The welds should be checked for evidence of lack of joint fusion, undercutting, and similar visual observable defects. An experienced inspector will detect many faults by visual inspection. The AWS is now establishing a testing program to *certify* welding inspectors that pass minimum established requirements.

A borescope or fiberscope can be used in hard-to-see vessel areas, such as in tubing in a heat exchanger. The borescope is a slender, handheld tube consisting of a series of lenses and internal light sources, which permit viewing an area of a vessel that normally cannot be seen. The advent of fiber optics has led to the development of the fiberscope, which consists of bundles of lights and image-transmitting fibers within a long flexible sheath having an eyepiece lens at one end. Cameras can be attached to the eyepiece to obtain a permanent record of what is seen. Defects found with such instruments include corrosion, blockage of fluid and gas passages, dents, bulges, and similar problems that would be noted if unaided visual inspection were possible.

Radiographic examination. Radiographic inspection includes x-ray and gamma rays obtained from isotopes, such as cobalt 60 and iridium 192 where the resultant radiant energy can be safely controlled. The radiographic method of testing basically involves passing rays through materials to be tested. The rays impinge on a film or screen, and by noting the contrast of the film, it is possible for an experienced radiographer to detect and detail the internal structure of the object under test. The focal spot is a small area in the x-ray tube from which radiation emanates. In gamma radiography, an isotope like cobalt 60 is the radiation source. When radioactive isotopes are used, strength in curies is important, as is the physical size of the source. The smaller the radiation source, the closer to the material it can be placed; at the same time, the smaller the size, the weaker the source in curies and the longer the exposure time needed. See Fig. 7.7.

Intensity of radiation is inversely proportional to the square of the distance from its source. However, it is not possible to bring the

Maximum voltage (kV peak)	Screens	Applications and approximate practical thickness limits
50	None	Extremely thin metallic sections. Wood, plastics, biological specimens, etc.
150	None or lead foil	Light alloys, 5-in. aluminum or equivalent; 1-in. steel, or equivalent
	Fluorescent	1½-in. steel, or equivalent
250	Lead foil	2-in. steel, or equivalent
	Fluorescent	3-in. steel, or equivalent
400	Lead foil	3-in. steel, or equivalent
	Fluorescent	4-in. steel, or equivalent
1000	Lead foil	5-in. steel, or equivalent
	Fluorescent	8-in. steel, or equivalent
2000	Lead foil	8-in. steel, or equivalent
15–24	Lead foil	16-in. steel, or equivalent
MeV*	Fluorescent	20-in. steel, or equivalent

* Million electronvolts.

(a)

Material	Half-life	Million electronvolts of rays	Steel thickness application, in.
Cobalt 60	5.3 yr	1.17, 1.33	1.5–5
Cesium 137	33 yr	0.66	1.0–4
Iridium 192	70 days	0.137–0.651	0.5–2.5

(b)

Figure 7.7 (a) Typical x-ray machines used in radiography to check for metal flaws. (b) Radioactive gamma-ray sources used in radiography for detecting metal flaws.

source and the material as close together as possible. It will be found that the shorter the source-to-film distance, the greater the parallax, or halo, effect on the radiographic film. This can blur or dissipate the image of any inclusion that might be in the test material and thus mask a defect. The minimum source-to-film distance, according to type of source, size of focal spot, and the thickness involved, is determined by a standard formula for geometrical unsharpness:

$$Ug = \frac{Ft}{D}$$

where Ug = geometric unsharpness
   F = source size, in.
   D = distance from source to object being radiographed, in.
   t = thickness of weld or object being radiographed, in.

Generally, geometric unsharpness shall be no more than 0.020 in. for material with a thickness under 2 in., no more than 0.030 in. for material with a thickness from 2 to 4 in., and no more than 0.050 in. for thickness greater than 4 in. A qualified radiographer is required to demonstrate the ability to meet these conditions.

The exposure time for any radiograph must be sufficient to meet film-density standards, which is a measure of the darkening effect on exposed film. Radiographs that are too light may not reveal defects such as inclusions, and those that are too dark may make it impossible to separate the defects from the background. Density requirements per Section V of the ASME Code stipulate that "transmitted film density through the radiographic image of the body of the appropriate penetrameter and the area of interest" shall be

1.8 for single film viewing of x-ray film

2.0 for radiographs of gamma-ray source

2.6 for double film exposures

4.0 for maximum density permitted for single or composite viewing

On welded areas that are to be radiographed, the surfaces must be ground so as to eliminate any surface irregularities that might interfere with proper interpretation of the film.

In all radiographic work, a health hazard exists for all personnel in the working area unless safety rules are followed. These include safety regulations of government bodies as well as those of plant radiation safety officers. Safety regulations generally specify the use of proper monitoring devices for the detection of radiation; the need to cordon off working area and the posting of warning signs; the establishment of dosage limits and the recording of the dosage received by workers; and the action to be taken when the limits are exceeded or violated.

Radiographic inspection requires an interpretation of pictures on a film. See Fig. 7.8.

Defects such as cracks, slag inclusions, lack of penetration, voids, and others appear as darkened areas on the film because they have a lower density than solid metal. Typical faults are noted as follows. *Cracks* appear as dark irregular lines. *Slag inclusions* show up as small, dark spots with irregular outlines. *Gas pockets* appear as small, dark spots with smooth outlines with occasional teardrop fails.

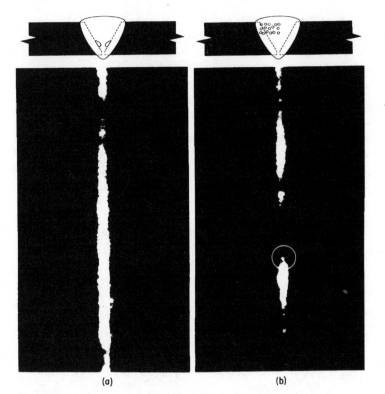

(a)   (b)

Figure 7.8 Skill is required in interpretation of radiographic film by studying typical pictures of defects. (*a*) Lack of side-wall fusion; (*b*) cluster porosity. (*Courtesy E. I. du Pont de Nemours & Co. Inc.*)

Lack of penetration is evidenced by a smooth, dark line most often located in the middle of a weld.

The Code provides guides or standards in order to judge whether an indication is acceptable. Some of the standards or requirements are as follows:

1. The first requirement is visual inspection of the weld. Joints must have complete joint penetration, and be free of undercutting, overlaps, or abrupt ridges and valleys. Weld reinforcement is specified not to exceed the following:

Plate thickness, in.	Maximum thickness of reinforcement, in.
Up to ½-in. inclusive	¹⁄₁₆
Over ½- to 1-in. inclusive	³⁄₃₂
Over 1- to 2-in. inclusive	⅛
Over 2-in. inclusive	⁵⁄₁₆

2. Welded joints to be radiographed must be free of ripples or weld surface irregularities to such a degree that the resulting radiographic contrast due to any irregularities cannot mask or be confused with the image of any objectionable defect.

3. Gauges called *penetrameters* must be used for every exposure of a film. The penetrameter serves as a comparison gauge on the film to compare faults in the weld. This is done by making a strip of metal for each exposure and drilling holes in the strip prior to exposure; these strips serve as guides to detect flaws within 2 percent of the plate thickness being welded. These holes are usually drilled with a minimum hole diameter of $\frac{1}{16}$ in. specified. The proper method of penetrameter location is shown in Fig. 7.9.

When an x-ray picture is taken, the penetrameter will be included, and the holes will serve as a guide for detecting faults right on the

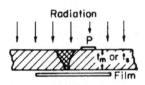

Single wall, no reinforcement, no backup strip

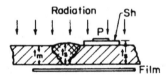

Single wall, weld reinforcements, no backup strip

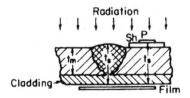

Single wall, weld reinforcement, stainless cladding redeposited over weld in base metal

(a)

(b)

Symbols

P = penetrameter
Sh = shim
t_m = design material thickness upon which the penetrameter is based
t_s = specimen thickness

Figure 7.9 Penetrameter and shim location in radiography. (*a*) For butt-welds; (*b*) for nozzle welds.

film. By comparing the holes on a penetrameter to the holes noted on a weld when the film is developed, an immediate comparison gauge is included on each film. The Code stipulates tolerances for weld rejects based on penetrameter data.

Power boilers constructed to Section I rules requires the following parts of the boiler to be radiographed.

Steel castings

1. If a 100 percent quality factor is to be used in the allowable stress per Code tables, all critical areas of the steel casting, such as gates, risers, and abrupt section changes, with thickness of 4½ in. or less must be radiographed.

2. Steel castings having a body greater than 4½ in. in thickness requires all parts of the casting to be radiographically examined.

Welded joints

1. Welded longitudinal joints require full-length radiographic examination for shells or drums of power boilers.

2. Circumferential welded joints of shells or drums require radiographic examination if the thickness of plate exceeds 1⅛ in.

3. For pipes, tubes, and headers, circumferential welded joints require radiographic examination when

   a. If containing steam, over 16-in. nominal pipe size or over 1⅝-in. thick. For those containing water, over 10-in. nominal pipe size, or over 1⅛-in. thick. In both cases, no contact with furnace gas exists.

   b. If in contact with furnace gases, but no furnace radiation, over 6-in. nominal pipe size, or over ¾-in. thick.

   c. If in contact with furnace gases and radiation, over 4-in. nominal pipe size, or over ½-in. thick.

If the tubes have at least five or more rows between them and the furnace, and if tubes are in contact with furnace gases that do not exceed 850°F, no radiographing of the weld is required.

Indications shown on the radiograph that are considered unacceptable per Code rules include:

1. Indications of a crack, and zones of incomplete fusion or penetration.

2. Other elongated indications which have lengths greater than those shown in this table:

Weld thickness	Maximum length of indication
To ¾ in.	¼ in.
¾ in. to 2¼ in.	⅓ thickness of weld
Over 2¼ in.	¾ in.

3. Aligned indications cannot have an aggregate length greater than the weld thickness in a total length of 12 times weld thickness.

4. Porosity charts in Section I are used to note if any group of aligned indications may not be acceptable.

Section I now requires that personnel performing NDT for a manufacturer or contractor must be qualified to SNT-1C-1A recommended practice of the American Society for Nondestructive Testing.

The Code requires steel castings that have weld repairs that exceed 1 in. or 20 percent of the section thickness to have a radiographic examination to make sure the weld repair is satisfactory. As can be seen, NDT is applied to repairs and not only to new construction.

Magnetic-particle testing. This test is used to detect surface faults by means of setting up a magnetic field, or magnetic lines of force, between two electrodes. Powdered magnetic material is sprinkled over the work to be tested. The magnetic field will affect the magnetic powder, and these particles will align themselves in a fault, as shown in Fig. 7.10. But the correct interpretation of the gatherings of the magnetic powder requires experience and practice. Magnetic-particle inspection is a practical means for spotting close-lipped discontinuities at or near the surface of a part. Both the wet and dry methods are presently available.

Discontinuities in a magnetized material give rise to localized leakage fields. And these fields attract finely divided magnetic particles. The latter point a finger at the defect and mark its extent on the surface of the part under inspection.

Both direct and alternating currents are used for magnetizing. Direct current (dc) is useful in finding subsurface discontinuities and is commonly used for inspecting welds and castings. Alternating current (ac) is usually employed when highly finished machined parts are checked.

Generally *dc magnetization* is considered where subsurface and surface cracklike defects must be found. With dc, the magnetic field extends within the part itself, and magnetic leakage fields are produced on the surface by interruption in the magnetic path below the surface.

Half-wave dc is a practical supply for subsurface defects, especially where the outer surface is rough as on casting or weldment. Current is rectified half-wave from a single-phase ac source at high amperage

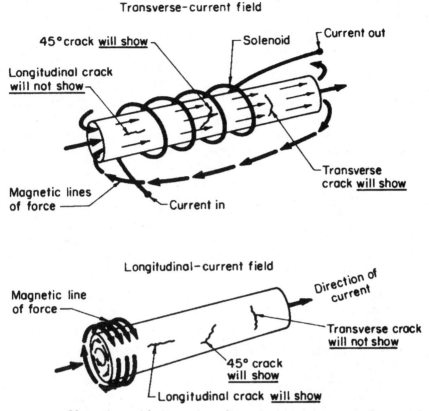

Figure 7.10 Magnetic-particle inspection—discontinuities in magnetized material develop localized magnetic leakage fields. (*Courtesy Power magazine.*)

and low voltage levels. Most fabricators of heavy industrial equipment use such tests to check for cracks on welds and castings. Heavy, thick welds can be inspected several times during the building up of the weld, after each welding pass if desired.

Dry magnetic powders are used for most portable types of inspection. Red, gray, or black powders give needed contrast with the color of the part being tested. Dry powders permit maximum sensitivity for deep, subsurface cracklike defects.

The *wet-method technique* uses a paste mixture of magnetic particles combined with oil or water. Either the part to be inspected is dipped into the mixture, or the mixture is flowed over the piece. In the dry method, which is more sensitive to subsurface defects, magnetic powder is dusted or blown over the part to be inspected.

For many applications, magnetic particles are coated with a *fluorescent* material and used in conjunction with black light. Thus, the resultant indication glows and attracts the eye. In effect, this increases the sensitivity of the inspection.

The principal limitation of the magnetic-particle method is that it applies to magnetic materials only and is not suited for very small, deep-seated defects. The deeper the defect is below the surface, the larger it must be to show up. Subsurface defects are easier to find when they have a cracklike shape, such as lack of fusion in the weld. On large, heavy objects, when extremely sensitive inspection is desired, the operation takes more time. This holds true on medium-sized critical parts such as aircraft propellers. With magnetic-particle testing, the surface to be inspected must be available to the operator. This means shafts or other equipment cannot be inspected without removing pressed wheels, pulleys, or bearing housings.

The advantages of magnetic-particle testing are many. Magnetic-particle testing can be used on magnetic-type material, which is very common in industrial products. It is a positive method of finding all cracks at the surface. And because cracks at the surface are most serious and may lead to failure, this is important. The method is flexible and permits effective use of portable equipment. The cost of both equipment and worker hours required for the test compares favorably with that of a simple visual-inspection method.

Liquid-penetrant (dye) inspection. This method of testing is used somewhat as magnetic-particle testing, except that it is used primarily on nonmagnetic material. But it can be used on magnetic material. The dye penetrant contains a visible dye, usually red. Indications of defects appear as red lines or dots against the white developer background. It is primarily a surface-defect indicator, and it is applied as follows: A dye penetrant is applied to the part by dip, brush, or spray and is allowed to sit for a specified time. After suitable penetration time, the excess penetrant is removed from the surface, and a developer is applied. The penetrant becomes entrapped in a defect and is brought to the surface by the action of a developer. See Fig. 7.11. Cracks are detected by noting the contrast between the white color of the developer and the red penetrant.

Another penetrant method used is the fluorescent penetrant method, which contains a material that fluoresces brilliantly under black lights. Indications of defects appear as fluorescent lines or dots against a nonfluorescent background.

The advantages of the dye-penetrant method are as follows: It provides fast, on-the-spot inspection during overhaul or shutdown periods; the initial cost of the test is relatively low. A perfectly white or blank surface indicates freedom from cracks or other defects that are open to the surface. The disadvantages are that it is not practical on very rough surfaces and color contrast is limited on some surfaces. Also it detects only defects open to the surface.

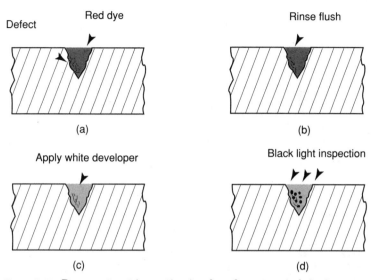

Figure 7.11 Dye-penetrant inspection involves four steps in bringing out a defect. (*a*) A red dye is applied and allowed to penetrate a suspected defect; (*b*) the excess dye is removed flush with the surface; (*c*) a white developer is applied; (*d*) under black light inspection, red dots show location of defect.

Ultrasonic inspections. Ultrasonic testing makes use of high-frequency sound waves of 0.5 to 10.0 megahertz (MHz) for inspection of material for flaws and for measuring plate or wall thickness. The basic principle used in an ultrasonic system is the transformation of an electric impulse into mechanical vibrations and then the retransformation of the mechanical vibrations into electric pulses that can be measured or displayed on a screen called a cathode-ray tube (CRT) screen. The transfer of mechanical energy to electric energy is done by means of a transducer. A transducer is a device which transforms energy from one form to another. Ultrasonic transducers transform high-frequency alternating electric energy into high-frequency mechanical energy through magneto restrictive elements up to 100 kilohertz (kHz); and above this range they are usually of the piezoelectric type. In the ultrasonic NDT field, piezoelectric transducers are chiefly used. Piezoelectric transducers depend on the expansion and contraction of certain crystalline and ceramic materials under the influence of an applied fluctuating electric potential. If mechanical pressure is applied to certain faces of specific crystals, then an electric potential is generated across them which is proportional to the applied pressure; conversely, an electric potential of appropriate frequency will cause a crystal to vibrate and so generate pressure waves in any medium with which it is in intimate contact. Figure 7.12 shows the

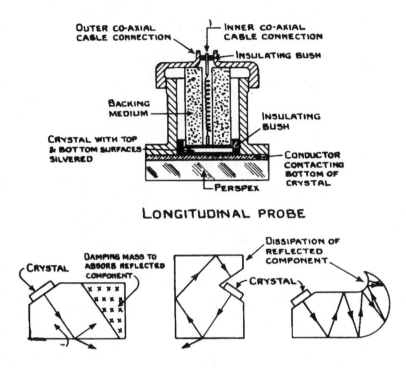

Figure 7.12   Details of ultrasonic probe detector crystals for longitudinal and angle detection methods.

arrangement of a probe in which the crystal or transducers are mounted to generate longitudinal or angle mechanical energy (sound) into a piece to be tested.

The basic component of a typical pulse-echo type of flaw detector is a short electric pulse generated and applied to the electrodes of the probe, which converts the pulse from electric to mechanical vibration energy. This ultrasonic pulse travels through the material, and part is reflected to the receiver probe by a reflector which could be the end of the material under test or an internal defect in the material. The returning pulse in the receiver generates an electric pulse which is amplified and displayed by a cathode-ray tube. See Fig. 14.19.

The CRT is a familiar component of a television set. It is employed in many applications to render visible any wave that can be converted to an electron beam. In order to evaluate the information displayed on the screen of the CRT, calibration techniques must be followed so that defects can be compared or picked up in relation to standard reference blocks. The main purposes of these blocks are to:

1. Calibrate the instrument for a time base for longitudinal or transverse waves.

2. To measure certain characteristics of probes and the apparatus.

3. To adjust sensitivity and periodically check to make certain it remains constant throughout a test.

4. To have a written procedure so that the settings and calibration can be duplicated at a later date.

Ultrasonic tests are grouped into three basic categories: *pulse-echo testing, through transmission,* and *resonance testing.* The pulse-echo method involves transmitting a short burst of high-frequency sound through the piece being tested and then detecting the echoes that are received from either a construction detail, such as a shoulder or hole, or a defect in the material with a separation or void in the material sufficiently large that sound cannot be transferred across the interface. In operation, a pulse-echo unit will produce, through an electronic pulser, a short burst of high-frequency electric signal. This is transmitted to the transducer which is forced to vibrate, usually at its resonant frequency. The probe must be coupled to the test piece with oil, water, or some other liquid or grease. The sound wave train then travels through the test piece until some form of discontinuity or boundary is encountered. This interruption in the medium then causes the sound wave to be reflected to the receiver transducer. The vibrational energy of the sound wave sets the transducer in motion to produce an electric impulse which is fed into an amplifier. The output of the amplifier is displayed on a cathode-ray tube which shows the signals on a linear time baseline. If a linear amplifier is used, the amplitude of the returned echoes can be used as a measure of the area producing the reflected signal.

Resonance testing makes use of a tunable, continuous wave system. This method is usually employed for measuring small or thin walls to 2 or 3 in. The resonance of the crystal is tuned to the piece under test. In practice, a loud pip is heard and can also be seen because the electronic circuit is also in resonance electrically. By proper calibration, direct thickness readings can be made.

Crystals are made for longitudinal waves, shear waves, and surface waves. High-frequency sound waves obey the same physical laws as light waves and can be reflected, refracted, absorbed, and polarized. Using an ultrasonic beam to scan materials such as acoustically transparent steel is analogous to using a light beam to scan optically transparent materials such as glass or water. Since most sonic testing equipment is portable, ultrasonic testing is well suited for field application. There are many battery-operated instruments on the market capable of performing many qualitative tests.

As with all other methods of nondestructive testing, there are distinct advantages and disadvantages associated with ultrasonic tests. The advantages of ultrasonic tests are:

1. Highly sensitive, permitting detection of minute defects
2. Penetrating power, allowing inspection of thick sections
3. Accuracy in locating and measuring flaw size
4. Fast response, permitting rapid inspection

The disadvantages of ultrasonic tests are that the test can be limited by unfavorable part geometry, such as size, contour, complexity, and relative defect orientation; and material structure, such as grain size, porosity, inclusion content, or fine dispersed precipitates.

Eddy-current testing. The underlying principle of eddy-current testing is the measurement of impedance of electron flow in the part being tested. Weak electric currents or eddy currents are induced by a probe containing inducing and sensing coils. Any changes in the geometry of the part such as a pit, crack, or thinning will affect the flow of the eddy currents. This change in the flow of the eddy currents will be detected by the sensing coil and displayed on a strip chart, a CRT display, or both. The accuracy of characterization of the detected flaw depends on the quality of the standard used in calibrating the eddy-current instrument. A piece of tubing of material identical to the tubes being inspected should be used for calibration. The standard should have a range of defects similar to those expected to be found in the tubes being tested.

Eddy-current testing can indicate tubes that are not leaking, but which may be deteriorated as a result of thinning, erosion, or rubbing. By periodically checking tubes and keeping accurate records, tube replacements can be conveniently scheduled in advance during turn-around plant-maintenance periods, and not on a breakdown basis during production runs. Eddy-current inspections from the inner diameter of exchanger tubing allows detection of external and/or internal pits, thickness reduction, cracks, voids, inclusions, dezincification, and carbonization. These conditions can be identified by experienced testers and located along the tube wall. The severity of the test defect can be judged from the results by comparing with the defects in a standard specimen.

Acoustic emission. When a material stretches or "gives," some sound is emitted, and by placing sensors on the material it is possible to pick up these sounds. It has been found that the acoustic sounds emitted by most structural material are above the hearing range of human beings and are generally above 100 kHz. This makes it possible to instrument for sound levels above most spurious noise sources, which will be at lower frequencies. For example, if we assume a small subcritical crack growth exists in a thick steel section, as the crack propagates forward, the atoms are displaced by a small distance in a very short time. This displacement leads to a transient elastic stress wave which then travels outward from the crack-tip area. This stress wave in the material is called the *acoustic emission*. Very large crack extensions that travel to the surface would be heard. However, by placing transducers on the surface, crack propagation can be detected as the crack develops and progresses. An acoustic emission monitoring system provides a means to continuously scan a loaded structural member for changes in acoustic emissions in order to detect any developing flaw. When the stress wave reaches a transducer, an electric signal is generated which can then be processed to provide information about the acoustic-emission event. By detecting and processing the information contained in the acoustic-emission stress wave, the growth of flaws can be tracked and studied in real time as they occur. Present systems include data acquisition for evaluating the severity of flaws. These data are stored in a computer for present analysis and for future reference. Acoustic emission has great potential in continuously monitoring critical pressure vessels for flaw formations such as exist in nuclear power plants, where, because of radiation hazards, access for inspection may be restrictive. This method is still in the developmental stage for such applications.

Proof and leak tests. These are also considered NDTs provided the tests are kept within prescribed limits in order to avoid permanent deformation. A hydrostatic test at 150 percent allowable working pressure is considered nondestructive. During this test visual inspections are made of welds, joints, and connections for leakage, while the pressure is watched for any drop that may indicate a hidden leak.

Other leak tests are bubble emission tests, halogen (Freon) leak detection and mass spectrometers.

Nondestructive examination while a piece of equipment is still in service is a long-term development goal. A recent development is for inspecting high-pressure steam pipes using *gamma-ray computed tomography* which measures photon transmission. When the photons enter the material, some of the energy is converted into light through excitation and ionization or the process of fluorescence. The light drives the photoelectric effect in a photomultiplier tube (or photovoltaic effect in a solid-state device) that produces a measurable electric pulse proportional to the energy of the incident gamma-ray photon. A discriminating circuit detects only those gamma rays that have been transmitted through the object to be scanned, eliminating the detection of scattered photons. By the use of computer analysis and special analytical cathode tube software, a three-dimensional density map is used to detect flaws.

There will be other nondestructive examination methods developed in this rapidly growing field that tries to assist our ability to inspect by limited visual methods.

Selecting NDT method. The wide choice of NDT requires considering the strength and weakness of each method, and what may be the defect that is of concern. Frequently more than one NDT must be used to find and evaluate discontinuities.

NDT examination and inspection require extensive experience in the method of NDT to be applied for flaw detection. This specialized experience is detailed in SNT-TC-1A, "Recommended Practice for Nondestructive Testing Personnel Qualification and Certification," published by the American Society for Nondestructive Testing (ASNT). Three levels or grades of qualifications are possible in each of the NDT methods previously described. The level 3 person is the most qualified, which generally requires not only knowledge of operating an instrument or applying a method but also theoretical knowledge of the NDT methods, their advantages and shortcomings, and the interpretation of test results. The ASME Boiler and Pressure Vessel Codes refer to the ASNT as the source for details on qualifying and certifying NDT "examiners." The ASME Code inspector must still make sure that the written procedures can detect the discontinuities by the NDT method to be used which are not acceptable in the section of the Code to which a boiler or pressure vessel is being built. For example, there are radiographic standards in Section I that may differ somewhat from those involving a nuclear pressure vessel. The "authorized" inspector making the Code inspections will review the NDT results in order to make sure discontinuities found by NDT are within Code-permissible limits. Close coordination is thus essential between the NDT examiner and the authorized inspector.

Quality Control System

ASME stamp holders, such as the "S" for power boilers, must have a quality control system that demonstrates that all Code requirements on design, materials, fabrication, NDT examination, and inspection requirements will be satisfied. The quality control system per item must be in writing, usually in a manual form, but is restricted for review by outside people only to the Authorized Inspector or ASME designee who reviews the stamp holders procedure for meriting the ASME stamp. Paragraph A-300 details the requirements of the quality control in Section I. Included in the outline are: authority and responsibility for the quality control; organizational chart showing how the quality control will be carried out throughout the manufacturer's shop; Code design calculations and drawings; material control procedures; examination and inspection methods and procedures; Code welding procedures to be followed; Code nondestructive examinations to be used; heat treatment to be applied; calibration of measurement and test equipment procedures; and records to be made and distributed (data reports) and retained per Code rules.

Questions and Answers

Welding

1 Name some typical dimensional defects that can occur in a welded joint.

ANSWER: Dimensional defects possible in a welded joint are:

1. Warpage of plate
2. Incorrect joint preparation or misalignment of the joint
3. Incorrect weld size for the thickness of plate to be welded
4. Incorrect weld profile such as overlap, excess convexity, excess weld reinforcement
5. Incorrect final dimension

2 Name some typical structural discontinuities that can be found in a welded joint.

ANSWER: Structural discontinuities that may be found in welded joints are: (1) porosity; (2) slag inclusion; (3) tungsten inclusion; (4) incomplete fusion; (5) undercut; (6) cracks; and (7) surface irregularities.

3 (*a*) What is electric arc welding? (*b*) What are the base or parent metals?

ANSWER: (*a*) Welding two pieces of steel together, with the fusion brought about by heating the metal by the electric arc. (*b*) The metals which are to be welded.

4 What should be the condition of the base metals at the time of welding?

ANSWER: Metals should be clean and free from grease, dirt, dust, paint, or other foreign substances except for a thin coating of linseed oil if this has been specified.

5 What precautions must be taken by the welder while welding or the inspector of welding while watching the welding process?

ANSWER: The face and neck of the welder or inspector should be covered by a helmet. The glass in the helmet should be darkened to protect the eyes. The welder should also wear gloves to protect the hands.

6 Which is better, the long or short arc?

ANSWER: The short arc because it makes a purer weld, gives better fusion and penetration, and causes less puttering of the molten metal.

7 How should a condemned weld be removed in order to permit the joint to be rewelded?

ANSWER: The weld metal should be chipped or ground out to good metal before the weld is redone.

8 What is meant by the terms *forehand* and *backhand* as applied to oxyacetylene welding?

ANSWER: In forehand welding, the welding flames are pointed in the direction in which the weld is progressing or is being made. In backhand welding, the welding flame is pointed in the direction opposite to that in which the weld is being made.

9 What is meant by double-butt welding and single-butt welding?

ANSWER: In double-butt welding, the metal is deposited in the welding groove from both sides of the plate, whereas in single-butt welding the metal is deposited from only one side with the plates to be welded butted together instead of being lapped one over the other.

10 After the first side of double-welded butt joint has been welded, what should be done to the joint before the second side is welded?

ANSWER: The second side should be chipped, ground, or melted out to secure a clean surface of proper shape to ensure fusion of the weld metal on that side.

11 What is meant by reinforcement of a weld?

ANSWER: The amount of weld above the surface of plate on the face side. This ensures full penetration for full thickness of plates to be joined.

12 Why is it common practice to use heavily coated electrodes in metallic arc welding?

ANSWER: The coating shields the arc from the atmosphere and produces a slag covering on the weld, both of which give the weld metal good tensile strength and ductility.

13 Is single-butt welding permitted in the fabrication of fusion-welded boiler drums?

ANSWER: No, unless a backing strip is used. If a backing strip is used to ensure complete penetration of the weld metal, the weld may be made from one side only.

14 (*a*) what is the least elongation acceptable in applying a free-bend test?

(b) How many sets of test plates are required in a test of welders?

(c) How often should welders be required to take the qualification test?

ANSWER:

(a) The specimen shall be bent cold under free-bending conditions until the least elongation measured within or across approximately the entire weld on the outside fibers of the bend-test specimen is *30 percent* or 700,000 divided by *U* plus 20 percent, whichever is less. Here *U* represents the minimum ultimate specified tensile strength of the material to be welded, in pounds per square inch, from Code stress tables.

(b) Two sets, one face-bend and one root-bend test per the thickness of the material, position, and welding process to be used. See Section IX of the ASME Code.

(c) When a welder has not used the specific welding process such as metal arc, submerged arc, etc. to weld ferrous on nonferrous materials for a period of six months or more, or when there is a specific reason to question the welder's ability to make welds that meet specifications.

15   What welds under Section I rules require radiographic examination?

ANSWER: On power boilers, the following:

1. For steel castings, if 100 percent quality factor is to be applied, all critical abrupt sections under 4½ in. thick. For steel castings over 4½ in. thick, all parts of the casting.

2. For welded longitudinal joints of shells or drums, the entire length of the weld.

3. For circumferential welded joints of shells or drums, for plate thickness over 1⅛ in.

4. For circumferential welds on pipes, tube or headers,
    a. If containing steam, and not in contact with furnace gases, over 16 in. nominal pipe size, or over 1⅝ in. thick. If containing water, 10-in. nominal pipe size, or over 1⅛ in. thick.
    b. If in contact with furnace gases, but no furnace radiation, over 6-in. nominal pipe size, or over ¾ in. thick.
    c. If in contact with furnace gases and radiation, over 4-in. nominal pipe size and over ½ in. thick.
    d. Tubes five rows or more back from the furnace are excluded as are tubes not in the gas path of furnace gases that do not exceed 850°F.

16   What are locked-up stresses in welding? What causes them?

ANSWER: Locked-up stresses are those internal stresses remaining in the weld metal and adjoining base material when a weld has been made. They are caused because the weld metal and the base material are, of necessity, at considerably different temperatures when a weld is made and therefore do not expand and contract uniformly, thus setting up internal stresses from HAZ effects.

17   What is meant by stress relieving and how is it accomplished?

ANSWER: Stress relieving or postweld heat treatment is the uniform heating of a structure or parts thereof (welded joint) to a temperature just below the critical metal range. The purpose of stress relieving is to relieve the major residual stress that may exist in a welded joint from the heat effects caused by the welding process.

The Code specifies preheating temperatures for the different types of P-number metals that may be welded. Generally, the preheat temperature is below 450°F. Postweld heat-treatment temperatures to be applied to the different types of metals are also specified in the Code by P-number groupings. For plain carbon steels, this is generally 1100°F to be held one hour per inch of plate thickness.

18   What is an "approved" or "qualified" process of welding?

ANSWER: One that has been written out in detail and followed in the making of test plates as required by the ASME Code, such test plates having been subjected to the tests specified in the Code and found to meet the requirements of the Code for the class of work to be done. This includes considering material (P numbers), thickness, welding process, electrodes, welding position, and heat treatment to be applied.

19   What is meant by the terms *lack of fusion, lack of penetration,* and *slag inclusions*? How are they caused and how should such conditions be remedied?

ANSWER: Lack of fusion is the lack of fusion or melting together of the base metals or weld metal being deposited. It is caused by improper welding techniques, among which could be the heat applied being too low because of low amperage, too fast welding and travel, improper surface preparation, and similar causes. When lack of fusion is discovered or noted, it is necessary to grind or chip out to sound weld metal and reapply weld metal to get fusion of the base metal.

Lack of penetration means that the welding or fusion of plates was not made for the entire thickness of the plate, leaving parts of the plates not joined. It is caused by improper joint preparation or fit-up, poor welding techniques, or improper surface preparation. It is corrected by chipping and grinding back to sound metal and filling up the area with a good penetrating weld per Code requirements.

Slag inclusions mean there is foreign matter in the weld, which can weaken the joint. This is generally caused by poor joint penetration or cleaning. To correct, the affected area is ground or chipped out and rewelded per Code requirements.

20 What is a backing strip as used in metallic welding? When a joint is welded from one side only, can it be considered the equivalent to a double-welded butt joint?

ANSWER: It is a strip of metal usually of the same material as that being welded, from 1- to 3-in. wide and ⅛- to ¼-in. thick, placed on the underside of the joint to be welded, to enable the welder to obtain penetration of the weld throughout the entire thickness of the plate or pipe and avoid the dripping through of molten metal. Welding from one side only can be considered a double-butt joint when complete penetration and reinforcement on both sides are achieved, as by the use of a backing strip or chill ring.

21 What is the maximum permissible offset of the abutting edges of welded joints at any one point in longitudinal joints and in girth joints?

ANSWER:

Thickness of plate, T	Maximum offset permitted	
	Longitudinal	Circumferential
To ½ in. inclusive	$\frac{1}{4}T$	$\frac{1}{4}T$
Over ½ to ¾ in. inclusive	⅛ in.	$\frac{1}{4}T$
Over ¾ to 1½ in. inclusive	⅛ in.	$\frac{3}{16}$ in.
Over 1½ to 2 in. inclusive	⅛ in.	$\frac{1}{8}T$
Over 2 in.	Lesser of $\frac{1}{16}T$ or ⅜ in.	Lesser of $\frac{1}{8}T$ or ¾ in.

22 What preparations are made before joining together by a butt weld a ⅞-in. head flange with a ½-in. shell plate? Why are those preparations necessary?

ANSWER: A 3-to-1 taper is required in the thicker plate for the purpose of avoiding stress concentration at the transition section between the plates of unequal thickness.

23 A boiler drum 5-in. thick has a fusion-welding longitudinal joint. What is the minimum temperature under which it must be stress-relieved and for how long a period?

ANSWER: At least 1100°F for a period of five hours if we assume a low P number for the material.

24 Has an authorized inspector any right to disagree with the manufacturer's records and call for a requalification of a welder or to have physical tests made?

ANSWER: If the inspector has any doubts regarding the ability of a welder or of the process, he or she can require a requalification test of the process or of the welder.

25 In what position must the test plates be when a welder makes test welds for the purposes of qualifications?

ANSWER: In whatever position the welder will encounter in actual work. If the welder makes the test plate in a flat position, she or he is not permitted to do work in other than that position, such as "vertical" or "overhead" welding.

26 (*a*) Is it permissible to locate a hole in a welded joint? (*b*) Is it permissible to roll and expand tubes into holes located in a welded joint?

ANSWER: (*a*) Yes, provided that magnetic-particle examination indicates the weld is sound after the hole is made. (*b*) Yes, as in part *a*.

27 When an unqualified welder is available for making a fusion-weld boiler repair, what two principles would determine the type of repair that may be made?

ANSWER: (1) The strength of the boiler must not be dependent on the weld. (2) The effect of welding heat and welding stresses must not be detrimental.

28 What is the maximum permitted length of a course in a horizontal boiler of either riveted or welded construction?

ANSWER: 12 ft.

29 (*a*) May the longitudinal riveted seams of older power boilers be seal-welded?

(*b*) May circumferential seams be seal-welded?

(*c*) When repairs are made to the caulking edges of girth seams by fusion welding, what precautions should be taken regarding the rivets?

ANSWER: (*a*) No. (*b*) Yes. (*c*) The rivets should be removed and the holes reamed after welding, and new rivets should be driven. This is to prevent the rivets being distorted from the heat of welding and possibly cracking. Consult with jurisdictional inspector for approval.

30 Describe briefly the necessary hydrostatic and other physical tests to be applied to a welded boiler drum or vessel before it can be accepted.

ANSWER: Welded joints on boiler drums must be radiographed in sections of the drum specified by the Code as well as stress-relieved. A hydrostatic test is required after the boiler has been completed with water at a temperature not less than 70°F. The pressure shall be applied gradually up to 1½ times the maximum allowable pressure to be stamped on the boiler. Control of the pressure must be such that the test pressure is never exceeded by more than 6 percent. The boiler is carefully examined for leakage when the pressure is dropped back to the maximum allowable pressure to be stamped on the boiler.

31 (*a*) Who is responsible for conducting tests of welding procedures and for qualifying welders?

(*b*) How long does a qualified welder's approval test remain in effect?

(*c*) Who keeps the qualification records of procedures and welder's approval and acceptance?

ANSWER:

(*a*) The manufacturer is responsible for the tests, subject to an inspector's approval.

(*b*) Indefinitely if regularly employed and employed in the same type of welding, thickness, and material.

(*c*) The manufacturer.

32 What is meant by preheating and stress relieving?

ANSWER: Preheating is the application of heat to the base metals prior to a welding or cutting operation. Stress relieving or postweld heat treatment is the uniform heating of a structure or parts thereof (welded joint) to a temperature just below the critical metal range. The purpose of stress relieving is to relieve the major residual stress that may exist in a welded joint from the heat effects caused by the welding process.

33 What temperature should be used if it is found necessary to preheat and stress-relieve the welding of low-carbon steel?

ANSWER: The Code specifies preheating temperatures for the different types of P-number metals that may be welded. Generally, the preheat temperature is below 450°F. Postweld heat-treatment temperatures to be applied to the different types of metals are also specified in the Code by P-number groupings. For plain carbon steels, this is generally 1100°F to be held one hour per inch of thickness of plate.

34 (*a*) What is meant by a nondestructive test of a welded joint? Give some typical examples.

(*b*) What would cause you to reject a welder's test piece?

ANSWER:

(*a*) A test which does not destroy the boiler component, such as radiograph, hydrostatic, ultrasonic, dye-penetrant, and magnetic-particle tests.

(*b*) Porosity, lack of penetration, slag inclusions, and cracks on the surface of the weld, or any other such defects not meeting Code requirements.

35 State the essential requirements that must be met to permit the welding of pipe or tube connections which do not require radiographic examination or the hammer test.

ANSWER: If the procedure and welder have been qualified per Code require-ments, circumferential welded joints on pipes or tubes do not require radio-graphic examination provided:

1. If containing steam, not over 16-in. nominal pipe size, and not over 1⅝ in. thick. If containing water, not over 10 in. nominal pipe size, and not over 1⅛ in. thick. In both cases, not in contact with furnace gases.

2. If in contact with furnace gases, but not radiation, not over 6-in. nom-inal pipe size, or over ¾ in. thick. If in contact with furnace gases and radiation, not over 4 in. nominal pipe size, or over ½ in. thick.

3. If the tubes have five or more rows between them and the furnace, and if the tubes are in contact with furnace gases that do not exceed 850°F, no radiographic examination is required of the welds.

Nondestructive testing

36 Define the half-life of an isotope.

ANSWER: The term *half-life* refers to the decrease in strength of radiation of an isotope; the half-life is the number of years it takes for the isotope to decrease one-half its present strength (time of reference).

37 If an isotope has a strength of 80 curies (Ci) now with a half-life of 10 years (yr), what will its strength be in 30 yr?

ANSWER: At the end of the first 10 yr, the strength will be 40 Ci. At the end of 20 yr, the strength will be 20 Ci. At the end of 30 yr, the strength will be 10 Ci.

38 Name two survey meters that are used to check on the radiation dosage to which an individual may be exposed.

ANSWER: Pocket dosimeters and film badges are used to check on the cumula-tive dosage of an individual.

39 (*a*) How is the energy level of an x-ray machine rated?
  (*b*) What penetrates material better, short or long radiation waves?
  (*c*) Does high or low voltage produce short waves?

ANSWER: (*a*) The energy level of an x-ray machine is rated by the peak voltage at which it can drive electrons into the tube target, and it is expressed as fol-lows: kVP = peak kilovolts (1000 V) and MV = megavolt (1 million V) (*b*) and (*c*) Short waves have greater penetrating power, and high voltages of the radiation source produce shorter wavelengths; therefore high voltage has greater penetrating power.

40 (*a*) Name two factors that control contrast in radiography.
  (*b*) Name the two principal types of screens used in radiography.

ANSWER:

(a) The voltage or energy level of the source of radiation that the radiography will be subjected to and the type of film used.

(b) Fluorescent and lead foil.

41 (a) Name two radiation sources permitted for radiographic examination in the Power Boiler Code.

(b) What class of film shall be used for radiographic examination of power boilers?

ANSWER:

(a) X-rays and radioactive isotopes.

(b) Type 4 film may be permitted for power boilers as well as type 2, as stipulated in Section V.

42 What must appear on a radiographic image in order to prove the technique and sensitivity of the radiographic examination?

ANSWER: The radiographic image must display the penetrameter and the specified hole in the penetrameter in order to prove that the technique used was of the proper sensitivity. In addition, the radiographs should display the radiograph identifying number and letters so the area radiographed can be traced.

43 (a) What is the purpose of a penetrameter? What is penetrameter sensitivity?

(b) From what type of material should penetrameters be fabricated when they are to be used to radiograph welds?

ANSWER:

(a) A penetrameter is used to obtain evidence on a radiograph that the technique used in exposing a joint to radiography will show certain minimum defects considered acceptable, but will show defects considered unacceptable by Code requirements. Penetrameter sensitivity refers to the image quality and smallest hole that can be "sensed" or seen; therefore, it is a measure of the lowest or minimum discontinuity that may be measured or observed by radiography on a particular thickness of metal.

(b) Penetrameters should be fabricated of material radiographically similar to the object being inspected.

44 Why are nonconsumable backing ring joints on the water side of welds not recommended?

ANSWER: The crevice formed by the boiler component and the backing ring can cause corrosion to develop at the crevice, which can lead to premature boiler part failure. This is especially applicable to backing rings used on the inside diameter of tubes.

45 (*a*) Where are penetrameters normally placed in relation to the weld and the source of radiation?

(*b*) If the penetrameter is placed on the film side of the joint, how is this identified?

ANSWER:

(*a*) Penetrameters are normally placed adjacent to the weld seam and on the source side of the seam to be radiographed.

(*b*) A lead letter "F" at least as high as the identification number must be placed adjacent to the penetrameter for those cases where the penetrameter is placed on the film side, and not on the source side.

46 How large must the diameter of the hole be in a penetrameter in relation to penetrameter thickness *t*?

ANSWER: 2*t* or twice the thickness of the penetrameter, except for penetrameters under 0.010-in. thickness, where the hole diameter must be a minimum of 0.020 in.

47 (*a*) Why is it necessary to remove weld ripples when a weld is to be radiographed?

(*b*) List some typical blemishes that must be avoided on radiographic film.

ANSWER:

(*a*) It is necessary to remove irregularities on weld surfaces so that any radiographic image taken of the weld does not mask or hide, or in other ways make it difficult to identify, any unacceptable weld discontinuity that may appear on the radiographic image.

(*b*) Radiographs must be free from the following blemishes so that discontinuities can be properly identified: fogging; streaks, water marks, stains; scratches, finger marks, crimps, dirt, smudges, tears; loss of detail on the image.

48 What two imperfections in a weld are considered unacceptable by the Power Boiler Code and which appear on a radiographic film?

ANSWER: Any type of crack or zone of incomplete fusion or penetration is considered unacceptable.

49 How long must a manufacturer retain in file radiographs taken on a power boiler?

ANSWER: A set of radiographs taken for each power boiler must be kept in file by the manufacturer for at least 5 yr.

50 (*a*) To what type of material may magnetic-particle examinations be applied?

(*b*) In prod magnetic-particle testing, what is the maximum prod spacing permitted?

(*c*) In magnetic-particle testing, what direction is the magnetic field in relation to current flow?

ANSWER:

(*a*) Only to ferromagnetic materials.

(*b*) A maximum of 8 in.

(*c*) At right angles, or perpendicular, to current flow.

51 (*a*) When are the best results obtained in magnetic-particle inspection— when the magnetic field is parallel to a discontinuity or when the magnetic field is at right angles to the discontinuity?

(*b*) Differentiate or describe the difference between the continuous method and residual method as used in magnetic-particle inspection.

ANSWER:

(*a*) The best results are obtained when the magnetic field is at right angles to the discontinuity or when the magnetizing current is flowing parallel to the discontinuity.

(*b*) In the continuous method, the dry powder is applied to the surface of the work under test while the magnetizing current is flowing. In the residual method, the dry powder is applied to the surface of the work after the magnetizing current has been switched off. This method requires the work to be one that retains the magnetism, whereas the continuous method can be used on material which has low magnetic retentivity.

52 How many magnetic-particle examinations are required of an area to be examined in a part under test?

ANSWER: At least two separate examinations shall be carried out on each area. The second test should produce lines of magnetic flux approximately perpendicular to the first test. This is necessary in order to pick up discontinuities which might be parallel to the magnetic field and thus not become visible if only one test were performed.

53 Name the three types of penetrant recognized by the Code in making liquid-penetrant examinations.

ANSWER: Water-washable, postemulsifying, and solvent-removable.

54 How does liquid penetrant work so that it can be used for examination for discontinuities?

ANSWER: Liquid-penetrant examination is used to detect discontinuities which are open to the surface of nonporous material. A liquid called a penetrant is first applied to the surface to be examined. The penetrant is allowed to stand for awhile on the part so that it can enter any discontinuity. The part is then wiped free of surface penetrant. A drier is then applied to the surface. Finally, a developer is sprinkled over the part. Any penetrant trapped in a discontinuity below the surface will now wet the developer so it sticks out and thus exposes the discontinuity that needs to be evaluated for acceptance of the part or rejection unless it is repaired.

55 What is type A and type B liquid-penetrant inspection?

ANSWER: Type A uses fluorescent penetrants while type B uses a visible-to-the-eye dye penetrant.

56 Name the two types of ultrasonic principles used in nondestructively testing material to be used for pressure-vessel parts.

ANSWER: The first type of ultrasonic principle used in nondestructive testing is pulse echo. In this system, sound pulses are sent through a material, and any reflecting surfaces echoes this sound. The signals are displayed on a CRT screen, and with proper adjustment, the echo principle is used to detect flaws at the interface of materials which produce echoes.

The second ultrasonic principle used in ultrasonic testings is the resonance principle. In this system, the frequency of sound from an instrument is varied until it matches the material under test or is in resonance with it. By electronic and electric circuit principles, the resonant frequency can be converted to thickness measurements of the material under test. This is the chief application for resonant ultrasonic-sound instruments.

57 (a) What method of ultrasonic examination of welds is permitted by the Code?

(b) When is it permissible to use ultrasonic examination on welded joints for power boilers?

ANSWER:

(a) Ultrasonic examinations shall be conducted with a pulse-echo type of system with rated frequency ranges of 1 to 5 MHz.

(b) When it is impractical to use a combination of radiography parameters so that a geometric unsharpness of 0.07 will not be exceeded, then ultrasonic examination shall be used to check the weld and heat-affected zones of those joints requiring radiographic examination by Section I rules.

58 Describe how the following surfaces shall be prepared for ultrasonic examination: (*a*) contact surfaces, (*b*) welded surfaces, (*c*) base material.

ANSWER:

(*a*) The finished contact surfaces shall be free from weld spatter and any roughness that would interfere with free movement of the search unit or impair the transmission of ultrasonic vibrations.

(*b*) The finished weld surfaces, where accessible, shall be of adequate smoothness to prevent interference with the interpretation of the examination. The weld surface shall merge smoothly into the surfaces of the adjacent base material.

(*c*) After the weld is completed but before the angle-beam examination, the area of the base material through which the sound will travel in angle-beam examination shall be completely scanned with a straight beam search unit to detect reflectors which might affect the interpretation of angle-beam results. Consideration must be given to these reflectors during interpretation of weld examination results, but their detection is not a basis for rejection of the base material.

59 (*a*) Name three types of waves used in ultrasonic inspections.

(*b*) What does the term CRT mean as applied to ultrasonic testing?

(*c*) What does a piezoelectric element do as applied to ultrasonic testing?

ANSWER:

(*a*) Straight-through waves, angle-beam waves, and surface waves.

(*b*) Cathode-ray tube. This is the tube on which the ultrasonic signals appear. If it is adjusted properly, they will show "beeps" where sound enters and where it bounces off either a fault or an opposite surface.

(*c*) The piezoelectric element is a carefully manufactured crystal which has the property of converting electric signals into mechanical vibrations or sound signals. The crystals also have reverse capability, namely, of converting vibrations or sound signals into electric signals.

60 (*a*) What nominal frequency should be used to calibrate by the ultrasonic angle-beam method?

(*b*) What is considered the primary reference response of a pulse echo using the angle-beam method when one is calibrating ultrasonic equipment?

(*c*) Describe the ultrasonic transfer method used to correlate the response from the basic calibration block and from the production material when the angle-beam method is used.

ANSWER:

(*a*) The nominal frequency shall be 2.25 MHz unless a variable, such as production-material grain structure, requires the use of other frequencies to ensure adequate penetration.

(*b*) Seventy-five percent of full screen on the cathode-ray tube.

(c) Transfer methods are used to correlate or check the response from the basic calibration block and from the component under examination. Readings are taken on the calibration block, and differences in readings from the component under test and the reference block are then corrected by recalibrating the instrument.

61 (a) When one is examining welds by the ultrasonic method, what shall serve as basic calibration reflectors and what are the material requirements for these calibration reflectors?

(b) In ultrasonically examining welds by the angle-beam method, how often must the transfer method be used to check technique on vessel welds and pipe welds?

ANSWER:

(a) Drilled holes shall be used as basic calibration reflectors, and these holes may be located in the base metal or extension of it or may be located in a calibration block of similar metallurgical structure as the metal being welded.

(b) For vessels, the transfer method must be used at least once for each 10 ft of weld or less per plate, and shall be performed at least twice for each type of welded joint. For pipe, the transfer method must be used, as a minimum, once for each welded joint on pipe sizes of 10-in. diameter and over and once for each 5 ft of weld for pipe less than 10 in. in diameter.

62 (a) Who has the responsibility of preparing an ultrasonic examination report for power boilers?

(b) What information shall the ultrasonic examination report contain?

(c) What minimum length of discontinuity is not permitted by the Power Boiler Code on welds examined by the ultrasonic method and where the base plate being welded is 3-in. thick?

ANSWER:

(a) The manufacturer.

(b) (1) All procedures and equipment shall be identified sufficiently to permit duplication of the examination(s) at a later date. This shall include initial calibration data for the equipment and any significant changes in subsequent recheck. (2) A marked-up drawing or sketch indicating the weld(s) examined, the item or piece number, and identification of the operator who carried out each inspection or part thereof are required. (3) A record of repaired areas shall be kept as well as the results of the re-examination of the repaired areas.

(c) Discontinuities with lengths of ¾ in. or over for plate thickness over 2¼ in.

63 (*a*) In ultrasonic examination of steel castings by the angle-beam method, how often must the transfer method be used to recheck the calibration of the ultrasonic instrument?

(*b*) When must both angle-beam and straight-beam methods be used in ultrasonically examining welded joints?

(*c*) When may flat basic calibration blocks be used for calibrating ultrasonic instruments that are to be used in examining circumferential welds on curved contact surfaces?

ANSWER:

(*a*) The transfer method shall be used for calibration checks against the standard at least each ½ hr.

(*b*) If the geometry of the welded joint does not make it possible to make examination by the angle-beam method for both sides of a weld from a single surface or a combination of surfaces, then both the angle-beam and the straight-beam methods shall be used to examine the welded joint.

(*c*) For contact surfaces with curvatures greater than 20-in. diameter, flat basic calibration blocks may be used or blocks of essentially the same curvature as the part to be examined.

64 (*a*) Name the two types of induced electric or magnetic fields that a part under examination may be subjected to when being tested by eddy current.

(*b*) Describe the term *inductor* as applied to *eddy-current* testing.

(*c*) When the *eddy-current* tests are applied, how are the magnitude and direction of the induced eddy currents altered?

ANSWER:

(*a*) The effect of the eddy current on the part being tested may be to induce electromagnetic *currents* or, if the material is magnetic, to induce or set up *magnetic* fields.

(*b*) The inductor is an electromagnetic coil to be operated with alternating current which is positioned close to the article under test and which, when energized, induces eddy currents or sets up magnetic fields or does both in the part under test.

(*c*) The magnitude and direction of the induced eddy currents are altered by discontinuities in the metal being tested.

65 (*a*) How is the change in the induced eddy current noted?

(*b*) How is the change in induced magnetic flux noted when eddy-current testing is used?

(*c*) What frequency of alternating current is used in eddy-current testing?

ANSWER:

(a) The change in the induced eddy current is detected by a detector coil which is connected to electronic circuitry. This circuit registers the discontinuity in several different electrical variables, depending on the design, but these may be voltage, current, or impedance. These then are related in magnitude to the discontinuity through reference blocks.

(b) The distribution of the induced magnetic flux is affected by discontinuities, too, and this is revealed by the change in magnetic-flux distribution as it is affected by the defect.

(c) The frequency chosen to excite the electromagnetic field depends on the depth of penetration beneath the surface of the part being tested. Frequencies vary from 500 to 20,000 Hz (cycles per second). The higher frequencies are used for small-diameter high-wall nonmagnetic tubing.

66 How would you define a leak test as applied to checking welds on a pressure vessel?

ANSWER: Leak tests of welds of pressure vessels are used to determine discontinuities that extend through a weld and thus result in a loss of contents—either liquid or gas. The tests used in leak testing may range from a hydrostatic test using water to sophisticated methods using low-pressure gas and electronic equipment for the detection of the leaking gas through a discontinuity.

67 (a) Who is responsible for drafting a nondestructive examination (NDE) procedure?

(b) What shall it contain?

(c) How is the procedure proved?

(d) To whom shall the procedure be available?

ANSWER:

(a) The manufacturer or assembler per Section V, ASME Codes; for nuclear in-service inspection, the owner of the nuclear facility is responsible for qualifying the NDE personnel.

(b) For nuclear work, all nondestructive examinations shall be performed to a written procedure, which will clearly indicate the method and technique to be used in order to properly inspect the vessel under test. The exception to this rule on power boilers is detailed in article 3 of Section V for radiographic examination in which procedure qualification is not required as long as density and penetrameter image requirements are met. As an example for good radiographic procedures for nuclear vessels, the following must be detailed: (1) material to be radiographed; (2) range of thickness to be radiographed; (3) type of radiation, source, voltage rating, and manufacturer of the radiation source equipment; (4) film brand or type to be used; (5) type and thickness of intensifying screens and filters to be used; and (6) minimum source-to-film distance.

(c) The procedure must be such as to prove the method to be used will show defects not permitted by the Code. In all cases, the procedure must demonstrate this capability to the satisfaction of the authorized Code inspector. This is a good rule to follow in inspection work.

(d) To the nondestructive examination personnel of the manufacturer and the authorized inspector.

Chapter

8

Material Testing, Stresses, and Service Effects

There is much emphasis in design and operation today on whether fitness for service is present. As power equipment becomes older, life prediction methods are employed to try to predict the future life of a boiler or turbine if worn parts are replaced.

Operators, inspectors, and maintenance personnel are also interested in material deterioration, because it can affect the future service of the equipment. The causes of material failure are many: distortion from thinning or overheating, ductile and brittle fractures initiated by cracking, residual stress cracks in welds, cycling fatigue cracks, corrosion causing thinning or pits that act as stress-concentration points, bulging caused by metal overheating, stress-corrosion cracks, and in some cases, hydrogen embrittlement cracks.

Material testing develops the properties that a material may have for boiler application. Many of these tests are required by ASME, Section II specification requirements, and operators, inspectors, and boiler maintenance personnel will come in contact with them in their job functions.

Boilers and Stresses

Boilers and many unfired pressure vessels have inherent dangers that can cause extensive property damage, injuries, and loss of life. These arise from the pressures that are confined in boilers, the high temperature in which they may operate, the cycling imposed in service that can lead to "fatigue" failures, and, last, the wear and tear that occurs on the material components with the passage of time. These are some of the reasons jurisdictions require periodic inspections of boilers.

In addition, modern boiler systems are now programmed to integrate, in most cases automatically, output flows or load with fuel, air, water, condensate, and exhaust products of combustion flows in a very narrow set range of desired values for each for efficiency purposes. If not detected in time on these integrated flow items, any malfunction can cause disruption in output, but also possible damage to the equipment, such as fire-side furnace explosions, tube ruptures, pressure part overheating, with bulging and even explosions, even though the pressure may be within design. See Fig. 8.1.

Boiler failures are preventable; and, since every potential failure involves some kind of stress, a working knowledge of stresses should be had by all concerned with boiler operation, testing, inspection, or maintenance. The cause of stresses in boilers should be understood before an attempt is made to calculate or analyze them.

Internal pressure and imposed stresses.   Internal pressure is the first cause to consider. The U.S. measure of steam pressure is in units of pounds per square inch as read on the pressure gauge; it is above the pressure of the atmosphere (14.7 psi at sea level). This unit measurement is known as *gauge pressure*. In some operation formulas and steam tables, absolute pressure is used. This means the pressure in

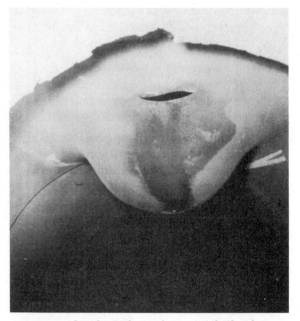

Figure 8.1 Overheated scotch marine boiler furnace sheet is bulged inward by boiler pressure. (*Courtesy Royal Insurance Co.*)

pounds per square inch above a perfect vacuum, that is, the gauge pressure plus approximately 14.7 at sea level.

The foregoing facts relate to unit pressure, that is, the force acting on 1 in.^{2}. If the area of a surface in a steam boiler is 20 in.^{2} and if it is exposed to a steam pressure of 50 psi, the total force or load on that surface will be $20 \times 20 \times 50 = 20,000$ lb. Thus, it is easily seen that the pressure acting on a large exposed area may cause a tremendous load, even with a comparatively low unit pressure. Also, this example makes clear the fallacy in feeling that there is no danger possible with a low-pressure boiler.

Other factors. In addition to pressure acting on the structural components of a boiler, other factors to consider are temperature, cycling, corrosion, creep of the material used in high-temperature service, and a multitude of factors involving the properties of the materials used in the fabrication of the boiler. With the advent of nuclear power, vast amounts of technical progress have been made in systematically analyzing the forces that can act on a pressure vessel's material and the methods or stress analysis that can be used to ensure a safe operating life. This includes the use of fracture mechanics to predict crack growth rates and determining cycles of operation before repairs are needed.

Material Testing—Force and Stress

The properties of a material that are important in analyzing its strength to resist loadings must be first reviewed.

Any body or material subject to external forces on it will resist these external forces. This resistance of the material comes from *within* the material. The internal structure of the material is subject to intercrystalline loading when an external force is applied. Thus, *stress* is defined as the internal force per unit area on the material which resists external forces on the material. It is expressed in pounds per square inch, but the notation *psi* is *not* used for stress as it is in pressure notation. Stress is always expressed by engineers with the notation *lb/in.^{2}* to differentiate it from the psi designation of pressure, which is an *external force* per unit area on the material.

There are several general classifications of stresses that affect materials. A *normal stress* is a *stress* on an area of a material produced by a force at a right angle to the area acted on. Normal stresses are further classified as *tension stresses* or *compression stresses*. In Fig. 8.2*a*, a 1-in.-diameter bar is pulled by a force *F*. This force produces a normal stress at a right angle to the cross-sectional area of the bar. Since the force tends to pull (stretch) the bar apart, it is called a *tensile* stress. Within the material, the intercrystalline struc-

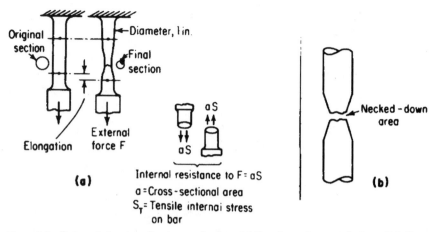

Figure 8.2 Determining tensile stress of a bar. (a) Specimen is stretched until failure occurs; (b) necked-down area after rupture.

ture (assuming it is steel) is also being stressed. The tensile stress of the bar is found by the following equation, based on the definition of stress as pounds per square inch, or internal force per unit area within in the material:

$$F = aS_T$$

where F = external force
a = cross-sectional area of material resisting F
S_T = tensile (internal) stress on the material

From this, the equation for stress S_T is

$$S_T = \frac{F}{a}$$

This equation shows that the tensile stress S_T is found by dividing the external force F, acting normal to the cross-sectional area, by the cross-sectional area of the material resisting the force.

If the force were acting in the opposite direction, a compressive stress (known also as a *bearing stress*) would be imposed on the material. But the compressive stress would be found by the same equation,

$$F = a \times S_c$$

where S_c = compressive stress.

See Fig. 8.2a. A 1-in.-diameter round rod has a force of 44,000 lb pulling it apart in a testing machine when it breaks in two as shown.

What is the ultimate tensile strength per Code? Using the tensile equation above and substituting,

$$F = 44,000 \text{ lb}$$

$$a = \frac{\pi(1)^2}{4} = 0.7854 \text{ in.}^2$$

$$S_T = \frac{44,000}{0.7854} = 56,024 \text{ lb/in.}^2 \text{ ultimate tensile strength}$$

Section II of the ASME Boiler Code lists the *tensile strength* of material by its ultimate failure stress under tension. This stress is used in calculating the strength of a boiler component, but the *allowable stress* is based on dividing the ultimate stress by a safety factor, usually 4 for normal shells, drums, and tubes calculations. Section II lists allowable stresses for various temperatures. The allowable stress will decline at upper temperatures per Code listing.

If a force acts tangent (sideways) to the area of a material, a shear stress is produced. This is illustrated in Fig. 8.3, in which a force F acts on the rivet area tangent to its cross-sectional area, thus producing a shear stress on the rivet. The shear stress is found by the equation $F = a \times s$, but a is the cross-sectional area resisting shear. In Fig. 8.3a, since only one area of the rivet is resisting the external force F, it is in *single shear*.

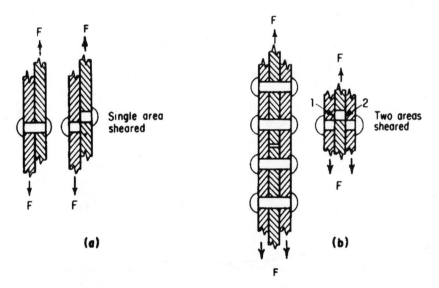

Figure 8.3   (*a*) Rivet in single shear; (*b*) rivet in double shear.

In Fig. 8.3*b,* a butt-riveted joint is shown with butt straps on each side of the butting plates. The rivets in this illustration are in double shear, as *two* areas of the rivet are resisting the load *F.* An example will illustrate the significance of single shear and double shear.

Assume that in Figs. 8.3*a* and 8.3*b* only one rivet is being considered for analysis, but one is in single shear, the other in double shear. Assume the rivet to be 1-in. diameter in each case and the load *F* to be 15,000 lb on each rivet. What is the shear stress for each rivet?

Single-shear rivet   Double-shear rivet

$$S_s = \frac{F}{a} \qquad\qquad S_s = \frac{F}{a}$$

$a = 0.7854$ in.^{2} $a = 1.5708$ in.^{2}

$$S_s = \frac{15,000}{0.7854} \qquad\qquad S_s = \frac{15,000}{1.5708}$$

$= 19,099$ lb/in.^{2} $= 9549.5$ lb/in.^{2}

This shows that a rivet in double shear is *twice* as strong as a rivet in single shear.

Another stress is that due to bending. A beam supported on each end and loaded in the middle will develop a *bending stress.* The beam, when bending, will actually be under a tension stress on one side and a compression stress on the opposite side. This is illustrated in Fig. 8.4. Flat plates and stayed surfaces in boilers are some elements that are subjected to bending stresses.

Note. Stress due to torsion, such as in an axle being rotated and transmitting power, is another stress considered in an analysis of resistance of materials. Stress analysis is beyond the scope of this book. However, a knowledge of *tension, compression, shear,* and *bending* stresses is essential in understanding how pressure is contained in a pressure vessel, pipe, or any other apparatus made of material designed to confine that pressure within safe limits.

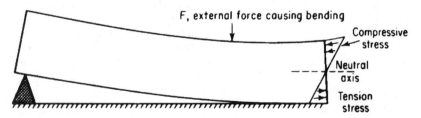

Figure 8.4 Bending causes compressive stress on top of the beam and tension stress on the bottom of the beam.

Strain. When a body or material is subjected to external forces, internal stresses resist these forces, but there is always some deformation with load. For example, a steel rod will stretch when pulled on by an external force. The *total stretch* is expressed in a length measurement such as inches or centimeters. *Strain* is defined as the stretch *per unit length,* or deformation of a body per unit length, and is always expressed as inches per inch, centimeters per centimeter, etc. For example, assume that a steel rod 10 in. long stretches 0.010 in. with load; then the unit strain will be 0.010/10 = 0.001 in./in.

Some of the fundamental properties of all structural materials that must be determined are found by means of *stress-strain diagrams.* From the stress-strain diagram, structural properties determined are *proportional limit, yield point, ultimate strength,* and *modulus of elasticity.*

Modern engineering practice requires testing of materials so as to specify and identify their physical properties. This is particularly true for materials intended to be used in boilers, pressure vessels, and nuclear reactors. Steel manufacturers' laboratories run tests, and the Boiler Code shows sketches and requirements for preparing samples to run tensile tests and bending tests on boiler materials.

Tensile tests. A test specimen of a specified grade of steel is cut out from a rolled stock, then machined to fit a test machine. A test specimen for a tensile test is shown in Fig. 8.5a. The square ends are clamped in jaws of a tension-test machine. The ½-in. round center piece is marked off in a 2-in. gauge length. An extensometer is attached to the 2-in. gauge length. This tension-test machine has a dial gauge indicating the force F applied to pull the rod apart. Incremental loading is applied. At each increment, the force F is recorded and the total amount of length l is taken at that incremental loading. This procedure is followed until rupture occurs. The data are tabulated, and then by the relationship F/a, where a is the original ½-in.-diameter area, stresses are found for each increment of load.

The strain, or stretch, is found by calculating the amount of stretch from the original 2-in. length to obtain the unit strain ε. These values are then plotted as in Fig. 8.5b, showing a stress-strain diagram for ductile steel.

The stress-strain diagram shows a sloping straight line extending from zero upward to the point marked *PL.* The reason is that as the load is increased, the imposed stress S increases, as does the strain ε. The increase for both is in the same ratio, meaning that if the load is doubled, the stress is doubled, and so is the strain. This is why a straight line is drawn, not a curve. The *proportional limit* of a material is thus the maximum unit stress that can be developed in the material without causing a deviation from the law of proportionality

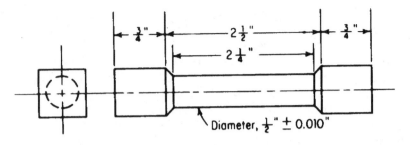

(a)

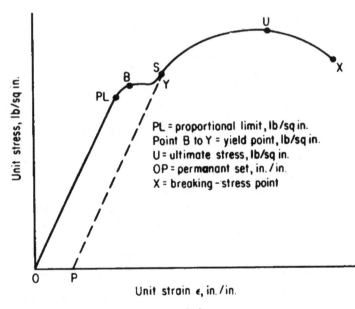

PL = proportional limit, lb/sq in.
Point B to Y = yield point, lb/sq in.
U = ultimate stress, lb/sq in.
OP = permanent set, in./in.
X = breaking-stress point

Unit strain ε, in./in.

(b)

Figure 8.5 (*a*) Specimen is prepared for tensile testing; (*b*) Stress-strain diagram is obtained from tensile test for a ductile material.

of unit stress to unit strain. Of even more significance, it implies that if the load is decreased, the material will return to its original length without having a permanent *set* as a result of the loading. The material will not be permanently deformed, but will return to its original shape as long as the proportional limit is not exceeded.

As the load on our test specimen is increased further, causing a stress greater than the proportional limit, a unit stress is reached at which point the material continues to stretch *without* an increase in load, assuming it is ductile steel. The unit stress at which this stretch-without-load occurs is called the *yield point* and is represented by the short horizontal line B to Y on the stress-strain diagram. The *yield point* of a material is defined as the minimum unit stress in the material at which the material deforms or stretches appreciably without an increase of load.

If a material is stretched or loaded slightly beyond the yield point, a permanent set or deformation occurs in the material. For example, in Fig. 8.5*b*, if the load is reduced to zero after just passing the yield point, the extensometer will show a permanent stretch or deformation. This is found by drawing a line parallel to the proportional-limit line, and the set will be length 0 to P per inch of test specimen.

If the loading on our test specimen is increased, as indicated by the curve S to U, a point of maximum unit stress is reached. Then the unit stress declines with slight additional loading and stretches until it breaks. This is particularly true of ductile material, which *necks down* very rapidly after reaching its maximum unit stress because of reduced area in the neck section, which requires less load to cause rapid stretching to complete breakage. See Fig. 8.2*b*.

The *ultimate strength* of a material is defined as the maximum unit stress that can be developed in the material as determined from the *original* cross section of the material. It is point U in the stress-strain diagram. The curve from U to X is a rapid, unstable testing condition, with point X called the *breaking-load point*. The maximum unit stress is at point U, and this is the ultimate stress designated for the material.

Example Data supplied on a tested specimen are: total length, 19 in.; sections at each end of a specimen are 1 in. $\times$ 1 in. $\times$ 6 in. long; center section $\frac{7}{16}$-in. diameter by 7 in. long with center section concentric with the square sections at each end; center 7-in. section has a 2-in. gauge length marked.

The specimen being tested here broke when the load reached a maximum of 11,274.75 lb, and the break was through the original $\frac{7}{16}$-in. diameter. But the diameter was now $\frac{1}{4}$ in. The cross-sectional area of a $\frac{7}{16}$-in.-diameter bar is 0.15033 in.^{2}. The length of the 2-in. gauge section had stretched to 2.55 in.

1. Find the ultimate strength of the material.
2. What is the ultimate strength in pounds per square inch of the 1 in. $\times$ 1 in. section at each end?
3. What is the percentage elongation of the 2-in. gauge section?

Solution

1. In this problem the break was in the $\frac{7}{16}$-in.-diameter section, so the stress S is

$$S = \frac{F}{a}$$

where a = original cross-sectional area.

$$S = \frac{11,274.75}{0.15033}$$

$$= 75,000\text{-lb/in.}^2 \text{ ultimate stress}$$

2. The ultimate stress at the 1 in.$\times$1 in. section is the same (75,000 lb/in.^{2}) because it is still the same material. It did not break at this section because the cross-sectional area is larger than at the $\frac{7}{16}$-in.-diameter section.

3. The percentage elongation is found as follows:

$$\frac{(\text{Final length} - \text{original length}) \times 100\%}{\text{Original length}} = \% \text{ elongation}$$

$$\frac{(2.55 - 2) \times 100\%}{2} = \frac{55\%}{2} = 27.5\%$$

The *modulus of elasticity,* also known as Hooke's law, states that the unit stress in a material is proportional to the accompanying unit strain, provided that the unit stress does not exceed the proportional limit. In different words, it states that the ratio of stress to strain for a certain material is always a constant, called E, the modulus of elasticity, or in equation form,

$$E = \frac{\text{stress}}{\text{strain}} = \frac{S}{\varepsilon} = \text{constant}$$

For steel, the modulus of elasticity is usually taken as 30,000,000 and written 30×10^6 lb/in.^{2}. This is the modulus of elasticity for normal or axial loads. There is also a shear modulus of elasticity. For steel it is 12,000,000 lb/in.^{2}.

Strain gauges are used to determine stresses at critical areas of boilers, nuclear reactors, and pressure vessels for which exact calculations cannot be made. With the following relationship among stress, strain, and the modulus of elasticity of the material, the stress can be calculated. It is much easier to measure strain, or the deformation of a material under load, than to measure stress.

We explained that stress for normal loads is F/a, where F = imposed load and a = original area of material resisting the load. We also explained that unit strain ε is e/l, where ε = strain in inches per inch, e = amount of strain from original length l, and l = original length.

Now,

$$E = \frac{\text{stress}}{\text{strain}}$$

Substituting the above values gives

$$E = \frac{F/a}{e/l} = \frac{S}{e/l}$$

as

$$S = \frac{F}{a}$$

Rewriting this in terms of stress S,

$$S = \frac{Ee}{l} = E\varepsilon$$

as

$$\frac{e}{l} = \varepsilon$$

Example An abnormal steel section of a boiler was shaped so that exact calculations of the stress imposed on it could not be made. The steel had a modulus of elasticity of 30,000,000 lb/in.^{2}. A length of 8 in. was marked off. From no load to full load, this length increased to 8.009 in. What was the stress at this section of the part?

Solution Use $S = E\varepsilon$ with

$$\varepsilon = 0.009/8 \text{ and } E = 30,000,000 \text{ then}$$

$$S = 0.009/8(30,000,000) = 33,750 \text{ lb/in.}^2 \text{ stress.}$$

It can be seen that if strain is measured, stress can be calculated by knowing the modulus of elasticity of the material, which is usually a constant for the class of material being considered.

The modulus of elasticity is a measure of the *stiffness* of a material. For example, if one material has a modulus of elasticity twice as large as that of another material, the elastic unit strain in the one material for a given unit stress is one-half as large as that in the other material. Thus, one material is considered twice as stiff as the other. Some common E values are steel, 30 million; cast iron, 15 million; aluminum, 12 million; concrete, 3 million.

The *elastic limit* is the maximum unit stress that can be developed in the material without causing a permanent set. Test results show that for most structural metals the elastic limit of the material has about the same value as the proportional limit, and in most technical literature the elastic and proportional limits are considered identical. A small difference is apparent in testing work, but for practical purposes they can be treated as identical quantities.

Stresses on cylindrical shells and pipes. Internal pressure in a cylindrical shell closed at each end tends to burst the vessel along two distinct axes. First, the total pressure acting on the shell tends to cause rupture along a longitudinal axis. The total pressure acting on the heads tends to cause fracture of the shell around its circumference.

Thin-walled cylinders, meaning those where the thickness of the shell does not exceed one-half the inside radius, have two stresses, *longitudinal stress* and *circumferential stress*. The latter sometimes is called the *transverse stress*. Thick-walled cylinders have these stresses also, but they are determined differently. Both stresses are known by these names because of the loading they resist in a cylinder. Both are fundamentally tensile stresses.

Figure 8.6*a* shows a seamless cylinder with an inside diameter *D*, shell thickness *t*, length *L*, and with a uniform pressure *P* acting inside the cylinder. Pressure acts on the cylinder walls, so the resultant force created tends to split the cylinder along its long axis. Thus the first stress to be considered is the longitudinal stress resisting this force tending to split the cylinder along this axis. The pressure

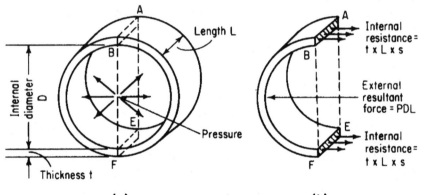

(a)   **(b)**

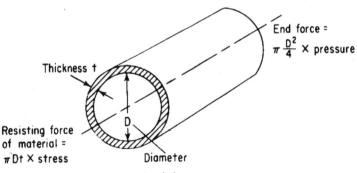

(c)

Figure 8.6 (*a*) and (*b*) Longitudinal force with pressure acting on the side of the cylinder tends to split the cylinder lengthwise. (*c*) Pressure on the end of the cylinder tends to split the cylinder circumferentially.

acts in all directions. But if we cut the cylinder as in Fig. 8.6*b*, which shows the external force on one side and also the internal material stress resisting this external force, the following is developed for a condition of equilibrium to exist.

The force tending to split the cylinder is area times pressure. This is

$$D \times L \times P = \text{force acting on one side}$$

where $D \times L$ = projected effective area. The internal force of the material resisting this force is

$$\text{Stress} \times \text{material area}$$

or

$$S_L \times t \times L \times 2 = \text{resisting force}$$

where $t \times L$ = one area of the material. But since there are two material areas resisting the force, it is multiplied by 2. Equating the two forces gives

$$D \times L \times P = S_L \times t \times L \times 2$$

From this,

$$\text{Longitudinal stress } S_L = \frac{DP}{2t}$$

The force tending to split the cylinder endwise, or around its circumference, is shown in Fig. 8.6*c*. Pressure acting on each end creates a force which is equal to the end area (circle) times the pressure, or

$$\frac{\pi D^2}{4} \times \text{pressure} = \text{end force}$$

The material resists this by a force equal to the end area of the material times the stress, or

$$\pi D t S_c = \text{resisting force}$$

where S_c is circumferential stress. Equating the two forces for equilibrium gives

$$\frac{\pi D^2}{4} P = \pi D t S_c$$

By elimination, solving for S_c, circumferential stress, gives

$$S_c = \frac{DP}{4t}$$

If we compare this with the longitudinal stress, we find the circumferential stress is one-half the longitudinal stress.

The two equations for longitudinal and circumferential stresses are fundamental strength-of-material equations. They are modified somewhat by the Boiler and Pressure Vessel Code to take into account manufacturing and experience factors.

The equations developed are for seamless construction, meaning that no welded, riveted, or ligament joint is present. Later chapters show how the joint efficiency has to be considered to modify these equations. Note that equations for both longitudinal and circumferential stresses (due to pressure) are independent of the length of the vessel. But if a vessel is very long, the bending stress will have to be added to the stress due to pressure. This is especially true of a vessel filled with a substance of considerable weight.

The significance of the circumferential stress being one-half the longitudinal stress in a cylinder enters many problems in boiler design and calculation. For example, on older boilers riveted circumferential joints did not have to be as strong in this direction as they did longitudinally. But in many calculations it is extremely important to check a cylinder both longitudinally and circumferentially, so as to make sure that the strength circumferentially is at least one-half the strength longitudinally. This is brought out in other chapters.

Temperature effects.   Temperature above designed limits has the immediate effect of lowering the permissible stress on a material. For example, SA-30 grade-A quality carbon-plate firebox steel has an allowable stress of 12,000 lb/in.^{2} for temperatures from -20 to 400°F. At 900°F the allowable stress is only 5000 lb/in.^{2}. By assuming the same pressure at both temperatures, it can be seen that a boiler designed for 12,000-lb/in.^{2} normal stress will be weakened to 5000/12,000, or 41.7 percent of its original strength with a temperature increase to 900°F.

Certain parts of boilers, particularly tubes, tube sheets, furnaces in scotch marine boilers, and cast parts in cast-iron boilers are very susceptible to temperature or overheating damages. A large temperature increase in a material, with accompanying *lower permissible stress levels,* is one of the most common causes of boiler damage. Low water, poor circulation, and scale are some causes of overheating of the material beyond safe stress levels. Let us not forget that the firing side of boilers is hot enough to melt steel. And with existing pressure on the water or steam side, it does not require much overheating to cause ruptures, bulges, and other deformation. Thus if the material is stressed well beyond the yield stress at high temperatures, permanent deformation will take place. In severe cases, the ultimate stress of the material is reached at the elevated temperature level, leading to complete rupture of the affected parts of the boiler.

Stress on boiler parts can also be caused by expansion due to temperature rise, and it is pronounced if the parts are restrained because of restrictions or uneven metal thicknesses being joined abruptly with no transition sections. Temperature causes expansion of steel, which can be calculated as follows:

$$e = nl(T_2 - T_1)$$

where e = change in length
l = original length
T_1 = original temperature, °F
T_2 = final temperature, °F
n = coefficient of expansion (change in length per unit of length per degree change in temperature)

Example Steel has a coefficient of thermal expansion of 0.0000065 in. per in. per °F. To show the possible rate of expansion to be considered, assume that a stay in a horizontal-return-tubular (HRT) boiler running from tube sheet to tube sheet is 30-ft long. How much will this rod expand with a temperature change from 70 to 300°F, assuming free expansion?

Substituting into the equation, we get

$$e = 0.0000065 \, (30)(12)(300 - 70)$$

$$= 0.538\text{-in. stretch, which is over } \tfrac{1}{2} \text{ in.}$$

If we assume that the stay rod was fixed at each end and that the tube sheets would *not give*, what compressive stress would be imposed on the rod, neglecting the column effect of a long rod?

This is calculated from the modulus-of-elasticity equation

$$S = E\varepsilon$$

where ε = stretch/inch. So,

$$S = 30{,}000{,}000 \, \frac{0.538}{30 \times 12}$$

$$= 30{,}000{,}000 \, (0.001494)$$

$$= 44{,}820 \text{ lb/in.}^2$$

This example illustrates the importance of considering temperature effects in boiler design and the rapid stress buildup when a part becomes accidentally overheated above design conditions. Remember that the stress developed is not calculated as simply as shown by the illustration. For example, we assumed that the shell and tube sheets would *not* expand because of temperature.

Strain gauges are used to measure the stretch, or "give" in a certain length of a material. By using the modulus of elasticity equation for the material, stress on a component can be calculated by using

$$S = E\varepsilon$$

If a tube leaks at the rolled joint but does not become bowed, it is an indication that the expansion force is greater than the rolled joint's holding power. Rolled joints are equivalent to *press fits,* depending on the friction of contact areas to hold the tubes tight in a tube sheet. The exception, of course, is welded-in tubes, where a shear stress is imposed by expansion.

Stress concentration effects.   If a structural material has an abrupt change in a section, for example, a flat plate containing an opening or a sharp corner as shown in the rod in Fig. 8.7a the stress distribution is not uniform over the cross-sectional area of the material. Near the abrupt change the stress is much higher than calculated. The affected section is said to have a *stress concentration* section, or area, and the ratio by which the normal stress has to be multiplied, K in Fig. 8.7a is called the *stress concentration factor.*

Stress concentration plays an important part in structural members subject to repeated type of loadings, for the stress concentration can lead to cracks and fatigue failures. If the stress concentration is severe enough (even in normal loading), stresses may be induced far above the normal expected stress. Sharp corners in welded joints and other sharply formed shapes must be avoided. Thus openings cut into plates must be reinforced to strengthen the edges around the opening against stress concentration.

The Boiler Code specifies permissible joint connections to avoid stress concentrations. Fillet radii are specified on formed shapes. Openings must be calculated by Code rules. In analytical and design work, stress concentrations are determined by the *photoelastic method, stress-coat method,* and *strain-gauge method* using the electrical resistant wire gauge.

Endurance limit.   In nuclear vessels one must carefully design the elements by the endurance limit and other stress-analysis methods. The *endurance limit* (also known as *fatigue limit*) is the maximum unit stress that can be imposed and repeated on a material through a definite cycle, or range of stress, for an indefinitely large number of times without causing the material to rupture.

How is the endurance limit determined? By testing a material through a complete reversal of stresses. When stressed nearly to its ultimate strength, the specimen will rupture after a few cycles. If a second sample of the same material is again tested but stressed

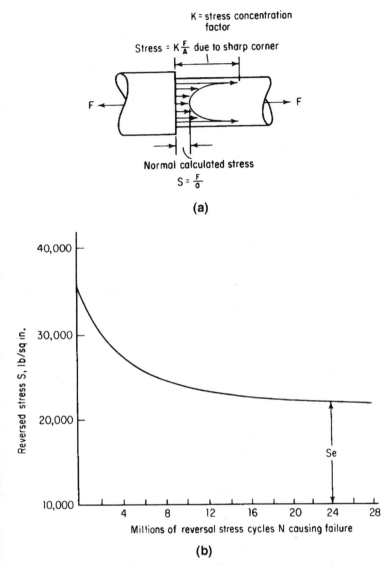

Figure 8.7 (*a*) Sharp corners produce stress concentrations which magnify normal calculated stresses. These must be multiplied by a stress concentration factor to determine the true stress at the abrupt section. (*b*) *S-N* diagrams are used to determine the endurance limit of different materials.

slightly less than before, a larger number of reversals, or cycles, can be imposed. This is continued until a stress value, known as the *endurance limit,* is reached where an almost indefinite cycle of stress can be imposed without causing rupture.

Figure 8.7b shows an S-N diagram, where stress to rupture is plotted on one side and number of cycles to failure on the other. The horizontal line obtained is the endurance stress for the material. In Fig. 8.7b this is 22,500 lb/in.^{2}. Endurance limits are widely used in machine design work, and with the adoption of the Nuclear Vessel Code, Section III, it will receive increasing attention in the Code. Endurance limits of materials can be modified considerably by environment, temperature swings, corrosion effects, hydrogen embrittlement, and similar other conditions which generally involve the study of the fatigue of metals. For example, discontinuities or stress concentration factors such as sharp corners or cracks can lower or modify endurance limits, as can corrosive surroundings on a stressed part.

Crack growth—fracture mechanics. Fatigue crack growth rate has been quantified as a result of intense developmental work in the space and nuclear stress analysis field. If a crack is assumed to exist in a material, under stress this material, if it is steel, will undergo plastic deformation about the crack tip. The crack can grow under plastic deformation as the applied stress or load is increased. The distance that the crack front advances with each cycle of loading is a function of the stress intensity applied to the tip of the crack, which is expressed as the stress intensity factor ΔK. The crack growth rate, in inches per cycle of stress application, is expressed as $d(a)/d(N)$, where a = crack size in inches and N = fatigue life or number of cycles of stress before failure. An equation for crack growth rate used in failure analysis is

$$\frac{d(a)}{d(N)} = C(\Delta K)^m$$

where C = material constant determined by test for class of material (for steels at room temperature, 4×10^{-24} is used)

m = material constant determined from tests (4 is usually used for ferrous material)

See Fig. 8.8a. Crack propagation data can be used to determine the stress intensity ΔK that can be tolerated for a particular design. Also, the number of cycles required to extend an initial crack size a_1, to what is considered a critical crack size, a_2, can be established by using the fatigue crack propagation equation, $d(a)/d(N)$.

Fracture mechanics and NDT methods are being used to determine if a flaw needs immediate repair or whether many more cycles of stress can still be applied until the defect has grown to a size that requires repair or replacement. In the last few decades, there has been a great deal of research activity directed to this failure-prediction method as applied to the basic mechanism of fracture phenomena

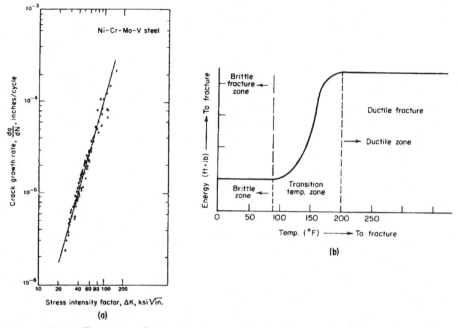

Figure 8.8 (*a*) Fatigue crack growth rate as a function of stress intensity at the tip of the crack for a nickel-chrome-moly steel. (*b*) Transition-temperature zone is important on certain steels to determine brittle and ductile fracture properties with change in temperature.

in solids. The determination of rate of crack growth helps in establishing the number of cycles that can be tolerated before corrective actions will be needed. Nondestructive testing methods are used to periodically check a developing defect in conjunction with fracture-mechanics methods.

Prevention of fatigue failures relies on containment of crack growth within what are considered safe limits. Where in-service inspections are possible, a safe service interval is established between inspections. Where in-service inspections are not feasible, a more conservative design is needed so that any anticipated defect will not grow to a dangerous size during the expected life of the component.

Stress corrosion. The simultaneous action of a part being subjected to repetitive stresses and some form of attack from the medium in which the part must operate can cause a part to fail as a result of *corrosion fatigue.* The combined effect of these two factors is much greater than the effect of either one alone. The cracking usually begins at surface defects, pits, or irregularities. These points act as stress-concentration points with the corroding medium intensifying the pitting action. The more repetitive the stress, the faster will be the pitting action, which again will increase the stress concentration. A cumulative effect can materialize that will cause a failure well below a predicted endurance stress.

Pits formed under simultaneous stress and corrosion are always sharper and deeper than pits formed in the same time under stress-less conditions. The more repetitive the stress, the faster will be the pitting action. At low-cycle repetitive stress, pitting will proceed entirely by normal corrosion, and the section will fail by the normal tension failure or when the resisting area is thinned down so the stress rises proportionately, assuming a constant load.

Stress corrosion, then, is primarily caused by repetitive stresses of high frequency in a corroding medium, leading to pitting caused by stress and corrosion. The pitting then develops high-stress concentration points which lead to fatigue failure.

No experimental data are available for determining the extent to which the endurance limit is reduced for most materials in combination with corroding solutions or media. But for boilers and pressure vessels, the obvious precaution to take against stress corrosion is to make sure the water or medium being confined is free of any corrosive tendencies. This is determined by analysis of the water at regular intervals by personnel experienced in water analysis. This also points out the value of periodic internal inspections of pressure-containing parts and examination of surfaces for evidence of pitting and corrosion.

Charpy V-notch tests. *Brittle materials* are those which are comparatively weak in tension. Brittleness is a characteristic of a material that is opposite to ductility and toughness. A material is generally considered brittle if its elongation at rupture from a tension test is less than 5 percent in a test specimen of 2 in. Brittle materials will fail with very little give or stretching before failure because of their lack of ductility. Brittle materials have low toughness or low resistance to impact loading, sometimes called lack of resiliency. As a result, impact testing is used as a measure of brittleness or toughness (described as the ability to absorb energy). When a material has a notch and is brittle, failure can be unexpected and well below calculated allowable stresses.

Charpy V-notch tests are used to measure resistance to impact loading or brittleness. The test uses a pendulum-type apparatus to strike a specimen, usually notched, and the foot-pounds of energy needed to cause a fracture are correlated to whether the material is considered brittle or not. A complete description of notched-bar impact testing can be secured from studying the E-23 standard as developed by the American Society for Testing and Materials (ASTM).

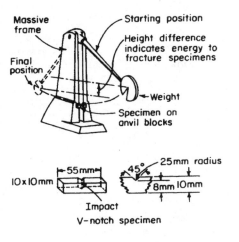

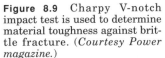

Figure 8.9 Charpy V-notch impact test is used to determine material toughness against brittle fracture. (*Courtesy Power magazine.*)

Figure 8.9 shows how a Charpy V-notch impact test is made. A V notch is cut out per standards on one face of a small specimen. When the heavy pendulumlike weight strikes the specimen's face opposite the notch, the foot-pounds needed to cause fracture are calculated by weight-lever arm relationships. Carbon steels with cold-weather resistance to impact failure are usually killed steels, or a steel that was deoxidized in manufacture so that no evolution of gas occurred during pouring and solidification. Nickel steels of 3½ or 9 percent are considered superior impact-resistance steels, as are austenitic stainless steels.

Nil-ductility temperature.   The effect of temperature altering or changing the brittle nature of a material was dramatically displayed in World War II when merchant ships cracked unexpectedly. It was noted that this occurred in cold waters, but not in warm waters. It has been established since these failures that materials may exhibit ductile behavior in a normal environment but act in a brittle manner if the environment, such as temperature, is changed. Stress corrosion cracking can also be of a brittle nature as a result of the combined action of stress, susceptible material, and a corrosive surrounding.

Low-carbon and low-alloy steels frequently exhibit a transition zone from ductile behavior to brittle failure over a small temperature range called the *nil-ductility temperature*. (See Fig. 8.8b.) Any flaws in the material will aggravate the tendency of a material with a nil-ductility characteristic to act brittle below the transition temperature.

HAZ.   Welding can also cause changes in a material that may make it act brittle in the weld metal or heat-affected zone (HAZ). It is thus important to determine and consider nil-ductility transition temperatures of material to be used in boilers and pressure vessels in order to make sure this temperature is not experienced during any service temperatures that the boiler or pressure vessel may be exposed to, even perhaps under possible test or operating abnormalities. One such possibility is *hydrostatic testing* with water that will be below the nil-ductility temperature. The other is testing when the surrounding temperature may be too low, as may be the case in winter. Failures due to brittleness from not considering the effect of temperature transitions on the property of the material have occurred during hydrostatic tests of pressure equipment and others such as water towers.

Other effects on materials. *Creep* is defined as the slow deformation of material with time, with the deformation taking place at elevated temperatures with no increase in stress. Modern material testing includes the determination of how much creep or permanent deformation can be expected in what length of time so that design can reflect these factors in life prediction for the pressure vessel. Creep properties are obtained by imposing a constant tensile load on a specimen at a constant temperature and then measuring the strain at intervals of time. Data are plotted to obtain a variety of constant-stress creep-strain curves or constant-time creep-stress curves. In using creep data, the designer tries to establish expected service life and the corresponding amount of creep deformation that can be tolerated. With these established, a stress can be selected that will satisfy these conditions.

Hydrogen embrittlement can occur under high-pressure and high-temperature operations of boilers and pressure vessels. Hydrogen attack on steels can result in severe loss of ductility with cracks developing that can lead to unexpected brittle fracture. The effect is most severe on high-strength steels. The hydrogen under high pressure and temperature diffuses into the metal as atomic hydrogen. It recombines in the metal as molecular hydrogen in grain boundaries and causes high pressure in these "voids," resulting in bulging and blistering. Hydrogen also combines with carbon in steel under high pressure and temperature and decarbonizes the steel, with embrittlement taking place. Frequently chromium-molybdenum steels are used where resistance to hydrogen embrittlement is required.

Nuclear pressure vessels may be subjected to *neutron irradiation* of steel. It is necessary to avoid impurities in the steel as well as large grain sizes in order to limit any vacancy or voids in the material. It is theorized that the radioactivity present in nuclear vessels promotes the formation of helium which, with time, collects at metal grain boundaries to weaken it by a decrease in ductility. Quality control is thus more stringent on nuclear vessels, so cavities may be eliminated.

Allowable Stresses and Safety Factor

Allowable stresses are used to design structures or machine components. The allowable stress is also sometimes called the *allowable working stress*. It is the maximum stress considered to be safe when the material is subjected to resisting loads assumed to be applied in service. In boiler applications, the term *allowable pressure* is often used. Actually, the allowable pressure is determined by applying the forces acting on a material and then calculating the allowable pressure from the allowable stress on the material.

The allowable stress is determined from the ultimate strength of the material, which is divided by a safety factor. The safety factor used in modern boilers is 4. However, certain elements of older boilers, such as riveted joints had to be designed with a safety factor of 5. Other parts had to be designed with a safety factor as high as 12.5, such as the rivets holding lugs on brackets on an HRT boiler to be suspended from a beam.

In boiler usage, the factor of safety is the ultimate strength divided by the allowable loadings, or the ultimate stress S_u divided by the allowable stress Sa. In equation form,

$$\text{Safety factor} = \frac{S_u}{Sa}$$

Another method of expressing the safety factor is by dividing the bursting pressure by the allowable pressure. This method is used on state inspection reports and on ASME data reports. In equation form, it is

$$\text{Safety factor} = \frac{\text{bursting pressure}}{\text{allowable pressure}}$$

In boiler design and usage and other critical structures where life and property may be at stake, there is a definite need for selecting working stresses and loadings considerably less than the ultimate, or the yield stress, for these reasons:

1. There is always some uncertainty in materials being used, how they were made, how they were assembled, and how they were joined or fabricated with other materials.

2. There is always some uncertainty as to the exact loading that a structure, or part of it, may have to resist and how it is abused in operation.

3. Calculations of all stresses possible in a fabricated structure are never that exact when one considers the variables to be encountered in service through the years.

Certain brittle materials such as cast iron do not have a *definite* yield point. Thus early Codes used the ultimate stress because it was much easier to work from ultimate stress and apply a safety factor to this to obtain an allowable stress. In some countries of Europe, the *yield point* is the basis of design. The Nuclear Vessel Code uses the yield stress or endurance stress or both as a basis of design. With increasing technological changes, the present Boiler Code may also be changed on this matter in time.

On existing installations, the question of safety factor and allowable pressure arises quite often. For example, can a boiler that was originally designed with a safety factor of 5 be operated with a safety factor of 4, since the latest Power Boiler Code allows a safety factor of 4? Usually, the original safety factor of 5 remains. The safety factor of 4 was drawn up principally for seamless-steel or welded boilers that met stiffer quality-control and inspection requirements on welding. Older boilers may not meet this requirement; thus the original safety factor of 5 should govern the allowable pressure.

If a boiler is stamped for an allowable pressure of 275 psi and the safety valve is set at 150 psi, what is the safety factor? Assume a Code-welded boiler meeting latest Code requirements and an original design safety factor of 4. The bursting pressure of this boiler would be $4 \times 275 = 1100$ psi. So,

$$\text{Safety factor} = \frac{\text{bursting pressure}}{\text{allowable pressure}} = \frac{1100}{150} = 7.33$$

This question brings up an important consideration as to which pressure to use in calculating safety factors on existing boiler installations. For example, in this question, should the safety-valve setting be used or the stamped allowable pressure? If the safety-valve setting is *below* the stamped allowable working pressure, use the safety-valve *setting*. If by chance the safety-valve or valves are set *higher* than the allowable pressure (above Code limits), then a dangerous condition exists because safety valves must always be set *at* or *below* the allowable, or working, pressure stamped on the boiler.

The safety-valve setting is used where the stamped allowable pressure is above the safety-valve setting in calculating the safety factor because the allowable pressure on the boiler is the *safety-valve setting*. The boiler is *not* supposed to operate above this safety-valve setting. If an increase in pressure is needed, other items such as valve and connection ratings will have to be checked before the pressure can be raised to the maximum allowable pressure stamped for the boiler. Then, a new safety valve will be required. Also, Code specifications on valve ratings and water-column connections will have to be checked to see if they meet the Code requirements for the new pressure.

The question of a safety factor continuing on an older boiler must be considered quite often, especially where the expense of repair or alteration must be evaluated in comparison to new equipment. The safety factor continuing on a boiler depends on the type of boiler, the condition of the boiler, and even the state in which it is located. Assuming that internal and external inspections are satisfactory, the National Board regulations state:

1. Lap-riveted longitudinal-joint boilers operating over 50 psi can be operated at this pressure for 20 yr. After that, 50 psi or less is permissible; but if the boiler is relocated, only low-pressure service is permitted.

2. For boilers of butt construction, at the end of 25 yr and every 5 yr thereafter, the safety factor must be increased by 0.5 unless a hydrostatic test of 1½ times the allowable pressure is imposed; if this test is satisfactory, no increase in the safety factor is necessary.

The best rule to follow on any safety-factor changes is to check with an insurance company boiler inspector or legal jurisdiction inspector. *Never* assume that the boiler can be operated at a higher pressure by just changing safety valves, even if the boiler is stamped for the higher pressure. There are other requirements to be met on feedwater, blowdown, water-column connections, low-water fuel cutoff, and service-valve ratings that must be considered. In determining the allowable pressure on a boiler, stress calculations must be made as shown in next chapters.

Questions and Answers

1 How would you define the ultimate tensile strength of a material mathematically if S = ultimate tensile strength, in pounds per square inch, F = maximum load when failure occurs, a_f = final cross-sectional area of a round specimen at the time of failure, and a_i = initial cross-sectional area of a round specimen?

ANSWER:

$$S = \frac{F}{a_i}$$

The initial area is used in material testing for the ultimate tensile strength.

2 Define ductility mathematically, given the following:

Percentage reduction in area = A
Initial area of specimen = a_i
Fractured area of specimen = a_f

ANSWER:

$$A = \frac{a_i - a_f}{a_i} \times 100\%$$

3 Differentiate between low-cycle and high-cycle fatigue material failure.

ANSWER: Low-cycle fatigue is a fatigue-type failure that occurs after a relatively low number of repeated stresses; below 10,000 cycles is the usual accepted range. High-cycle fatigue failure requires longer cycles of repetitive stress, usually well above 100,000 cycles of stress.

4 What is stress?

ANSWER: Stress is the resistance to an external force that a material provides within its crystalline structure; it is expressed as pounds per square inch of the material's cross-sectional area.

5 What is the test most commonly used for measuring a metal's strength, elasticity, and ductility?

ANSWER: The tension test is the mechanical test usually applied to determine these properties of a material.

6 What is strain as applied to material testing?

ANSWER: Strain is a measure of deformation of a material under load, and it is expressed as the increase in length per inch of original length.

7 What is the proportionality constant called that exists between stress and elastic strain?

ANSWER: This constant for a class of material is called the modulus of elasticity E.

8 How would you define ultimate tensile strength?

ANSWER: The ultimate tensile strength is the stress which a material experiences in a testing machine under maximum load during the test. The ultimate stress is obtained by dividing this maximum load by the original cross-sectional area of the material.

9 What is notch sensitivity?

ANSWER: Notch sensitivity or notch toughness of a metal is its resistance to the start and propagation of a crack at the base of a standardized notch. Notch sensitivity is measured by the amount of energy absorbed in foot-pounds by a specimen as it fractures under impact of a hammer blow that is delivered by a standard weighted pendulum. Brittle materials will absorb little energy in fracture.

10 What is meant by fatigue?

ANSWER: Fatigue is the tendency of a metal to fracture under conditions of repeated cyclic-type stresses that are considerably below the ultimate strength of the material.

11 How is ductility defined?

ANSWER: The ductility of a metal is the amount of permanent deformation or strain that it can undergo before fracture occurs. Two methods are used to define or measure ductility: the percentage of elongation and the reduction in area of a specimen tested under tension. Welds are tested for ductility by a bend test. This consists of bending the sample by a specific amount about a plunger of given radius. The increase in distance between gauge marks on the tension side of the specimen is noted, and the change is expressed as the percentage of elongation.

12 What are brittle materials?

ANSWER: These are materials that can be deformed very little without rupture taking place. This is usually characterized by a sudden, shattering type of failure. Cast iron, concrete, brick, and glass are examples of brittle materials.

13 What are resilient materials?

ANSWER: These are materials that can absorb large amounts of energy without experiencing permanent deformation. To express it another way, they return to their original shape after the load is reduced. Materials with capability of low modulus of elasticity and high elastic limit would produce high resilience.

14 Define tough materials.

ANSWER: Tough materials can absorb large amounts of energy before rupturing. This quality is related to a material's high strength, high ductility, or flexibility. Toughness is a useful measure of the ability of a material to absorb shock loading or sudden blows without rupturing.

15 Name some environmental impurities that can enter a boiler's steam or water system, which could produce stress corrosion cracks on a high-strength turbine steel.

ANSWER: The following have been identified as possibly causing stress corrosion cracks on turbine blades: (1) use of coordinated phosphate treatment with improper control, which can produce free caustic that can attack metals; (2) sodium getting into the boiler water from a leaking condenser tube; and (3) improperly treated water used in attempering sprays for the control of superheat or reheat.

16 What causes thermal stress?

ANSWER: Thermal stress is caused when a material is not free to expand or contract because of temperature changes to which a material may be subjected in service. The resultant stress that is developed can be found by the equation

$$s = Ea(t_2 - t_1)$$

where E = modulus of elasticity of material
a = coefficient of expansion for material
t_1, t_2 = initial and final temperatures

17 What can the effect of creep be on a structural metal?

ANSWER: Creep causes planes of slow movement in a material's crystalline structure that are also a function of time. This slow movement or slip can cause sufficient deformation to cause a sudden fracture even when the applied stress is much lower than that which normally could produce a failure under normal loadings.

18 The term *concentration factor* defines what ratio?

ANSWER: The stress concentration factor is the ratio of the actual stress existing on a given plane of a member under loading to the calculated stress that is needed to resist that loading without taking into account the discontinuity which causes normal stresses to be magnified by the stress concentration factor.

19 How is Poisson's ratio described?

ANSWER: Below the proportional limit, a material under load will stretch in tension lengthwise and be thinned at right angles to the length. The ratio of the unit strain at right angles to the stress to the unit strain in the direction of the stress is called *Poisson's ratio,* expressed as follows:

$$u = \frac{e_t}{e_s}$$

where e_t = unit strain in transverse direction
e_s = unit strain in direction of stress

20 A ½-in. round bar is stretched 0.00195 in. in a length of 2 in. and decreased 0.000162 in. in diameter when a load in tension of 2000 lb is imposed. What is Poisson's ratio?

ANSWER:

$$\text{Unit strain in direction of stress} = \frac{0.00195}{2} = 0.000975$$

$$\text{Unit strain in transverse direction} = \frac{0.00016}{0.5} = 0.00032$$

$$\text{Poisson's ratio} = \frac{0.00032}{0.000975} = 0.328$$

21 Calculate the tensile strength of the specimen in the above question.

ANSWER:

$$s = \frac{P}{a} = \frac{2000}{\pi(0.25)^2} = 10{,}191 \text{ lb/in.}^2$$

22 A specimen has a modulus of elasticity of 6.5 million. What is the maximum load axially that may be applied on a ½-in. rod without stretching the rod $\frac{1}{16}$ -in. in a length of 6 ft?

ANSWER: Use

$$E = \frac{F/a}{e/l} \qquad \text{and solve for } F$$

where F = load

a = rod area = $(\frac{1}{4})^2\pi = \pi/16$

$e = \frac{1}{16}$

$l = 6 \times 12 = 72$ in.

$F = 6{,}500{,}000 \times 0.0625/72 \times 16/3.14$

$\quad = 28{,}751$ lb.

23 How much will the diameter of a steel rod's 2-in. diameter change if an axial load of 60,000 is imposed and Poisson's ratio is 0.30?

ANSWER:

$$\text{Axial stress} = \frac{60{,}000}{\pi(1)^2} = 19{,}091$$

Use $E = 30$ million for steel.

$$E = \frac{\text{stress}}{\text{strain}}$$

$$\text{Strain axially} = \frac{19{,}091}{30{,}000{,}000} = 0.000636 \text{ in./in.}$$

$$\text{Transverse strain} = 0.3 \times 0.000636 \text{ in./in.} = 0.0001808$$

$$\text{Change in 2-in. diameter} = 2 \times 0.0001808 = 0.0003616 \text{ in.}$$

24 How is the endurance limit of a material determined?

ANSWER: The endurance limit of a material is determined by applying to several test specimens repeated loads, which cause complete reversed stresses of known values, and recording the number of stress reversals each specimen endures before it fails or breaks. Each specimen is subjected to a lower stress than the preceding one and also breaks after a large number of cycles of stress. Finally, a stress is reached that does not cause failure regardless of how many reversed cycles are applied. This stress is called the endurance limit for the material under test.

25 What loading besides internal pressure is required to be evaluated by the ASME high-pressure Boiler Code?

ANSWER: The loadings that must be considered include the effect of the weight of water and stresses imposed on a boiler during a hydrostatic test, as well as any additional loading that will increase the average stress above the working pressure and induced stresses by more than 10 percent of the allowable working stress. These loads can include wind, snow, and earthquake loads where specified as in the nuclear Codes. It also includes the reaction of supporting lugs, rings, saddles, and similar types of support.

26 Who must prepare stress reports for nuclear pressure vessels?

ANSWER: The manufacturer of the pressure vessel is responsible for the preparation of the stress report, and it must be reviewed by the owner of the plant and in most cases by an inspection agency.

27 What broad outline is stipulated for making a stress report?

ANSWER: The stress report must be prepared and signed by professional engineers experienced in pressure-vessel design. Generally it has three sections: thermal analysis, structural analysis, and fatigue evaluation. Of interest is the fact that areas of severest stress condition at any transient must be listed in the report, including the values of the stresses.

28 What is the purpose for the reduced-section tensile test, the free-bend test, the root-bend test, the face-bend test, and the side-bend test?

ANSWER: These tests are required to qualify welding procedures and welders per ASME Code requirements:

(a) A reduced-section tensile test is a qualifying procedure. When the section is broken in tension, it shall have a tensile strength at least that of the minimum of the range of the plate which is welded, and elongation or stretch shall be 20 percent minimum in 2 in.

(b) A free-bend test is a qualifying procedure. It consists of bending a specimen cold; the outside fibers of weld shall elongate at least 20 percent before failure occurs.

(c) A root-bend test is for qualifying welders. It consists of bending a specimen against the bottom of the weld.

(d) A face-bend test is for qualifying welders. It consists of bending a specimen against the surface of the weld.

(e) A side-bend test is for qualifying welders. It consists of bending a specimen against the side of the weld.

29 How is corrosion fatigue defined?

ANSWER: This is a failure produced by the combined action of repetitive stresses on a material and a corrosive environment but at lower stress levels or fewer cycles than would be required to produce a fatigue failure without the presence of a corrosive environment.

30 Name three hardness tests applied to metals that are welded.

ANSWER: Brinell, Rockwell, and Vickers.

31   What type of fracture normally is analyzed by the notched-bar impact test?

ANSWER: Notched-bar impact tests are used to analyze ferritic steels for possibilities of failure by *brittle fracture.*

32   Name the destructive test normally carried out on ferritic steels in order to determine the nil-ductility transition temperature in order to determine the low-temperature ductility properties of the ferritic steel.

ANSWER: The drop-weight test is used to determine the *fracture resistance* of ¾-in. and over thick steel at *different temperatures* with the steel having a notch that can start a fracture.

33   What is a Charpy V-notch test?

ANSWER: This test is to determine the toughness of material that has a notch (or defect) for resisting shock impact loads or failure by brittle fracture. The test determines the energy required to break a standard notched specimen that is supported at two ends, and this is compared to established standards. The ASME Code requires this test for many boiler materials as specified in Section II.

34   How is stress-corrosion cracking defined?

ANSWER: The cracking is caused under sustained tensile load with accompanying action of a corrodent that causes a failure well below the normal stress on the material.

Code Strength, Stress, and Allowable Pressure Calculations

Chapter 8 reviewed some basic strength-of-material principles and the basic stresses that a material under load must resist. This chapter is devoted to ASME Code strength calculations as applied to high-pressure boiler components. Equations expressing the strengths of shells, stayed surfaces, and like components of low-pressure boilers and pressure vessels are similarly calculated, but specific equations are available in the appropriate section of the ASME Code. The boiler-strength calculations demonstrate methods on how to do this. It must be recognized that new materials, better manufacturing quality control, and increased knowledge of how materials behave under load will continue to cause Code changes in the equations and allowable stresses. Therefore, the latest ASME-published Code should be consulted for specific design details.

Applicable Code sections. In order to obtain the latest equations and rules for calculations, the following sections of the Code were used as reference material:

Section I: Rules for the Construction of Power Boilers

Section II: Materials, Part A, C, and D

ASME Power Piping Code: B31.1

The National Board Inspection Code

Component analysis. Any boiler and its parts confining pressure must be analyzed per component by carefully considering the strength of the material being used, its physical characteristics as to type and grade, allowable stress, thickness, etc. The forces acting on this material must then be analyzed. This force is usually created by pressure, but may also include temperature, the weight it is supporting, and stress concentration, such as around an opening. The problem then evolves to comparing the forces acting on the material and determining whether the material is being stressed beyond the allowable stresses governed by the Boiler Code rules. Elements to be considered depend on the type of boiler but will generally include shells or drums, tubes, tube sheets, heads, flat surfaces, stays, stay bolts, openings, furnaces, welded joints, (rivets in the past), structural supports, and connected piping and valves. Each of these is governed by Boiler Code rules as to allowable material, allowable stresses, and method of calculating forces to obtain the allowable pressure. Finally in boiler and pressure-vessel application, the weakest element producing the lowest pressure then determines the *allowable pressure* for the boiler.

Existing installations.   There is an old rule that the allowable pressure of a boiler is based on the Code requirements prevailing when the boiler was built. With regard to calculating allowable pressure in this chapter, it is the aim of the author to show existing installation equations used up to 1986, and the *new construction* Code requirements since the changes that were made by the Code of 1986. This will assist the reader to determine allowable pressures for the vintage boiler question that may arise.

Revisions in calculation, based on the 1986 changes, include:

1. Longitudinal welded joints must have backing strips removed, and the weld reinforcement removed substantially flush with the plate. Prior Codes permitted these to stay as welded, but the welding efficiency allowed was only 90 percent.

2. Calculations on firetube components such as tubes and furnaces now require the use of *external pressure charts* and new equations for determining allowable pressure.

3. Riveted boiler construction is still permitted, but the present Code makes reference to the 1971 edition of Section I for further details on Code requirements for riveted boilers. (This indicates that riveted construction is basically obsolete.)

Gauge pressure is used in most Code-allowable pressure calculations, because this is the pressure noted on gauges in operation. Gauge pressure is the pressure above atmospheric pressure, usually expressed as 14.7 psi at sea level. Absolute pressure is the *total* pressure above atmosphere, i.e., 100 psi gauge is equal to 114.7 psi absolute pressure.

There is also a difference between allowable pressure and operating pressure. Boilers are always operated below the maximum allowable pressure in order to avoid the safety valve opening since one or more safety valves must be set at the maximum allowable pressure. Experience indicates that the following differential between operating and allowable pressure will avoid the safety valve lifting:

Maximum allowable pressure	Operating pressure differential
To 300 psi	10%, but not less than 10 psi
Over 300 to 1000	7%, but not less than 35 psi
Over 1000 to 2000	5%, but not less than 80 psi
Over 2000	Manufacturer's design recommendation

Code external boiler piping. ASME Code certification or inspection and approval of the Authorized Inspector is required on the boiler proper, and the Code-defined external piping. This includes data forms and Code symbol stamp stamping for these components. The external piping is defined in the Preamble to Section I and generally includes all piping from the boiler proper connections to the valve(s) required by the Code on that piping, such as steam outlet, feedwater, blowoff, drains, vents, surface blowoff, water column, gauge glass, pressure gauge, and the recirculation return line for a high-temperature water boiler.

This classified external piping is designed as respects material, fabrication, installation, inspection, and allowable pressures by the requirements as detailed in ASME B31.1, titled Power Piping Code, but may also have reference material in Section I. Other piping within the boiler proper, or for tubes over 5 in OD, requires the use of Section I equations for cylindrical drums or shells.

The Power Piping Code provides details on acceptable pipe joints; however, *threaded joints* on piping cannot be used where severe corrosion, crevice corrosion, shock or vibration may occur, nor at temperatures over 925°F. The maximum nominal threaded pipe size is 3 in., and the maximum permissible pressure per nominal threaded pipe size for steam and hot water or other fluid with temperatures above 220°F is

Maximum nominal threaded pipe size, in.	Maximum pressure, psi
3	400
2	600
1	1200
¾ and smaller	1500

Figure 9.1 provides data on American Standard Association steel pipe. A B31.1 pipe equation will be illustrated on how to determine the allowable pressure on a Code-defined external pipe.

Figure 9.1 American Standard Association steel pipe data.

Nominal Size	External Diam. In.	Standard Weight Pipe A.S.A. Schedule 40											Extra Strong A.S.A. Schedule 80		Double Extra Strong	
		Internal Diam. In.	Wall Thickness In.	Weight per Ft Plain Ends Lb	Threads per In.	Circumference In. External	Circumference In. Internal	Transverse Area Sq. In. External	Transverse Area Sq. In. Internal	Length of Pipe per Sq Ft External Surface	Length of Pipe per Sq Ft Internal Surface	Wall Thickness In.	Weight per Ft Plain Ends Lb	Wall Thickness In.	Weight per Ft Plain Ends Lb
1/8	0.405	0.269	0.068	0.244	27	1.272	0.845	0.129	0.057	9.431	14.199	0.095	0.31	......	......
1/4	0.540	0.364	0.088	0.424	18	1.696	1.144	0.229	0.104	7.073	10.493	0.119	0.54	......	......
3/8	0.675	0.493	0.091	0.567	18	2.121	1.549	0.358	0.191	5.658	7.748	0.126	0.74	......	......
1/2	0.840	0.622	0.109	0.850	14	2.639	1.954	0.554	0.304	4.547	6.141	0.147	1.09	0.294	1.71
3/4	1.050	0.824	0.113	1.130	14	3.299	2.589	0.866	0.533	3.637	4.635	0.154	1.47	0.308	2.44
1	1.315	1.049	0.133	1.678	11 1/2	4.131	3.296	1.358	0.864	2.904	3.641	0.179	2.17	0.358	3.66
1 1/4	1.660	1.380	0.140	2.272	11 1/2	5.215	4.335	2.164	1.495	2.301	2.768	0.191	3.00	0.382	5.21
1 1/2	1.900	1.610	0.145	2.717	11 1/2	5.969	5.058	2.835	2.036	2.010	2.372	0.200	3.63	0.400	6.41
2	2.375	2.067	0.154	3.652	11 1/2	7.461	6.494	4.430	3.355	1.608	1.847	0.218	5.02	0.436	9.03
2 1/2	2.875	2.469	0.203	5.793	8	9.032	7.757	6.492	4.788	1.328	1.547	0.276	7.66	0.552	13.70
3	3.500	3.068	0.216	7.575	8	10.996	9.638	9.621	7.393	1.091	1.245	0.300	10.25	0.600	18.58
3 1/2	4.000	3.548	0.226	9.109	8	12.566	11.146	12.566	9.886	0.954	1.076	0.318	12.51	0.636	22.85
4	4.500	4.026	0.237	10.790	8	14.137	12.648	15.904	12.730	0.848	0.948	0.337	14.98	0.674	27.54
5	5.563	5.047	0.258	14.617	8	17.477	15.856	24.306	20.006	0.686	0.756	0.375	20.78	0.750	38.55
6	6.625	6.065	0.280	18.974	8	20.813	19.054	34.472	28.891	0.576	0.629	0.432	28.57	0.864	53.16
8	8.625	7.981	0.322	28.554	8	27.096	25.073	58.426	50.027	0.443	0.478	0.500	43.39	0.875	72.42
10	10.750	10.020	0.365	40.483	8	33.772	31.479	90.763	78.855	0.355	0.381	......	......	......	......
12	12.750	12.000	0.375	49.562	8	40.055	37.699	127.676	113.097	0.299	0.318	......	......	......	......

Pipe sizes 14 in. and above are designated by outside diameter, and wall thickness is specified.

Problem   A steam pipe from the boiler proper to the required two stop valves with plain ends to be welded to the 1200 psi boiler connections is a nominal 6 in., schedule 80, size, and, as Fig. 9.1 shows, has a thickness of 0.432 in. The pipe material is A210C per B31.1 listing. Operating temperature is not to exceed 700°F. What is the allowable pressure if a $\frac{1}{16}$-in. corrosion allowance is to be provided?

solution   The Piping Code requires the allowable stress for the listed material to be not greater than the expected steam temperature; therefore, from the stress tables in B31.1, the allowable stress for A210C material is 16,600 lb/in.^{2} The equation to be used per B31.1 is

$$P = \frac{2S(t_m - A)}{D_o - 2y(t_m - A)}$$

where P = design or allowable internal pressure, psi
  t_m = minimum required wall thickness, in.
  t_m = 0.432 − 0.0625 = 0.3695 in.
  D_o = outside diameter of pipe, in.
  D_o = 6.625 in. from Fig. 9.1
  S = maximum allowable stress for material, lb/in.^{2}
  S = 16,600 lb/in.^{2}
  y = a Code factor for ferritic or austenitic steel based on temperature of design
  y = 0.4
  A = additional thickness for erosion or superimposed loads from supports
  A = 0.050 as estimated

Substituting in the above equation,

$$P = \frac{2(16,600)(0.3695 - 0.050)}{6.625 - 2(0.4)(0.3695 - 0.050)}$$

$$P = 1665 \text{ psi}$$

Therefore, the piping is thick enough for the expected 1200 psi operating pressure.

Boiler Tubes

Three common methods of boiler-tube fabrication are used. (1) The seamless tube is pierced hot and drawn to size. (2) The lap-welded (forge-welded) tube consists of metal strip ("skelp") curved to tubular shape with the longitudinal edges overlapping. Heat is applied and the joint forge-welded. (3) The electric-resistance butt-welded tube is formed like the second type, but as its name implies, the joint is butt-welded.

It is considered good practice by some to place the weld on welded tubes away from the radiant heat of the fire. Tubes for bent-tube-type boilers are bent usually by machine.

The diameter of boiler tubes always refers to nominal outside diameter while pipe diameter refers to nominal inside diameter.

Tube ends: expanding, flaring, beading.   Practically all boiler tubes have the ends expanded into the tube hole of the shell or drum. This is to make the tube tight against leakage and to give it a firm grip on the tube hole so that the tube may have a definite holding or staying effect.

The edges of the tube holes are chamfered about ¹⁄₁₆ in. after the holes are drilled so that there will be no sharp edges to cut into the tube when it is expanded.

Tube holes are finished ¹⁄₃₂ in. larger in diameter than the outside diameter of the boiler tube, except in the tube sheet of firetube boilers. Through this, tubes must be drawn during retubing, and therefore its holes are finished ¹⁄₁₆ in. larger in diameter so as to permit a tube that is coated with scale to be removed without damage to the tube sheet.

Thick drums may be counterbored in order to have a reasonably narrow circumferential strip of tube to expand. The diameter of the counterbore should be sufficient to allow for flaring the tube end according to requirements.

The counterbore may be from either the inside or the outside. When a drum of a watertube boiler has tubes expanded in its upper side, it is best practice not to use the outside counterbore, for pockets for soot would thus be formed.

For watertube boilers the tubes and nipples should extend through the tube hole ¼ to ¾ in. and be flared to at least ⅛ in. larger than the tube-hole diameter.

Firetube boilers have the tube ends exposed to heat and products of combustion, and therefore the tube ends might soon be burned off if they were flared. In these boilers the tube ends are driven back into a bead after expanding the tubes, in order to protect them against overheating, although the bead does not increase the holding power of the tube appreciably.

AWP for tubes in watertube boilers.   To calculate the allowable pressure for tubes in watertube boilers up to 5-in. OD, the following Code equation is used:

$$P = S\left[\frac{2t - 0.01D - 2e}{D - (t - 0.005D - e)}\right]$$

or

$$t = \frac{PD}{2S + P} + 0.005D + e$$

where P = maximum allowable pressure, psi
D = outside diameter of tubes, in.
t = minimum required thickness, in.
S = maximum allowable stress, lb/in.^{2}
e = thickness factor for expanded tube ends

Note. For selecting the S value of *tubes,* the operating temperature of the metal shall be not less than the maximum expected mean wall temperature (the sum of the outside and inside surface temperatures divided by 2) of the tube. This in no case shall be taken as less than 700°F for tubes absorbing heat. For tubes which do not absorb heat, the wall temperature may be taken as the temperature of the fluid within the tube, but not less than the saturation temperature.

Note. Over a length at least equal to the length of the seat plus 1 in., e is equal to 0.04 for tubes expanded into tube seats. However, e = 0 for tubes expanded into tube seats, provided that the thickness of the tube ends over a length of the seat plus 1 in. is *not* less than the following:

0.095 in. for tubes 1¼-in. OD and smaller

0.105 in. for tubes above 1¼-in. OD and up to 2-in. OD

0.120 in. for tubes above 2-in. OD and up to 3-in. OD

0.135 in. for tubes above 3-in. OD and up to 4-in. OD

0.150 in. for tubes above 4-in. OD and up to 5-in. OD

For tubes strength-welded to headers and drums, e = 0.

Figure 9.2 shows a typical allowable stress table for different tube material at different temperatures. Section II, Part D, of the ASME Code lists the allowable stresses under SA numbers, not to be confused with A numbers listed in the Power Piping Code. Figure 9.3 lists the typical allowable stresses for steel plate material, now listed in Section II, Part D, allowable stress tables. Note that tubes absorbing heat and shells and drums are designed by Section I equations and rules.

Example of a tube problem for a watertube boiler   A seamless tube in a watertube boiler is made of SA-210 C material, is 2¼-in. OD, 0.188-in. thick, operating where it absorbs heat at 650°F. The tube has been expanded into the drum. What is the allowable pressure for this tube?

solution   Use the P equation with

S = a value for 700°F from Fig. 9.2: 16,600 lb/in.^{2}
t = tube thickness = 0.188 in.
D = outside tube diameter = 2.25 in.
e = 0 (because thickness of tube is greater than 0.120 in.)

Spec no.	Composition	Form	Ultimate	Allowable stress		
				−20 to 650°F	700°F	800°F
A Carbon steel—tubes						
SA 192	C-Si	Seamless	47.0	18.8	11.5	9.0
SA 178A	C	Welded	47.0	11.8	11.5	7.7
SA 226	C-Si	Welded	47.0	11.8	11.5	7.7
SA 210 A-1	C	Seamless	60.0	15.0	14.4	10.8
SA 178 C	C	Welded	60.0	15.0	14.4	9.2
SA 210 C	C-Mn	Seamless	70.0	17.5	16.6	12.0
B Low-alloy steel—tubes						
SA 209 T1b	C-½ Mo	Seamless	53.0	13.3	13.2	13.1
SA 250 T1b	C-½ Mo	Welded	53.0	11.3	11.2	11.1
SA 250 T1	C-½ Mo	Welded	55.0	11.7	11.7	11.7
SA 209 T1	C-½ Mo	Seamless	55.0	13.8	13.8	13.7
SA 213 T2	½ Cr-½ Mo	Seamless	60.0	15.0	15.0	14.4
SA 423-1	¾ Cr-½ Ni-Cu	Seamless	60.0	15.0	15.0	—
SA-213-T12	1 Cr-Mo	Seamless	60.0	15.0	15.0	14.8
SA-213-T11	1¼ Cr-½ Mo-Si	Seamless	60.0	15.0	15.0	15.0
SA-213-T3b	2 Cr-½ Mo	Seamless	60.0	15.0	15.0	14.7
SA-213-T22	2¼ Cr-1 Mo	Seamless	60.0	15.0	15.0	15.0

				−20 to 100	300	500	700	800
SA 213-T21	3 Cr-1 Mo	Seamless	60.0	15.0	15.0	15.0	14.8	14.5
SA 213-T5	5 Cr-½ Mo	Seamless	60.0	15.0	15.0	15.0	13.4	12.8
SA 213-T7	7 Cr-½ Mo	Seamless	60.0	15.0	15.0	14.5	13.4	12.5
SA 213-T9	9 Cr-Mo	Seamless	60.0	15.0	15.0	14.5	13.4	12.8
C High-alloy steel—tubes								
SA 268-TP405	12 Cr-1Al	Seamless	60.0	15.0	13.3	12.9	12.1	—
SA 268-TP446	27 Cr	Seamless	70.0	17.5	15.6	14.5	14.1	—
SA 213-TP304	18 Cr-8 Ni	Seamless	75.0	18.8	16.6	15.9	15.9	15.2
SA 213-TP316	16 Cr-12 Ni & 2MO	Seamless	75.0	18.8	18.4	18.0	16.3	15.9
SA 213-TP321	18 Cr-10 Ni & Ti	Seamless	75.0	18.8	17.3	17.1	15.8	15.5
SA 213-TP347	18 Cr-10 Ni & Cb	Seamless	75.0	18.8	15.5	14.9	14.7	14.7

Figure 9.2 Typical allowable stresses for tube material per ASME Code, Section II, Part D. (*Courtesy American Society of Mechanical Engineers*)

Substituting in the tube equation,

$$P = 16,600 \left[\frac{2(0.188) - 0.01(2.25)}{2.25 - (0.188 - 0.005(2.25)} \right]$$

$$= 16,600 \left[\frac{0.3535}{2.2323} \right]$$

$$= 2628 \text{ psi}$$

Example What thickness is required on 2-in. OD tubes, made of SA-178A material and which are located in a heat-absorbing area of a 600 psi water-tube boiler? The tubes were expanded into the drums.

solution Use the equation

$$t = \frac{PD}{2S + P} + 0.005D + e$$

where $P = 600$ psi
$\quad\quad S = 11,500$ lb/in.^{2} from Fig. 9.2 at 700°F
$\quad\quad D = 2$ in.

Spec no.	Designation	Ultimate tensile strength, kips/in.^{2}	Representative allowable stresses not exceeding metal temperature, °F	
			−20 to 650	800
A Carbon steels—plate				
SA 285A	Carbon, C	45.0	11.3	8.3
SA 285B	Carbon, C	50.0	12.5	9.0
SA 285C	Carbon, C	55.0	13.8	10.2
SA 442-Gr55	C-MN-Si	55.0	13.8	10.2
SA 515 Gr55	C-Si	55.0	13.8	10.2
SA 516 Gr55	C-Si	55.0	13.8	10.2
SA 442 Gr60	C-MN-Si	60.0	15.0	10.8
SA 516 Gr60	C-Si	60.0	15.0	10.8
SA 515 Gr60	C-Si	60.0	15.0	10.8
SA 515 Gr65	C-Si	65.0	16.3	11.4
SA 516 Gr65	C-MN-Si	65.0	16.3	11.4
SA 515 Gr70	C-Si	70.0	17.5	12.0
SA 516 Gr70	C-MN-Si	70.0	17.5	12.0
SA 299	C-MN-Si	75.0	18.8	12.0
B Low-alloy steels—plate				
SA 204A	C-½ Mo	65.0	16.3	16.2
SA 204B	C-½ Mo	70.0	17.5	17.5
SA 204C	C-½ Mo	75.0	18.8	18.8
SA 302A	MN-½ Mo	75.0	18.8	17.7
SA 302B	MN-½ Mo	80.0	20.0	18.8
SA 302C	MN-½ Mo-½ Ni	80.0	20.0	18.8
SA 302D	MN-½ Mo-¾ Ni	80.0	20.0	18.8
SA 225A	MN-V	70.0	17.5	14.8
SA 225B	MN-V	75.0	18.8	12.0
SA 202A	½ Cr-1¼ MN-Si	75.0	18.8	12.0
SA 202B	½ Cr-1¼ MN-Si	85.0	21.3	12.0
SA 203A & D	2½ Ni & 3½ Ni	65.0	16.3	11.4
SA 203B & E	2½ Ni & 3½ Ni	70.0	17.5	12.0
SA 387 2Cl.1	½ Cr-½ Mo	55.0	13.8	13.5
SA 387 12Cl.1	1 Cr-½ Mo	55.0	13.8	13.8
SA 387 11Cl.1	1¼Cr-½ Mo-Si	60.0	15.0	14.8
SA 387 22Cl.1	2¼Cr-1 Mo	60.0	15.0	15.0
SA 387 21Cl.1	3 Cr-1 Mo	60.0	15.0	13.9
SA 387-5	5 Cr-½ Mo	60.0	15.0	12.8
C High-alloy steels—plate				

			−20 to 100	300	500	700
SA-240-405	12 Cr-1Al	60.0	15.0	13.3	12.9	12.1
SA-240-304	18 Cr-8Ni	75.0	18.8	16.6	15.9	15.9
SA-240-316	16 Cr-12Ni-2 Mo	75.0	18.8	18.4	18.0	16.3
SA-240-321	18 Cr-10Ni-Ti	75.0	18.8	17.3	17.1	15.8
SA-240-347	18 Cr-10Ni-Cb	75.0	18.8	15.5	14.9	14.7

Figure 9.3 Typical allowable stresses for plate material per ASME Boiler Code, Section II, Part D. (*Courtesy American Society of Mechanical Engineers*)

$e = 0$ (assumed; if necessary add 0.04 if tube is below Code-specified thickness for expanded tubes)

Substituting,

$$t = \frac{600(2)}{2(11500) + 600} + 0.005(2) + 0$$

$$= 0.061 \text{ in.}$$

Therefore the tube thickness required is

$$t = 0.061 + 0.04 = 0.101 \text{ in.}$$

AWP for Tubes in Firetube Boilers

Old method—pre-1986. For firetube boilers using the common SA-83 and SA-178A tube material,

$$P = 14,000\left(\frac{t - 0.065}{D}\right)$$

where P = maximum allowable pressure, psi
$\quad$ t = minimum required thickness, in.
$\quad$ D = outside diameter of tube, in.

For firetube boilers using copper tubes of SB-75 specification,

$$P = 12,000\left(\frac{t - 0.039}{D}\right) - 250$$

Example What pressure is allowed on a steel tube of SA-192 material, expanded into tube seats, 3 in. in diameter, with wall thickness 0.115 in., to be used in a firetube boiler?

$$P = \frac{14,000(0.115 - 0.065)}{3.0} = 233 \text{ psi}$$

New Code method. It is necessary to use the external pressure charts in Section II, Part D of the ASME Boiler Code. There are two factors to be obtained. Factor A is obtained from a geometric chart as shown in Fig. 9.4. With Factor A known, Factor B is obtained from another chart, which is based on the material used with yield strength as a criteria for each chart. Figure 9.5 is for carbon and low-alloy steel with a minimum yield strength of 24,000 psi but under 30,000 psi. The above problem is solved by the new method as follows.

The new method requires using external pressure as the force acting on the tube for a certain length, which for firetube boilers is tube sheet to tube sheet. For this example, assume this distance is 10 ft. The first step per the new ASME method is to obtain the following ratios for the same tube as given in the previous problem:

$$\frac{\text{Outside diameter}}{\text{thickness}} = \frac{D_o}{t} = \frac{3}{0.115} = 26$$

$$\frac{\text{length}}{\text{Outside diameter}} = \frac{L}{D_o} = \frac{10 \times 12}{3} = 40$$

See Fig. 9.4. Use above values and move vertically on the L/D_o line to 40; then move horizontally to the D_o/t value of 26. Read down to obtain the factor $A = 0.0012$.

See Fig. 9.5. This is the external pressure chart to use for a SA-178A tube material. Using the A value determined previously, move vertically to the minimum 700°F temperature line; then move horizontally right to obtain the B factor of 7400.

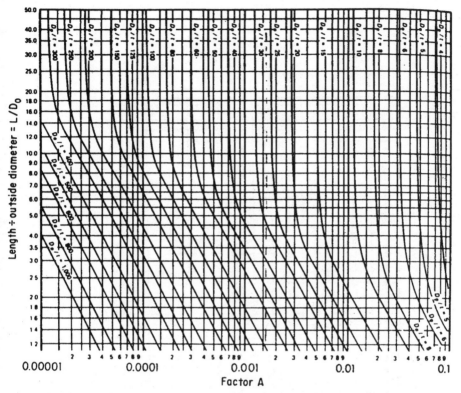

Figure 9.4 ASME Section II, Part D chart for determining Factor A in calculating allowable external pressure for firetube boiler tubes, flues, and furnaces. (*Courtesy American Society of Mechanical Engineers.*)

The Code provides the following equation for this case:

$$\text{Allowable pressure} = P = \frac{4B}{3 \times D_o/t}$$

Substituting,

$$P = \frac{4 \times 7400}{3 \times 26} = 379.5 \text{ psi}$$

As can be noted, the external pressure chart method considers tube length, which the older method did not consider.

Shells and Drums

To calculate the allowable pressure on shells or drums and Code listed pipes and tubes over 5-in. diameter, two factors must be taken into consideration:

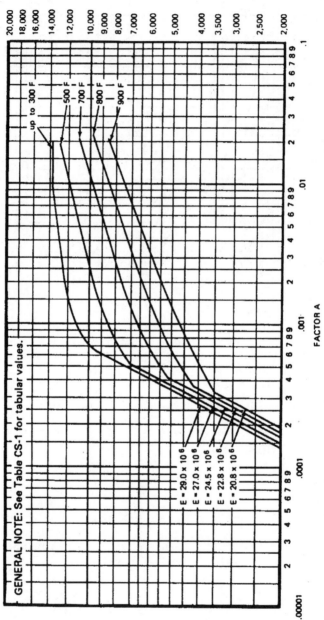

Figure 9.5 ASME Section II, Part D chart for determining Factor B for firetube boiler tubes, flues, and furnaces under external pressure, and constructed of carbon or low-alloy steels with specified minimum *yield strength* of 24,000 psi, but under 30,000 psi. (*Courtesy American Society of Mechanical Engineers.*)

1. Longitudinal-weld joint efficiency. If all weld reinforcement is removed flush with the plate, a 100 percent efficiency can be used, which has been required by the Code since 1986. Otherwise, a joint efficiency of 90 percent was used. If the boiler was of riveted construction, the appropriate riveted joint efficiency was used. See 1971 ASME Code. If the shell is seamless, a 100 percent efficiency is used. See Fig. 9.6a.

2. Since watertube boilers are arranged so that the drums have tube holes in them, the weakening effect is calculated in terms of a ligament efficiency, and this efficiency is used with the corresponding shell thickness to obtain an allowable working pressure for the shell or drum by the tube-hole area.

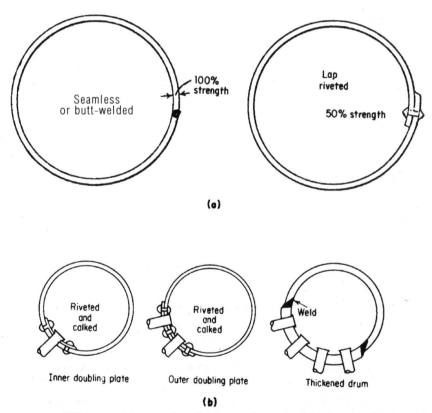

Figure 9.6 Efficiency of drum and shell joints is determined by the construction. (a) Seamless and butt-welded joint has 100 percent efficiency compared to an old riveted lap joint. (b) Ligament sections of drums or shells were made thicker to counteract the weakening caused by the tube holes. Older riveted boilers used doubling plates as shown. Welded drums have a thin and thick section of the drum welded together.

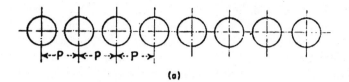

(a)

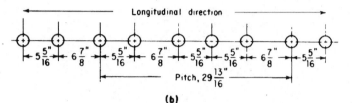

(b)

Figure 9.7 Pitch determination in calculating ligament efficiency. (a) Tube holes spaced equally with same hole diameters; (b) unequal tube-hole spacing, same hole diameters, and pitch to be used in calculating ligament efficiency.

The efficiency of a ligament is found as follows: Where the pitch between the tube holes is equal (Fig. 9.7a):

$$\frac{p - d}{p}$$

where p = pitch, or longitudinal distances between holes, in.
d = tube-hole diameter, in.

If the pitch of the tube holes is unequal (Fig. 9.7b), a unit longitudinal length in which all the unequal pitches are included should be selected. Then,

$$\frac{p - nd}{p} = \text{efficiency of the ligament}$$

where p = length selected to include all variations in pitches, in.
d = tube-hole diameter, in.
n = number of tube holes in longitudinal line in selected length

The ASME Code, Section I, covers methods of finding efficiency of *diagonal* ligaments by means of charts.

The efficiency of the usual tube ligament is quite low, usually 35 to 50 percent. In a past practice with riveted boilers often the ligament was strengthened by riveting a reinforcing strap or doubling plate over the tube-hole section. The tube holes are cut through the entire section; thus, although the efficiency of the ligament is not increased because the tube spacing remains unchanged, the thickness is increased. See Fig. 9.6*b*.

As Fig. 9.7*b* illustrates, another method of increasing the strength of the ligament is to make the drum in two longitudinal halves with two longitudinal welded seams and the side with the tube holes made thicker. The edges of unequal thickness usually are machined on the thicker section by a 3 to 1 taper to match the thinner plate thickness prior to welding. This is done to prevent a stress concentration at the joint.

Calculating Allowable Pressure

1. For old riveted boilers, see 1971 Code. The following equation was used for calculating the allowable pressure for shells or drums:

$$P = \frac{0.8SEt}{R + 0.6t} \quad \text{or} \quad t = \frac{PR}{0.8SE - 0.6P}$$

2. For welded or seamless construction, with the weld reinforcement removed substantially flush and the backing strip removed, use the equation from Section I:

$$P = \frac{SE(t - C)}{R + (1 - y)(y - C)} \quad \text{or} \quad t = \frac{PR}{SE - (1 - y)P} + C$$

where P = maximum allowable pressure, psi

   S = maximum allowable stress for operating temperature of metal, lb/in.^{2}

   t = minimum required thickness, in.

   R = inside radius of cylinder, in.

   E = efficiency of joint (E = efficiency of longitudinal welded joints or of ligaments between openings, whichever is lower. $E = 1.00$ for seamless cylinders, and $E = 1.00$ for welded joints, provided all weld reinforcement on the longitudinal joints is removed substantially flush with the surface of the plate. For older welded boilers, $E = 0.9$ if joint was not ground smooth. E = efficiency for riveted joints, E = efficiency for ligaments between openings.)

   C = factor depending on whether a pipe or shell is threaded (for usual drum or shell $c = 0$; see Code for specifics)

   y = factor which depends on whether steel is ferritic or austenitic as follows:

	y value	
Temperature, °F	Ferritic steel	Austenitic steel
Below 900	0.4	0.4
950	0.5	0.4
1000	0.7	0.4
1050	0.7	0.4
1100	0.7	0.5
1150	0.7	0.7

Example 1: Old riveted boiler problem   A riveted boiler is 66 in. in diameter and 16 ft long. The shell plate is $\frac{7}{16}$-in. thick, the allowable stress is 13,750 lb/in.^{2}, and the allowable pressure is 125 psi. What is the least permissible circumferential efficiency of the girth joint?

solution Using the formula

$$P = \frac{0.8SET}{R + 0.6t}$$

$$125 = \frac{0.8(13,750)(E)(0.4375)}{33 + 0.6(0.4375)}$$

$$E = \frac{125(33.26)}{11,000(0.4375)} = 86.6\%$$

$$\text{Circumferential } E = \frac{86.6}{2} = 43.3\%$$

The circumferential strength of the shell must be always at least half that of the longitudinal joint. This was illustrated in Chap. 8.

Example 2: Unequal shell thickness because of ligaments problem
What is the maximum allowable working pressure on a welded drum that has 1.469-in.-thick shell plate and 2.406-in.-thick tube sheet? The outside diameter of the shell plate is 57.75 in. The outside diameter of the tube sheet is 58.688 in. Material is SA-515-70, and the metal temperature is not more than 650°F. Tube ligament efficiency is 0.429 and welded-joint efficiency is 100 percent. Allowable stress for this ferritic steel is 17,500. Calculations for allowable pressure must be made twice as follows.

solution Based on welded joint and shell-plate thickness, and inside radius of shell or tube sheet:

$$P = \frac{SE(t - C)}{R + (1 - y)(t - C)}$$

$$R = \frac{57.75}{2} - 1.469 = 27.406$$

$$y = 0.4 \quad C = 0$$

So

$$P = \frac{17,500(1)(1.469)}{27.406 + 0.6(1.469)} = 908.8 \text{ psi}$$

Based on the tube-sheet thickness and ligament efficiency:

$$R = \frac{58.688}{2} - 2.406 = 26.938$$

The same equation is used:

$$P = \frac{17,500(0.429)(2.406)}{26.938 + 0.6(2.406)} = 636.1\text{-psi allowable pressure}$$

The lowest pressure governs, AWP is 636.1 psi.

Example 3: Ligament efficiency and shell allowable pressure problem
The mud drum of a watertube boiler has a tube sheet $\frac{5}{8}$-in. thick with a 43.250-in. OD, and contains $3\frac{9}{32}$-in. diameter tube holes pitched horizontally $5\frac{5}{16}$ in. in banks of three (Fig. 9.7b) and two tubes with $6\frac{7}{8}$ in. between banks. The shell plate is $\frac{1}{2}$-in. thick. The tube sheet and shell are welded per Code requirements. The material for both is SA-285C with maximum temperatures below 650°F. What is the maximum allowable pressure for this drum if the shell's OD is 43.125 in.?

solution Use the equation

$$P = \frac{SE(t - C)}{R + (1 - y)(t - C)}$$

with

$$S = 13,800 \text{ from Fig. 9.3}$$

$$E = 1 \text{ or calculated ligament efficiency}$$

$$C = 0$$

$$y = 0.4$$

Based on the shell:

$$R = \frac{43.125}{2} - 0.5$$

$$= 21.0625 \text{ in.}$$

$$t = 0.5 \text{ in.}$$

Substituting,

$$P = \frac{13,800(0.5 - 0)}{21.0625 + (1 - 0.4)(0.5 - 0)}$$

$$= 323 \text{ psi}$$

Based on the tube sheet (it is necessary to calculate the ligament efficiency first):

$$E = \frac{p - nd}{p}$$

$$= \frac{29.8125 - 5(3.28125)}{29.8125}$$

$$= 0.448$$

$$R = \frac{43.250}{2} - 0.625$$

$$= 21.00 \text{ in.}$$

Substituting as previously in the shell equation,

$$P = \frac{13,800(0.48)(0.625 - 0)}{21.00 + (1 - 0.4)(0.625 - 0)}$$

$$= 194 \text{ psi, and this is the allowable pressure}$$

Thick shells. Thick shells are defined by Section I as those where the thickness of the shell exceeds one-half the inside radius; under those conditions, the following equation is used to calculate the allowable pressure for the shell or header.

$$P = SE\left[\frac{Z - 1}{Z + 1}\right]$$

where S = allowable stress from Section II, Part D
  P = allowable maximum pressure
  E = efficiency of longitudinal joint or ligament efficiency

$$Z = \frac{(SE + P)}{(SE - P)}$$

or

$$= \left[\frac{(R + t)}{R}\right]^2$$

or

$$= \left[\frac{R_o}{R}\right]^2$$

where t = minimum thickness of shell plate, in.
  R = inside radius of weakest course of shell, in.
  R = outside radius of weakest course of shell, in.

Example of thick shell allowable pressure calculation   A cylindrical header on a watertube boiler is made of SA-204B material operating under 650°F. The OD is 12 in., and it is 3-in. thick. Calculate the allowable pressure.

solution

$$S = 17,500 \text{ lb/in.}^2$$

$$R = \frac{12 - 6}{2} = 3 \text{ in. Thickness exceeds one-half the inside radius}$$

$$E = 1 \text{ (welded joint)}$$

$$Z = \left[\frac{6}{3}\right]^2 = 4$$

Substituting in the equation,

$$P = 17,500(1)\left[\frac{4 - 1}{4 + 1}\right]$$

$$= 10,500 \text{ psi allowable pressure}$$

Dished heads

The ends of drums of most conventional-type watertube boilers are closed with a dished-out head. The shorter the radius of the curvature, that is, the nearer the dish approaches a semispherical shape of reduced radius, the greater will be the resistance to internal pressure. Conversely, the ASME Code specifies that the radius shall not be greater than the diameter of the shell or drum to which the head is attached—otherwise, the head requires bracing.

There are four types of blank unstayed heads permitted by the Power Code: *segment of a sphere, semiellipsoidal, hemispherical,* and *flatheads.* The first three are bumped heads (Fig. 9.8*a*). Some Code flatheads and methods of attachment are shown in Fig. 9.8*b*. Bumped heads are flanged, with a corner radius on the concave side of the head of not less than 3 times the head thickness and in no case less than 6 percent of the diameter of the shell for which the heads are to be attached.

Flanged-in manhole openings in dished, or bumped, heads must be flanged to a depth of not less than 3 times the required thickness of the head for plate of up to 1½-in. thickness. If thicker, the depth of the flange must be the thickness of the plate plus 3 in. The minimum width of the bearing surface for a gasket on a manhole must be ¹¹⁄₁₆ in., and the gasket thickness when compressed must be less than ¼ in.

Dished heads are calculated for allowable pressure by the following methods, depending on what type of head is involved and whether it has a manhole. For *segments of a sphere head without a manhole,* use the formula

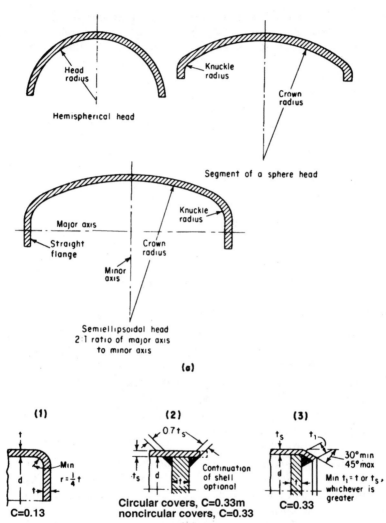

Figure 9.8 Code dished heads and flatheads. (*a*) Three types of dished heads: hemispherical, segment of a sphere, and semiellipsoidal; (*b*) Code-acceptable methods of flathead attachments determines the *C* factor in the flathead equation.

$$t = \frac{5PL}{4.8S}$$

where S = maximum allowable stress, lb/in.^{2} (Section II, Part D)
t = thickness of head, in.
P = maximum allowed pressure, psi
L = radius on concave side to which head is dished, in.

For *semiellipsoidal heads* with pressure on the concave side (without a manhole), use the shell formula for cylinders, but assume that the shell is seamless (no joint efficiency). For *hemispherical heads* with no manhole and pressure on the concave side, use the following formula:

$$t = \frac{PL}{1.6S} \qquad (9.1)$$

or

$$t = \frac{PL}{2S - 0.2P} \qquad (9.2)$$

where t = required thickness, in.
P = maximum allowable pressure, psi
S = maximum allowable stress, lb/in.^{2} (Section II, Part D)
L = radius to which head was formed, in.

Equation (9.2) may be used for heads over ½-in. thick to be used with shells or drums over 5-in. diameter and that are integrally formed on seamless drums or are attached per Code fusion welding with no staying required.

If the required thickness of the head in the above equation exceeds 35.6 percent of the inside radius, the following formula must be used for head thickness:

$$t = L(Y^{1/3} - 1)$$

where

$$Y = \frac{2(S + P)}{2S - P}$$

Dished heads with manholes. When any of the heads—segment of a sphere, semiellipsoidal, or hemispherical—has a flanged-in manhole or an access opening that exceeds 6 in. in any dimension, it is computed on the following basis:

1. By the formula for a segment of a sphere head.

2. The thickness of the head must be increased by 15 percent, but in no case less than ⅛ in. after the thickness is obtained by the formula.

3. If the radius to which a head is dished is less than 80 percent of the diameter of the shell, the thickness of the head with a flanged-in manhole opening must be found (or calculated) by making the dish radius equal to 80 percent of the diameter of the shell.

Example dished head calculations A semiellipsoidal 42-in.-diameter dished head has a flanged-in manhole, is ⅞-in. thick, has a crown radius of 84 in. with pressure on the concave side. The material is SA-285C. What is the allowable pressure for this head?

solution Because of the flanged-in manhole, the segment of a sphere equation must be used to calculate the allowable pressure. Transposing this equation for P,

$$P = \frac{4.8St}{5L}$$

with

$$S = 13,800 \text{ lb/in.}^2 \quad \text{(from Section II, Part D)}$$

$t = 0.875 - 0.131 = 0.744$ in.   [thickness must be reduced first by $0.15 \times$ (0.875) or 0.125, whichever is greater]

For semiellipsoidal heads, the Code requires using a dish radius equal to 0.8 times diameter of shell, or

$$L = 0.8(42) = 33.6 \text{ in.}$$

Substituting,

$$P = \frac{4.8(13,800)(0.744)}{5(33.6)}$$

$$P = 293 \text{ psi}$$

Example A hemispherical head of a 54-in.-ID drum is welded to the shell of a watertube boiler with pressure on the concave side. Welding meets Code requirement. Head material is SA-387 Grade 2 with Code allowable stress of 13,800 lb/in.^{2} per Section II, Part D. The boiler has been operating at 750 psi at 600°F superheat. What should the head thickness be without a manhole and with a manhole?

solution *Without a manhole:*

$$t = \frac{PL}{2S - 0.2P} \quad \text{with}$$

$$P = 750 \text{ psi}$$

$$S = 13,800 \text{ lb/in.}^2$$

$$L = 27 \text{ in.}$$

Substituting,

$$t = \frac{750(27)}{2(13,800) - 0.2(750)}$$

$$t = 0.738 \text{ in.}$$

With manhole:

Required $t = \dfrac{5PL}{4.8S}$ plus $0.15t$ or 0.125 in. (whichever is greater)

Substituting as before,

$$t = \frac{5(750)(27)}{4.8(13,800)} = 1.53 \text{ in.}$$

$$\text{Required } t = 1.53 + 0.15(1.53)$$

$$\text{Required } t = 1.76 \text{ in.}$$

Calculating dish radius with bump and chord data.   See Fig. 9.9a. Refer to triangle ABO:

$$R = \text{hypotenuse of right triangle}$$

$$R^2 = \left(\frac{C}{2}\right)^2 + (R - b)^2$$

Clearing and multiplying out $(R - b)^2$

$$2Rb = \frac{C^2}{4} + b^2$$

$$R = \frac{C^2}{8b} + \frac{b}{2}$$

Example   Chord $C = 38$ in. on a head; bump $b = 4$ in. So

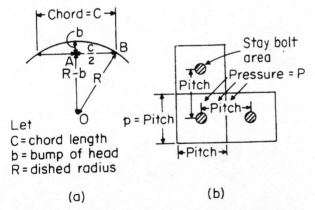

(a)   (b)

Figure 9.9 The dish radius can be calculated by measuring chord length C and bump b. (b) A stay bolt must resist the force created by the pressure acting on the area encompassed by the pitch dimension.

$$R = \frac{(38)^2}{8(4)} + \frac{4}{2} = 47.1 \text{ in.}$$

Flat heads

The ASME Code has several equations for flat heads, depending on whether the head is round, rectangular, or square. For a typical round head, the equation used is

$$t = d \sqrt{\frac{CP}{S}}$$

where t = minimum required thickness, in.
   d = diameter, measured as indicated in Code
   C = a factor, depending on method of attachment (Fig. 9.8b)
   S = maximum allowable stress value, lb/in.^{2} (Section II, Part D)
   P = maximum allowable pressure, psi
   t_r = thickness required for shell to which head is attached, in.
   t_s = actual thickness of shell to which head is attached, in.
   m = ratio of t_r / t_s

(See PG-31 of Section I and Fig. PG-31 for attachments allowed by the Code.)

Example   An unstayed flat head is attached to a shell as shown in Fig. 9.8b(2) with welding and heat treatment meeting Code requirements. The flat head is circular with a 16-in. diameter and is 1½-in. thick. The material is SA-285C, the operating temperature is 560°F. The seamless shell to which the head is attached is ⅜-in. thick, and calculations indicate that only a ⁵⁄₁₆-in. thickness for the shell is required. What is the allowable pressure for the flat head?

solution   The flat head equation must be used with S = 13,800, d = 16, t = 1.5, and m = 0.3125/0.375 = 0.833, and from Fig. 9.8b(2), C = 0.33 m; therefore, C = 0.33(0.833) = 0.27,

$$t = d \sqrt{\frac{CP}{S}}$$

Substituting,

$$1.5 = 16 \sqrt{\frac{0.27P}{13,800}}$$

$$P = \frac{13,800}{0.27} \left[\frac{1.5}{16} \right]^2$$

$$= 449 \text{ psi}$$

For *noncircular flat heads* a Z factor is introduced into the flat head equation:

$$Z = 3.4 - \frac{2.4d}{D}$$

where d = short span
D = long span

Example If the above head were rectangular with dimensions of 10 in. by 16 in. and all other conditions were the same, except that $C = 0.33$, what would be the allowable pressure on this rectangular flat head?

solution With $d = 10$ in. as shorter span of rectangle is used per Code, $S = 13,800$ lb/in.^{2}.
For noncircular flat heads

$$t = d \sqrt{\frac{ZCP}{S}}$$

$$C = 0.33$$

$$Z = 3.4 - \frac{2.4(10)}{16}$$

$$= 1.9$$

Substituting, and solving for P

$$1.5 = 10 \sqrt{\frac{1.9(0.33)P}{13,800}}$$

$$P = \frac{13,800}{1.9(0.33)} \left[\frac{1.5}{10} \right]^2$$

$$P = 495 \text{ psi}$$

Bracing and Staying

The first point to remember in all problems dealing with bracing or staying is that the stress set up in a stay is due to the unit pressure in pounds per square inch acting on the area of plate supported by that stay. This *total pressure* is resisted by the internal resistance of the brace (unit stress) times the net area of the brace. These facts are the basis for all bracing formulas.

Stay bolts and stays are used in boilers to reinforce flat or other surfaces exposed to pressure loading, because the plate surfaces would have to be made too thick to resist this loading if stays or stay bolts were not used.

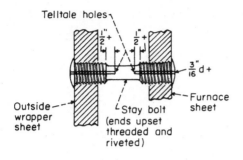

Figure 9.10 Code-threaded stay bolt has telltale hole extending ½ in. beyond threads on both ends.

To calculate stay-bolt problems, it is necessary to establish the Code requirement, which states that the required area of a stay bolt at its minimum cross section shall be found by calculating the load on the stay bolt, dividing this by the allowable stress on the stay bolt, and increasing the resultant area by a factor of 1.1. See Fig. 9.9b and 9.10. In equation form, the following can be developed to use on stay-bolt problems:

$$\frac{\text{Load on stay bolt}}{\text{Allowable stress}} = \text{resisting area of stay bolt}$$

Let S = allowable stress, lb/in.^{2} (Section II, Part D)
  a = area of stay bolt (usually round), in.^{2}
  p = pitch of stay-bolt spacing, in.
  P = allowable working pressure, psi

Then

$$\frac{(p^2 - a)P}{S} = \frac{a - (\text{telltale hole area})}{1.1}$$

In addition to the strength of the stay bolt, the strength of the plate between the stay bolts must be adequate, or the plate might buckle between the stay bolts. The Code requires this to be checked by one of the following equations:

$$t = p \sqrt{\frac{P}{CS}} \qquad \text{or} \qquad P = \frac{St^2C}{p^2}$$

where S = maximum allowable stress, lb/in.^{2}
  t = required thickness of plate, in.
  p = maximum pitch, in.
  P = maximum allowable pressure, psi
  C = factor, depending on construction

For example $C = 2.1$ for welded or stays screwed through plates not over $\frac{7}{16}$-in. thickness with ends riveted over, $C = 2.2$ if plate is over $\frac{7}{16}$-in. thick.

Section I of the Code lists other C values and illustrations of the pitch to be used in corner-welded construction, illustrated in Fig. A-8 of Section I. The pitch is generally from stay bolt to stay bolt, or from point of tangency on curved surfaces to the first stay bolt as illustrated in Section I.

Example of stay-bolt problem   Question (a): What is the maximum allowable square pitch of stay bolts in the flat furnace sheet of a Code firebox boiler if the sheets are supported by $\frac{7}{8}$-in. screwed stay bolts with $\frac{3}{16}$-in. telltale holes in the outer end as illustrated in Fig. 9.10? The stay bolts have 12-V threads per inch, and the pressure carried is 115 psi. Stay-bolt material is SA-675 with an allowable stress of 12,500 lb/in.² from Section II, Part D, while plate material is SA-299 with an allowable stress of 18,800 lb/in.² for under 650°F operation.

Question (b): What is the least thickness of plate required between stay bolts?

solution   Use this equation for question (a):

$$(p^2 - a)P = \left(\frac{a - 0.0276}{1.1}\right) S$$

where $a = 0.419$ in.² (cross-sectional area of $\frac{7}{8}$-in. V-threaded bolt), telltale hole area of $\frac{3}{16}$-in. diameter $= 0.0276$ in.²

$P = 115$ psi
$S = 12,500$ lb/in.²

Substituting, and solving for pitch, p

$$(p^2 - 0.419)115 = \left(\frac{0.419 - 0.0276}{1.1}\right) 12,500$$

$$p^2 = \frac{4,495.9}{115} = 39.09$$

$$p = 6.25 \text{ in.}$$

For question (b), use

$$t = p \sqrt{\frac{P}{CS}}$$

where $p = 6.25$ in.
$P = 115$ psi
$C = 2.1$ (on the assumption that the plate is under $\frac{7}{16}$ in.)
$S = 18,800$ lb/in.²

Substituting, and solving for t

$$t = 6.25 \sqrt{\frac{115}{2.1(18,800)}}$$

$$= 0.337 \text{ in.}$$

Stay bolts may be used to stay furnaces to the outside shell or wrapper sheet (Fig. 9.10). The size and the pitch of the stay bolts have much to do with the maximum allowable pressure on the boiler.

It will be noted in Fig. 9.10 that the ends are upset slightly so that the area of the stay bolt at the root of the threads will not be less than that of the body. However, some stay bolts are made without upsetting the ends; thus, in calculating the net area, the diameter at the root of the threads should be used.

Upset ends of stay bolts should be annealed in order to reduce any tendency to brittleness. The length of the stay bolt must be such that at least two threads extend over the plates. The ends are then riveted over.

Stay bolts sometimes break because of furnace-sheet expansion and contraction; the point of breakage is usually near the inside surface of the shell. Telltale holes are required in stay bolts not over 8-in. long, for these bolts are considered less flexible and more susceptible to breakage than the longer bolts. The telltale hole is at least $\frac{3}{16}$-in. diameter and is drilled in from the outside to a depth at least $\frac{1}{2}$ in. past the inner surface of the plate or, if the stay bolt is reduced in diameter, to at least $\frac{1}{2}$ in. beyond this reduction. It is obvious that when the stay bolt cracks halfway through its cross section, leakage through the telltale hole should give warning.

Diagonal stays. To stay the flat portions of heads that are not supported by tubes, diagonal stays are used above the tubes. This stay is not as direct as the through-stay, and it throws stress on the shell plates as well. But the diagonal stay leaves more room above the tubes for inspection, repair, and cleaning. A common form of diagonal stay is shown in Fig. 9.11a, b now welded on the attachments per Code. For calculating the strength of a diagonal stay, the Code requires the following to be considered:

1. What is the slant of the diagonal or its angle to the flat surface being supported?

2. Is it welded or riveted to the shell and head?

3. What is the construction on the ends of the stay where it is fastened to the shell or head: riveted or welded, pins, split palms, or blades (crowfoot type)? See Fig. 9.11a(2) and a(3).

The Code permits most diagonal stays to be calculated as straight stays similar to the stay-bolt method. This method calls for multiplying pressure times area on one side, with the holding power of the

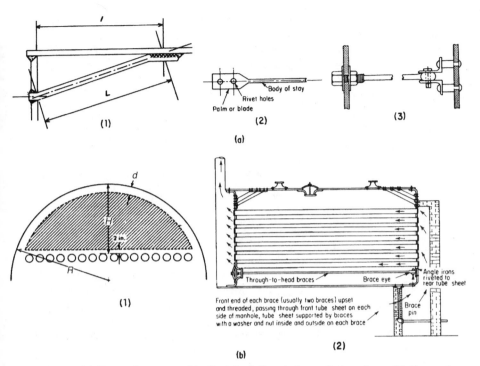

Figure 9.11 (*a*) Types of stays used in firetube boilers to brace flat surfaces: (1) diagonal stay, (2) palm stay, and (3) through-stay. (*b*) Flat surfaces above the tubes 2 in. from the tube tangent line require staying, usually by diagonal stays as shown, while the flat surfaces below the tubes are stayed with through-stays.

stay on the other side of the equation. For example, in Fig. 9.11*a*(1) if the ratio of L/l is 1.15 or less (on an HRT boiler), the body of the stay is calculated as a straight stay. But the allowable stress to be used is 90 percent of that allowed for a straight stay. If L/l is over 1.15, the body of the stay is calculated by increasing the area required on the body of the stay by L/l. In equation form, this is expressed as follows:

$$A = \frac{aL}{l}$$

where a = cross-sectional area of direct stay body
A = cross-sectional area of diagonal stay body
l = length of right angles to area to be supported [see Fig. 9.11*a*(1)]
L = diagonal length of stay

The Code rules on palms that were riveted on diagonal stays required the cross-sectional area of this part of the stay to be at least 25 percent greater than the body of the stay.

Example The net area of a segment to be stayed is 504 in.² and is supported by seven 1¼-in.-diameter diagonal braces (1.227-in.² net area). The length of these diagonals is less than 1.15 times the length of a direct pull and does not exceed 120 diameters. SA-285C is used for Code approved welded construction.

What pressure is allowable on the stayed segment?

solution SA-285 Grade C material per Section II, Part D has an allowable stress S of 13,800 lb/in.² All temperatures are below 400°F.

$$AP = 0.9(7aS)$$

$$a = \text{area of 1 brace} = 1.227 \text{ in.}^2$$

$$A = \text{area to be stayed} = 504 \text{ in.}^2$$

Substituting, and solving for P

$$P = 7(1.227)(13,900)(0.9)/504$$

$$= 212 \text{ psi}$$

Example The area to be stayed on the front tube sheet of a boiler is 136 in.². If it is braced with two diagonal stays of the welded type, what diameter of brace would be required to safely carry 165 psi, $L = 29\frac{1}{4}$ in., $l = 28\frac{5}{8}$ in.? SA-285C material for stays under 600°F. The allowable stress is 13,800 lb/in.²

solution

$$\frac{L}{l} = \frac{29.25}{28.625} = 1.02, \text{ less than 1.15, can be calculated as straight stay.}$$

Use

$$AP = 0.9naS$$

where $A = 136$ in.
 $n = 2$ braces
 $P = 165$ psi
 $S = 13,800$ lb/in.²

Substituting, and solving for a

$$136(165) = 2(13,800)(a)(0.9)$$

So

$$a = \frac{22,440}{24,840} = 0.903 \text{ in.}^2$$

Thus a 1⅛-in.-diameter brace must be used, as $\pi D^2/4 = 0.903$, and solving for $D = 1.1$ in. The nearest standard size is 1⅛ in.

Areas of tube sheets to be stayed. The Code has two equations for determining the area to be stayed on tube-sheet flat surfaces, or the area with no tube holes called segments.

Segments of a flanged head. The area to be stayed is enclosed 2 in. from the tubes and a distance d from the shell as shown in Fig. 9.11b(1). The d distance is the larger of the following:

d = the outer radius of the flange, but not exceeding 8 times the thickness of the head, or

$$= 80 \ \frac{t}{\sqrt{P}}$$

where d = unstayed distance from shell, in.
t = thickness of head, in.
P = maximum allowable working pressure, psi

The net area, A, to be stayed for a flanged head is then

$$A = \frac{4(H - d - 2)^2}{3} \sqrt{\frac{2(R - d)}{(H - d - 2)} - 0.608}$$

where A = area to be stayed, in.^{2}
H = distance from tubes to shell, in.
d = distance (as defined previously), in. For *unflanged heads,* $d = 0$.
R = head radius, or diameter, in.^{2}

Example (a) A 66-in. HRT boiler is operating at 140-psi working pressure. The flanged heads are $\%_6$-in. thick. The distance from the upper tubes to the shell is 24 in., and $d = 3$ in. [Fig. 9.11b(1)]. What is the area to be stayed?

(b) Head meets Code welding requirements. If this head is to be stayed by 1¼-in.-diameter diagonal braces, how many braces will be required when L does not exceed l more than 1.15 times and 9500 lb/in.^{2} stress at cross-sectional area is allowed for a straight brace?

solution (a)

$$A = \frac{4(24 - 3 - 2)^2}{3} \sqrt{\frac{2(33 - 3)}{24 - 3 - 2} - 0.608}$$

$$= 481.3 \ \sqrt{2.550}$$

$$= 768 \ \text{in.}^2$$

(b)

$$768P = 0.9naS$$

where n = number of braces
a = area of one brace = 1.2272 in.^{2}
S = allowable stress, lb/in.^{2}
P = allowable pressure, psi

$$768(140) = n(1.2272)(9500)(0.9)$$

$$n = 10.2$$

Thus 11 braces must be used.

Example   What would be the area to be stayed per above data if the head were *unflanged*?

solution   Use the same equation, but d = 0. Substituting,

$$A = \frac{4(24-2)^2}{3} \sqrt{\frac{2(33)}{24-2} - 0.608}$$

$$= 645.33 \sqrt{2.392}$$

$$= 998 \text{ in.}^2$$

In older HRT boilers, the segment of the tube sheets above the top row of tubes required staying. For this, there were three common methods. In a boiler of this type that did not exceed 36-in. diameter or 100-psi working pressure, structural shapes such as angle irons or channel irons were riveted to the segment. The outstanding legs were proportioned to have sufficient strength in bending to resist the pressure load.

For boilers exceeding 36-in. diameter or 100-psi working pressure, the flat segment of the tube sheets requires staying either by diagonal stays between the tube sheet and shell or by through-stays running the entire length of the boiler [Fig. 9.11*b*(2)]. The former are usually preferable, for they leave more room inside the boiler for cleaning and inspection. The through-stays may make it quite difficult for a worker or an inspector to move around over the tubes in the boiler.

There were three general types of diagonal stay for attachment by rivets: the Huston, the MacGregor, and the Scully. It is important to have them all under tension.

The diagonal stays are stretched into position before the tubes are installed in the boiler. In order to prevent distortion of the head through the tension of the stays, one or more heavy steel bars are clamped across the tube sheet as a beam, the tube sheet thus being held against the tension of the stays until some of the tubes are installed. The beam is known as a *strongback*.

Modern diagonal stays are welded as shown in Fig. 9.11*a*(1).

The staying of the section of the tube sheet below the tubes of HRT and similar boilers is usually effected by through-to-head stays. These stays are "spooled off" at the rear tube sheet [Fig. 9.11*a*(3) and *b*(2)]. The front ends of the through-to-head stays—usually two in number—pass through the front tube sheet with inside and outside nuts and washers. They are usually made tight against leakage by grommets or soft metallic packing of various types beneath the outside nuts and washers.

The reason why the rear ends of the stays do not pass through the rear tube sheet but are spooled off inside is that the heat of the fire would damage the nuts and threaded ends.

If a flanged-in manhole is provided below the tubes in the front tube sheet, the stiffening effect of the flange is sufficient to allow 100-in.^{2} deduction from the area to be stayed. However, if through-to-head stays are used, the full-sized stays required to brace the rear tube sheet must be used (unless diagonal stays between the bottom of the rear tube sheet and the shell supplement the smaller through-to-head stays that would be sufficient to brace the front segment), for no deduction in area to be braced is permitted for the rear tube sheet.

Figure 9.12a illustrates the net area to be stayed when the tube pattern is irregular. Areas adjacent to cylindrical furnaces as illustrated in Fig. 9.12b may not require staying as long as the distance is at or below 1½ times the pitch distance where the pitch is calculated by the stay-bolt equation and using the Code recommended C value in the equation.

Girder and radial stays.   The girder stay (Fig. 9.13a) was formerly used very extensively to support flat crown sheets in locomotive firebox units. But it has been largely superseded by the radial stay (Fig. 9.13b)

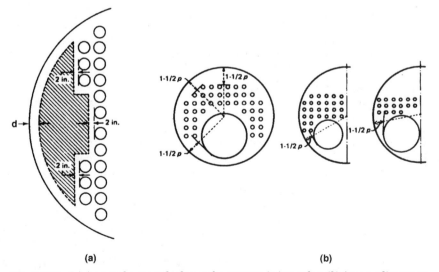

(a)   (b)

Figure 9.12   (a) Area to be stayed when tube pattern is irregular. (b) Areas adjacent to cylindrical furnaces can be designed by the stay-bolt equation with maximum pitch allowed as illustrated.

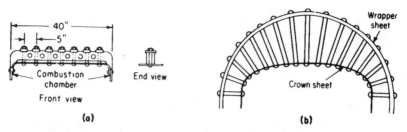

Figure 9.13 (a) A girder stay is used to support the crown sheet of locomotive firebox and scotch marine boilers. (b) Radial stays.

for this purpose. It is still used to support the tops of combustion chambers in boilers of the scotch marine type. The girder stay consists of a cast steel or built-up girder with its ends resting on the side, or end sheets, of the firebox or combustion chamber. It supports the flat crown sheet (the top of the combustion chamber) by means of bolts.

The radial stays are more flexible and tend to hold less scale from circulation than do girders. About the only advantage of the girder stays is that they pass straight through the sheet rather than at an angle. The Power Boiler Code has examples on the use of special equations for girder stays.

Staying Furnace Against Collapse

The furnace sheets of a boiler and of other internally fired boilers must resist the pressure on the external surfaces that tends to cause collapse. This collapsing tendency is resisted either by the stiffness of the furnace or by staying it to the shell with stay bolts.

A furnace not exceeding 38-in. OD may be self-supporting, and the use of stays may be eliminated provided that the thickness of the furnace is sufficient for necessary stiffness and that the span of furnace length is not too great.

Pre-1986 method. The Power Boiler Code gave the following two formulas for self-supporting, unstayed circular furnaces not over 4½ diameters in length. When the length does not exceed 120 times the thickness of the furnace sheet, use

$$P = \frac{51.5(300t - 1.03L)}{D} \tag{9.3}$$

When the length exceeds 120 times the thickness of the sheet, use

$$P = \frac{1.09(10^6)(t^2)}{LD} \tag{9.4}$$

where P = maximum allowable pressure, psi

D = outside diameter of furnace, in.

L = total length of furnace between centers of head seams, in.

t = thickness of furnace walls, in.

Example Determine the allowable working pressure for an unstayed furnace in a vertical tubular boiler when the furnace is 26-in. OD, $\frac{7}{16}$-in. thick, and 42-in. long.

solution

$$120 \times 0.4375 = 52.38$$

Equation (9.3) applies (length is below $120t$).

$$P = \frac{51.5(300t - 1.03L)}{D}$$

where t = 0.4375, L = 42, D = 26.

$$P = \frac{51.5[300(0.4375) - 1.03(42)]}{26}$$

$$= \frac{51.5(87.99)}{26}$$

$$= 174.3 \text{ psi}$$

Post-1986 method. This method requires using external pressure charts to determine allowable pressure, somewhat similarly to the firetube method illustrated previously.

Using the same data as in the previous example and referring to Fig. 9.4,

Length of furnace, L = 42 in.

OD of furnace, D_o = 26 in.

Thickness of furnace, t = 0.4375 in.

The material of the furnace is SA-210C with a yield stress of 40,000 lb/in.[2] per Section II, Part D. The following ratios must be determined in order to obtain the A factor from Fig. 9.4:

$$\frac{L}{D_o} = \frac{42}{26} = 1.26 \qquad \frac{D_o}{t} = \frac{26}{0.4375} = 59.4$$

Refer to Fig. 9.4 and use the above ratios, factor A = 0.00124. With this factor known (Code geometric chart), refer to Fig. 9.14 for the material chart to find Factor B at a minimum material temperature of 700°F. Factor B is 9900 from the right of the chart. The Code provides this equation for determining the allowable working pressure:

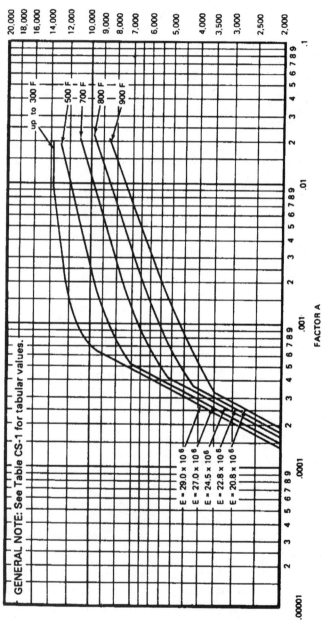

Figure 9.14 ASME Section II, Part D chart for determining Factor B for firetube boiler tubes, flues, and furnaces under external pressure, and constructed of carbon or low-alloy steels with specified minimum *yield strength* 30,000 psi and over. (*Courtesy American Society of Mechanical Engineers.*)

$$\text{Allowable } p = \frac{4B}{3D_o/t}$$

Substituting,

$$\text{Allowable } P = \frac{4(9900)}{3(59.4)} = 222 \text{ psi}$$

If a furnace does not meet the requirements for an unstayed unit, one of the following three methods of support may be used:

1. A corrugated furnace may be used. A common type of corrugated furnace is known as a Morrison furnace. It may be used in either vertical tubular or horizontal types of firebox boiler such as the scotch marine.

2. The Adamson ring is a device used to stiffen a circular furnace against collapse under external pressure. It is used primarily in horizontal furnaces, for sediment might lodge on the flanges on the waterside if it were installed in a vertical axis which would cause overheating. Figure 9.15 shows a cross-sectional view of a past and present Adamson ring joint and the ASME Code specifications for proportions.

3. Stay bolts may be used to stay the furnace to the outside shell or wrapper sheet. The size and the pitch of the stay bolts have much to do with the maximum allowable pressure on the boiler.

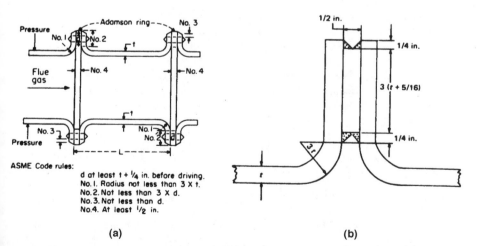

Figure 9.15 Adamson rings are used to stiffen furnaces against collapse from the external pressure imposed on the sheet. (*a*) Older riveted Adamson ring details per then-existing Code. (*b*) Adamson ring installed by welding and Code details for this type of construction.

The reader should consult the Code for more complete specifications for unstayed furnaces, for it gives details not permitted by the space or intent of this book, which is to emphasize methods of analysis to determine allowable pressure.

Reinforcement of Openings in Shells

In cutting a manhole in the shell of a boiler, it is necessary to compensate for the metal removed. This is done by installing a manhole frame if needed.

The minimum-size elliptical manhole permitted by the ASME Code is 11 in.×15 in. In cutting the shell for a frame having an opening of this size, the shorter dimension is placed along the longitudinal axis of the boiler so that less frame material will be required for replacement in this weaker directional axis.

In considering a cross-sectional plane of the boiler shell plate in the vicinity of a manhole "cutout," it is necessary to find the total area of metal removed, including rivet holes, and to provide a manhole frame having an equal cross-sectional area in the same plane if the shell does not have excess thickness.

The Power Boiler Code has detailed requirements on openings cut into shells or headers and how to calculate if reinforcement around the opening is necessary. The following procedure generally applies. See Fig. 9.16. The area required to be restored by the finished opening d is

$$A = d \times t_r \times F$$

where d = diameter of finished opening in given plane, in.
t_r = required thickness of seamless shell for the pressure
F = factor that considers axis of the nozzle, usually 1.00

To determine if enough metal is available from the shell, nozzle, welds, or reinforcement, the Code provides Eqs. (9.5) and (9.6). (See Fig. 9.16 for the meaning of the symbols and limits.)

To determine the metal available from the *shell,*

$$A_1 = (t_s - Ft_{rs})d \quad \text{or} \quad A_1 = 2(t_s - Ft_{rs})(t_s + t_n) \tag{9.5}$$

The larger value is used.

To determine the metal available from the *nozzle,*

$$A_2 = (t_n - t_{rn})5t_s \quad \text{or} \quad A_2 = (t_n - t_{rn})(5t_n) \tag{9.6}$$

Use the smaller value.

Example A 5-in. extra heavy pipe nozzle of SA-52 Grade S/A material is welded into a shell similar to that shown in Fig. 9.16 with ½-in. welds. The .

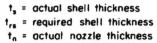

t_s = actual shell thickness
t_{rs} = required shell thickness
t_n = actual nozzle thickness
t_{rn} = required nozzle thickness

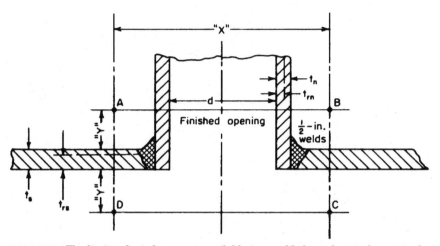

Figure 9.16 The limits of reinforcement available in a welded nozzle attachment is the excess material available from shell, nozzle, any reinforcement ring and welds encompassed in rectangle ABCD. The distances X and Y are specified in the Code and are illustrated by example in the text.

shell has an inside diameter of 30 in., is ⅞₆-in. thick, and has a working pressure of 200 psi. All welds meet Code requirements. This includes weld strength. The 5-in. pipe is 0.375-in. thick with an OD of 5.563 in. Using Section I calculations for reinforcement openings, demonstrate if the reinforcement meets Code requirements for the 200 psi working pressure.

solution From Section II, Part D, the allowable stress for the shell material is 13,800 lb/in.^{2} and the allowable stress for the nozzle material is 12,000 lb/in.^{2}

The same equation is used to determine the required shell and nozzle thickness by using the equation for a seamless cylinder, but with the C factor omitted, or

$$t = \frac{PR}{SE} - (1 - y)P$$

t = required thickness, in.

P = allowable pressure, psi

S = allowable material stress, lb/in.^{2}

$E = 1$

$y = 0.4$

For shell:

$$t_{rs} = \frac{200(15)}{13,800(1) - 0.6(200)} = 0.219 \text{ in.}$$

For nozzle:

$$t_{rn} = \frac{200(2.407)}{12,000(1) - 0.6(200)} = 0.037 \text{ in.}$$

The area of reinforcement required is

$$A = d \times t_{rs} \times F \text{ with } F = 1 \text{ as nozzle is inserted square.}$$

$$A = 4.813(0.219)(1.0) = 1.054 \text{ in.}^2$$

Areas of reinforcement available:

From shell: $A_1 = 4.813[1.0(0.438) - 1.0(0.219)] = 1.054 \text{ in.}^2$
From nozzle: $A_2 = 5(0.375)[0.375 - 0.037] = 0.634 \text{ in.}^2$
From welds: $A_3 = \frac{1}{2} \times \frac{1}{2} \times \frac{1}{2} \times 2 = 0.25 \text{ in.}^2$
Total reinforcement area available $= 1.938 \text{ in.}^2$

This exceeds the required area of reinforcement, therefore the opening reinforcement meets Code requirement.

Identifying Code Boilers

An identifying stamp on a power boiler is required by states and municipalities. The National Board standard form of stamping is sufficient for the boiler to pass rules and regulations for construction in practically all sections of the United States. This stamp consists of the ASME symbol above the manufacturer's serial number, the manufacturer's name or approved abbreviation, the maximum pressure for which the boiler was built, the heating surface and maximum capacity, and the year built.

"National Board" followed by a serial number stands for the National Board of Boiler and Pressure Vessel Inspectors. It is an enforcement body for the ASME Power Boiler Code, and a copy of the manufacturer's constructional data is filed with this board under the serial number. The National Board stamp indicates that the boiler is of ASME standard construction and that its construction was followed in the shop by a qualified inspector.

On horizontal firetube boilers (externally fired), the stamping should be in the middle of the front tube sheet, above the top row of tubes. On horizontal firetube boilers of the firebox type, the stamping should be located above the center or right-hand furnace or above a handhole at the furnace end. On vertical firetube boilers, the stamping should be located over the furnace door. Watertube boilers have the stamping on the drum heads above the manhole flange.

It is very important when purchase or relocation of a boiler is contemplated to ascertain the boiler laws of the state. Many states require filing a special form and following a definite procedure before relocating a boiler.

The Use of Code Equations

The Code calculations reviewed in this chapter were designed to show how the allowable pressure is obtained for the different boiler components. Those readers preparing for operator or inspector examinations should note the method of attack required in solving boiler component strength problems based on Code formulas or equations. It must also be realized that the method illustrated does not apply only to new construction. Operators and inspectors as well as repairers should review the Code equations and requirements when faced with a problem on existing installations. By using the appropriate equation for the component that requires review, a decision can be reached on appropriate Code-type repairs that will restore the strength of the component. It is also possible to determine wear rates on components by comparing actual thickness, for example, to those required by the methods illustrated in this chapter. This will assist in calculating remaining life for the component(s), and may even involve changing the allowable pressure until repairs are made. The intent of this chapter is to show how allowable pressures are calculated so that an engineering decision on facts can be made on how to proceed when a questionable condition arises.

Questions and Answers

1   What are the major disadvantages of a lapped longitudinal joint?

ANSWER: The major disadvantage of a lap-riveted or welded longitudinal seam is that the plate is "lapped"; thus, the drum or shell does not form a true circle. A certain degree of bending stress may occur along the lap of the seam when the shell is subjected to pressure. This causes a stress concentration that eventually may result in fatigue and cracking of the plate, thus producing a disastrous boiler explosion.

About the only "advantages" of the lap-riveted or welded seam are low cost, simplicity, minimum width of space, and minimum total thickness where the seam is exposed to the fires, as in girth seams of some firetube boilers.

2   What is the maximum diameter and pressure that can be allowed on a boiler whose longitudinal joint is fusion-welded and is not x-rayed or stress-relieved?

ANSWER: 16-in. ID and 100-psi pressure. Build to miniature-boiler specifications.

3 A Code-approved welded watertube boiler is 52-in. ID with 2-in. drum-plate thickness in the ligament area of the drum. The plate material is SA-515 Gr.70 with an allowable stress for under 650°F operation of 17,500 lb/in.² What is the minimum required tube ligament efficiency for an allowable pressure of 600 psi?

ANSWER: Transpose the equation for E:

$$P = \frac{SE(t - C)}{R + (1 - y)(t - C)}$$

with

$$R = 26 \text{ in.}$$

$$t = 2 \text{ in.}$$

$$C = 0$$

$$y = 0.4$$

$$P = 600 \text{ psi}$$

$$S = 17{,}500 \text{ lb/in.}^2$$

Transposing and substituting,

$$E = \frac{P[R + (1 - y)(t - C)]}{S(t - C)}$$

$$E = \frac{600[26 + (1 - 0.4)(2)]}{17{,}500(2)} = 0.466 = 46.6\% \text{ efficiency required}$$

4 What is the largest opening on a shell permitted without calculation for reinforcement on a welded attachment?

ANSWER: 2-in. pipe size.

5 A dished head concave to pressure, ⅝-in. thick, and 26 in. in diameter. What is the minimum length of flange permitted where the seam is to be welded to the shell or drum?

ANSWER: The corner radius must be not less than 3 times the head thickness or in no case less than 6 percent of the diameter or shell, or

$$3 \times \tfrac{5}{8} = 1\tfrac{7}{8} \text{ and } 6\% \text{ of } 26 = 1.56$$

The flange must be at least 1⅞ in. (larger value).

6 In the building of a miniature boiler, which is of all welded construction, is not x-rayed or stress-relieved, and is used for a maximum working pressure of 100 lb, what pressure must be used in testing this boiler hydrostatically?

ANSWER: 300 psi per Code.

7 A boiler drum is drilled for tubes in a series of rows parallel to its axis; the tubes circumferentially are in alignment. The diameter of tubes holes is 4.03125 in., and they are spaced equally in every row in groups 13½ in. center to center (tubes alternately pitched 7½ and 6 in.). Determine the horizontal ligament efficiency E and the minimum pitch circumferentially.

ANSWER: Use

$$E = p - \frac{nd}{p}$$

with

$$p = 13.5 \text{ in.}$$

$$n = 2$$

$$d = 4.03125 \text{ in.}$$

Substituting,

$$E = \frac{13.5 - 2(4.03125)}{13.5} = 0.403$$

For the pitch in the circumferential direction, transpose equation to solve for p, with $E = 0.403/2 = 0.202$ and $n = 1$ as tubes are in-line, or

$$p = \frac{nd}{1 - E} = \frac{1(4.03125)}{1 - 0.202}$$

p circumferentially = 5.05 in.

8 What is the minimum required thickness for a tube sheet of a Code water-tube boiler steam drum of the following specifications?

Working pressure, 675 psi
Inside radius, 27 in.
Longitudinal tube-hole efficiency, 46.6 percent
Circumferential tube-hole efficiency, 31.1 percent
Circumferential joint efficiency, 100 percent
Material, SA-515-70
Temperature, 650°F
Allowable stress, 17,500 lb/in.² for material

ANSWER:

$$t = \frac{675 \times 27}{17,500(0.466) - 0.6(675)} = 2.35 \text{ in.}$$

based on longitudinal tube-hole efficiency

9 (a) Why are manhole and handhole plates usually made oval?

(b) Why is water used instead of steam or air to test a boiler?

(c) What precautions are necessary to take on safety valves and steam pipes when the hydrostatic test is applied to one boiler in a battery?

ANSWER:

(a) To enable the operator to remove them from the boiler and replace them. An oval manhole also provides the largest entrance opening with the removal of a minimum amount of material. The long axis of the manhole should be across, or girthwise of, the shell.

(b) Incompressibility avoids explosions.

(c) Use proper test clamp or remove safety valves and blank flanges or cap. Disconnect stop valves and blank off steam-pipe ends.

10 (a) What causes a plate formed into a cylinder to retain its shape?

(b) What is the maximum permissible distortion in a welded boiler drum?

ANSWER:

(a) The stretching of the outside fiber of the plate beyond its elastic limit.

(b) 1 percent of mean diameter.

11 (a) On a boiler, why is the hydrostatic test limited to 1½ times the maximum allowable working pressure?

(b) What good does this test do?

(c) Why is there a minimum and maximum temperature set for the water during a hydrostatic test? What are the minimum and maximum temperatures?

ANSWER:

(a) Because a greater pressure is likely to damage the boiler plate by causing a permanent deformation.

(b) It tests for tightness of parts and strength.

(c) So that boiler parts and water will be approximately the same temperature. The temperature must be not less than 70°F to prevent sweating. That may mask a leak through a defect, and the metal temperature cannot exceed 120°F in order to permit close examination of welds and similar joint-type areas.

12 What is the allowable working pressure on an unstayed full-hemispherical head of welded construction with the following specifications? 48-in. ID; pressure on the concave side; thickness of plate = ⅜ in.; efficiency of weld is 100 percent; SA-285C material with an allowable stress of 13,800 lb/in.².

ANSWER: Use the equation

$$t = \frac{PL}{(1.6)S}$$

and transposing and substituting,

$$P = \frac{1.6(13,800)(0.375)}{24} = 345 \text{ psi}$$

13 What is the minimum required thickness permitted for a Code flat unstayed circular head forged integral with a header and complying with other requirements of the Code? The diameter D is 22 in., and the required pressure is 150 psi. The maximum allowable stress value S is 11,000 lb/in.[2].

ANSWER: Note: $C = 0.17$,

$$t = 22 \sqrt{\frac{0.17(150)}{11,000}} = 1.056 \text{ in.}$$

14 (a) Name two important features to be noted in connection with a manhole opening formed by flanging inward the head plates of a boiler.

(b) Why is it preferable that heads butt-welded to shells have a skirt or flange when they are attached by the fusion-welding process?

ANSWER:

(a) The plate must be flanged to a depth of not less than 3 times the thickness, and the gasket bearing surface must not be less than $1\frac{1}{16}$ in.

(b) To prevent concentration of stresses in the knuckle of the flange.

15 To what percentage may the knuckle of a dished head be thinned in forming?

ANSWER: Not over 10 percent.

16 The stay bolts in the firebox of a locomotive-type boiler are spaced 7 in. horizontally and $6\frac{1}{2}$ in. vertically, and they are of $1\frac{1}{4}$-in. diameter with 12V threads per inch. The plate thickness is $\frac{1}{2}$ in. The ends of the bolts are drilled with $\frac{3}{16}$-in.-diameter telltale holes, and the area of telltale hole is 0.0276 in.[2]. What is the allowable working pressure on the furnace? These data are given: SA-285 Grade C plate under 600°F; allowable stress, 13,700 lb/in.[2]; area of stay bolt at bottom of threads is 0.960 in.[2].

ANSWER: Use the stay-bolt equation

$$(p^2 - a)P = \frac{(a - 0.0276)S}{1.1}$$

and transposing and substituting,

$$P = \frac{(0.960 - 0.0276)13,700}{1.1(7 \times 6.5 - 0.960)} = 261 \text{ psi}$$

It is also necessary to calculate the allowable pressure, based on plate thickness between the stay bolts. Use

$$P = \frac{St^2C}{6.5 \times 7} \quad \text{with } C = 2.2$$

Substituting,

$$P = \frac{13,700(0.5)^2(2.2)}{6.5 \times 7} = 165.6 \text{ psi}$$

This is the allowable pressure.

17 What hydrostatic pressure is required on a watertube boiler drum of Code-approved welding with longitudinal weld efficiency of 100 percent, plate thickness of 1%₁₆ in. at this joint with an inside radius of 27.25 in.? The tube sheet is 1¹⁵⁄₁₆-in. thick with an inside radius of 26¹⁵⁄₁₆ in. Tube hole ligament efficiency is 0.304 circumferentially and 0.429 longitudinally. The plate material is SA-515 Gr.70, operating under 650°F, with an allowable stress of 17,500 lb/in.².

ANSWER: Note that the circumferential ligament efficiency is greater than one-half the longitudinal ligament efficiency of 0.429; therefore, the allowable pressure is based on this longitudinal ligament efficiency and tube-sheet data. Using shell equation, with $C = 0$,

$$P = \frac{SE(t - C)}{R + (1 - y)(t - C)}$$

where $C = 0$ and $y = 0.4$.

$$P = \frac{17,500(0.429)1.938}{26.938 + 0.6(1.938)} = 518 \text{ psi}$$

Per Code, hydrostatic pressure required is $1.5 \times 518 = 777$ psi.

18 What would be the required knuckle radius of an unstayed dished head if the plate is ½-in. thick and the diameter of the drum is 48 in.?

ANSWER: $48 \times 0.06 = 2.88$ in.

19 Why are flat surfaces in a boiler stayed or braced?

ANSWER: A boiler plate under pressure tends to assume the shape of a sphere, so braces and stays are used to hold it in place.

20 How should telltale holes be drilled in stay bolts?

ANSWER: Solid stay bolts 8 in. and less in length should be drilled with telltale holes at least ³⁄₁₆-in. diameter to a depth at least ½ in. beyond the inside of the plate.

21 Why is the pitch of stay bolts in a vertical tubular boiler taken from the inside of the firebox rather than the outside of the shell?

ANSWER: The furnace sheet is the sheet that normally requires staying; therefore, the pitch is taken from inside the firebox.

22 Name three methods used to strengthen a furnace to resist collapse.

ANSWER: On a vertical tubular boiler by stay bolts; on horizontal furnaces by the use of Adamson rings; and by corrugations.

23 What is an Adamson ring and where is it used and for what purpose?

ANSWER: It is a flanged-type joint between two or more sections of a plain furnace. It is located somewhere in the length of the furnace; there may be one in the center or several spaced at intervals, not less than 18 in. apart. It serves as a reinforcement, stiffening the furnace against collapsing pressure. The ring between the flanges also serves as a filler for caulking.

24 A 16-in.-ID circular nozzle manhole (welded type) is located on a welded drum. These data pertain:

Thickness of welded neck, ¾ in.
Thickness of drum shell, 1⅛ in.
Diameter of drum shell, 48 in. inside
Working pressure, 500 psi
Working temperature, not above 600°F
Welding used complies with the Code for power boilers
Nozzle extends uniformly ¾ in. below the inside drum shell
Nozzle is welded to the drum on the inside and outside with ¾ in. × ¾ in. fillet welds
Reinforcement ring on the inside of the drum is 1 in. thick, with 21½-in. ID and 36-in. OD
Inside edge welded to the drum with 1 in. × 1 in. fillet welds
Material for the drum nozzle and the reinforcing ring have a design stress of 15,000 lb/in.²

Assume the welding meets the Code requirements, and show by calculations that this reinforcement is adequate to meet Code requirements.

ANSWER: For the shell,

$$t_{rs} = \frac{500(24.0)}{15,000 - 0.6(500)} = 0.817 \text{ in.}$$

For the nozzle,

$$t_{rn} = \frac{500(8.0)}{15,000 - 0.6(500)} = 0.272 \text{ in.}$$

The area of reinforcement required is

$$A = d \times t_{rs} \times F$$

$$= 16.0(0.817)(1.0) = 13.072 \text{ in.}^2$$

The area of reinforcement provided is

A_1 (shell) = [1.0(1.125) − 1.0(0.817)]16.0   =   4.928 in.^{2}

A_2 (nozzle) = 2(2.5)(0.75)(0.75 − 0.272) + 2(0.75)^2 =   2.918 in.^{2}

A_3 (welds) = 2(0.75)^2 + 1(1.00)^2 =   2.125 in.^{2}

A_4 (ring) = 1.0(32 − 21.5)   = 10.500 in.^{2}

Total   20.471 in.^{2}

So construction complies with Code requirements.

25 A high-temperature boiler weighs 50,000 lb and is supported by columns carrying beam girders. The boiler has proper attachments at four points for the reception of U bolts, which are supplied with washers and nuts; the allowable stress is 6000 lb/in.^{2} on these bolts. What is the diameter of bolts at the weakest section?

ANSWER: Each bolt must carry 50,000 divided by 4, or 12,500 lb. The area of the bolt required equals 12,500 divided by 6000, or 2.083 in.^{2}. The diameter of the bolt required equals the square root of 2.083 divided by 0.7854, or 1.6 in. The bolts should be 1⅝-in. diameter.

26 A riveted old HRT boiler, 72-in. diameter, original ½-in.-thick shell plate with an ultimate 55,000 lb/in.^{2} tensile strength, longitudinal joint efficiency of 87.5 percent, factor of safety, FS = 5 had been operating at 133 psi. Inspection revealed that the bottom of the shell on the first course measured only 5/16-in. thickness due to extensive corrosion. (a) What would be the allowable pressure based on the weakest course, and (b) at what safety factor, FS, was this boiler operating?

ANSWER: (a) Since the corrosion is in the solid plate, use

$$E = 1$$

$$S = 55,000/5 = 11,000 \text{ lb/in.}^2$$

$$R = 36 \text{ in.}$$

$$t = 0.3125$$

$$y = 0.4$$

then

$$P = \frac{11,000(1.0)(0.3125)}{36.0 + 0.6(0.3125)} = 95 \text{ psi allowable}$$

(b) Use 55,000 lb/in.^{2} in the same shell equation, and solve for FS, or

$$FS = \frac{55,000(1)(0.3125)}{133[36 + 0.6(0.3125)]} = 3.57 \text{ instead of 5}$$

27 A stoker-fired 66 in. × 16 ft horizontal-return-tubular boiler, carrying 125-psi pressure, has sixty-six 3½-in. no. 11 gauge tubes. What is the total heating surface if one uses one-half the shell and ignores the heads?

ANSWER:

$$\text{Shell heating surface} = \frac{66(3.14)(16)}{2(12)} = 138 \text{ ft}^2$$

$$\text{Tube heating surface} = \frac{3.26(3.14)(16)(66)}{12} = 901 \text{ ft}^2$$

$$\text{Total heating surface} = 1039 \text{ ft}^2$$

Therefore, two or more safety valves are required.

28 Who has the responsibility to establish and maintain a quality-control system for building power boilers?

ANSWER: The manufacturer and/or assembler holding or applying for an ASME certificate of authorization.

29 Briefly explain what information the manufacturer must maintain in the quality-control record file.

ANSWER: In general, a file should include the following:

1. Drawings or blueprints of the vessel being made or assembled.
2. Calculations to show Code compliance on thickness, materials, and similar requirements.
3. Identification of material to be used, including dimensions and quality to be supplied by the steel mills or other suppliers.
4. Inspections made on material and fabrication during designated checkpoints.
5. Qualification of any nondestructive examination (NDE) personnel who performed NDE.
6. Qualification of any welders who performed welding and applicable written welding procedures.
7. NDE records on radiographs and nondestructive testing.
8. Repairs made, hydrostatic test performed, and a record of data sheets for the vessel.

30

   (*a*) Who is responsible for maintaining records for a vessel constructed in accordance with ASME Code?

   (*b*) How long must these records be kept?

ANSWER:

   (*a*) The manufacturer or assembler holding the ASME certificate of authorization, except with nuclear vessels, where the responsibility is with the owner of the nuclear facility per NRC rules as well as ASME Code requirements.

   (*b*) By Section I, for at least 5 yr by the manufacturer; for nuclear vessels, indefinitely by the owner of the nuclear facility.

31   What is the allowable pressure on a *convex* head that is unstayed with a 12 in. × 16 in. flanged-in-manhole, 48-in. diameter, radius of bump 36 in., allowable stress is 13,750 lb/in.^{2}, and thickness of head is 1.0625 in.?

ANSWER: Segment of sphere equation must be used because of manhole, and thickness must be reduced by 15 percent, or $t = 1.0625/1.15 = 0.924$ in. Also, by Code, convex heads are allowed 0.6 the pressure on a concave head. Then substituting

$$P = \frac{4.8(13,750)(0.924)(0.6)}{5(38.4)} = 275 \text{ psi}$$

with $R = 0.8(48)$, because head is dished less than 0.8 times diameter of head and Code requires a dished head to be made at least equivalent to a head with a dish radius of 0.8 times diameter of the head.

32   A firetube boiler requires 374 in.^{2} of flat surfaces to be stayed with stay bolts. Layout shows spacing of stay bolts 5¼ in.×5½ in. would cover the area. If stay-bolt material has an allowable stress of 11,300 lb/in.^{2}, and bolts are drilled with ³⁄₁₆-in. telltale holes, how many and what size stay bolts are required for an allowable pressure of 150 psi?

ANSWER: Number of stay bolts required is

$$\frac{374}{5.25 \times 5.5} = 13$$

With a = net area on bottom of threaded stay bolt, and telltale hole area = 0.0276, use stay-bolt equation to solve for a as follows:

$$[(5.25)(5.5) - a]\,150 = \left[\frac{a - 0.0276}{1.1}\right] 11,300$$

and solving for a,

$$a = 0.460 \text{ in.}^2$$

A ¹⁵⁄₁₆-in., 12-V threaded stay bolt has an area at bottom of threads of 0.494 in.², and this should be used.

33 An unflanged head on a firetube boiler has a diameter of 72 in. with a distance from the upper tubes to the inside shell of 30 in. What is the area to be stayed?

ANSWER:

$$\text{Area to be stayed} = \frac{4(30-2)^2}{3} \sqrt{\frac{2(36)}{30-2}} - 0.608$$

$$\text{Area to be stayed} = 1464.7 \text{ in.}^2$$

10

Boiler Connections, Appurtenances, and Controls

Figure 10.1 illustrates the many connections and appurtenances that are required on a high-pressure boiler that may be connected to a common steam main. All the pipe and valve connections serve a purpose in operation and maintenance, and most are governed by ASME Section I Code requirements as respects size and installation. The Code requires as a minimum: (1) a pressure gauge and test connection, a safety valve, a blowdown valve, gauge glass, gauge cocks, a

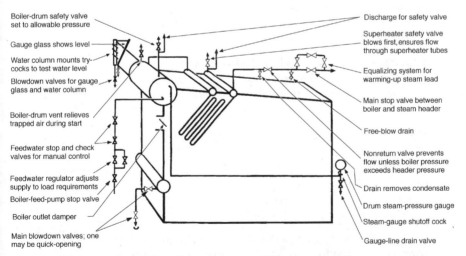

Boiler-drum safety valve set to allowable pressure

Gauge glass shows level

Water column mounts try-cocks to test water level

Blowdown valves for gauge glass and water column

Boiler-drum vent relieves trapped air during start

Feedwater stop and check valves for manual control

Feedwater regulator adjusts supply to load requirements

Boiler-feed-pump stop valve

Boiler outlet damper

Main blowdown valves; one may be quick-opening

Discharge for safety valve

Superheater safety valve blows first, ensures flow through superheater tubes

Equalizing system for warming-up steam lead

Main stop valve between boiler and steam header

Free-blow drain

Nonreturn valve prevents flow unless boiler pressure exceeds header pressure

Drain removes condensate

Drum steam-pressure gauge

Steam-gauge shutoff cock

Gauge-line drain valve

Figure 10.1 Connections and appurtenances on a high-pressure boiler serve an important function in operation and inspection and most are required by the ASME Code. (*Courtesy Power magazine.*)

stop valve in the steam line, and stop and check valves in the feed line. (2) In addition to the minimum mentioned above, boiler feed pumps and/or injectors are needed. Low-water fuel cutoffs on automatically fired boilers are now required in many jurisdictions on not only low-pressure but also high-pressure boilers. Combustion safeguards are also becoming a necessity on boilers firing fuel in suspension. These are covered in later chapters. The term *boiler fitting* is applied to valves, gauges, and other connections or devices that are attached directly to the boiler so that the unit or units can be operated safely and efficiently.

Applicable definitions.   See also Appendix 1, "Terminology and Definitions."

1. *Safety valve:* Prevents boiler pressure from rising above the setting of the valve by relieving excessive steam pressure, guarding against hazards of over pressure.

2. *Steam supply stop valve:* The valve installed at the steam outlet of the boiler to shut off the flow of steam.

3. *Steam pressure gauge:* Indicates the steam pressure in the boiler in pounds per square inch.

4. *Steam gauge siphon:* The device installed between the steam gauge and the boiler to provide a water seal, so that live steam will not enter the gauge to cause a false reading or damage to the gauge.

5. *Inspector's test gauge connection and cock:* Provides the necessary connection to check the accuracy of the steam pressure gauge on the boiler.

6. *Water column:* The hollow casting or forging connected at the top to the boiler's steam space and at the bottom to the water space. The water gauge glass and water test cocks are installed on the column.

7. *Water glass and gauge fixtures:* To show the water level in the boiler.

8. *Water test gauges or try cocks:* For testing the water level in the boiler should the water glass be out of service temporarily for any reason.

9. *Drain valve under the water column and low-water fuel cutoff switch:* To provide a means for daily flushing under the water column and water level controls to keep the chamber and lines clean, so the water will register accurately in the glass. Also provides a means of testing the low-water cutoff.

Recommended instrumentation. Any installation will be improved by the use of instruments when trained operating personnel are in attendance and make intelligent use of the data provided. Instrumentation of larger packaged boilers should include an apparatus for obtaining flue-gas analysis and determining combustion efficiency.

As a minimum, the ASME Boiler Code for high-pressure boilers recommends the following instruments: (1) steam pressure gauge; (2) feedwater pressure gauge; (3) furnace draft gauge: (4) an outlet pressure gauge on the forced-draft fan and an inlet pressure gauge on the induced-draft fan; (5) steam-flow recorder for checking boiler output; (6) CO_2 recorder to check on combustion; (7) superheater inlet and outlet temperature recorder; (8) inlet and outlet temperature recorders for air heaters; (9) thermometers indicating inlet and outlet steam temperatures for boiler reheaters; (10) feedwater-temperature recorders for checking degree of deaeration and economizer operation; (11) pressure gauges on pulverizers to check differential pressure for fuel-air mixtures to burners; (12) pressure gauges for oil-fired boilers on oil lines to burners and temperature gauges before and after any oil preheaters; and (13) pressure gauges for gas-fired boilers on the main gas line to burners and on individual burners.

Pressure and gauges. *Pressure* is the unit force imposed on a unit area by a fluid or gas, this force acting on the confining walls of a vessel. In English units, this is expressed as pounds per square inch, psi. *Gauge pressure* is that pressure indicated on gauges that measure the internal pressure of pressure vessels. This is a pressure that is above the atmospheric pressure surrounding the pressure vessel on the outside. *Absolute pressure* is the sum of the gauge pressure and atmospheric pressure. At sea level, the standard atmospheric pressure is 14.696 psi. It has also been defined as a 760-mm (29.921-in.) height of a mercury column at 32°F. In laboratory work requiring close precision, a mercury barometer must be used to obtain the exact mercury height and weight at different temperatures, because the weight of the column of mercury in a barometer also gives the pressure per unit area of cross section. At 32°F, mercury weighs 0.491 lb/in.[3] Thus, each inch of mercury height will exert 0.491 psi.

Pressures *below* atmospheric pressure are defined as vacuum conditions and are usually measured in inches of mercury.

Example What is the absolute pressure on an air tank when the pressure gauge reads 148 psi and a barometer indicates 29.45 in. of mercury?

solution

$$\text{Atmospheric pressure} = 29.45(0.491) = 14.46 \text{ psi}$$

$$\text{Absolute pressure} = 148 + 14.46 = 162.46 \text{ psia}$$

Example If the vacuum gauge on a condenser shows 28.20 in. of mercury, and a barometer indicates 29.58 in. of mercury, what is the absolute pressure in the condenser?

solution

$$\text{Absolute pressure (in inches of mercury)} = 29.58 - 28.20$$
$$= 1.38 \text{ in. of mercury}$$

Therefore,

$$\text{Pressure} = 1.38(0.491) = 0.678 \text{ psia}$$

The two main types of pressure gauges are the Bourdon tube and the diaphragm type. Figure 10.2a shows the interior mechanism of the single-tube Bourdon gauge with the dial removed. The bent tube of oval cross section is closed at one end and connected at the other to boiler pressure. The closed end is attached by links and pins to a toothed quadrant, which in turn meshes with a small pinion on the central spindle. As pressure builds up inside the oval tube, it attempts to assume a circular cross section, thus tending to straighten out lengthwise. This action turns the spindle by the links and gearing, causing the needle to move and register the pressure on a graduated dial.

Pressure gauge Code requirements. The boiler must have at least one pressure gauge so located and of such size that it is easily readable and which at all times indicates the boiler pressure. A valve or cock must be placed in the gauge connection (Fig. 10.2b) adjacent to the gauge so it can be removed for repairs. The gauge must be connected to the steam space or to the water column or its steam connection. For a steam boiler the gauge or its connection must have a

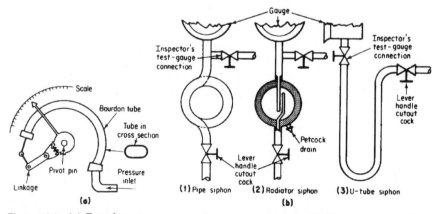

Figure 10.2 (a) Bourdon steam pressure gauge movement. (b) Siphons are used on pressure gauges to protect the gauge from exposure to high-temperature steam.

siphon for maintaining a water seal to prevent steam from entering the gauge tube. The connection of a pressure gauge must be a minimum of ¼-in. ID.

For temperatures over 406°F, no brass or copper tubing should be used. The pressure gauge dial should be graduated to twice the safety-valve setting, but in no case less than 1½ times this setting. A valve connection of at least ¼-in. pipe size must be installed on the boiler for the exclusive purpose of attaching a test gauge and, when the boiler is operating, for checking the accuracy of the boiler pressure gauge. This connection is known as the inspector's connection. The pressure gauge must be illuminated, free from objectionable glare or reflection that can in any way obstruct an operator's view while noting the setting on the gauge. The pointer on the gauge must be in a near-vertical position when indicating the normal operating pressure. This is also true of other pressure gauges in the boiler room that are used on auxiliaries. Pressure gauges must not be tilted forward more than 30° from vertical and then only when it is necessary for proper viewing of the dial graduations.

The siphon is simply a pigtail or drop leg in the tubing to the gauge for condensing steam, thus protecting the spring and other delicate parts from high temperatures. Three forms are shown in Fig. 10.2b. If there is danger of freezing during long periods of shutdown, the siphon should be removed or drained.

Temperature and gauges. Temperature can be classified into three scales:

1. *Fahrenheit:* On this scale the freezing point of water is 32°F, and the boiling point of water is 212°F, at atmospheric pressure. As can be seen, this gives a 180° spread on the scale.

2. *Centigrade:* The metric system uses this scale, with 100° between the freezing and boiling points of water. Conversion from one scale to the other involves this difference in the spread between freezing and boiling of water, and can be calculated using:

$$°F = \frac{9}{5}°C + 32 \qquad °C = \frac{5}{9}(°F - 32)$$

3. *Absolute temperature:* This temperature scale was derived from experiments that revealed that a perfect gas expands at a rate $\frac{1}{491.7}$ of the gas's original volume for each change in temperature of 1°F. On the centigrade scale, the expansion rate is $\frac{1}{273}$ of the original volume per °C. By lowering the temperatures of various substances, it has been established that all molecular activity would cease at 491.7°F below the freezing point of water, or −459.7°F.

This absolute temperature is called "absolute zero." Thus, to convert a Fahrenheit or centigrade temperature to absolute temperature, use the following equations: Absolute temperature, in Fahrenheit degrees is

$$°F_a = °F + 459.7$$

Absolute temperature, in Centigrade degrees is

$$°C_a = °C + 273$$

Example What is the absolute temperature of a substance at $-10°F$?

answer

$$°F_a = -10 + 459.7 = 449.7°F_a$$

Division of temperature-measuring devices is into mechanical devices (bimetallic and filled-system thermometers) and electronic devices (resistance temperature detectors, or RTDs; thermocouples; thermistors; semiconductors; and noncontact infrared instruments). Mechanical temperature-measuring devices are good for displaying temperatures, but they cannot transmit output signals and thus are not used with electronic or digital control systems. See Fig. 10.3a for the operating temperature ranges of temperature-measuring devices. This figure also shows a bimetallic thermometer (Fig. 10.3b) and a filled-system thermometer (Fig. 10.3c).

Mechanical Temperature Devices

Bimetallic thermometers. If a composite metal strip, made of two dissimilar metals welded or riveted together, is heated, it bends in the direction of the metal with the lower coefficient of expansion. The amount of bend for a given temperature change is repeatable and can be used to indicate temperature, provided the purity of the two metals is controlled. Usually the strip is formed into a coil, and a pointer is attached to the inner end. As the coil expands, the pointer indicates temperature on a scale (Fig. 10.3b).

Bimetallic thermometers are inexpensive and moderately accurate—typically 1 percent of reading for industrial types. The upper temperature limit is around 1000°F, but 500°F is the usual limit. Because of the scale layout, close resolution is difficult to obtain except for narrow-range thermometers. In industrial models the elements are usually sealed in a tube, since corrosion can degrade the metal strip.

Filled-system thermometers. Devices of this type (Fig. 10.3c) are made with an expandable gas or liquid in a hermetically sealed enclo-

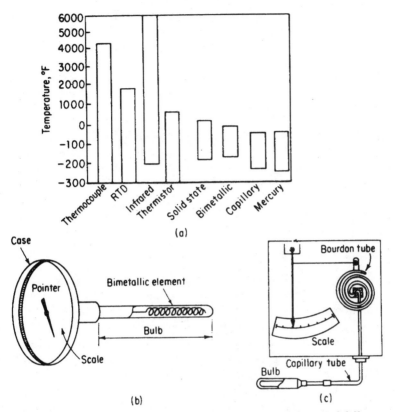

Figure 10.3 Temperature measurement. (*a*) Temperature limits of different temperature sensors; (*b*) bimetallic thermometer; (*c*) filled-system thermometer. (*Courtesy Power magazine*)

sure. When heat is applied, the pressure of the expanding fluid causes a helical tube to stretch, moving a pointer over a scale. Other variations use the fluid to operate a piston connected to an indicator, which displays temperature on a linear scale. This type is not often used for temperature measurement, but may be used to operate a switch for thermostatic control.

Electronic Temperature Devices

Resistance temperature detectors.   RTDs operate on the principle that an electrical conductor's resistance changes with temperature. Platinum is usually used because of its stability at high temperatures.

Thermocouples.   Thermocouples depend on the principle that when two dissimilar metals are joined, an electrical voltage is generated and this varies with temperature. This voltage can be measured with electronic circuitry. Thermocouples have a "hot" junction, which is the measuring point, and a "cold" junction, which is the reference point. Thermocouple sensors can be made in a variety of configurations and sizes and are extensively used for the continuous measurement of the temperatures of hot streams. These are widely used in boilers to check on circulations.

Thermistors. Thermistors are resistors that vary in their electrical resistance with temperature changes and thus are similar to RTDs. This quality can be used to measure temperature with the appropriate electrical circuitry. Shortcomings of thermistors in comparison with RTDs and thermocouples is a narrower operating range and the fact that the change of resistance with temperature is not linear.

Instrumentation Development

There has been a tremendous growth in the use of sensors and instrumentation in boiler plants for the purpose of better automatic control and efficient operation. Among the items measured in power plants are: pH, conductivity, dissolved oxygen, silica, hydrazine and sodium ion all for the purpose of controlling water quality in the boiler. Each loop in a boiler system has its own sensors, instrumentation and display needs. An example is the gas stack analyzer for emission control, shown in Fig. 10.4a. These instruments measure CO, CO_2, SO_2, and NO_x. With these readings, adjustments can be made on air-fuel ratios and other adjustments, if needed, to bring the values to design set points.

Among the most widely measured quantity is pH and conductivity. These are now measured at convenient points in a boiler plant on a continuous basis as a result of instrument and sensor development in industry. *Conductivity* is used to detect solids content in a solution, or boiler water, and the instrument is based on determining the flow of current across set points, and relating this to solid content of the solution. See Fig. 10.4b.

pH is the measurement in a solution of the hydrogen ion concentration, and this is related to the acidity or alkalinity of the solution, with 7.0 being the neutral point on a scale of 0 to 14. Values below 7 indicate an acidic condition, while values above 7 are alkaline. See Fig. 10.4c. The instrument is based on the electromechanical potential at the surface of glass electrodes. The readings are used to analyze water-treatment results.

The growth in power plant instrumentation and control has also been accelerated by the evolution of semiconductor technology, or "computer on a chip," commonly also called the microprocessor. This

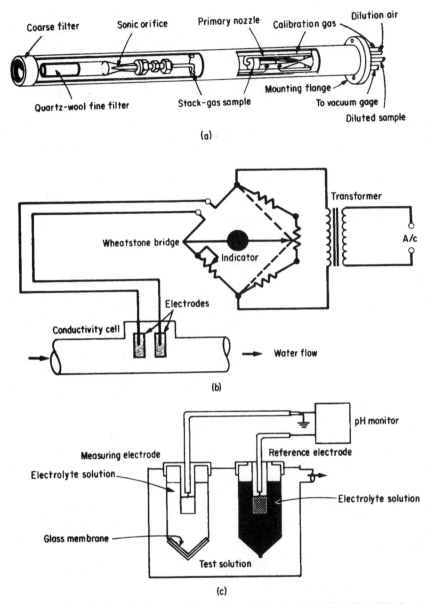

Coarse filter   Sonic orifice   Primary nozzle   Calibration gas   Dilution air

Quartz-wool fine filter   Stack-gas sample   Mounting flange   To vacuum gage

Diluted sample

(a)

Transformer

A/c

Wheatstone bridge

Indicator

Electrodes

Conductivity cell

Water flow

(b)

pH monitor

Measuring electrode   Reference electrode

Electrolyte solution

Electrolyte solution

Glass membrane

Test solution

(c)

Figure 10.4 (a) Stack probe for sampling and conditioning CO, CO_2, SO_2, NO_x from stack gases. These are transported to remote analyzers and display boards. (b) Conductivity electric circuit for measuring solids in solution. (c) pH monitor. (*Courtesy Power magazine*)

growth in microprocessors has resulted in the growth of advanced instrumentation and accompanying controls to achieve automatic operation, and also has the following advantages:

1. By programming starting and shutdown procedures through a computer, power plant equipment can be started up and shut down faster.

2. By data logging and trending of readings, detections of impending malfunctions is possible, and the operator can take corrective actions.

3. By setting operating limits closer to theoretical design limits, higher thermal efficiency can be achieved.

4. Emission control can be improved by more rigid control of air-fuel ratios and other factors controlling the combustion of fuel.

Admittedly, this trend to more automatic operation also reduces staff requirements in power plants. Thus, the skill level must be increased as less personnel are required in a modern plant. Operators must study the many automatic features that are being incorporated into instrumentation and controls in a modern boiler system.

However, there are certain ASME Code requirements on connections, valves, and appurtenances that are still required by jurisdictional authorities in order to maintain good operating practices for safety reasons.

Instrumentation and control calibration. The growth in instrumentation and controls now requires technicians to periodically test and calibrate the instrument against a standard of reference, to permissible deviations that are established for the field calibrator's guidance. Usually stickers are applied to show the date of calibration, and next due date. Larger plants have personnel specializing in instrument calibration, while smaller plants with fewer instruments may expect an operator to develop this skill or may farm it out to instrument servicing organizations.

Water Columns, Gauge Glasses, and Gauge Cocks

The gauge glass and gauge cocks are essential Code appliances for indicating the level of the boiler water. The water column (often omitted in railroad and marine practice) is installed between the gauge glass and the boiler. It serves to eliminate excessive fluctuations of water-level indication in the glass due to rapid boiler circulation or ebullition and thus acts as a steadying medium. See Fig. 10.5a, b, and c.

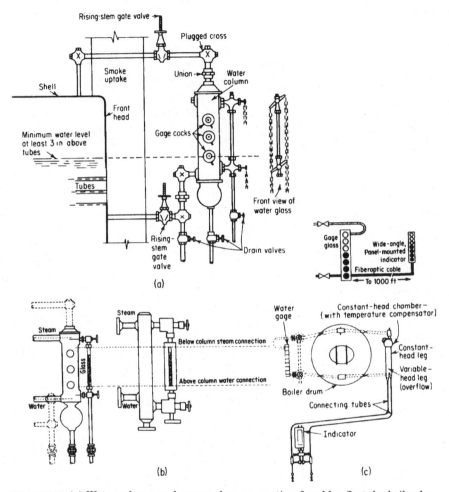

Figure 10.5 (*a*) Water-column and gauge glass connection for older firetube boiler has rising stem gate valves which must be sealed or locked open when boiler is operating. (*b*) Water-column connection has valves only to gauge glass connection. Note required location of top and bottom of gauge glass. (*c*) Remote water level indicators permit operator to check water level at a location away from the column connection to the boiler shell or drum.

Mirrors are sometimes used where, because of extremely high elevation, gauge glasses are not visible from the operating floor. Other styles of glass make use of remote gauge-glass indicators which may be located at the operating floor level.

The gauge-glass connections between the glass and the water column (or boiler if the column is omitted) should be at least ½ in., and a drain valve or cock should be provided in the bottom of the lower connection chamber. A shutoff valve is installed in both connections to permit replacement of a defective glass while the boiler is under pressure.

Gauge cocks are installed to serve as a check on the accuracy of the gauge glass and to determine the water level in the event of glass failure. If two independent gauge glasses are installed at the same elevation and at least 2 ft apart, the gauge cocks are usually dispensed with. Most codes require but two gauge cocks for locomotive boilers within 36-in. diameter or on firebox boilers within 5 hp. Larger boilers are fitted with three gauge cocks. These cocks may be mounted on the water column or directly on the boiler and are equally spaced, within the visible range of the gauge glass.

Water-column material is usually cast iron up to 250 psi. The ASME Code specifies that malleable iron may be used up to 350 psi, but steel is required for higher pressures. The piping connecting a water column to the boiler should be at least 1-in. diameter. Brass pipe for the lower connection is preferred up to about 150 psi. Where any bend or turn is necessary in the lower connection, crosses should be used instead of 90° elbows. The two unused openings should be fitted with pipe plugs, and these should be removed at each annual internal inspection (more often if required) to see that the connecting pipes are clear and to clean them if necessary.

Water-column drains are required at the bottom and in the lowest part of any pocket in the lower connection. A ¾-in. or larger gate-valve drain is used for sediment removal. A globe valve may be used on the nonpressure side of the gate valve for tightness if desired. Use of the gate valve permits a wire to be run through an obstructed valve and pipe without removing the boiler from service.

The steam connection should pitch toward the column and the water connection toward the boiler so that a false indication of level will not be shown by water trapped in the column when the boiler water level is declining.

Leveling a water column or a gauge glass is important. The *lowest visible point* in the gauge glass should be at least 2 in. above the elevation of the lowest safe water level as determined and specified by the boiler manufacturer.

In firetube boilers, the lowest safe water level is determined by the type of boiler. For scotch marine boilers, it is 2 in. above the highest point of the tubes or crown sheet. For horizontal-return-tubular boilers, it is 3 in. above the highest point of the tubes, flues, or crown sheet. For locomotive boilers under 36-in. diameter, it is 2 in. above the highest point of the crown sheet, and for boilers over 36-in. diameter, it is 3 in. For vertical tube (VT) boilers of the submerged-tube type, it is 2 in. above the upper tube sheet. For the dry-type VT boiler, it is one-third the height of tubes. In most standard types of watertube boilers, it is at least 2 in. above the lowest permissible water level as determined and specified by the boiler manufacturer.

Valves are not absolutely essential on steam and water connections to the water column (see Fig. 10.5b), but if used, they must be outside screw and yoke (see Fig. 10.5a), lever-lifting gate valves, stopcocks with lever handles, or other valve types that offer a straight-way passage and show by the position of the operating mechanism whether they are open or closed. Lock these valves or cocks open or see that they are sealed open, as required by the Code.

The water-gauge glass with its steam, water, and drain valves is placed on the water column as shown, along with the required number of gauge cocks. Damper regulators, feedwater regulators, steam gauges, and other pieces of apparatus that do not require or permit escape of an appreciable amount of steam or water may be connected to the pipes leading from water column to boiler.

Boilers operating at 400 psi and above. The ASME rule on gauge-glass connections for high-pressure boilers requires that each boiler have at least one water-gauge glass, except boilers operated over 400 psi, which must have two water-gauge glasses, connected to a single water column or directly to the drum. For power boilers with all drum safety valves set at or above 400 psi, two independent remote level indicators may be used instead of one of the two gauge glasses for boiler drum water-level indication. When both remote level indicators are in reliable operation, the one gauge glass may be shut off but must be maintained in serviceable condition. When the direct reading of the gauge-glass water level is not readily visible to the operator in her or his working area, two dependable indirect indications must be provided, either by transmission of the gauge glass or by remote level indicators.

Figure 10.5c shows a typical high-pressure remote level indicator. It is mounted on the boiler room instrument panel or installed at any other eye-level location convenient for the operator. The indicator is connected to fittings on the boiler drum by two small tubes. Changes in the boilerwater level cause corresponding change in static head in one of these tubes; static head in the other tube remains constant. Variations in the differential pressure at the indicator cause movement of the pointer which accurately indicates water level. The indicator is operated by the boiler water itself, using the pressure differential between a constant head of water and the varying head of water in the boiler drum. By means of a diaphragm-operated mechanism, with the diaphragm sides connected to high and low levels by tubing connection, the water level is shown by a graduated scale on the instrument.

Developments on remote indicators include using green glass ports for water, and red for steam showing through ports of a water level gauge. By means of fiberoptic cable, remote viewing is possible on a second port-hole type level gauge. Transmission of over 1000 ft is possible with fiber-optic cables. (See Fig. 10.5c.)

Safety Valves

ANSI B95.1 has an extensive listing of terminology related to safety valves. This is the most important safety device on a boiler, and can be the last defense against an overpressure explosion. A few definitions on this most important device will help in differentiating the different types that are available:

A *pressure relief device* is designed to relieve pressure or open to prevent a rise in internal pressure of a closed vessel in excess of the allowable working pressure.

A *pressure relief valve* is actuated by inlet pressure having a gradual lift proportional to the increase in pressure and is primarily used for liquid service in preventing over pressure.

A *safety valve* is a pressure relief device actuated by set pressure, but characterized by a rapid or pop opening action to immediately release the steam from a closed vessel. It may also be used for air release service.

A *pilot-operated pressure relief valve* is a pressure-relieving device where the major relieving device is actuated and controlled by a self-actuated auxiliary "trigger" pressure relief valve.

There are many Code requirements on this important safety device.

Safety-valve construction. It is of the utmost importance that a safety valve be correctly constructed. Such construction may be ensured by specifying that the construction must conform to ASME or National Board approved and registered direct spring-loaded pop type, properly marked as to pressure and capacity and equipped with a testing lever. The pressure setting must match either the maximum allowable pressure for which the boiler is designed or, an older boilers, the maximum pressure allowed by state or city law. The capacity of the safety valve should be at least equal to the maximum steam that can be generated by the boiler. An ASME standard safety valve bears the following information stamped on the valve body or name plate (see Fig. 10.6a):

Manufacturer's name or trademark

Manufacturer's type or design number

Size, in.

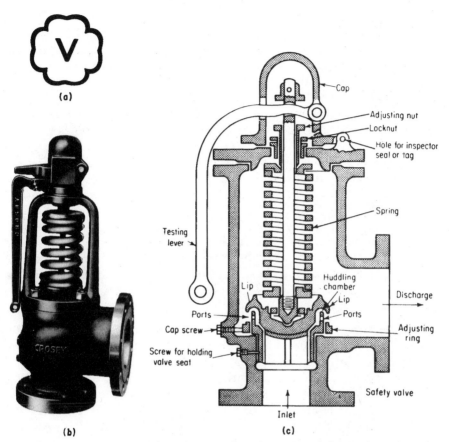

Figure 10.6 Pop safety valves. (*a*) ASME symbol for approved safety valve. (*b*) Superheated-steam safety valve has exposed spring. (*Courtesy Crosby Valve & Gage Co.*) (*c*) Huddling chamber provides pop action for safety valve.

Seat diameter, in.

Pressure at which valve is set to blow, psi

Blowdown, psi

Discharge capacity, lb/hr

Capacity lift, in.

ASME standard symbol

When a safety valve bears the ASME or National Board stamping, it is the manufacturer's guarantee that the rules of the ASME Code have been followed in the construction of the product.

In brief, the major constructional requirements are that the disk and seat be of noncorrosive material, the seat being fastened to the body so that it cannot lift with the valve disk. All parts should be constructed so that no failure of any part will interfere with full discharge capacity of the valve. The seat may be inclined at any angle between 45° and 90°.

The safety valve must be of the direct spring-loaded type. Code states do not allow the installation of weight and lever types or deadweight safety valves, for the adjustment of such valves is too easily tampered with. Their use is not recommended anywhere.

The safety-valve *spring* is usually of round or square stock, for maximum clearance between the coils. If the coils come in contact, the valve cannot lift. It is for this reason principally that the maximum range of adjustment permitted with a spring is 10 percent of its rated setting. This rule is for safety valves set at up to 250 psi. For higher pressures, the allowable range of adjustment is 5 percent of the spring rating. If the setting is changed to a greater deviation, a new spring and nameplate should be installed by the manufacturer's representative.

Connections. Safety valves should be connected directly to an independent nozzle on the boiler without any intervening valves of any description. Threaded connections may be used up to and including 3-in. diameter. For boilers operating at over 15 psi, all safety valves over 3-in. diameter should have flanged inlet connections.

It is important that the nozzle opening to and the discharge piping from the safety valve be at least as large as the safety-valve connection. If two or more safety valves are connected on a common nozzle or fitting, the area of this nozzle or fitting should at least equal the combined areas of all safety valves served.

A *lifting lever* is required in order to lift the valve from its seat when there is 75 percent of the popping pressure in the boiler. Lifting levers that can lock the valve in raised position are not approved.

Blowback, or *blowdown,* is the number of pounds of drop in boiler pressure from the value at which a safety valve pops to the point where the valve closes. For pressure up to 100 psi, the blowback should be not over 4 percent, but not less than 2 lb. Higher pressures call for a minimum blowback of 2 percent of the popping pressure. Safety valves used on forced-circulation boilers of the once-through type may be set and adjusted to close after blowing down not more than 10 percent of the set pressure. The valve for this special use must be so adjusted and marked, and the blowdown adjustment must be made and sealed by the manufacturer. A lesser blowback may result in a destructive chattering (rapid popping and seating) action. Too great a blowback wastes steam and fuel. Although the Code is silent regarding maximum blowback, it is usually good practice to adhere to the minimum allowed.

The ASME Code specifies that the blowback must be adjusted and sealed by the manufacturer or an authorized representative. The adjustment is effected by the *blowback ring,* or *adjusting ring,* in valves of the type shown in Fig. 10.6c. The adjusting-ring access screw is removed, and with a pointed tool or screwdriver the ring is rotated part of a turn on its threaded sleeve. This raises or lowers the ring, changing the area of the huddling chamber.

The action of the huddling chamber and blowback ring is to expose a greater area to escaping steam when the valve lifts slightly. The steam pressure acting on an increased area gives a greater total lifting pressure against the spring, which results in the valve opening with a pop, with the abrasive cutting action of the steam (known as "wire drawing") which might be caused on the valve and seat by slow opening being thus eliminated.

The Crosby nozzle-type safety valve (Fig. 10.7) dispenses with the blowback-ring principle. When the valve first lifts, steam strikes the adjusting ring and is deflected downward. The reaction of the diverted steam flow causes the valve to pop open. This construction results in high lift and high capacity.

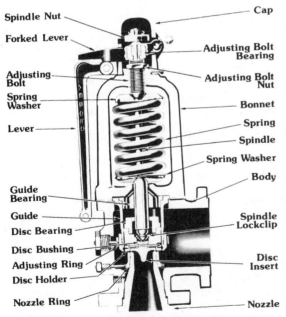

Figure 10.7 Cross-sectional view of top-guided, full-nozzle reaction-type safety valve for steam temperatures to 1020°F. (*Courtesy Crosby Valve & Gage Co.*)

Discharge from safety valve. Discharge pipes should be used if the discharge is located where workers might be scalded. A proper discharge pipe is as essential to the safety of plant personnel as the safety valve is to the boiler. Too often a worker has been opening a stop valve when a safety valve, having no discharge pipe and pointing directly at the person, pops. To be standing in the path of a high-pressure 3- or 4-in. jet of steam is usually fatal.

Every discharge pipe should be at least 6 ft high. If headroom makes it impossible to terminate the discharge pipe within a reasonable distance from the ceiling, it should extend out through the building wall or roof. If it is a flat roof where workers may be, the discharge pipe should extend at least 6 ft above it. If a horizontal discharge pipe is more practical, it should discharge at a safe location.

It is essential that the discharge pipe diameter be at least equal to the size of the safety valve. If a length of over 12 ft is necessary, it is better to use a diameter ½ in. larger for each 12 ft in length. A long line with no increase in diameter will cause a back pressure because of flow friction and may cause serious chattering of the safety valve. All 90° bends should be avoided if possible.

The discharge pipe should be supported independently of the safety valve. Serious stresses may be set up in the safety-valve body, connection, or boiler nozzle by the weight of a heavy, unsupported discharge pipe.

After a safety valve has blown many times, it is not uncommon for slight leakage to develop. Condensation of this leakage may gradually fill an undrained escape pipe with water. This condition alone prevents the safety valve from blowing at its set pressure. The popping point will be increased 1 lb for every 2.3-ft elevation of water in the escape pipe. Also, in an outdoor discharge pipe exposed to severe winters, ice may form and seriously interfere with proper safety-valve operation. Every discharge pipe should have a ⅜- or ½-in. open drain at its lowest point. This drain should be conducted off the boiler top in order to prevent external corrosion induced by dampness. Figure 10.8 shows a correctly installed safety valve.

Capacity. The number and capacity of safety valves required on boilers are governed by Code rules. The following rules on boilers must be followed:

1. The safety-valve capacity on a boiler must be such that the safety valve (or valves) will discharge all the steam that can be generated by the boiler (this is assumed to be the maximum firing rate) without allowing the pressure to rise more than 6 percent above the highest pressure at which any valve is set, and in no case more than 6 percent above the maximum allowable pressure.

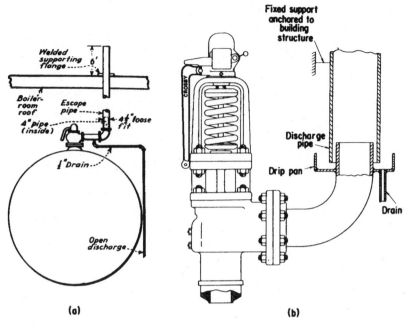

Figure 10.8 Discharge pipe from the safety valve should have a drain to remove condensate. (*a*) Low-pressure pipe; (*b*) high-pressure pipe.

2. The boiler manufacturer is now required by Code rules to stamp the nameplate and also to show on the master data report what the maximum-designed steaming capacity of the boiler is. The old minimum safety-relieving capacity rules based on heating surface and fuel fired, as shown in Fig. 10.9, are now placed in the nonmandatory appendix paragraphs of Section I of the ASME Code and can be used as a guideline if the nameplate does not show the maximum capacity as now stipulated in the Code.

3. For electric boilers, the relieving capacity is determined by multiplying the kilowatt input by 3½ to obtain the pounds per hour of steam-relieving capacity.

4. For HTHW boilers, the required steam-relieving capacity in pounds per hour is determined by dividing the maximum Btu output (for the fuel being fired) of the boiler by 1000.

Popping tolerance and number of safety valves. The popping-point tolerance that a valve must meet is the following, on a plus or minus basis:

1. Two psi for pressures up to and including 70 psi.

2. Three percent for pressures for 71 to 300 psi.

Minimum Pounds of Steam per Hr per Sq Ft of Surface on HP Boilers

Surface	Fire-tube boilers	Water-tube boilers
Boiler heating surface:		
Hand-fired................................	5	6
Stoker-fired..............................	7	8
Oil-, gas-, or pulverized-fuel-fired........	8	10
Waterwall heating surface:		
Hand-fired...............................	8	8
Stoker-fired..............................	10	12
Oil-, gas-, and pulverized-fuel-fired......	14	16

NOTE: When a boiler is fired only by a gas having a heat value not in excess of 200 Btu per cu ft, the minimum safety-valve relieving capacity may be based on the values given for hand-fired boilers above.

Figure 10.9 The old rule of determining safety relief valve capacity by the heating surface and fuel fired has been replaced by manufacturer's stamping of the capacity on the boiler nameplate per Code requirements. The above table is now a guide for those boilers not stamped with the maximum capacity of output.

3. Ten psi for pressures over 301 to 1000 psi.

4. One percent for pressures over 1000 psi.

One or more safety valves must be set at or below the maximum allowable pressure. The highest pressure setting of any safety valve cannot exceed the maximum allowable working pressure by more than 3 percent. The range of pressure settings of all the saturated steam safety valves on the boiler cannot exceed 10 percent of the highest pressure setting to which any valve is set.

Each boiler requires at least one safety valve, but if the heating surface exceeds 500 ft^2 or the boiler is electric with a power input over 500 kW, the boiler must have two or more safety valves. When not more than two valves of different sizes are mounted singly on the boiler, the smaller valve must be not less than 50 percent in relieving capacity of the larger valve.

Superheater safety valves. Safety valves discharging steam over 450°F from superheaters should have a flanged or welded inlet connection for all sizes. Also, such valves should be constructed of steel or alloy steel throughout, suitable for heat resistance at maximum steam temperatures. The spring in superheater safety valves should be fully exposed (Fig. 10.6b) so that it will not be in contact with high-temperature steam.

Every superheater attached to a boiler with no intervening valves between the superheater and boiler requires one or more safety

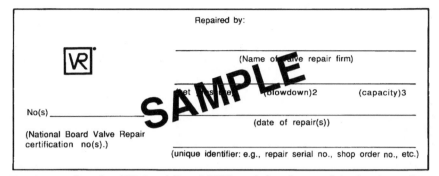

Figure 10.10 Sample of National Board "VR" symbol and nameplate showing safety valve was repaired by a NB-authorized organization.

valves on the superheater outlet header. With no intervening stop valves between the superheater and the boiler, the capacity of the safety valves on the superheater may be included in the total required for the boiler, provided the safety-valve capacity in the boiler is at least 75 percent of the aggregate safety-valve capacity required for the boiler.

The superheater safety valves should always be set at a lower pressure than the drum safety valves so as to ensure steam flow through the superheater at all times. If the drum safety valves blow first, the superheater could be starved of cooling steam, leading to possible superheater tube overheating and rupture.

Testing of safety valves. Corrosion and deposits on valve and valve seat are caused by the safety valve not having been lifted for a long period. To avoid this most dangerous condition on automatic-fired (especially low-pressure) boilers, the safety valve should be periodically raised by using the hand lever, or preferably by raising the steam pressure to the popping point. The latter practice should be done only with constant attendance at the boiler, and then only under the supervision of trained personnel who will carefully watch boiler pressure and immediately shut down the boiler if the pressure starts exceeding the maximum allowable. The lever testing of safety valves should be done with at least 75 percent boiler pressure on the safety valve.

Other problems that may be encountered on safety relief valves besides corrosion and scale deposits include damaged seating surfaces from corrosion, foreign particles (rust and scale) and weld beads, obstructions in the piping going to or from the safety valve, causing chatter, pipe strains on the discharge side causing the safety valve to leak sporadically, and improper blowdown ring setting which can also cause chatter. Other more obvious defects are broken or corroded springs and settings too close to operating pressure.

Repairs to safety valves. The NB now has a procedure to qualify safety relief valve repairers and then issues a "VR" stamp to the qualified organization. The procedure requires the maintaining of a quality-control program by the repair organization and also the demonstration of repair capability to the satisfaction of a representative of the National Board. This includes acceptance testing by an approved laboratory to show that repaired valves meet the functional and operating criteria (set pressure and capacity) as stipulated by the section of the ASME Boiler Code. Repaired valves may have a nameplate attached showing the NB "VR" repair symbol as shown in Fig. 10.10. Many jurisdictions now require safety relief valves to be repaired only by NB-qualified "VR" stamp holders; therefore, to comply with jurisdictional requirements, operators of boiler plants should check with the jurisdictional inspector on any safety valve repairs. Additional information regarding the National Board "VR" stamp for safety valves can be obtained by contacting:

The National Board of Boiler and Pressure Vessel Inspectors
1055 Crupper Ave.
Columbus, OH 43229
(614) 888-8320

Valves and Piping

Valves on boilers, as shown in Fig. 10.1, include steam valves on the main headers; feed valves on the water feed to a boiler; drain valves on water columns, gauge glass, and drain connections; blowdown valves for both surface blowoff and bottom sediment blowoff; check valves on feed lines; and nonreturn valves on steam mains. Materials for power plant piping are mostly carbon steels or austenitic stainless steel. In nuclear applications, changes in metal properties as a result of radiation must also be considered in piping design. Factors affecting piping in power plants include thermal transients due to rapid changes of temperature from operating problems, water hammer due to entrapped water in piping, creep and vibration from connected machinery requiring close attention to balancing of machinery. Large-amplitude and low-frequency seismic vibration must be designed for in nuclear plants. Figure 10.11 shows some pipe support systems.

Each main or auxiliary discharge steam outlet, except the safety valve and superheater connections, must have a stop valve placed as close to the boiler as possible. When the outlet size is over 2 in. (pipe size), the valves must be the outside-screw-and-yoke (O S & Y) type to

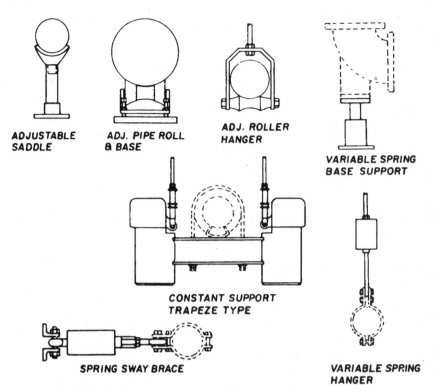

Figure 10.11 Methods of supporting power-plant piping.

indicate by the position of the spindle whether the valve is open or closed. When two or more boilers are connected to a common steam main, the steam connection from each boiler having a manhole must have two stop valves in series, with an ample free-blowing drain between them. The discharge of the drain must be in full view of the operator when opening or closing the valves. Both valves may be of the O S & Y type, but one should be an automatic nonreturn valve. This should be placed next to the boiler so that it can be examined and adjusted or repaired when the boiler is off the line. Steam mains going into a plant from the boiler should be adequately supported.

A *nonreturn valve* (Fig. 10.12e) is used sometimes as a stop valve on the main steam line next to a steam boiler. The function of stop-checks, or nonreturn valves as they are frequently called, is as important as pop safety valves on boilers where two or more units are connected to the same header. They automatically prevent backflow from the header should a boiler fail. They simplify the work of cutting out a boiler or bringing a cold boiler into operation. They protect boiler repair or inspection crews against steam backflow should the header

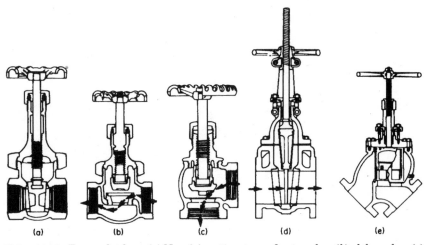

(a)   (b)   (c)   (d)   (e)

Figure 10.12 Types of valves. (*a*) Nonrising-stem type of gate valve; (*b*) globe valve; (*c*) angle valve; (*d*) outside screw-and-yoke type of gate valve; (*e*) nonreturn valve for steam line. (*Courtesy Crane Co.*)

valve be accidentally opened. No multiboiler plant should be without stop-checks. They are available in straight-way Y patterns, angle Y patterns, globe, and angle patterns, depending on working pressure.

Valves of various types are shown in Fig. 10.12. As a rule, the *globe valve* (Fig. 10.12*b*) is used where positive tightness against leakage is desired and where the fluid controlled is practically free from suspended solids. It should be noted in the figure that both the valve disk and the seat are renewable. If they become worn, all that is necessary is to remove pressure from the pipe line, unscrew the valve bonnet, and renew these parts. The valve body does not have to be disturbed from its piping connections.

When a globe valve is used in the feedwater piping to a boiler, it is important for the flow to enter under the valve disk. If it entered from above and the disk became detached from the stem, the valve would automatically close, the feed to the boiler being thus prevented. As shown in Fig. 10.12*b*, fluids change direction when flowing through a globe valve. This seating construction increases resistance to—and permits close regulation of—fluid flow.

The disk and seat can be quickly and conveniently reseated or replaced. This feature makes them ideal for services that require frequent valve maintenance. Shorter disk travel saves operators time when valves must be operated frequently.

Angle valves (Fig. 10.12*c*) have the same operating characteristics as globe valves. Used when making a 90° turn in a line, an angle valve reduces the number of joints and saves makeup time. It also gives less restriction to flow than the elbow and globe valve it displaces.

The *gate valve* may be one of two types: the nonrising-stem inside-screw and the rising-stem outside-screw-and-yoke types (Fig. 10.12*a* and *d*). Gate valves are best for services that require infrequent valve operation and where the disk is kept either fully opened or closed. *They are not practical for throttling.* With the usual type of gate valve, close regulation is impossible. Velocity of flow against a partly opened disk may cause vibration and chattering and result in damage to the seating surfaces. Also, when throttled, the disk is subjected to severe wire-drawing erosive effects.

The plug gate valve, however, is excellently suited for throttling service. The gate valve operates on the wedge principle, with considerable seat contact. Also, since there is straight-through flow, no dam is offered to trap sediment or pieces of scale. Hence, the gate valve should be used for water-column drains and for similar services.

The rising-stem-type gate valve is used where it is urgent that there be a visible indication that the valve is in the open position. This type is required when shutoff valves are used in the connecting pipes between a water column and a boiler.

Many other modifications of the globe and gate valves are available for special services with various fluids.

Blowdown valves and fittings. A blowdown connection is required at the lowest waterspace of a boiler to serve three purposes:

1. To remove precipitated sludge or loose scale

2. To permit rapid lowering of the boiler water level if it has become too high accidentally

3. As a means of removing water from the boiler system so that fresh water may be added to keep concentration of solids in the boiler water below the point where difficulties may be experienced

Since at least one of these functions may be performed under emergency conditions, it is essential that the blowdown valves conform to rigid specifications. Globe valves are not permitted under any condition, for inherent in their design is a tendency for a dam or pocket to form, where sediment may build up.

Blowdown valves and fittings should be designed for at least 25 percent greater pressure than the maximum allowable pressure on the boiler.

Cast-iron fittings may be used between the blowdown valve and the boiler for pressures of 100 psi or less, but steel is preferred. Steel fittings are an ASME Code requirement for pressures over 100 psi, and steel valve construction for over 250 psi.

At least two blowdown valves (Fig. 10.13a) should be used if the pressure exceeds 100 psi. At least one of these valves should be a slow-opening valve. The ASME Code defines a slow-opening valve as one requiring at least five full 360° turns between full-open and closed position. Where there is more than one blowdown connection

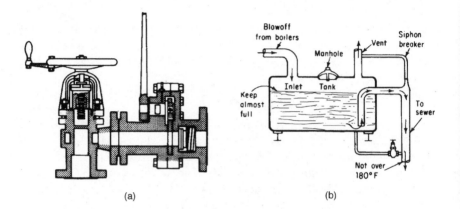

(a)   (b)

Minimum pipe size		
Boiler blowoff size, in.	Outlet size, in.	Vent size, in.
Minimum ¾	¾	2
1	1	2½
1¼	1¼	3
1½	1½	4
2	2	5
2½	2½	6

	Minimum tank size		
Boiler rating, HP	Diameter, in.	×	Height, in.
2–20	18		24
21–50	24		30
51–100	30		36
101–200	36		36
201–400	36		42
401–800	42		48
801–1000	48		60

Note: Use 34.5 lb/(hr/HP) to convert to horsepower.

(c)

Figure 10.13 Blowdown valves and blowdown tanks are used to remove sediment from boilers. (a) Angle and quick-opening blowdown valve where two blowdown valves are requried if boiler pressure is over 100 psi. (b) Blowoff tanks prevent hot water and steam being blown into sewers or open space in order to prevent scalding of personnel. (c) NB-recommended pipe sizes and blowoff tank sizes based on boiler horsepower rating.

from a boiler, joining a common blowdown header, one valve on each independent line and a master valve on the header meet ASME Code regulations. The Code specifies further that a double blowdown valve in one casing is permissible provided that failure of one valve does not affect the other.

A *blowoff tank* is necessary when no open space is available into which blowoff from the boilers can discharge without danger of accident or damage to property. For example, discharging to a sewer would probably damage the sewer by blowing hot water under high pressure directly into it. A good blowoff tank installation is always nearly full of water (Fig. 10.13*b*). The National Board (NB) of Boiler and Pressure Vessels Inspectors has published recommended rules for boiler blowoff equipment, including methods of calculating the size of blowoff tank needed for the pressure and capacity of the boiler under consideration. Figure 10.13*c* shows the NB-recommended pipe-connection sizes for blowoff tanks, and minimum-size tank based on boiler capacity. The tank must be designed in accordance with Section VIII of the ASME Boiler Construction Code, for a working pressure of at least one-fourth the maximum working pressure of the boiler to which it is connected. In no case, however, can the plate thickness be less than ⅜ in.

Check valves. *Check valves* are used where unidirectional flow is essential, as when feedwater flows into a boiler. The swinging disk in the valve (Fig. 10.14*a*) closes against its seat if the flow tends to reverse. By bleeding pressure from the piping and removing the bonnet and side plug, the valve and seat may be ground to a new face when worn. Also, all these parts are renewable. Being nonreturn valves, check valves are used to prevent backflow in lines. In operating principle, all check valves conform to one of two basic patterns. Shown is the swing check type (Fig. 10.14*a*). Flow moves through these valves in approximately a straight line comparable to that in

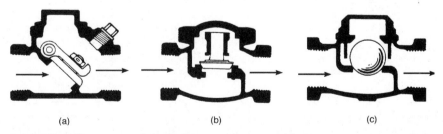

(a)   (b)   (c)

Figure 10.14 Some types of check valves. (*a*) Swing-disk has replaceable composition washers on disc that absorbs most wear. (*b*) Piston check valve is used for high-temperature service, providing tight shutoff. (*c*) Free-floating ball check valve provides good seating life as ball never seats in the same place.

gate valves. In lift or piston and ball check valves (Fig. 10.14b and c) flow moves through the body in a changing course, as in globe and angle valves. In both swing and lift types, flow keeps the valve open while gravity and reversal of flow close it automatically.

Primary caution to be observed when installing any check valve, whether swing or lift check type, is to see that flow enters at the proper end, i.e., that the disk opens with flow. If you follow the marking on the body of the valve, you can be sure the valve is properly installed and that the disk will be properly seated by backflow or by gravity when there is no flow.

Steam Traps

Steam traps should be installed in lines wherever condensate must be drained as rapidly as it accumulates and wherever condensate must be recovered for heating, for hot-water needs, or for return to boilers (Fig. 10.15). They are a "must" for steam piping, separators, and all steam-heated or steam-operated equipment.

The purpose of a steam trap is to have the steam first give up the latent heat of vaporization, about 970 Btu/lb, by stopping the steam flow through the trap until it condenses, and then opening to let the condensate flow into the boiler water loop. This permits the steam to give up its energy to do useful work and thus contributes to saving on fuel costs.

Another feature of steam traps is vents incorporated in their design so that entrapped air or gases can be vented from the steam-condensate piping system. This venting prevents

1. Loss of heat transfer because air acts as an insulator within the steam piping system
2. Oxygen corrosion within the steam piping system
3. Accumulation of gases such as carbon dioxide that could form carbonic acid, which attacks and corrodes steam metal piping

There are many types of steam traps as illustrated in Fig. 10.15. They all require periodic inspections and maintenance, because the following detrimental conditions will affect their proper operation.

1. Dirt, rust, weld beads and similar contaminants inside a trap will affect its internal movements. One way to prevent this is to install strainers ahead of the trap. The strainers will have to be cleaned periodically to remove the contaminants. If corrosion products are found, a review should be made of the feedwater and boiler water treatment methods so that corrosion is minimized. This may also require a check to make sure the trap vents are functioning.

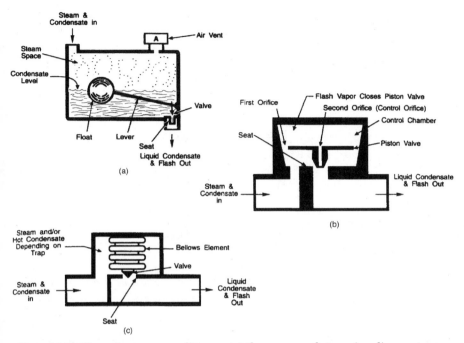

Figure 10.15 Steam traps are used to prevent the passage of steam in a line or steam-using apparatus until the steam gives up its heat of condensation, and then the trap opens to allow the condensate to flow to collection tanks or back to boiler condensate tanks. Their mode of operation varies. (*a*) A closed float trap opens and closes by changes in the condensate level. (*b*) Piston traps utilize the heat energy in hot condensate and the kinetic energy in steam to open and close the trap. (*c*) The bellows trap is a thermostatic trap that has a bellows with a fluid that responds to temperature changes, and thus opens and closes the trap, usually based on steam saturation temperature for the pressure involved. (*Courtesy Yarway Corp.*)

2. It is important to place traps at low points in the steam line, and at proper intervals so that condensate separates out and thus prevent the occurrence of water hammer. Steam and water traveling at high speed in a pipe can cause large impact forces if suddenly stopped by a trap or valve. The shock caused by water hammer can break valves, traps, and elbows of a piping system.

3. It is also necessary to guard against condensate that is in traps possibly freezing in subfreezing temperatures and causing trap damage. This is especially true for outdoor installations or unheated buildings. There are self-drain traps available to prevent this. Also used is heat tracing of steam and condensate lines that may be exposed to the freezing hazard.

4. To avoid traps closing and causing condensate to back up into the steam line, modern practice is to install traps that fail in the open position. This can cause energy loss, but prevents the risk of equipment "upstream" being damaged by water.

Detecting faulty traps. Leaking or defective traps cause an energy loss besides possibly endangering process equipment. Periodic inspections are, therefore, needed to make sure they are operating as designed. Methods used:

1. If the condensate pressure is above atmospheric pressure, the vent valve can be opened after the trap to note if water or steam is coming out. High-temperature condensate may flash into steam; however, this can be recognized because flashing steam usually has a cloudlike appearance, whereas live steam coming out of the vent is not as visible, and it also escapes at high velocity. Live steam, of course, means that the trap is defective and needs checking for repairs.

2. Noise change detection or sound inspection methods used include careful listening with a rod placed on the trap or using available sonic detectors. Most traps, if they are operating properly, give an opening and closing sound as steam comes in and condensate is allowed to flow out. If not operating properly, there is the pronounced sound of steam blowing, or "whistling" through the trap without giving up the heat of condensation.

The piston trap illustrated in Fig. 10.15 is used for high-pressure service and is rated as a thermodynamic trap. On start-up such as on a steam line going to a steam turbine, pressure created by the cold condensate lifts the piston valve, thus discharging the condensate in the steam line. When the steam temperature and condensate temperature approach saturation temperature, this causes the condensate in the control chamber to flash into steam and chokes the flow through the control orifice, increasing the pressure in the control chamber. The pressure increase, acting on the larger effective area of the piston than the inlet pressure, causes it to snap shut, and this prevents steam flow through the trap. When cooler condensate reaches the trap, the control chamber pressure drops, flashing ceases, and the trap reopens to let the condensate out, then the cycle is repeated.

Feedwater and Water Level Regulation

Regulating water level and feedwater flow can be performed by a self-contained regulator as shown in Fig. 10.16. More modern level control may be by control valves that depend on sensors to detect the level, then transfer this measurement by pneumatic or electric signals to an actuator or to a microprocessor, which then sets the level to desired or required set points.

Older boilers are still equipped with self-contained feedwater and level regulators. There were three general classes of feedwater regulator

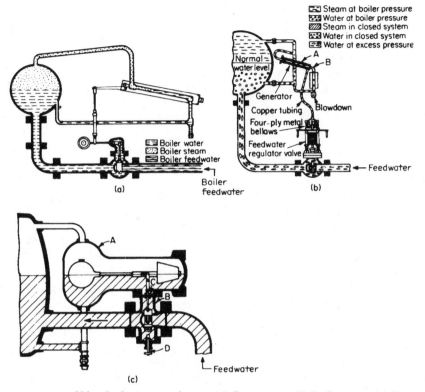

Figure 10.16   Older feedwater regulators. (*a*) Copes type; (*b*) Bailey type; (*c*) Stets type. (*Courtesy Bailey Meter Co.*)

in use for all sizes of power boilers: (1) the thermostatic-mechanical regulator; (2) the thermostatic-fluid type; and (3) the float-operated style.

The Copes feedwater regulator (Fig. 10.16*a*) is a thermostatic-mechanical type. Each end of the inclined tube is connected to the boiler; thus, when the boiler water level is normal, the water level will be about midway in the inclined tube. Owing to the inclined position of the thermostatic tube, a 1-in. vertical change in the boiler water level will cause several-inches change in the water position in the tube. This fact makes the regulator extremely sensitive in operation. One end of the tube is connected through a bell crank and linkage to the feedwater control valve. As the boiler water level falls, more of the tube length is exposed to steam; the tube expands and opens the feedwater control valve. As the water level rises, the tube cools, contracts, and closes the feedwater valve.

The Bailey regulator represents the thermostatic-hydraulic type (Fig. 10.16*b*). An inclined thermostatic tube is used. This tube is surrounded by a jacket having fins to dissipate heat to the atmosphere and make it rapidly responsive to temperature changes within the thermostatic tube. The jacket contains water which through a closed piping system connects with a metal bellows on top of the feedwater controlling valve. A spring tends to balance the valve against the fluid pressure on the siphon or bellows. The feedwater valve position is controlled by the pressure of the fluid as caused by temperature changes of water-level (and steam-space) variation in the thermostatic tube.

The Stets regulator represents the third type (Fig. 10.16c). Here a float chamber is installed and connected to the boiler at the same elevation as the normal water level in the boiler. A ball float rises and falls with the boiler water level and, through a linkage system, controls the position of the feedwater valve.

Three-element feedwater regulator. In a three-element feedwater control, as shown in Fig. 10.17, steam flow, feedwater flow, and water level are measured and recorded by mechanically operated meters. Measurements of steam flow and water flow are balanced against each other with differential linkage. A pilot control is connected to the linkage so that any difference between the amounts of steam flow and of water flow causes a change in the pneumatic output signal. This signal is transmitted to an air relay where it is combined with the pneumatic signal from the water-level recorder.

A change in boiler load unbalances the differential linkage, thus producing a change in the output of the pilot control. That in turn changes the output of the air relay. This new signal repositions the feedwater control valve, admitting required water into the boiler

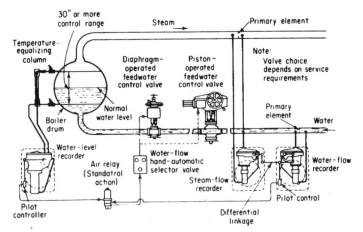

Figure 10.17 Three-element regulator controls steam and water flow to boiler and also maintains water level.

equal to the steam flow out of the boiler. The resulting change in feed-water flow rebalances the differential linkage and brings the pilot-control signal back to its neutral point. As a final check, and to ensure having the proper drum level, the signal from the pilot control in the water-level recorder readjusts the feedwater control valve, if required. The selector valve in the system provides automatic or remote manual control. Under normal operating conditions, the control pressure gauge on the selector valve is an indication of valve position.

Low-Water Fuel Cutoffs

A low-water cutoff (Fig. 10.18), separate from the programming-sequence control, immediately shuts down the boiler if the water drops to a dangerously low level. Three types are as follows:

1. The float-magnet type (Fig. 10.18a) has a ferrous plunger on one end of a float rod. The plunger slides within a nonferrous sleeve. A permanent magnet, with a mercury switch affixed, is supported by a pivot adjacent to the nonferrous sleeve. Under normal water conditions, the ferrous plunger is above and out of reach of the magnetic field. In this position, the mercury switch is in a horizontal plane, keeping the burner circuit closed. But if the boiler water level drops, the float also drops, bringing the ferrous plunger within the magnetic field. Then the magnet swings through a small arc toward the plunger; the mercury switch tilts, opening the burner circuit.

2. The float-linkage type (Fig. 10.18b) has a float connected through linkage to a plate supporting a mercury switch. Because the plate is horizontal in the normal water-level position, the switch holds the burner circuit closed. If the water level drops, the float drops, tilting the plate so the switch opens the circuit.

3. The submerged-electrode type (Fig. 10.18c) uses boiler water to complete the burner circuit. If the water level drops below the electrode tip, current flow is interrupted, shutting down the burner. On firetube boilers, the low-water cutoff generally includes an intermediate switch that controls the feed pump.

The automatically operated boiler firing fuel in suspension can be exposed to low-water damage if there is an interruption in the water feed to the boiler. The low-water fuel cutoff is the last defense against possible serious overheating damage to the boiler; therefore, it should be tested and maintained properly. For unattended boilers, the testing should be done at least once per week. Except for the probe type, the testing will also clear any sediment out of the bowl in which the sensing element, such as a float, is located. With the burner firing, the low-water cutoff drain valve is opened, and the burner should

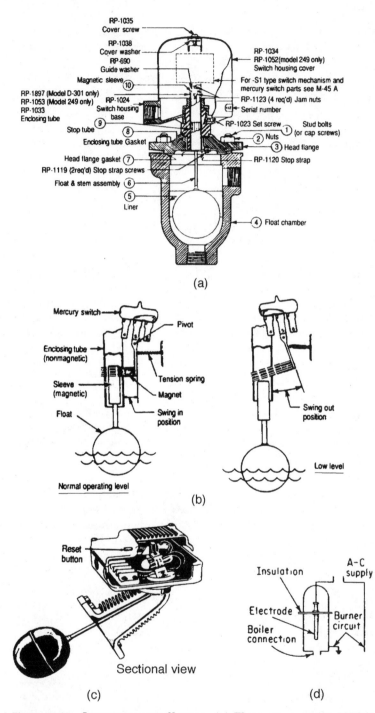

(a)

(b)

(c)   (d)

Sectional view

Figure 10.18   Low-water cutoff types. (a) Float-magnet type LWCO.
(*Courtesy Magnetrol Corp.*); (b) mercury switch operation on float-magnet
type LWCO; (c) float-linkage type LWCO; (d) electrode probe-type LWCO.

shut down as the water level in the bowl drops below the line usually indicated by the low-water cutoff manufacturer. If the burner does not shut off, service is required on the low-water cutoff mechanism. Periodic testing and blowing down the low-water cutoff will prevent scale, mud, and rust from building up in the bowl, which can result in restricting the proper movement of the low-water cutoff, and thus prevent it from operating when a low-water condition does develop.

Combination water feeder control and low-water fuel cutoff operating off one sensor, or float, are very prone to developing low water if the sensing element becomes disabled or stuck in deposits. This condition interrupts feed and also the low-water protection. Most jurisdictions require a second independent low-water fuel cutoff to prevent one mechanism failing, thus endangering the boiler with possible serious overheating damage that may include an explosion from weakened metal parts.

Probe-type low-water cutoffs can also be tested if the probes are mounted in a separate chamber that can be drained independently from the boiler water. If mounted inside the boiler, the manufacturer's instructions should be followed on testing and maintenance.

Most low-water fuel cutoffs require an annual dismantling and cleaning program, and this is usually recommended by jurisdictional inspectors.

Pressure and temperature switches are extensively used on smaller automatically fired boilers to cycle boilers on and off, with pressure switches used on steam boilers and temperature switches used on hot-water boilers. See Fig. 10.19. Upper limit switches with manual resets are now required by the ASME Boiler Code, Section IV, Heating Boilers, and for high-pressure boilers with fuel inputs to 12,500,000 Btu/hr as stipulated in ANSI/ASME CSD titled, "Controls and Safety Devices for Automatically Fired Boilers." While the upper limit switch shown in Fig. 10.19c is for pressure, a similar upper limit switch is based on temperature as applied to hot-water boilers. The full requirements of ANSI/ASME CSD are covered in the next chapter as it also has requirements on fuel trains and combustion safeguards.

On-off controls are limited to firetube and small watertube boilers. As the name implies, a drop in pressure actuates a pressurestat or mercury switch to start the stoker or burner and open the air damper, or to reverse the process when pressure rises again. Since control is limited to varying the lengths of on and off periods, combustion efficiency is low. A typical on-off combustion-control circuit for a low-pressure steam-heating boiler has the low-water cutoff control and pressure control switches hooked up in electric series. Thus if either control opens, the current to the burner motor is interrupted. The primary control consists of an electromagnetical relay that is energized by the thermostat.

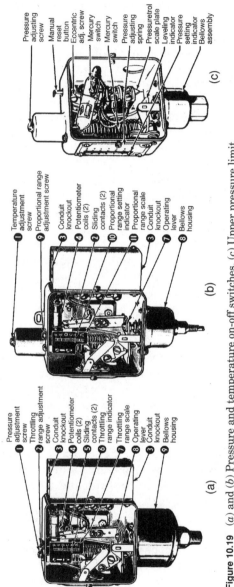

Figure 10.19 (*a*) and (*b*) Pressure and temperature on-off switches. (*c*) Upper pressure limit switch with manual reset. (*Courtesy Minneapolis-Honeywell Regulator Co.*)

(a)

- ❶ Pressure adjustment screw
- ❷ Throttling range adjustment screw
- ❸ Conduit knockout
- ❹ Potentiometer coils (2)
- ❺ Sliding contacts (2)
- ❻ Throttling range indicator
- ❼ Throttling range scale
- ❽ Operating lever
- ❸ Conduit knockout
- ❾ Bellows housing

(b)

- ❶ Temperature adjustment screw
- ❾ Proportional range adjustment screw
- ❸ Conduit knockout
- ❹ Potentiometer coils (2)
- ❺ Sliding contacts (2)
- ❿ Proportional range setting indicator
- ⓫ Proportional range scale
- ❸ Conduit knockout
- ❼ Operating lever
- ❽ Bellows housing

(c)

- Pressure adjusting screw
- Manual reset button
- Eccentric adj. screw
- Mercury switch
- Mercury switch
- Pressure adjusting spring
- Pressuretrol scale plate
- Leveling indicator
- Pressure setting indicator
- Bellows assembly

Flow Measurement

The principle underlying flow measurement is shown in Fig. 10.20a. A pressure drop across an orifice, in this case in the steam line, can be measured by tapping the pipe at each side of the restriction. The resulting pressure differential is proportional to the square of the fluid velocity. But correct location of the tapping points is important. By using nozzle restriction (Fig. 10.20b), the best result (largest pressure differential for a given flow) is obtained when connections are located about one pipe diameter upstream and one-half diameter downstream from the nozzle's inlet face.

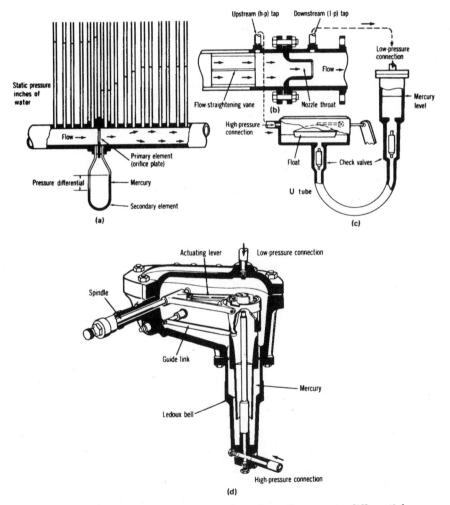

Figure 10.20 (a) and (b) Flow measurement depends on the pressure differential across an orifice; (c) mercury float manometer; (d) a Ledoux bell mechanism converts the differential pressure to an output signal.

Converting differential pressure to a usable output signal may be done in various ways. Two widely used secondary elements are a mercury float manometer (Fig. 10.20c) or a Ledoux bell mechanism (Fig. 10.20d). In Fig. 10.20c the two pressure tappings from the primary element are connected to two mercury chambers, joined by a U tube. Pressure variations raise and lower the float. A pressuretight shaft conveys this movement to mechanical linkage within the controller. Check valves in the two legs of the U tube prevent damage to the mechanism resulting from sudden changes or reversals in pressure differential.

Steam piping auxiliaries may include a pressure-reducing station and a desuperheating station where plant needs may require lower pressure or saturated steam supplied by high-pressure boilers.

Pressure-reducing station.   *Pressure-reducing valves,* sometimes known as *pressure regulators,* are used to supply steam at a desired constant pressure lower than that of the supply. Their applications include supply for manufacturing processes, low-pressure feedwater or fuel-oil heaters, and other auxiliaries. See Fig. 10.21a.

When high-pressure boilers are used for heating low-pressure apparatus through a reducing valve, it is necessary to guard the low-pressure system against overpressure. To guard against overpressure, it is

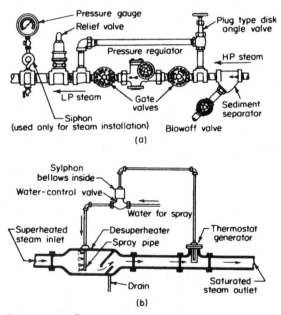

Figure 10.21   Steam piping auxiliaries. (*a*) Pressure-reducing valve arrangement; (*b*) desuperheating steam with treated water spray.

essential to install a safety relief valve on the low-pressure side of the reducing valve, with adequate capacity in pounds per hour to match the reducing-valve capacity. The pressure setting of the safety valve should always be based on the maximum allowable pressure of the weakest equipment on the low-pressure side of the reducing valve. The safety valve should be as near the reducing valve as possible. In the event of multiple vessels on the low-pressure side of the reducing valve, the total safety valves on each vessel should not be combined. It should be assumed that only one vessel may be operating, and, therefore, the total relieving capacity of all safety valves may be no protection.

The National Board has drawn up the following method to calculate the required relief-valve capacity in pounds per hour on the low-pressure side of a reducing station. These rules are drawn from thermodynamic principles involving flow through a nozzle. A coefficient of discharge had to be assumed (around 70 percent). Check with the reducing-valve manufacturer for the maximum valve flow and then install a safety valve with the recommended capacity. In lieu of data from the manufacturer, use the following formula:

$$RVC = \tfrac{1}{3} \times OC \times VSPA \tag{10.1}$$

where RVC = relief-valve capacity required, lb steam/hr
   OC = orifice capacity, lb steam/(hr/in.^{2}) (Fig. 10.22a)
   VSPA = valve-size pipe areas, in.^{2} (Fig. 10.22b)

Where a pressure-reducing valve is supplied by steam from the boiler, the capacity of the safety valve or valves on the low-pressure side of the system need not exceed the capacity of the boiler, for obvious reasons.

Most pressure-reducing valves are arranged with a valved bypass, which also acts as a potential steam-source hazard in case the bypass is left open. Where such a valved bypass is used, the following formula should be used to determine the steam flow rate through the bypass:

$$RVC = \tfrac{1}{2} \times OC \times BPA \tag{10.2}$$

where RVC = relief-valve capcity, lb steam/hr
   OC = orifice capacity, lb steam/(hr/in.^{2}) (Fig. 10.19a)
   BPA = bypass pipe area, in.^{2}

The larger of the relief-valve capacities calculated by Eqs. (10.1) and (10.2) should be used for selecting the relief-valve capacity.

Example. A reducing valve has no data on capacity on its nameplate. It is 3-in. in size with 125 psi on the inlet side and 30-psi steam on the outlet side,

Orifice Relieving Capacities, lb per hr per sq in. for Determining the Proper
Size of Relief Valves Used on Low-pressure Side of Reducing Valves

Outlet pressure, psi	Pressure-reducing valve inlet pressure, psi								
	125	100	85	75	60	50	40	30	25
110	4,550								
100	5,630								
85	6,640	4,070							
75	7,050	4,980	3,150						
60	7,200	5,750	4,540	3,520					
50	7,200	5,920	5,000	4,230	2,680				
40	7,200	5,920	5,140	4,630	3,480	2,470			
30	7,200	5,920	5,140	4,630	3,860	3,140	2,210		
25	7,200	5,920	5,140	4,630	3,860	3,340	2,580	1,485	
15	7,200	5,920	5,140	4,630	3,860	3,340	2,830	2,320	1,800
10	7,200	5,920	5,140	4,630	3,860	3,340	2,830	2,320	2,060
5	7,200	5,920	5,140	4,630	3,860	3,340	2,830	2,320	2,060

(a)

Nominal pipe size, in.	Standard		
	Actual ext diam, in.	Approx int diam, in.	Approx int area, sq in.
⅜	0.675	0.49	0.19
½	0.840	0.62	0.30
¾	1.050	0.82	0.53
1	1.315	1.05	0.86
1¼	1.660	1.38	1.50
1½	1.900	1.61	2.04
2	2.375	2.07	3.36
2½	2.875	2.47	4.78
3	3.5	3.07	7.39
3½	4.0	3.55	9.89
4	4.5	4.03	12.73
5	5.563	5.05	19.99
6	6.625	6.07	28.89
8	8.625	8.07	51.15
10	10.750	10.19	81.55
12	12.750	12.09	114.80

(b)

Figure 10.22 Charts for determining relieving capacities of reducing valves. (a) Relieving capacity in pounds per hour per square inch of reducing-valve area and inlet and outlet pressure. (b) Standard nominal pipe sizes, internal diameters, and areas in square inches.

which is the pressure rating of the low-pressure side connected equipment. Determine safety valve setting and capacity for the low side of the steam line if the reducing station has a 1-in. bypass line. Use Fig. 10.22 data.

solution Note that 3-in. diameter valve has 7.39-in.^{2} internal area, while 1-in. bypass has 0.86-in.^{2} internal area.

Through reducing valve, flow possible is

$$RVC = \tfrac{1}{3} \times 7200 \times 7.39 = 17{,}736 \text{ lb steam/hr}$$

Through bypass, flow possible is

$$RVC = \tfrac{1}{2} \times 7200 \times 0.86 = 3096 \text{ lb steam/hr}$$

Therefore, safety valve should have a set pressure of 30 psi and a relieving capacity of 17,736 lb steam/hr.

Desuperheaters. Large power-generating units are designed to operate more efficiently with a high degree of superheat. But small steam auxiliary units are often designed to operate with saturated-steam temperatures, for the use of superheated steam necessitates higher costs of construction with respect to close clearance and rotating expansion control than would be warranted, even when compared with the possible operating-expense reduction.

Rather than run a separate steam line from the boiler, independent of the superheater, it is often more practical (especially for temperature control) to pass all steam through the superheater and tap off a small line, from the superheater steam header, for auxiliary use. This small line passes steam through the desuperheater (Fig. 10.21b), which sprays a carefully proportioned amount of properly treated water into the flow. This proportion is regulated so that the amount of superheat to be removed will equal (or not quite equal) the amount of heat necessary to evaporate all water added to make saturated steam.

Maintaining Varied Instruments and Controls

Boiler plants can vary from older plants with simple and easy-to-maintain parts to newer "hi-tech" plants that use digital signals to display values and with microprocessors doing calculations to set controls for the best operating and economic results. It is essential for operators and maintenance personnel to understand their system, and to know whether adjustments are possible by plant in-house capability. The most common devices found in an older plant are analog devices and discrete controls, defined as follows:

1. *Discrete controls* activate devices from a single measured physical event, such as pressure, and this information is relayed to a solenoid, valve regulator, or coil for an action that is basically on-off, start-stop, or go-no go.

2. In *analog controls* a quantity is measured over a range of dynamic values, and the measurement is now tracking these sliding values, and transmitting signals to a controller to change its setting so that desired set values are obtained analogous to the sliding values. Analog controls were generally either hydraulic or pneumatic. Instruments for both

discrete and analog controls are mounted on panel boards, with round chart recorders and dial indicators to guide the operator in making adjustments if needed. The advent of electronic transmitters allowed greater lengths for transmitting measured values from the operating area to central control rooms. For example, a voltage indicator will show voltage between a range of 0 to 300 volts on a 250-volt system by having the pointer move smoothly as values change.

The maintenance requirements for operators with analog devices are usually limited to periodic minor adjustments, if devices are not sealed, and routine housekeeping cleaning. Sealed units must be replaced if they fail. Adjustments should follow manufacturer's instructions. Checking of the device should be against a standard reference source or test sample that gives the exact reading the device should have. If required, have the adjustments made by a qualified technician or return the device to the manufacturer for adjustment or replacement.

Pneumatic control and recording devices must also be checked and maintained to note if they function properly. This includes compressors, air dryers, storage tanks, and pressure regulators. The air must be kept clean and dry at all times for the pneumatic system to operate properly.

3. *Digital devices* change the value shown on an instrument face to a pulse from an external source. The value from an analog sensor is converted by a small microprocessor or similar computer-type device into a pulse that can be transmitted by low voltage to a display module in digital form. The maintenance requirements of digital equipment include periodic cleaning of the sensor device to make sure no binding can affect its accuracy or function. The accuracy should be checked to a standard or by reliable test instruments. The microprocessor or signal converter are not prone to drift, and thus do not require adjustments.

Maintenance of the boards or display sent by the microprocessor may require an electronic technician to analyze and locate defective board or display components. Some plants carry spare boards for immediate replacement of the component and return the defective board to the manufacturer for repair or exchange for a replacement. This way repairs are expedited.

4. *Transmission of data* may be by: (*a*) telemetry systems comprising a transmitter and receiver, which require a dc power supply. Some use batteries. The power supply needs periodic checking for proper operation. Telephone lines may be used between the transmitter and receiver. (*b*) By cable and fiber-optic lines paralleling the use of telephone lines; (*c*) by other remote operation control such as radio or satellite systems. Some of the problems are limitation of tower loca-

tions and interference from adjacent frequencies, which may require sophisticated search equipment if problems occur.

All telemetry-type equipment needs monitoring. However, because the equipment is electronic, very little adjustment is possible, and "chips" or boards are generally replaced. The units transmitting and receiving do require checks on the adequacy of cooling. Repairs to the equipment should be by qualified personnel. Failure of a component is usually noticeable immediately by failure to transmit signals.

Sensors and controls. All electronics and computer hardware that are involved with controlling a system depend on their performance on the availability of sensors to measure the variables, on their accuracy in getting this information to a computer or programmable controller. There are many physical measurements for which no sensors have yet been developed. One of these, for example, is a continuous readout of the heating value of a fuel. Other measurements must be made to compensate for this so that this information can be used to regulate fuel input to a boiler system.

The arrival of electronic control circuits has seen applications of this technology to boiler controls that are rapidly changing boiler control methods. This is being accelerated by the development of more reliable sensors to detect boiler operating conditions, instruments, and recorders that convert the sensors "reading" into electronic signals that are then transmitted to remote indicators and controllers, as well as being displayed at a central point, such as a control room. With the advent of fiber-optic cables, even longer distance of data transmission is now possible.

Sensors are devices that have the inherent ability to monitor changes in the medium being measured, such as temperature, pressure, flow, level, draft, percentage of CO_2, and similar quantities. Transducers are used to convert the changes noted by a sensor to an electric or a pneumatic signal. These signals are sent to controls, which are set to regulate a quantity within set points. The signals are thus compared to these set points, and if needed, a signal is then relayed to an actuator so that control of the medium is maintained within the established set points. Some controls incorporate the sensor, transducer, and actuator into a self-contained control device.

Figure 10.23b illustrates how one type of device may transmit pressure electrically to indicating, recording, and control equipment at close or remote points by using transmitting equipment and then show measured pressure at that location. The device illustrated has a Bourdon tube that positions a movable core of a transformer from zero pressure to the design pressure. The core position determines the magnetic flux linkage between the primary and secondary windings,

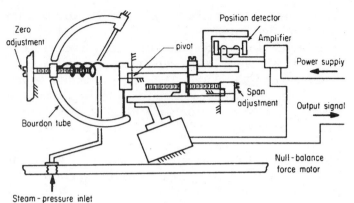

(a)

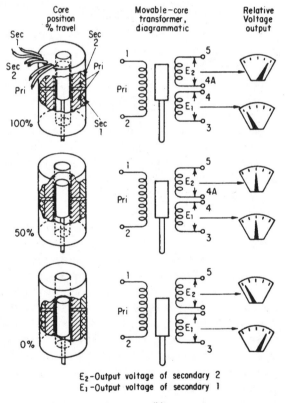

E_2 – Output voltage of secondary 2
E_1 – Output voltage of secondary 1

(b)

Figure 10.23 Transmitting pressure readings electrically. (*a*) Explosion-proof Bourdon-tube pressure gauge to electric output device. (*b*) Transformer designed for electrically transmitting pressure measurements.

which in effect changes the voltage in the secondary with pressure changes. This voltage can be transmitted to standard electronic receivers for further use in control, data recording, computer input, and similar applications.

In the figure, the voltage induced in each secondary winding is proportional to the displacement of the core from its center position; thus, the core position determines the signal voltage output. At 100-percent travel, voltage E_2 is larger than E_1 since the core is near the top of the transformer. At 50 percent travel, voltages E_2 and E_1 are equal since the core is centered between the two secondary windings. The output voltage on the secondary windings can be calibrated for proportional pressure reading on the receiving end.

Packaged boilers of the firetube type generally provide fully automatic operation because of the use of controls. The standard package will include a compact control cabinet containing motor starters, relays, switches, fuses, control transformers, and electronic flame safeguard controls. Indicating lights portray the sequence of operation and can be tied in with a remote station or even to a commercial surveillance system. (See Fig. 10.24.) Packaged watertube boilers are designed for gas or oil burning and are instrumented for basic automatic operation through control technology. However, complete reliance on automatic controls without periodic testing and maintenance can be dangerous. Figure 10.25 shows the result of neglectful operation. The tubes in the watertube boiler were melted from overheating resulting from a combination of scale in the tubes and failure of the low-water cutoff to stop the fuel input as the water level dropped to dangerous levels.

Controls on boilers are used extensively to regulate steam pressure and load within 1 to 2 percent of design points—steamdrum water levels within 1 in. of set points, fuel-air ratios within 5 percent of excess air requirements, and many more variables that can affect the operation of a boiler or its safety. A multitude of new sensors have been developed such as sodium analyzers, oxygen analyzers, and complete stack-gas analysis equipment in order to assist an operator or attendant in operating a boiler efficiently and safely.

Questions and Answers

1 What are the minimum appliances or appurtenances necessary for safe operation of a boiler?

ANSWER: Pressure gauge and test connection, safety valve, blowdown valve, gauge glass, gauge cocks, stop valve in steam line, and stop and check valves in the feed line.

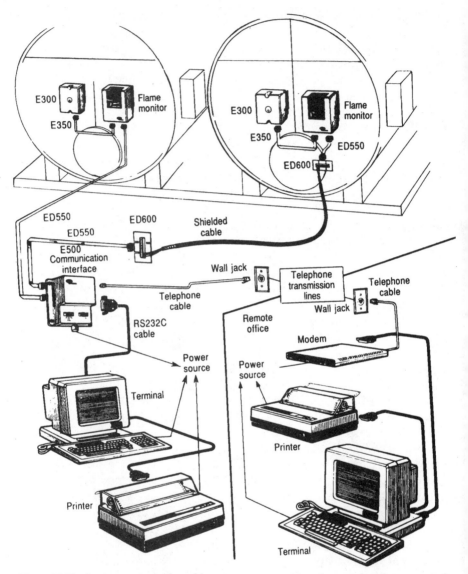

Figure 10.24 A microprocessor-based burner management system can be interconnected to a computer terminal, including a remote office, and also have a message center for displaying various limits. (*Courtesy Fireye Products, subsidiary of Allen-Bradley*)

2 What is the most important boiler appliance?

ANSWER: The safety valve.

3 What type of safety valve should you install?

ANSWER: ASME standard direct spring-loaded pop type.

Figure 10.25 Scale and low water caused tube melting from lack of testing and maintenance of basic boiler controls. (*Courtesy Factory Mutual Engineering.*)

4 How should a safety valve be connected to the boiler?

ANSWER: Directly to an independent nozzle with no intervening valve of any description. If over 3-in. diameter and over 15 psi, the valve should have a flanged connection.

5 What should be the maximum diameter for safety valves?

ANSWER: Heating Boiler Code stipulates 4½ in.

6 How do requirements of superheater safety valves discharging steam at over 450°F differ from requirements of those on the boiler drum?

ANSWER: They should have a flanged connection for all sizes. They should be constructed of steel or alloy steel suitable for the maximum temperature. The spring should be exposed so that it will not come in contact with high-temperature steam.

7 With a 4-in. safety valve, what is the minimum-size direct-connected nozzle and escape pipe which should be used?

ANSWER: 4 in. in each case.

8 What testing attachment is required on safety valves?

ANSWER: A lifting lever.

9 What is meant by blowback?

ANSWER: It is the number of pounds per square inch of steam pressure drop from the point at which a safety valve pops to the pressure at which it reseats.

10   What controls the amount of blowback?

ANSWER: The blowback adjusting ring.

11   What is a huddling chamber?

ANSWER: It is a chamber exposing the underside of the valve disk in a safety valve to increased pressure area on its primary lift. Pressure acting on the increased area results in the pop, or secondary, lift.

12   What principle do some safety valves use in place of the increased area exposed by the huddling chamber?

ANSWER: The reaction principle.

13   Why go to the expense of installing huddling chambers or reaction flow to pop a safety valve?

ANSWER: Without such installation, a gradual lifting and seating of the valve would rapidly ruin the valve seat by cutting or wire-drawing action of the steam.

14   When is more than one safety valve required on a boiler?

ANSWER: If the boiler has over 500-ft^2 heating surface or over 500-kW input for an electric boiler.

15   What is the purpose of a water column?

ANSWER: To steady the turbulence of boiler water between the drum and the gauge glass so that its level may be determined more accurately.

16   What attachments are permitted to pipe connections of a water column? Why limit the number of attachments?

ANSWER: Pressure gauge, damper regulator, feedwater regulator, drains, level indicators, or such connections as take negligible flow. Any appreciable flow would cause a false water-level indication.

17   Where should a gauge glass be located?

ANSWER: In an easily seen location with its lowest visible point at least 2 in. above the lowest safe water level in the boiler.

18   (1) What are gauge cocks, or try cocks? (2) How many are there and where should they be?

ANSWER: (1) They are valve cocks used to show the water level in a boiler as a check on the gauge glass. (2) Two are required on locomotive boilers up to 36-in. diameter or on firebox boilers up to 5 hp. Three are required on all other boilers over 15-psi pressure. They should be located equidistant within the visible range of the gauge glass. Try cocks are not required if two gauge glasses are installed at the same level at least 2 ft apart on a boiler.

19 Why is a drain required on a water column, and why should it be at least ¾-in. diameter?

ANSWER: To permit removal of sediment which might block the lower connection and cause a false water-level indication. Smaller sizes might become obstructed easily.

20 Why is a globe valve not desirable for water-column drain control?

ANSWER: Because the dam or pocket in this type of valve forms a natural trap for sediment and scale.

21 What three purposes does a blowdown valve serve?

ANSWER: Removal of sludge and loose scale, control of boiler-water concentration, and emergency control of abnormally high water levels.

22 Would you select cast iron or steel for elbows in a blowdown line between the boiler and the valve?

ANSWER: Cast iron is permitted up to 100 psi. Steel is required for higher pressures. Steel is preferred for all pressures over 15 psi.

23 On what principle do the majority of pressure gauges operate?

ANSWER: On the Bourdon-tube principle; that is, a curved tube tends to straighten when subjected to internal pressure.

24 Describe a fusible plug.

ANSWER: It is a threaded bronze or brass casing having a tapered core of nearly pure tin. A fusible plug is installed at the lowest safe water level in some low- and moderate-pressure boilers with the small end of the tapered core exposed to gases in the primary pass. Should the water level approach a dangerously low level, the core is designed to melt and escaping steam will sound the alarm.

25 Where and why are self-locking door latches required on firing doors?

ANSWER: On watertube boilers, in order to prevent the door from being blown open from positive furnace pressure resulting from tube ruptures, gas explosions, etc.

26 Where and why is a siphon required in pressure-gauge lines?

ANSWER: It is a pigtail or drop leg in the piping to the gauge, designed to trap condensate and to prevent live steam from entering the Bourdon tube. It prevents the tube, springs, and other delicate parts from being subjected to high temperatures.

27 A high-pressure steam boiler has two safety valves. What is the Code permissible range of pressure settings for these safety valves?

ANSWER:

1. One of the safety valves must be set at or below the maximum allowable pressure for the boiler.

2. The highest pressure setting of any safety valve cannot exceed the maximum allowable working pressure of the boiler by more than 3 percent.

3. The range of pressure settings of all the saturated steam safety valves on the boiler cannot exceed 10 percent of the highest pressure setting any valve is set.

28 A safety valve on a low-pressure steam boiler has no safe discharge outlet pipe. What is the danger?

ANSWER: The operation of the valve could scald a person standing near the outlet from the safety valve, and it may also damage equipment, such as electrical, located near the safety-valve outlet.

29 How may the water-gauge glass on a boiler indicate a false water level?

ANSWER: There may be obstructions in the water gauge fixture, water column or the piping connecting the column to the boiler proper.

30 For what type of boiler installation may the gauge glass be shut off in operation?

ANSWER: For boilers with drum safety valves set at or above 400 psi and equipped with two independent remote and reliable water level indicators, the gauge glass may be shut off, but it must be maintained in serviceable condition at all times.

31 What may cause water in a gauge glass to fluctuate up and down in an erratic manner?

ANSWER: The boiler water may be dirty and have floating scale that has broken loose. Blowdown the boiler water column and open gauge glass drain to remove possible obstructions to the gauge glass connections. Also check for boiler leaks that may be causing loss of water and makeup problems. Overload on the boiler that is sporadic may also cause rapid water fluctuation. If none of these is the cause, the boiler may require shutting down for an inspection of internal damage.

32 What is a steam trap and where should you expect to find one in a boiler installation?

ANSWER: It is a device designed to remove condensate from steam space with minimum loss of steam. A steam trap is very often used on pockets or separators of steam lines.

33 What is the purpose of a pressure-reducing or pressure-regulating valve? Where might you find one?

ANSWER: It serves to reduce an available pressure to a lower, cosntant, desired pressure. For example, a steam line operating at 400 psi for power generation has to supply a branch line to a heater shell designed for 75 psi. A reducing valve would be installed on such a line and set to maintain heater pressure within the prescribed limits. Also needed on the low-pressure side would be a safety valve set at 75 psi with a capacity equal to the capacity of the reducing valve in lb steam/hr, and a pressure gauge to note if the regulator is working.

34   For what type of boiler may power-actuated pressure-relieving valves be used, and what is the definition of a power-actuated pressure-relieving valve?

ANSWER: See Section I of the ASME Boiler Code. Power-actuated pressure-relieving valves are permitted on forced-flow steam generators with no fixed steam and waterline. These valves are defined as valves whose movement to open or close are fully controlled by an external source of power, such as air, steam, or hydraulics. They may discharge into intermediate pressure sections of the once-through steam generator. They are required to open by a control impulse signal when the maximum allowable pressure at the superheater outlet is reached.

35   Can superheater safety valves be included in the total relieving capacity required for a steam generator if it is considered part of the boiler flue gas flow?

ANSWER: With no intervening valve between the superheater and the boiler, the capacity of the superheater safety valves may be included in the total required for the steam generator; however, the maximum capacity allowed from the superheater is 25 percent of the total required for the boiler.

36   Under what conditions are shutoff valves permitted in connecting pipes between boiler and water columns?

ANSWER: If they are of the rising-stem outside-screw-and-yoke type or are straightaway valves or cocks marked plainly for their open and closed positions. They should be locked or sealed open.

37   Why are check valves required in the feedwater line?

ANSWER: So that boiler pressure will not force the boiler water back in case of piping failure; also, to assist the functioning of the feed pump or injector.

38   What are the minimum and maximum sizes permitted for blowdown valves or connections?

ANSWER: 1 in. minimum, 2½ in. maximum except for 100-ft^2 or less heating surface boilers where ¾-in. minimum size is permitted.

39   What are deflectors and when are they required on explosion doors?

ANSWER: They are sheet-metal plates placed in front of explosion doors in a boiler setting to divert any blast from operating floors, stairways, platforms, or anywhere anyone might be passing.

40 What force operates bucket and tilt traps?

ANSWER: Gravity.

41 Of what material are trap buckets or floats made?

ANSWER: Corrosion-resistant metal.

42 Why must a return trap be vented to the atmosphere after each discharge?

ANSWER: Because the body is at a pressure higher than that of the condensate returns, and the return check will be held closed. By venting the trap to atmosphere, the returns may flow freely into the trap until it tilts and closes these valves.

43 (*a*) In what units is a vacuum gauge graduated?

(*b*) What is the relation between these units and pounds per square inch?

ANSWER:

(*a*) In inches of mercury (Hg) vacuum.

(*b*) Each inch of mercury equals 0.49 psi below atmospheric pressure (14.7 psi at sea level).

44 What is a compound gauge?

ANSWER: A gauge reading positive pressure in pounds per square inch on one side of the zero reading and negative pressure (vacuum) in inches of mercury on the other side.

45 Provision for what four details of installation must be made in a steam line transmitting steam a considerable distance from the boiler?

ANSWER: Provision for insulation, support, expansion, and drainage.

46 May reheater safety-valve capacity be included in the total required capacity for a steam generator?

ANSWER: The safety-valve capacity cannot be included. Note, however, that the reheater safety-valve capacity in total must be equal to the total design flow through the reheater. One safety valve must be at the reaheater outlet with a capacity not less than 15 percent of the total required for the reheater.

47 An ASME boiler has 650 ft^2 of heating surface. In a boiler built for 150 psi, the safety valve is set to pop at 100 psi. Two means of feeding are used, city pressure at 110 psi and a pump. It is desired to operate the boiler at 150 psi. What would you recommend to do and why?

ANSWER: Recommend new springs for present safety valves. Recommend an additional means of feeding the boiler (pump or injector) since city pressure at 110 psi is not adequate. The fittings should be satisfactory for 150 psi. Two blowoff valves are required when the pressure is over 100 psi. The steam gauge should be graduated to approximately 250 psi.

48   In a boiler operating at a pressure of 65 lb, what is the smallest size of feed and blowoff connections if the safety valve is set at 90 psi?

ANSWER: ½-in. feed, ¾-in. blowoff.

49   Under what conditions does the Code permit stop valves or cocks in the connections to a water column?

ANSWER: They must be either the outside-screw-and-yoke type of gate valve or stop cocks with levers permanently fastened thereto and marked in line with their passage.

50   (a) What are the minimum and maximum amounts of blowdown permitted on a safety valve?

   (b) How is this adjusted when needed?

   (c) What tolerance, either plus or minus, would you allow on the opening pressure on a valve set for 150 psi?

ANSWER:

   (a) Minimum blowdown shall be not less than 2 lb and the maximum not lower than 96 percent of the safety valve set pressure.

   (b) By the manufacturer or the manufacturer's representative only.

   (c) Plus or minus 3 percent.

51   What is the allowable adjustment on the spring of an ASME safety valve set at (a) 290-psi pressure, (b) 190-psi pressure?

ANSWER: (a) 5 percent either way. (b) 10 percent either way.

52   Why should the safety valve not be placed on the same nozzle with the main steam line?

ANSWER: There would be a difference in pressure (because of the flow of steam in the pipe) between the boiler pressure and the pressure directly under the seat of the valve. The flow of steam in the pipe would cause the valve to chatter and damage the disk and seat when it blows. If the boiler has a dry pipe, it is liable to become obstructed, and then the valve could not blow to relieve overpressure. A stop valve is liable to be installed in steam line between the safety valve and boiler.

53   Does an economizer require a safety relief valve by the ASME Code?

ANSWER: If the economizer may be shut off from the boiler, one or more safety relief valves are required, set at the maximum allowable pressure allowed on the economizer, and with a capacity in Btu/hr calculated from the maximum heat absorption as determined by the manufacturer, and if the safety relief valve is stamped in steam capacity, divide the Btu output design by 1000.

54   What causes safety valves to leak below the popping pressure?

ANSWER: Leakage is usually caused by damaged seats, lodged scale, or an operating pressure too close to the popping pressure. Strains set up in the valve body, either by pipe expansion or by weight of unsupported discharge piping, may also cause leakage. Such leakage can be stopped by relieving piping strain on the safety valve. Valves also leak or fail to pop at the set pressure because scale or dirt becomes wedged between the disk holder and the guide. Hardened boiler compounds deposited on the nozzle under the seat may also cause leakage or valve sticking.

55   What type of boilers require no gauge glass or gauge cocks?

ANSWER: Forced-flow steam generators with no fixed steam line and waterline (once-through boilers) and high-temperature water boilers of the forced-circulation type. The same applies to once-through hot-water-heating and hot-water-supply boilers having no fixed steam line and waterline.

56   What area of the boiler should be computed as heating surface?

ANSWER: That side of the boiler surface exposed to the products of combustion, exclusive of superheating surface. The areas to be considered for this purpose are tubes, fireboxes, shells, tube sheets, and the projected area of headers. For vertical firetube steam boilers, compute only the portion of the tube surface up to the middle gauge cock.

57   Calculate the heating surface required for an oil-fired, firetube boiler of 100 tubes, each 2½ in. in diameter, no. 20 gauge in thickness, each 15 ft long. The remaining heating surface of fire sheet and tube sheet totals 130 ft² at a working pressure of 125 psi. How many and what size valves should be installed?

ANSWER: Use the ID of the tubes as heating surfaces. Heat transfer is from the inside of the tube through the tube thickness to the waterside. Number 12 gauge tube has a wall thickness of 0.105 in.; thus the ID of the tube equals $2.5 - (2 \times 0.105) = 2.29$ in.

The area in square feet of all tubes equals the circumference times the length times the number of tubes $= \pi\, 2.29/12 \times 15 \times 100 = 897$ ft².

Total heating surface $= 897 + 130 = 1027$ ft²

Because the boiler has over 500 ft² of heating surface, two or more safety valves are required.

58   Under what other name is the "computer on a chip" sometimes called, and how has this device affected instrumentation and controls of boilers?

ANSWER: The great growth in semiconductor technology has led to the development of miniaturized circuits to do certain control jobs; thus the term "computer on a chip" was applied and then replaced by the term "microprocessor." As these chips were further miniaturized, designers of instruments and controls could pack more information and instructions as well as data storing in the devices. This has caused a continuous trend in power plants to control the power equipment automatically by on-line computer systems, and the further development of distributed control, "smart" self-diagnostic controls, and predictive maintenance data analysis capability. As can be noted, the greatest concentration in the use of miniaturized instruments and controls has been in power plants.

59   What is a CRT graphics display?

ANSWER: CRT (cathode ray tube, or TV screen) graphics display to the power-plant operator a pictorial representation of the entire power-plant system for monitoring the plant and, if needed, for controlling a function by simply pressing the appropriate button. This has replaced the old power-plant boards. The easy-to-scan, on-screen schematics display to the operator pre-programmed values of monitored points in the power-plant loops, such as temperature, pressure, flows, extraction pressures, loads on generator, and similar operating data. Printouts of data can be obtained anytime by appropriate operator action on this type of computer-controlled operation.

60   How often should water column and water glass levels be checked on high-pressure boilers?

ANSWER: For plants with a three-shift operating schedule, at the beginning of each shift, and this should be done by the operator responsible for maintaining proper water level in the boiler. For other plants with automatic boiler operation, the level in the column and gauge glass should be checked at the beginning of the daily work effort.

61   For a boiler equipped with a quick-opening blowdown valve and a slow-opening valve in combination, what is the recommended opening and closing blowdown prcoedure?

ANSWER: It is recommended the quick-opening valve be opened first and blowdown rate be adjusted with the slow-opening valve. When stopping blowdown, close the slow-opening valve first and then the quick-opening valve. It is also good practice to open the valve nearest the steam generator last and close it first when stopping blowdown.

62   What is the valve arrangement on a boiler feedwater line next to the boiler and what precautions are needed on feedwater entrance inside the boiler?

ANSWER: A stop valve is required next to the boiler with a check valve placed next to the stop valve. The entire feed line can then be shut off with the stop valve if work is required on the check valve. At other times it is left open with the check valve preventing flow out of the boiler. Discharge of feedwater inside the boiler should be away from heating surfaces to avoid thermal shock on boiler components and should be below the minimum water level.

63   What is the minimum-size feed pipe required for a VT boiler with (a) 50 ft^2 of heating surface, (b) 125 ft^2 of heating surface?

ANSWER: (*a*) ½-in. pipe size. (*b*) ¾-in. pipe size.

64 What are the maximum- and minimum-size openings for a blowdown connection in a power boiler?

ANSWER: Maximum, 2½ in.; minimum, 1 in.; miniature boiler, ¾ in.

65 A VT firetube boiler has 95 ft² of heating surface. What is the minimum size of blowoff piping that can be used on this boiler?

ANSWER: ¾ in.

66 Why should low-water fuel cutoffs be inspected internally or dismantled at least once a year?

ANSWER: Dismantled inspections are made to check on the following possible places that could lead to this vital safety device not operating when needed: (1) binding of floats, rod, and associated pivot points; (2) linkage parts that may be broken, corroded, or worn; (3) excessive scale buildup in float chamber; (4) pipe connections to cutoff plugged with sediment; (5) float waterlogged; and (6) electric-type probe bridged with scale, giving false electric signal on low-water cutoff.

67 What type of feedwater control is especially vulnerable to possibly causing a low-water failure?

ANSWER: Units with combination pump return and/or feeder and low-water cutoff control, because a single failure of the float will make immediately inoperative the water feed to the boiler as well as nullify the low-water fuel cutoff, because the nonfunctioning float controls both feed and the cutoff. An independent second low-water fuel cutoff should supplement the combination feeder and low-water fuel cutoff.

11

Combustion, Burners, Controls, and Flame Safeguard Systems

Basic Combustion Process

The combustion process is a special form of oxidation in which oxygen from the air combines with fuel elements, which generally are carbon, hydrogen, and, though detrimental, sulfur. Important to combustion studies are the chemical thermodynamics and kinetics of flame travel and velocity of reactions. A proper mixture of fuel and air as well as an ignition temperature is required for the combustion process to continue. Fuel must be prepared so that thorough mixing of fuel and air is possible. The term *flammability* is used to describe a fuel's ability to be burned, or really its ability to be converted to a gas so that combustion can take place.

Three conditions must be satisfied for proper chemical reactions to take place in the combustion process:

1. Proper proportioning of fuel and oxygen (or air) with the fuel elements, as shown by chemical equations, is necessary.

2. The mixing of fuel and oxygen (or air) must be thorough, so a uniform mixture is present in the combustion zone and so every fuel particle has air around it to support the combustion. Solid fuels generally will be converted to gas first by the heat and presence of air. Liquid fuels will vaporize into gases and then burn. Atomization of liquids increases the mixing with air and increases the vaporization into a gas. Pulverization of coal will have the same effect.

3. The ignition temperature must be established and monitored so that the fuel will continue to ignite itself without external heat when combustion starts.

The chief heat-producing elements in fuels (except for atomic reaction and electricity) are carbon, hydrogen, and their compounds. Sulfur, when rapidly oxidized, is also a source of some heat energy, but its presence in a fuel has bad effects. The burning of coal, oil, or gas is a chemical reaction involving the fuel and oxygen from the air. Air is 23 percent oxygen by weight and 21 percent by volume. The remainder of air is mostly nitrogen, which takes no actual chemical part in combustion but does affect the volume of air required and the formation of NO_x. The table in Fig. 11.1 represents some typical combustion reactions for various fuel constituents. It is always the carbon, hydrogen, or sulfur that produces the chemical reaction for heat by combining with oxygen.

Since oxygen in the air is known to be 23.15 percent by weight and 21 percent by volume (from combustion equations), the amount of air required can be calculated. For example, in the complete combustion of carbon, it can be determined that 2⅔ lb of oxygen is required to burn 1 lb of carbon. The amount of air required to burn 1 lb of carbon would then be

$$A_R = \frac{\text{Amount of oxygen}}{\% \text{ oxygen in air by weight}} = \frac{2.67}{0.2315} = 11.52 \text{ lb}$$

This is shown in Fig. 11.1.

Incomplete combustion results in smoke and lowered operating efficiency. In order to obtain complete combustion, the furnace volume must be adequate to permit complete burning of fuel particles before

Combustible element	Symbol	Chemical reaction	Combustion product	Volumes	Oxygen, lb	Nitrogen, lb	Air, lb	Gaseous products, lb	ft³ O₂/ lb fuel	ft³ air/ lb fuel	Heating value, Btu/lb
					\multicolumn{8}{Weights per pound of combustible}							
Carbon	C	$C + O_2 \rightarrow CO_2$	Carbon dioxide	1 vol C + 1 vol O₂ = 1 vol CO₂	2.67	8.85	11.52	12.52	31.65	151.3	14,600
Carbon	C	$2C + O_2 \rightarrow 2CO$	Carbon monoxide	2 vol C + 1 vol O₂ = 2 vol CO	1.33	4.43	5.76	6.76	—	—	4,440
Carbon monoxide	CO	$2CO + O_2 \rightarrow 2CO_2$	Carbon dioxide	2 vol CO + 1 vol O₂ = 2 vol CO₂	0.57	1.90	2.47	3.47	6.79	32.5	10,160
Hydrogen	H	$2H_2 + O_2 \rightarrow 2H_2O$	Water	2 vol H₂ + 1 vol O₂ = 2 vol H₂O	8	26.56	34.56	35.56	94.8	453	62,000
Methane	CH₄	$CH_4 + 2O_2 \rightarrow CO_2 + 2H_2O$	Carbon dioxide and water	1 vol CH₄ & 2 vol CO₂ = 1 vol CO₂ + 2 vol H₂O	4	13.28	17.28	18.28	47.4	226.5	23,850
Ethylene	C₂H₄	$C_2H_4 + 3O_2 \rightarrow 2CO_2 + 2H_2O$	Carbon dioxide and water	1 vol C₂H₄ + 3 vol O₂ = 2 vol CO₂ + 2 vol H₂O	3.43	11.38	14.81	15.81	—	—	21,600
Ethane	C₂H₈	$2C_2H_8 + 7O_2 \rightarrow 4CO_2 + 6H_2O$	Carbon dioxide and water	2 vol C₂H₈ + 7 vol O₂ = 4 vol CO₂ + 6 vol H₂O	3.73	12.40	16.13	17.13	44.5	212	22,230
Sulfur	S	$S + O_2 \rightarrow SO_2$	Sulfur dioxide	1 vol S + 1 vol O₂ = 1 vol SO₂	1	3.32	4.32	5.32	11.87	56.7	4,050

Figure 11.1 Combustion reactions of fuel elements with combining weights and volumes and heating value of the fuel.

they enter heating surfaces and are cooled below their ignition temperature.

In order to thoroughly mix oxygen with burning fuel gases and particles, the flame action must produce turbulence. Flexibility of flame control may be affected by control of the primary air supply. Primary air is that which conveys fuel to burners or mixes with fuel at burners or through the fuel bed. See Fig. 11.2a. Secondary air is supplied to the burning fuel so that oxygen may unite in combustion at advantageous points.

If not enough oxygen or air is supplied, the mixture is rich in the fuel; thus the fire is reduced, with a resultant flame that tends to be longer and smoky. The combustion also is not complete, and the flue gas (products of combustion) will have unburned fuel such as carbon particles or carbon monoxide instead of carbon dioxide. Less heat will be given off by the combustion process. If too much oxygen or air is supplied, the mixture and burning are lean, resulting in a shorter flame and cleaner fire. Excess air takes some of the released heat

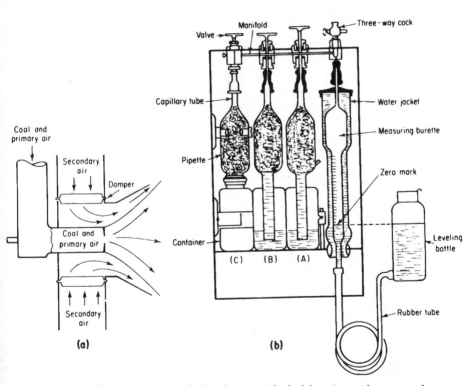

Figure 11.2 (a) Primary air in turbulent burner aids fuel burning at burner, and secondary air completes combustion in furnace (secondary air is two-thirds of total). (b) Orsat apparatus used for analyzing flue gases. New sensors permit continuous readout of flue gas constituents.

away from the furnace and carries it up the stack. Burning should always be with excess air to ensure that all the fuel is properly burned and thus attain better efficiency in heat release. This also reduces smoke formation and soot deposits, which today, with stricter pollution laws, is important.

When flue gas comes out of a stack as black smoke, it is an indication of insufficient air. Too much air usually causes a dense, white smoke. A faint, light-brown haze coming from the stack is a sign of a reasonably good air-fuel ratio. Of course, a more exact analysis is made with a flue-gas analyzer, such as an Orsat apparatus. From this analysis, the percentage of either excess or insufficient air can be determined. See Fig. 11.2b.

Flue-gas analyzers. The portable Orsat analyzer shown in Fig. 11.2b is used to determine the percentages by volume of carbon dioxide, oxygen, carbon monoxide, and nitrogen in the flue gas. A more complete analysis may include the percentages of hydrogen and hydrocarbons that may appear in the flue gas. However, the combustion of hydrogen produces water vapor, which does not appear in an Orsat analysis of dry gases in the flue gas. The basic mechanism of the Orsat apparatus involves bottles containing different chemical solutions for absorbing the different flue gases. For example, carbon dioxide is absorbed by the bottle containing caustic potash, oxygen in a bottle containing pyrogallic acid, and carbon monoxide in a bottle containing an acid solution of cuprous chloride. The nitrogen percentage is obtained by subtracting from 100 the sum of the other gases. Operators should follow the instructions that are usually provided for these type of instruments. Their main purpose is to fine-tune the air-fuel ratios.

Continuous emission monitoring. Governmental regulations on SO_2 and NO_x emissions have encouraged a growth in new burner systems as well as new electronic instrumentation and controls to control air-fuel ratios and emissions within jurisdictional restraints. Another goal is to improve the heat rate by monitoring the combustibles in the fuel gas, because poor mixing of air and fuel at the burner and in the furnace results in the formation of unburned carbon monoxide in the flue gas.

O_2 *analyzers* now can measure the percentage of O_2 in the flue gas; this permits fine-tuning the combustion process to obtain the best air-fuel ratio, thus improving burning efficiency. The most often used O_2 analyzer is the zirconium oxide unit, which consists of a probe and electronic unit. The zirconium oxide probe, constructed of stainless steel, houses a ceramic dust filter, flame trap, detector cell, cell heater, and thermocouple. A flexible conduit connects the probe to the electronic unit in which is located a transmitting system, probe heater control circuits, and reference air pump.

NO_x formation in the furnace is the combining of nitrogen in the fuel with oxygen, called "fuel NO_x," while that produced in the high temperature zone where nitrogen combines with the oxygen in the air is termed "thermal NO_x." When firing natural gas and light-distillate-oil, most NO_x is the thermal type, while heavy oils and coal burning may also produce fuel NO_x. Emission limits on NO_x are usually specified in terms of pounds of NO_x per million Btu of gross heat released, or pounds per hour. One state, New York, recently stipulated 4.5 ppm by volume for NO_x emission and 15 ppm per volume for CO emission for an independent power producer using the combined-cycle gas-fired gas turbine, waste-heat boiler combination. It is essential for operators to know their jurisdictional emission limits in order to make sure the firing equipment performs within these imposed governmental limits. The emission analyzers installed for this purpose, and for controlling the air-fuel ratio, should be carefully monitored for proper operation with any controls that may be connected to them. It is also important to maintain records of emission readings for any governmental review.

Air-fuel ratios.   Air-fuel ratios are being trimmed to as low a level as possible as a means of saving fuel. However, unless excess air is used in the combustion chamber, incomplete combustion can occur, which also wastes fuel, and there is the risk of a late-ignition-type furnace explosion occurring from unconsumed fuel. In order to ensure complete combustion, excess air must be supplied in amounts varying from 20 to 30 percent, depending on the fuel used, the boiler load, and boiler configuration. The correct amount of excess air is also influenced by the need to control NO_x and SO_2 emission. Combustion analyzers using infrared absorption techniques are being used to provide readings of CO and CO_2, which can be fed to a controller with input signals on boiler load. Feed-forward concepts are used to automatically control air dampers as a function of load, while the controller also alters the relationship between the air damper and the fuel valve as the boiler firing rate changes with load. The system is called an automatic *boiler combustion trim control.* As can be noted, its main purpose is to maintain a constant air-fuel ratio.

Combustion efficiency also can be affected by poor heat transfer due to soot deposits, or scale on the water side of a boiler. Therefore, close monitoring at fixed intervals of fuel consumed in Btus vs. load in Btus is another method of checking overall boiler efficiency.

Flue-gas analysis measurements can be used to calculate the weight of air used per pound of fuel burned by the following equation:

$$W_A = \frac{28N_2}{12(CO_2 + CO)(0.769)} \left(\frac{W_f C_f - W_r C_r}{W_f \times 100} \right)$$

where CO_2 = percentage of carbon dioxide in flue gas by volume
CO = percentage of carbon monoxide in flue gas by volume
N = percentage of nitrogen in flue gas by volume
W_f = weight of fuel fired, lb
C_f = carbon content of fuel, percentage from ultimate analysis
C_r = carbon content of ash and refuse, percentage
W_r = weight of ash and refuse from W_f pounds of fuel, lb
W_A = actual weight of air per pound of fuel burned, lb

Example   To illustrate the use of this equation, assume that 700 lb of coal is fired in a boiler. The carbon content of this coal is 68 percent. Ash and refuse after burning amount to 60 lb, with the carbon in the refuse being 7.8 percent. Flue-gas analysis showed the following percentages by volume:

$$CO_2 \text{ (carbon dioxide)} = 11.9\%$$

$$N_2 \text{ (nitrogen)} = 80.9\%$$

$$CO \text{ (carbon monoxide)} = 1\%$$

$$O_2 \text{ (oxygen)} = 7.3\%$$

What is the actual weight of air used to burn this coal? Substituting values, we have

$CO_2 = 11.9$	$W_f = 700$
$CO = 1.0$	$C = 68$
$N_2 = 80.9$	$C_r = 7.8$
	$W_r = 60$

Substituting in the equation gives

$$W_A = \frac{28(80.9)}{12(11.9 + 1.0)(0.769)} \left[\frac{700(68) - 60(7.8)}{700 \times 100} \right]$$

$$= 19.0(0.68)$$

$$= 12.9 \text{ lb of air per pound of fuel fired}$$

or

$$700 \times 12.9 = 9030 \text{ lb of air for 700 lb of coal}$$

Another useful equation is used to determine the weight of flue gas W_{fg} formed by burning a pound of fuel (as determined from a flue-gas analysis):

$$W_{fg} = \frac{4CO_2 + O_2 + 700}{3(CO_2 + CO)} \left(\frac{W_f C_f - W_r C_r}{W_f \times 100} \right)$$

Example The same coal is being burned with the same analysis as shown in the previous example.
Substituting, we get

$$W_{fg} = \frac{4(11.9) + 7.3 + 700}{3(11.9 + 1.0)} \left[\frac{700(68) - 60(7.8)}{700 \times 100} \right]$$

$$= 20.06(0.68)$$

$$= 13.6 \text{ lb flue gas per pound of fuel fired}$$

or

$$13.6 \times 700 = 9520 \text{ lb of flue gas for } 700 \text{ lb of coal}$$

By using the specific heat (Btu per pound per degree temperature rise for a gas), it is possible to calculate the heat lost up the stack. Use the equation

$$H_L = WC_p(T_2 - T_1)$$

where H_L = heat lost by flue gas
W = weight of flue gas going up the stack, usually lb/hr
C_p = mean specific heat of flue gas; can be taken as approximately 0.25 Btu (lb/°F)
T_2 = stack temperature, °F
T_1 = temperature of air entering furnace, °F

Example In the previous problem, 9520 lb/hr of flue gas went up the stack. Assume the stack temperature is 650°F and the air inlet temperature is 80°F. Then the heat lost up the stack is

$$H_L = 9520(0.25)(650 - 80)$$

$$= 1,356,600 \text{ Btu/hr}$$

Since 700 lb of coal was fired, with an assumed average of 14,500 Btu/lb, the total energy input is

$$700 \times 14,500 = 10,150,000 \text{ Btu/hr}$$

The percentage of heat input going up the stack is then

$$\frac{1,356,600}{10,150,000} = 13.4 \text{ percent}$$

This indicates a boiler efficiency of $100 - 13.4 = 86.6$ percent.

Draft. Draft provides the differential pressure in a furnace to ensure the flow of gases. Without draft, stagnation in the burning process would result, and the fire or process of combustion would die from lack of air. Draft pushes or pulls air and the resultant flue gas through a boiler and up into the stack. The draft overcomes the resistance to flow presented by the obstructions of tubes, furnace walls, baffles, dampers, and chimney lining (also slag).

Natural draft is produced by a chimney into which the boiler exhausts. The cool air admitted to a furnace (by means of damper openings) rushes in to displace the lighter hot gases in the furnace. Thus the hot gases rise (chimney effect), causing a natural draft.

Mechanical draft is produced artificially by means of forced- or induced-draft fans. The chimney is still necessary on mechanical-draft installations for venting the products of combustion high enough not to be offensive to the surroundings. Most modern boilers, including the domestic type, use some form of mechanical draft. Domestic burners may have a fan built into the burner unit.

Industrial Boiler Emission Standards

EPA programs, effective June 1989, require a 1.2-lb SO_2 per million Btu input emission limit for coal-fired boilers with a capacity above 10 million Btu/hr input. For boilers above 75 million Btu/hr input, SO_2 removal must additionally be 90 percent minimum.

NO_x emission standards applies to wood-, coal-, and oil-fired boilers. This requires NO_x to be limited to 1 lb per million Btu input regardless of boiler size.

SO_2 removal systems are especially applicable to the larger coal-burning plants. The new Clean Air Act of the federal government makes it mandatory in all fuel-firing systems to limit not only particulate emission normally handled by bag-house filters, or precipitators, but also sulfur dioxide and NO_x emissions, which are a gaseous product. Present emphasis in stack-gas cleanup is on flue-gas desulfurization, but nitrogen oxides have also become an emission subject to control. Gas-scrubbing systems for cleaning boiler and incinerator flue gases are being used. The three major types of scrubbing systems designed are the lime or limestone slurry type, the magnesium oxide slurry, and the double alkali system. Scrubbing systems are an additional expense, and in power generation they absorb 2 to 7 percent of total power plant output because of the additional pumps, fans, and similar equipment that is needed. Material of construction for scrubber systems must be acid- and abrasion-resistant and must also withstand high temperatures during plant-upset conditions. Stainless and nickel-alloy steels are used to avoid stress corrosion cracking in the presence of chlorides. Rubber-lined mild steel is also being used, as are flakeglass-reinforced polyester linings and similar acid-proof coatings or linings. Figure 11.3 shows a typical limestone scrubbing system that is added to the flue-gas flow system after the boiler precipitator. The flue gas, after being cleaned in the precipitator, enters the bottom of the scrubber tower and passes up through the limestone slurry, which is being sprayed into the tower from above. The chemical reaction between the SO_2 and limestone results in removal of about 80 to 85 percent of the SO_2 from the flue gas. The sludge formed must be disposed of in approved landfill sites.

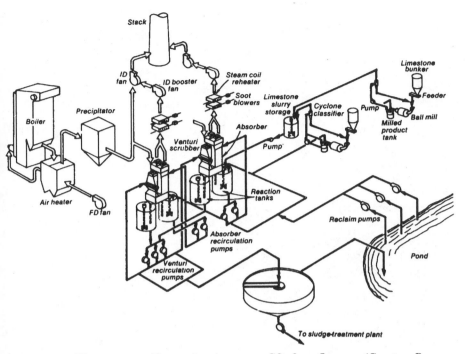

Figure 11.3 Limestone scrubber system to remove SO_2 from flue gas. (*Courtesy Power magazine.*)

Fluidized-bed burning. This type of burning has received attention as a means of controlling NO_x and SO_2 emissions into the atmosphere. See Fig. 11.4a. This is accomplished by introducing with the fuel, inert material such as sand, silica, and alumina as well as a sorbent such as limestone into an agitated fuel bed. The bed is kept suspended through the action of primary combustion air that is introduced below the combustion floor. See Chap. 4 for a description of the boilers used in this process. The use of in-bed steam-generating tubes requires special consideration in order to avoid abnormal erosion wear on the tubes. Composite tubes, which are shown in Fig. 11.4b, are used in these boilers, as are full stainless steel tubes, which are employed to reduce the effect of erosion and corrosion on the in-bed tubes.

NO_x *discharge control systems* are being applied not only to boilers, but also, by jurisdictional requirements, to fired heaters in the petrochemical industries, and to gas turbine emissions. Heavy smog formations in industrial areas is a result of NO_x emissions, which combine with reactive-type organic gases to form ozone (O_3). While ozone is required in the upper atmosphere to block the sun's harmful radiation, it is considered carcinogenic when mixed with ambient air, and

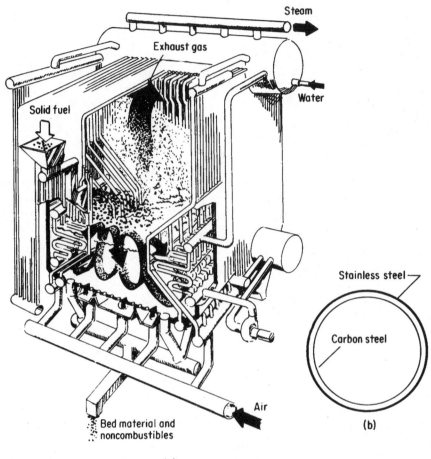

Figure 11.4 (*a*) Fluidized-bed burning exposes tubes to erosion and corrosion in the fluidized bed. (*Courtesy Power magazine.*) (*b*) Composite tube has stainless steel rolled over carbon steel for corrosion protection.

thus presents a health threat if concentrated above recognized threshold limits. NO_x reduction in the combustion process can take many forms, depending on the boiler or burning process involved. Among the methods used are:

1. Tight air-fuel ratio control, called *low excess air burners,* that work on the principle that low levels of excess air suppress NO_x formation.
2. Combustion staging in order to lower peak flame temperatures because nitrogen oxide formation and concentration are primarily influenced by fuel compositions (fuel NO_x), burner design, and fire-

box or flame temperatures and dwell burning time. This is especially true of thermal NO_x formation. Staged burners consist of two types:

a. Staged air burner which divides the combustion air into primary and secondary. The primary fuel-rich burning limits peak flame temperature, thus limiting NO_x formation. Secondary air is introduced to complete the burning of the fuel still left in the rich mixture.

b. Staged fuel burners inject a portion of the fuel gas into the combustion air, and this results in very lean combustion, which also reduces the peak flame temperature, thus resulting in low thermal NO_x formation. The remainder of the fuel is injected into a secondary combustion through secondary nozzles. The combustion products from the primary zone reduce the peak temperatures and oxygen concentration in the secondary zone, thus inhibiting NO_x formation.

3. Flue gas recirculation extracts a portion of the flue gas going to the stack and returns it to the burner with the combustion air, which results in a lower peak flame temperature that reduces thermal NO_x formation.

4. Water or steam injection with air-fuel mixtures is used with gas turbine burners to limit flame temperature, thus reducing NO_x formation.

5. Postcombustion methods that include selective catalytic and noncatalytic reduction methods and chemical scrubbing of the flue gas.

a. In the selective catalytic reduction process, ammonia is injected into the flue gas upstream of a catalyst bed, which can be titanium, vanadium, platinum, zeolites, and ceramics shaped in honeycombed plates, rings, or pellets. A chemical reaction converts the NO_x and ammonia to an ammonium salt, which decomposes to form elemental nitrogen and water discharge to the stack.

b. In the selective noncatalytic reduction process, ammonia or urea is injected into a thermally favorable location, and this also results in the reduction of NO_x to N_2 and H_2O.

Coal burning

Coal is a major source of energy in the United States and most probably will continue to be so for many years. However, one of the problems with coal as a source of energy is its sulfur content, present mostly as iron sulfide (pyritic sulfur) and in coal-borne organic systems, some containing the thiophene ring.

Coal quality is an important consideration when burning this abundant fuel. Because of the wide variety of coals (anthracite, bituminous, semibituminous) the quality of the coal may affect the rate of burning, capacity of the boiler, heat rate, slagging and fouling rate of heat transfer surfaces, corrosion and fly-ash erosion from the fireside, environmental impact, and type of burning equipment required, to name a few.

Upon combustion of the coal as mined, harmful sulfur-containing gases and ashes are emitted into the atmosphere with deleterious effects on animal and plant life. In order to make coal burning environmentally safe and acceptable, some method of desulfurization must be applied. Stack-gas scrubbing is most often used, but it is economical only for large industrial and power plants. Fluidized-bed burning is also being applied to control emissions.

Coal cleaning, and especially desulfurization prior to combustion, is another alternative. Many pollution controls and much monitoring can be omitted when desulfurized coal is available to plants carrying out relatively small operations.

Anthracite coal is very hard, is noncoking, and has a high percentage of fixed carbon. It ignites slowly, unless the furnace temperature is high, and requires a strong draft. The heating value is around 14,000 Btu/lb. Bituminous coal is soft, has a high percentage of volatile matter, burns with a yellow, smoky flame, and has a heating value of 11,000 to 14,000 Btu/lb. Semibituminous coal is the highest grade of bituminous. It burns with little smoke, is softer than anthracite, and has a tendency to break into small pieces when handled. The heating value is 13,000 to 14,500 Btu/lb. Subbituminous (black lignite) is a low grade of bituminous coal with a heating value between 9000 and 11,000 Btu/lb. Lignite is between peat and subbituminous coals, with a wood structure and claylike appearance. The heating value is 7000 to 11,000 Btu/lb.

Culm is a waste product left over in anthracite coal-mining operations with less than 3000 Btu/lb energy value and high ash content. The advent of commercially feasible circulating fluidized-bed burning and the rise of independent power producers under PURPA has encouraged going back to this marginal fuel at anthracite mine sites to produce electric power by burning this previously ignored fuel. Special fuel preparation is required as is disposing of the large amount of waste generated by this high-ash-content fuel (to 75%).

Coal analysis.   Two methods of analyzing coal are *ultimate analysis* and *proximate analysis*. Ultimate analysis gives the percentages of the various chemical elements of which the coal is composed. Proximate analysis determines the percentage of moisture, volatile matter, fixed carbon, and ash with a fair degree of accuracy.

This analysis requires a laboratory and a skilled chemist. If a sample of coal is separated into its elements, certain proportions of oxygen, hydrogen, carbon, etc., will be found. These proportions are generally expressed as percentages of the weight of the original sample, the unit weight being 100 percent. The heating value of coal is estimated from the ultimate analysis by getting the percentages of carbon, oxygen, hydrogen, and sulfur in the coal and by measuring the heat of combustion available in 1 lb of coal.

Other conditions reported in a coal analysis are: (1) as-received; (2) air-dried; (3) moisture-free; (4) moisture- and ash-free; and (5) moisture- and mineral-free.

Because coals range over a broad spectrum of properties compared to gas and oil, burner and furnace designs for coal burning will vary a great deal. In addition, coal handling to the burners, the crushing and pulverizing equipment to be employed, the ash disposal methods to be used, and the type of environmental control that is to be exercised must all be considered.

Figure 11.5a shows some average heating values of coal in comparison to other fuels. Figure 11.5b shows some typical coal properties that must be known when burning coal. The grindability factor determines the ease with which the coal can be pulverized, as indicated by the ASTM index numbers, also called the Hardgrove index. The base

Fuel	$H = $ Btu/lb
Semibituminous coal	14,500
Anthracite	13,700
Screenings	12,500
Coke	13,500
Wood, hard or soft, kiln dried	7,700
Wood, hard or soft, air dried	6,200
Wood shavings	6,400
Peat, air dried, 25% moisture	7,500
Lignite	10,000
Kerosene	20,000
Petroleum, crude oil, Pennsylvania	20,700
Petroleum, crude oil, Texas	18,500

	$H = $ Btu/ft^3
Natural gas	960
Blast-furnace gas	100
Producer gas	150
Water gas, uncarbureted	290

(a)

Type of coal	Source	Heating value, Btu/lb	Proximate analysis, as received				Ultimate analysis, dry and ash-free, percent					Grindability ASTM
			Moisture	Percent volatiles	Fixed C	Ash	S	C	H$_2$	O$_2$	N$_2$	
Anthracite	Pennsylvania	13,000	2	6.3	79.7	12	0.6	93.5	2.6	2.3	0.9	25
Bituminous	Pennsylvania	13,600	3	23.1	63.9	10	2.17	87.6	5.2	3.3	1.4	95
Bituminous	Ohio	12,450	6	34.8	49.2	10	2.44	82.2	3.5	7.7	1.7	66
Subbituminous	Colorado	9,200	24	30.2	40.8	5	0.36	75	5.1	17.9	1.5	58
Lignite	North Dakota	6,330	40	27.6	23.4	9	1.42	72.4	4.7	18.6	1.5	—

(b)

Figure 11.5 (a) Some typical heating values of coal in comparison to other fuels. (b) Typical coal properties that are used in analyzing coal burning and necessary burner and boiler auxiliaries.

is taken as 100; a coal is difficult to grind if its index is below 100, while easier if it is above 100.

Coal gasification. Under intense development, including demonstration plants, are integrated coal gasification, combined-cycle, and pressurized fluidized-bed coal combustion that will promote further use of coal in the future. See Fig. 11.6a for a schematic of an integrated gasification combined-cycle system. Fuel gas is generated in a gasifier by coal reacting with steam, air, and limestone as shown. The pressurized fuel gas is then cleaned by passing through a hot gas particulate remover, as shown in Fig. 11.6b. The clean coal gas is then supplied to a gas turbine combustor with the exhaust gas from the gas turbine going to a waste-heat boiler to make steam for the steam turbogenerator. The system shown uses the hot gas cleanup method in order for the hot gas to be compatible with gas turbine needs. Hot cleanup operation at elevated pressure reduces the volume of gases that require processing. Technical problems being studied are the effects of coal gasification systems on gas turbine blades, which require particulate-free gas in order to avoid (1) deposit accumulation on blades; (2) erosion of blades from high velocity particulate matter; and (3) corrosion of blades from the breakdown of blade coatings. Government-sponsored research and development have resulted in

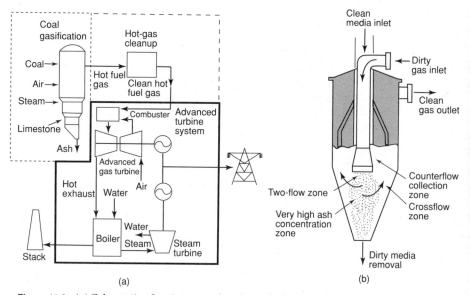

(a)   (b)

Figure 11.6 (*a*) Schematic of an integrated coal gasification combined-cycle and pressurized fluidized-bed combustion plant with associated gas turbine and steam turbine driven generators. (*b*) Combustion Power Co.'s hot gas particulate removal system features a granular-bed ceramic filter to clean the particle-laden gas coming from the coal gasification unit. (*Courtesy Mechanical Engineering magazine.*).

many particulate control concepts. The filter system shown in Fig. 11.6*b* from Combustion Power Co. has particle-laden inlet gas being discharged into a slowly moving bed of granular material, made of high-temperature ceramic. While the bed of granules moves down, the dirty inlet gas flows up past the bed material, and this makes it possible for the particles from the gas to be captured by the granular bed. The dirty granular bed material is pneumatically conveyed, cleaned, and returned to the filter vessel. The pneumatic gas in turn is cleaned by a conventional high-pressure baghouse to remove the collected dust.

Coal gasification will extend the use of coal as a fuel for electric power generation, once the inherent problems are solved in the demonstration plants.

Methods of coal burning

Stoker burning. Coal firing has progressed from simple hand shoveling to stokers and pulverized firing. Fuel not burned in suspension is burned on various stokers. Two broad classes of stokers are over-feed, in which the fuel is carried into the furnace above the stoker, and underfeed, where the fuel is carried by the stoker underneath. Overfeed stokers are further classified into spreader and chain-grate stokers. See Fig. 11.7.

In the overfeed spreader stoker, raw coal is blown or thrown by air or steam or rotating paddles in suspension above the burning bed. The dust coal particles tend to burn in suspension. In the traveling grate stoker the fuel is added above the grate by a coal hopper through a gate, which regulates fuel-bed thickness. Coke is formed and burned as the grate moves the fuel to the back of the furnace, so that by the time the end is reached only ash remains, which is dumped off the grates. These grates travel to the front of the furnace by means of a sprocket drive for a fresh load of fuel to keep the cycle going. The underfeed type of stokers shown in Fig. 11.7*d* has the coal reach the fuel bed from below. The fuel is pushed along a feed trough, or retort, into the furnace and spills over onto the fuel bed at each side. Several names are used for the underfeed stoker types. The name of each is determined by the mechanism used to move the coal, such as single retort, multiple retort, screw feed, or ram feed. Single-retort units handle up to 50,000 lb/hr; multiple-retort designs handle up to 500,000 lb/hr.

Pulverized coal firing. Pulverized coal firing is the most widely used method for burning coal in large boilers. The system requires coal to pass from feed bunkers through scales or feeders to the pulverizer. The grinding of the coal permits the fuel elements in the coal to

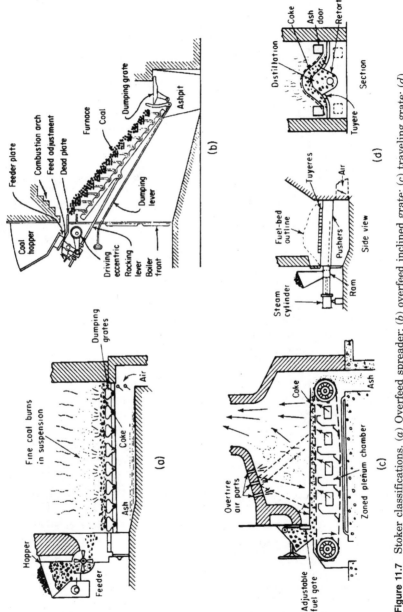

Figure 11.7 Stoker classifications. (a) Overfeed spreader; (b) overfeed inclined grate; (c) traveling grate; (d) underfeed retort.

rapidly oxidize (burn) as the ignition temperature is reached. More complete burning is thus possible than with fuel-bed burning.

As these fine particles enter the furnace and become exposed to radiant heat, the temperature rises, and the volatile matter of the coal is distilled off in the form of a gas. Enough primary air is introduced at the burner to intimately mix with the stream of coal particles, which thus support combustion. The volatile matter burns first and then heats the remaining carbon to incandescence. Secondary air is introduced around the burner, which supplies the oxygen to complete the combustion of carbon particles in flames several feet long. Figure 11.8a–c shows pulverized-coal furnaces arranged for long-flame-system firing, shelf-system firing, and corner- (tangential) system firing.

In general, *pulverizers* (sometimes called *mills*) may be classified as attrition or impact types. To these might be added the shearing type, which is a form of the attrition type, impact type, or both. The impact mills generally have some attrition action. And conversely, while attrition may be the primary action of a mill, impact is usually present as a secondary action. Thus we have impact mills, including ball mills and hammer mills, and attrition mills, including bowl mills and ball-and-race mills.

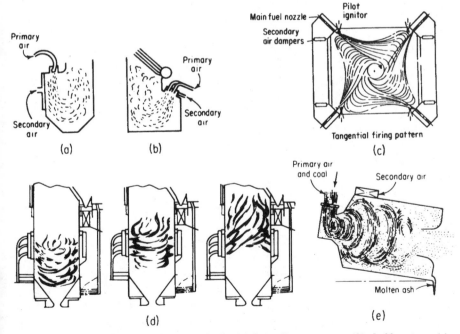

Figure 11.8 Pulverized-coal firing methods. (*a*) Long flame system; (*b*) shelf system; (*c*) corner or tangential firing; (*d*) tilting burner in tangential firing is used for variable load; (*e*) cyclone whirls coal and air against the walls of the burner.

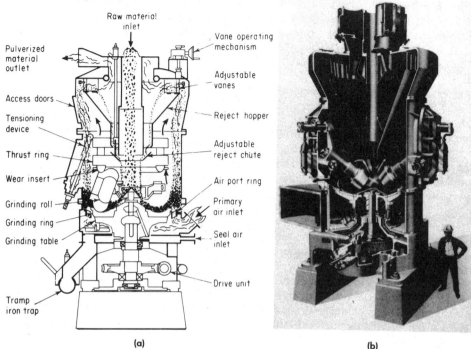

(a)   (b)

Figure 11.9 Bowl mill pulverizer. (*a*) Internal detail. (*Courtesy Foster-Wheeler Corp.*) (*b*) External view of a unit. (*Courtesy Combustion Engineering Inc.*)

Figure 11.9 shows a bowl-mill type of pulverizer. The grinding elements consist of three equally spaced, hollow toroidal rolls which run in a concave grinding ring. Force is applied to the rolls from above by a uniformly loaded thrust ring which fits smoothly into the necks of the rolls. The main drive shaft turns the table supporting the grinding ring, which transmits the motion to the rollers, driving them at about half the speed of the grinding ring. The rolls revolve about their own axes and simultaneously, in planetary fashion, they revolve about the axis of the mill, separated and spaced equally around the grinding ring by the satellite spacer.

Burners for pulverized coal must supply air and fuel to the furnace in a manner that permits stable ignition, effective control of flame shape and travel, and thorough and complete mixing of fuel and air. The air used to transport the coal to the burner forms the primary air; secondary air may be introduced in the burner (turbulent burners) or around or near the burner (nozzle burners). The turbulent burner (Fig. 11.10) imparts a rotary motion to the coal-air mixture in a central nozzle and the secondary air issuing from a chamber around that nozzle, all within the burner. This gives some premixing for coal and air and considerable turbulence. In some burners the coal-air mixture issues from a series of nozzles to mix within the furnace with

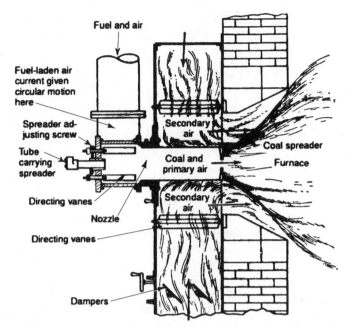

Figure 11.10 Pulverized-coal firing has coal-air mixture burning in a central nozzle with secondary air coming in around the nozzle to obtain complete combustion of the pulverized coal.

secondary air admitted through separate openings. Tangential firing has burners in furnace corners that direct their flames tangentially to an imaginary circle in the furnace space (Fig. 11.8c). Turbulence can be set up in this fashion, and there is a tendency for unburned combustible material in the tail of the flame to be caught up in the secondary air dampers. A special form of tangential firing appears in Fig. 11.8d. These burners are adjustable to shift the flame zone vertically and so regulate the temperature of the furnace exit gas according to load. This, in turn, controls superheat over a wide load range. Newer installations use fully automatic control of burner inclination.

The cyclone burner illustrated in Fig. 11.8e receives crushed (not pulverized) coal in a stream of high-velocity air tangent to the circular burner housing, which forms a primary water-cooled furnace. Coal thrown to the rim of the furnace by centrifugal force and held by a coating of molten ash is scrubbed by fast-moving air. Secondary air enters at high velocity also and parallel to the path of the primary coal-air mixture. The coal in the sticky slag film burns as if it were in a fuel bed. Volatiles are distilled off, and carbon is burned out to leave ash. Combustion of volatile matter begins in the burner chamber and is completed in the secondary furnace into which the burner chamber discharges. Molten ash, under centrifugal force, clings to the burner-chamber walls, and the slight inclination causes slag to discharge continuously. The nature of this burning tends to reduce greatly the amount of ash carried in suspension, and hence fly-ash emission is negligible.

Coal-burning boilers—waste disposal. Plants burning coal must dispose of a large amount of fly ash, bottom ash, and flue-gas desulfurization solids until this waste is reprocessed or used, for example, as a cement for road building. Jurisdictional requirements on waste disposal is one of the items that needs attention in the disposal of coal-burning wastes. Permits are required, as a minimum, for ash disposal in ponds or landfills. A major concern is the effect of any runoff or leaching on the Clean Water Act and the Safe Drinking Water Act passed by the federal government. In other jurisdictions, such as states, landfills (1) must be a sufficient distance above local groundwater table; (2) must have soil conditions not prone to settling; (3) must be sited away from limestone quarries, waterways, flood plains, wellheads, aquifers, or underground mines. There are other state requirements such as topping the landfill so that it will support vegetation growth and grading of the landfill to acceptable slope levels.

Other waste includes that coming from fire-side cleaning of tubes, and those coming from water-side cleaning. These may involve hazardous wastes because of the use of chemicals in the cleaning operation, which can require different jurisdictional disposal and testing requirements than that from ash disposal. It is thus essential in operating boiler plants to be aware of these waste disposal regulations.

Fuel Oils

Fuel oils differ from gas, with viscosity being the outstanding difference. In burning gas, intimate mixing with air can be achieved, and thus complete combustion occurs as long as there is no deficiency of air in the combustion process. Fuel oils are viscous; therefore, it is necessary to break up the oil by atomization so that the air can combine with the finer units of oil. However, the viscous nature of oil never permits the oil to reach the gaseous nature of gas; therefore, the temperature that can be attained in combustion is a bit lower than might be expected in comparison to the instant heat release when burning gas. The oil flame is further from the burner orifice. Better atomization equipment has been developed that disperses the fuel more completely, thus burning the fuel more rapidly, which produces higher flame temperatures. Petroleum products are listed in Fig. 11.11a and b. Refining of petroleum involves separating and recombining the carbon and hydrogen molecules into fractions having the same range of boiling points. Typical fractions from light to heavy are naphtha, gasoline, kerosene, and gas-oil.

Type of fuel	Specific gravity	API gravity	Weight, lb/gal	Heating value Btu/lb	Heating value Million Btu per barrel
Residual fuel	1.0	10	8.337	18,540	6.5
No. 4 fuel oil	0.966	15	8.053	18,840	6.35
Heavy distillate	0.910	24	7.587	19,190	6.1
Light distillate	0.865	32	7.215	19,490	5.95
Kerosene	0.825	40	6.879	19,750	5.7

(a)

API gravity	Specific gravity	Weight, lb/gal	Btu/lb	Btu/gal	lb/42-gal barrel	lb/ft³
3	1.0520	8.76	18,190	159,340	368.00	65.54
5	1.0366	8.63	18,290	157,840	362.62	64.59
7	1.0217	8.50	18,390	156,320	357.37	63.65
9	1.0071	8.39	18,490	155,130	352.46	62.78
11	0.9930	8.27	18,590	153,740	347.71	61.93
13	0.9792	8.16	18,690	152,510	342.88	61.07
15	0.9659	8.05	18,790	151,260	338.22	60.24
17	0.9529	7.94	18,890	149,980	333.64	59.42
19	0.9402	7.83	18,980	148,610	329.23	58.64
21	0.9279	7.73	19,060	147,330	324.91	57.87
23	0.9159	7.63	19,150	146,110	320.71	57.12
25	0.9042	7.53	19,230	144,800	316.59	56.39
27	0.8927	7.44	19,310	143,670	312.60	55.68
29	0.8816	7.35	19,380	142,440	308.70	54.98
31	0.8708	7.26	19,450	141,210	304.92	54.31
33	0.8602	7.17	19,520	139,960	301.18	53.64
35	0.8498	7.08	19,590	138,690	297.57	53.00
37	0.8398	7.00	19,650	137,550	294.04	52.37
39	0.8299	6.92	19,720	136,400	290.64	51.76
41	0.8203	6.83	19,780	135,090	287.23	51.16

(b)

Figure 11.11 (a) Typical fractions of petroleum fuels; (b) fuel-oil properties arranged by API gravity scale at standard 60°F.

The API scale is the American Petroleum Institute scale for showing specific gravity. The API scale fixes a reading of 10°F as equal to a specific gravity of 1.00. Readings greater than 10°F indicate a specific gravity of less than 1.0 or an oil which is lighter. To obtain the actual specific gravity in relation to water from the API reading, use the following equation:

$$\text{Actual specific gravity} = \frac{141.5}{131.5 + \text{API deg}}$$

Specific gravity in degrees Baumé (°Bé) is found in the same way except that the numbers are 140 and 130, respectively. For practical purposes, the two specific-gravity scales may be considered the same.

Fuel oils are sold in six standardized grades, under the numbers or grades of 1, 2, 3, 4, 5, and 6. Grades 1, 2, and 3 are light, medium, and heavy domestic fuel oils. These usually do not require heating prior to burning in a furnace. Grades 4, 5, and 6 correspond to federal specifications for Bunkers *A, B,* and *C,* respectively. These oils are heavy and viscous; thus they require heating prior to being sprayed into a furnace.

Viscosity is the relative ease, or difficulty, with which an oil flows. It is measured by the time in seconds a standard amount of oil takes to flow through a standard orifice in a device called a *viscosimeter.* The usual standard in this country is the Saybolt Universal, or the Saybolt Furol, for oils of high viscosity. Since viscosity changes with temperature, tests must be made at a standard temperature, usually 100°F for Saybolt Universal and 122°F for Saybolt Furol. Viscosity indicates how oil behaves when pumped and, more particularly, shows when preheating is required and what temperature must be held.

Figure 11.12 is a chart that can be used to determine the best viscosity and temperature for burning efficiently an oil with a certain temperature and viscosity. This assists in determining to what temperature a viscous oil must be heated prior to burning. The chart incorporates instructions.

Flash point represents the temperature at which an oil gives off enough vapor to make an inflammable mixture with air. The results of a flash-point test depend on the apparatus, so this is specified as well as temperature. Flash point measures an oil's volatility and indicates the maximum temperature for safe handling.

Pour point represents the lowest temperature at which an oil flows, under standard conditions. Including pour point as a specification ensures that an oil will not give handling trouble at expected low temperatures.

By centrifuging a sample of oil, the amounts of water and sediment present can be determined. These are impurities, and while it is not economical to eliminate them, they should not occur in excessive quantities (not more than 2 percent). Incombustible impurities in oil, from natural salts, from chemicals in refining operations, or from rust and scale picked up in transit, show up as ash. Some ash-producing impurities cause rapid wear of refractories, and some are abrasive to pumps, valves, and burner parts. In the furnace, they may form slag coatings.

The temperature required at the burner for numbers 4, 5, and 6 fuel oil is generally 10°F below the flash point, or for number 4: 150°F; number 5: 175°F; and number 6: 275°F.

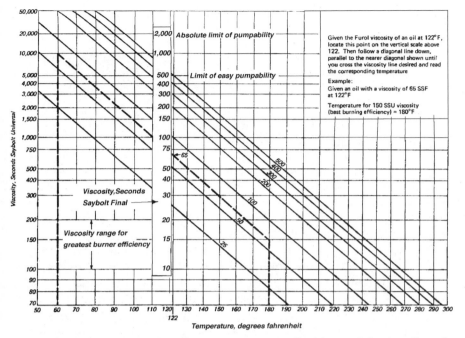

Figure 11.12   Chart showing to what temperature an oil with a certain viscosity and temperature must be heated to obtain greatest burner efficiency.

Oil Burners

In addition to proportioning fuel and air and mixing them, oil burners must prepare the fuel for combustion. Two ways (with many variations) are: (1) oil may be vaporized or gasified by heating within the burner or (2) oil may be atomized by the burner so vaporization can occur in the combustion space. Vaporizing burners (first group) are limited in range to fuels they can handle and find little use in power plants. If oil is to be vaporized in the combustion space in the instant of time available, it must be broken up into many small particles to expose as much surface as possible to the heat. Atomization is effected in three basic ways: (1) by using steam or air under pressure to break the oil into droplets; (2) by forcing oil under pressure through a nozzle; and (3) by tearing an oil film into drops by centrifugal force. All three methods are used. In addition, a burner must provide good mixing of fuel and air so complete combustion of the oil droplets may ensue.

The steam-atomizing burners possess the ability to burn almost any fuel oil, or any viscosity, at almost any temperature. Air is less extensively used as an atomizing medium because its operating cost is apt to be high. These burners can be divided into two types by their methods:

1. Internal mixing or premixing of oil and steam or air as shown in Fig. 11.13a for steam, and Fig. 11.13b for air. The mixing is inside the body or tip of the burner before being sprayed under pressure into the furnace.
2. External mixing, where all of the oil emerging from the burner is caught by a jet of steam or air (Fig. 11.13c).

Figure 11.13d shows an air register which controls the amount of combustion air that will surround the sprayed fuel into the furnace. Steam consumption for atomizing runs from 1 to 5 percent of the steam produced, with the average around 2 percent. The pressure required varies from about 75 to 150 psi.

In the burner of Fig. 11.13c, oil reaches the tip through a central passage, with the flow being regulated by the screw spindle. Oil whirls out against a sprayer plate to break up at right angles to the stream of steam, or air, coming out behind it. The atomizing stream surrounds the oil chamber and receives a whirling motion from vanes in its path. When air is used for atomizing, it should be at 10 psi for lighter oils and 20 psi for heavier. Combustion air enters through a register (Fig. 11.13d). Vanes or shutters are adjustable to give control of excess air.

Mechanical atomizers.   Mechanical atomizing oil burners depend on high oil pressure produced by an oil pump to force oil through nozzles that produce a fine mist for more complete fuel combustion in the furnace.

Good atomization results when oil under a pressure of 75 to 200 psi is discharged through a small orifice, often aided by a slotted disk. The disk gives the oil a whirling motion before it passes on through a hole drilled in the nozzle, where atomization occurs. For a given nozzle opening, atomization depends on pressure, and since pressure and flow are related, the best atomization occurs over a fairly narrow range of burner capacities. To follow the boiler load as steam demand goes up or down, a number of burners may be installed and turned on or off, or burner tips with different nozzle openings may be used. All nozzle openings must be changed to the same size in a given system, never fired with mixed sizes.

There are many burner designs to extend the usual 1.4:1 capacity range of the mechanical-atomizing nozzle. One has a plunger that opens additional tangential holes in the nozzle as oil pressure increases. This gives a 4:1 range. The burner in Fig. 11.14a uses a movable control rod which, through a regulating pin, varies the area

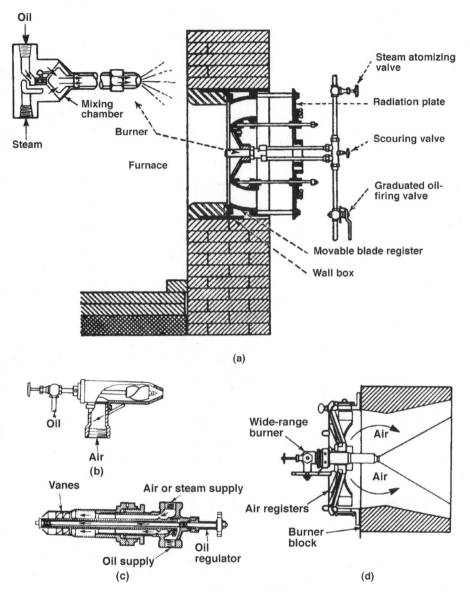

Figure 11.13 Steam and air atomizer oil burners. (*a*) Steam atomizer burner mixes oil and steam internally; (*b*) air atomizer also mixes oil and air internally; (*c*) burner mixes air or steam with oil externally; (*d*) air register controls combustion air around burner.

of tangential slots in the sprayer plate and the volume of the oil passing the orifice.

The wide-range mechanical atomizer (Fig. 11.14*b*) gives a capacity range of about 15:1 and much higher if needed. By use of either a con-

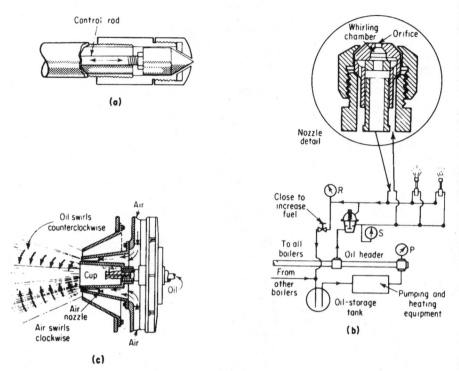

Figure 11.14 Mechanical atomizing burner types. (*a*) Movable control rod varies spray; (*b*) constant-pressure-differential pump or valve control spray; (*c*) rotary-cup burner atomizes fuel by centrifugal force.

stant-differential valve or pump, as shown, the difference in pressure between supply and return is held constant. This pump system offers advantages in many plants:

1. No hot oil is returned to the storage tank or pump suction.

2. Fuel enters the closed circuit at the same rate it is burned, thus simplifying fuel metering and combustion control.

3. The pump may be used to boost pressure on existing oil burner systems.

The *horizontal rotary cup* burner atomizes fuel oil by literally tearing it into tiny droplets. A conical or cylindrical cup rotates at high speed (usually about 3500 r/min) if motor-driven. Oil moving along this cup reaches the rim where centrifugal force flings it into an air stream (Fig. 11.14*c*). This system of atomization requires no oil pressure beyond that needed to bring oil to the cup. But high oil preheat temperatures must be avoided since gasification may develop. The rotary cup can satisfactorily atomize oils of high viscosity (300 seconds Saybolt Universal, or SSU) and has a wide range of about 16:1.

Two types of horizontal rotary burners used are the pump type with the pump mounted at the burner, and the separate pump type with remotely located pump, as in Fig. 11.14*b*, where the oil is taken from the storage tank to be pumped directly to the burner. Where heavier oils are used, a preheater is required before the oil goes to the burner in order to obtain good atomization. The heavier fuel oils for these burners are harder to ignite, and may have a spark-ignited gas pilot for igniting the main flame, and thus require a timed ignition period with flame proving.

Mechanical atomization has resulted in the growth of *low-excess-air* burners by raising the atomization pressure, metering carefully the fuel and air supplied to the burner and the temperature of the oil supplied to the burner. As can be noted, low-excess-air operation requires special controls and equipment, more precise adjustment of the controls, and continuous care that the proper air-fuel ratio for low-excess-air burning is maintained per the design of the specialized equipment necessary to achieve the desired results.

Oil burner operating guides.   Some precautions that must be observed in oil burner operation and maintenance include the following:

1. Avoid the burner flames striking any boiler heating surfaces directly, such as water legs, tubes, shells, and similar boiler surfaces to avoid metal overheating and bulges (blowtorch effect).

2. Burner tips and nozzles must be kept clean in order to maintain good atomization. The oil to the burners must be kept at the proper temperature for the same reason, as well as proper oil pressure on mechanical atomization burners.

3. Strainers in oil lines must be periodically cleaned or replaced to prevent rust, scale, and similar foreign material to be drawn from fuel tanks, and thus interrupt the flow of oil to the burners, cause a sporadic flame at the burner, or possibly flame extinguishment. Under these conditions, if the flame safeguard system does not detect this partial loss of flame, unconsumed fuel in the furnace could produce a furnace explosion on the fire side.

4. Good housekeeping is essential around any fuel-burning equipment to avoid the threat of a fire erupting around or near the burner. Leaks should be promptly cleaned and the source of any fuel leak traced for correction. Do not pile oily rags and waste around fuel lines or burners.

5. Maintain proper airflow in order to maintain designed air-fuel ratios, as well as to avoid air pollution fines from improper stack discharges, such as smoke.

6. Check safety controls, such as flame safeguards per the manufacturer's instruction and frequency, as well as other controls to maintain efficient and safe burning on the combustion equipment.

7. It is essential to keep air intakes to boiler rooms in good condition, especially on smaller boilers that receive combustion air from internal air in the boiler room. Also essential is to keep discharge ducts to chimneys free of leaks into the boiler room. Besides poor combustion, there is the risk from harmful gases, such as carbon monoxide, a deadly gas. A NFPA rule is to have at least one square inch of fresh outdoor air opening per 4000 Btu/hr fuel input, or

$$\text{Fresh outside opening, in.}^2 = \frac{1}{4000 \text{ Btu/hr input}}$$

Example A boiler receives combustion air from an enclosed boiler room and is rated 6000 lb/hr with an efficiency of 78 percent. What size opening to fresh outside air is needed?

Use 1000 Btu per lb of steam.

$$\text{in.}^2 \text{ opening required is } \frac{6000(1000)}{0.78(4000)} = 1923 \text{ in.}^2$$

8. Always purge the furnace before any light-off procedure. Make sure automatic firing systems follow the proper purge timing period according to design or instructions for the boiler burner system.

Steam-heated oil heaters with oil pumped at high pressure require special attention in order to avoid oil contaminating the steam-condensate system. Fuel-oil heaters are required when heavy, viscous oil is burned in order to facilitate flow and assist atomization. In these heaters, oil is pumped through tubes that are surrounded by steam and enclosed in a shell.

The maximum safe pressure of the shell should be ascertained definitely, and a safety valve set at not over this pressure should be installed on the shell or on the steam supply system.

The manufacturer's specifications of maximum oil pressure should be learned, and an oil-pressure relief valve installed between the oil pump and the first shutoff valve in discharge line to the heater. The discharge of this relief valve may be piped back to the oil storage tank for cleanliness. A second oil-pressure relief valve should be installed at the outlet from the heater, for, when valves are closed, expansion may create excessive pressures.

The condensate from the heater should be drained to waste unless a well-lighted and frequently observed gauge glass is on the trap body, or other suitable means of oil detection are employed. One cannot be too careful with this installation, for a split tube in the oil heater might allow the fuel oil to pass into the feedwater system, flooding the inside of the boilers with oil, an extremely hazardous condition.

Gas Fuels

Natural gas is the main fuel used in steam generation because manufactured gases run too high in cost. By-product gases usually have low heating values and are produced in relatively minor quantities, so they are ordinarily used at the production point and not distributed. Natural gas is colorless and odorless. Composition varies with source, but methane (CH_4) is always the major constituent. Most natural gas contains some ethane (C_2H_6) and a small amount of nitrogen. Gas from some areas, often called *sour gas* contains hydrogen sulfide and organic sulfur vapors. The heating value averages about 1000 Btu/ft^3 (20,000 Btu/lb), but may run much higher. Natural gas is usually sold by the cubic foot, but may be sold by the therm (100,000 Btu).

Coal gas and *coke-oven gas* (manufactured gases) are produced by carbonizing high-volatile bituminous coal in retorts that exclude air and are heated externally by producer gas. Usually a number of by-products result. Cleaned of impurities, these gases are roughly one-half hydrogen and one-third methane, plus small amounts of carbon monoxide, carbon dioxide, nitrogen, oxygen, and illuminants (C_2H_4 and C_6H_6). The heating value runs around 550 Btu/ft^3.

The gas served in a given area may be a mixture of two or more gases or a mixture of natural and manufactured gas. The heating value, usually held to 525 to 550 Btu/ft^3, is often fixed by state or local ordinance.

Commercial butane and propane are essentially by-products from the manufacture of natural gasoline and from certain refinery operations. As supplied, propane (C_3H_8) is essentially pure, while butane (C_4H_{10}) usually contains a small amount of propane. Both have high heating values, are easily liquefied at low pressure, and are widely used as bottled fuels.

Blast-furnace gas, a by-product of iron making, has the lowest heating value of any commercial gas, about 90 Btu/ft^3. It is close to three-quarters nitrogen and carbon dioxide, the only important combustible constituent being carbon monoxide. Raw gas, which usually contains a high concentration of solid impurities, is normally washed before use. But unwashed gas has been successfully burned in boiler furnaces.

Sewage-sludge gas runs about two-thirds methane and one-third carbon dioxide, with small amounts of hydrogen, nitrogen, and usually some hydrogen sulfide. The heating value is about 650 Btu/ft^3. Although used mostly in internal-combustion engines, this gas is also burner-fired. See Fig. 11.15*a* for properties of fuel gases, and Fig. 11.15*b* for properties of waste fuels.

Gas Burners

Burning gas requires no preparation of the fuel, as do other fuels. But proportioning with air, mixing, and burning can be handled in several ways. Also, the fuel's characteristics need to be known for sound selection of equipment and successful operation. Atmospheric burners are used for gas burning and differ mainly in the way air and fuel mix. The atmospheric burner is popular, as in home gas ranges. The momentum of the incoming low-pressure gas stream is used to draw in, or aspirate, part of the air needed for combustion. A shutter or similar device regulates the amount of air so induced. Gas and air together pass through a tube leading to the burner ports, mixing in the process. The mixture burns at the ports or openings in the burner head (with a blue, nonluminous flame). Secondary air is drawn into the flame from the surrounding atmosphere.

A single-port atmospheric burner is shown in Fig. 11.16*a*. A needle valve controls the gas flow through the spud; air is drawn in around the shutter at the end. With burner-port size and shape fixed, the nature of burning depends largely on the amount of primary air, or premix. With premix low, the flame is long and pale blue. It may have a yellow tip, indicating cracking and presence of free carbon.

Operation is usually satisfactory with 30 to 70 percent premix; in some special designs, 100 percent primary air is used. This premix range gives a turndown, or capacity, range of about 4:1. Usually premix and capacity ranges are somewhat narrower. Secondary air may be drawn in around the burner, the amount depending on the area of the opening and the draft. The high-pressure burner uses gas at about 20 to 30 psig and air at atmospheric pressure. Another type uses compressed air, with gas at atmospheric pressure.

A refractory gas burner is shown in Fig. 11.16*b*. It depends on natural or fan draft to draw in all the air required for combustion; hence draft conditions are important. One design uses multiple gas jets, which discharge into the airstream to cause violent agitation in a short mixing tube or tunnel of refractory. In the burner of Fig. 11.16*b*, turbulence vanes impart a swirling motion to the air entering the tunnel.

Large steam-generating units often use a high-pressure (2 to 25 psi) gas burner of the gas-ring, center-diffusion-tube, or turbulent, design. The gas ring has an annular manifold located between the air

(a)

Fuel	Source	Average composition	High heat value, Btu/ft³	Remarks
Blast-furnace gas	By-product of iron making	58% N_2, 27% CO, 12% CO_2, 2% H_2, some CH_4	90–100	Good fuel when cleaned—used mainly at source
Butane	By-product of gasoline making, also in casing-head gas	C_4H_{10} (usually has some butylene C_4H_8 and propane C_3H_8)	3200–3260	Liquefies under slight pressure, sold as liquid (bottled gas)
Casing-head gas	Oil wells	Varies, mostly butane, propane	1200–2000	Used mostly in oil fields
Carbureted water gas	Manufactured from coal, enriched with oil vapor	34% H_2, 32% CO, 16% CH_4, 7% N_2, 5% C_2H_4, 4% CO_2, 2% C_2H_6	500–600	Good fuel, but usually costly Part of most city gas
Coke-oven gas	By-product coke ovens	48% H_2, 32% CH_4, 8% N_2, 6% CO, 3% C_2H_4, 2% CO_2, 1% O_2	500–600	Good fuel when cleaned, often used at source
Natural gas	Gas wells	Varies, mostly CH_4, C_2H_6, C_3H_8	950–1150	Ideal fuel, piped to point of use
Oil gas	Manufactured from petroleum	54% H_2, 27% CH_4, 10% CO, 3% N_2, 3% CO_2, 3% C_2H_4	500–550	Used on West Coast, often mixed with coke-oven gas
Producer gas	Manufactured from coal, coke, wood, etc.	51% N_2, 25% CO, 16% H_2, 6% CO_2, 2% CH_4	135–165	Requires cleaning
Propane	By-product of gasoline	C_3H_8	2500	Similar to butane
Refinery gas	By-product of petroleum processing	Varies, mostly butane, and propane	1200–2000	Used mainly at refineries
Sewage gas	Sewage-disposal plants	65% CH_4, 30% CO_2, 2% H_2, 3% N_2, traces of O_2, CO, H_2S	600–700	Many disposal plants meet all power needs with this fuel

(b)

Waste	Average heating value (as fired), Btu/lb
Gases:	
Coke-oven	19,700
Blast-furnace	1,139
Carbon monoxide	575
Refinery	21,800
Liquids:	
Industrial sludge	3,700–4,200
Black liquor	4,400
Sulfite liquor	4,200
Dirty solvents	10,000–16,000
Spent lubricants	10,000–14,000
Paints and resins	6,000–10,000
Oily waste and residue	18,000
Solids:	
Bagasse	3,600–6,500
Bark	4,500–5,200
General wood wastes	4,500–6,500
Sawdust and shavings	4,500–7,500
Coffee grounds	4,900–6,500
Nut hulls	7,700
Rice hulls	5,200–6,500
Corn cobs	8,000–8,300

Figure 11.15 (a) Properties of fuel gases; (b) heating value of waste fuels.

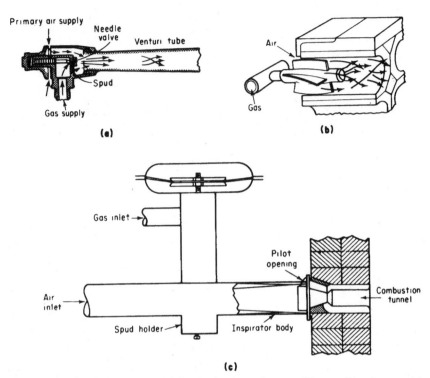

Figure 11.16 Gas burner types. (a) Atmospheric gas burner; (b) tunnel gas burner; (c) inspirator burner premixes air and gas for burning.

register and the furnace wall surrounding the burner opening. Orifices drilled in this ring spray gas angularly across an incoming airstream, controlled in quantity, velocity, and rotation by the resistor.

Figure 11.16c shows a low-pressure-type burner in which gas is at atmospheric pressure and air is at 1 to 2 psi. This unit, through the inspirator governor, prepares a burning mixture to several burners. It can provide a higher burner-head pressure, to overcome variable draft conditions in the furnace, and good overload capacity with uniform air-gas mix at all loads.

Waste Fuels

Combustion of industrial wastes and municipal refuse is gaining increased attention as fossil fuel becomes scarcer and more expensive. Among the solid fuels are the following: Wood from lumber and woodworking industries in the form of sawdust, slabs and shavings, and hog wood. Hog wood is wood refuse cut to uniform size before burning. Bark from pine, oak, and hemlock tress is burned in special furnaces. The heating value of wood varies from 2500 to 3000 Btu/lb. Bagasse is the crushed stalks of sugarcane from which the sap has been extracted. The heating value is from 3500 to 4500 Btu/lb. Coke is the solid remains after the destructive distillation of either petroleum oils or certain bituminous coals. The heating value of petroleum coke is from 11,500 to 15,000 Btu/lb. See Fig. 11.15b for waste-fuel heating values.

The boiler manufacturer should be consulted on any conversion of a fossil-fuel-burning boiler to a waste-fuel-burning one. There are many variables to consider in order to prevent future operating difficulties. In general, solid wastes are burned on a fuel bed, or by shredding, in suspension. Liquids are burned by atomizing-type burners.

The continuing emphasis on environmental control adds another design consideration: the problem of emissions from burning of waste fuel. Mechanical dust collectors are used as well as wet or dry scrubbers and electrostatic precipitators in order to comply with emission standards.

Combination-fuel burners. Many utilities and industrial plants employ combination-fuel burners in large boilers, as shown in Fig. 11.17, which can burn pulverized coal, oil, and gas. The usual reason for these combination-fuel burners is to take advantage of differences in fuel prices during yearly seasonal changes, such as summer and winter variations in natural gas or fuel oil prices. The design may also include capabilities for burning two fuels at the same time. Furnaces are usually designed to burn gas simultaneously with oil or

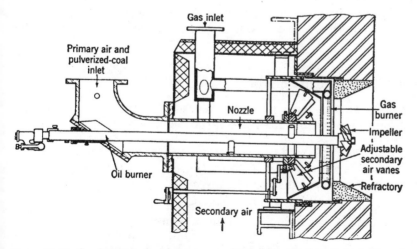

Figure 11.17 Combination-fuel burners can burn pulverized coal, fuel oil, and gas. These types of burners are used by utilities and large industrial plants in order to take advantage of seasonal variations in the price of one fuel from the other.

pulverized coal, or to burn gas only. Designers of combination-fuel burners and furnaces try to attain flame travel where there is no impingement on tubes, walls, or floor tubes, and where radiant heat is the chief heat transfer method in the furnace.

Pulse burners. Pulse combustion is primarily used in heating boilers up to about 5000 lb/hr capacity with gas as the primary fuel, although manufacturers claim liquid and solid fuel burners can also be used.

In ordinary gas burners, there is a steady flow of fuel and air into the combustion chambers of the boiler, while in pulse burning, the gas and air are introduced intermittently into a special geometrically proportioned combustion chamber that develops an oscillation, or cyclic burning as illustrated in Fig. 11.18. The advantage claimed for this method of burning is that full burning of fuel in the combustion chamber is possible due to the cyclic entrance of fuel into the combustion chamber, thus achieving higher efficiency of burning.

To start a pulse burner, an initial flow of air is provided by a small blower, fuel is introduced, and a spark plug is energized. The burner immediately is placed in a resonant operation, with positive and negative pressures produced as illustrated in Fig. 11.18. The starting blower and spark plug are automatically switched off once a detector signals combustion has been established. The cycling is timed so that each cycle of burning leaves a pocket of flame that will ignite the next charge of gas-air mixture.

The air-fuel ratio is carefully maintained by modern metering devices. The valves are set at the factory to provide minimum excess air for a clean and safe burn. Pulse combustion boilers use a pressure switch to verify that a flame exists instead of flame rods or ultraviolet flame detectors.

Combustion and Load Control

Load control on a boiler is generally geared to maintaining a constant pressure or temperature. Pressure variation is caused by:

1. *Load on the boiler:* An increase in load without additional fuel input causes a pressure drop. A decrease in load without an accompanying decrease in fuel input causes a pressure rise.

2. *Fuel input to the boiler:* Too high an input will cause a pressure rise, while too low an input will cause a pressure drop.

Thus pressure regulation and fuel regulation, or combustion controls, are directly related. For this reason combustion controls are geared for modulation by pressure variations within close limits. While air-

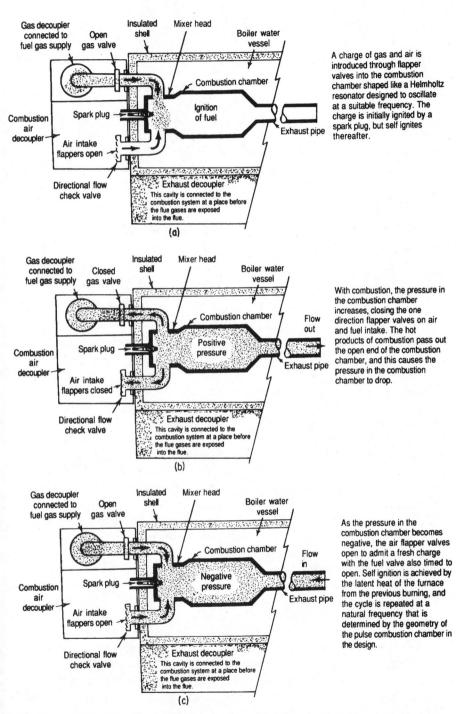

Figure 11.18 Pulse burning has three sequences. (*a*) The starting phase; (*b*) the positive-pressure phase; (*c*) the negative-pressure phase.

flow and exhaust flow usually follow fuel flow, the latter is determined by the pressure-set limits within which a boiler is to operate.

Although manufacturers differ in approach, the following factors must be considered in any control used on a boiler: (1) steam pressure and flow; (2) furnace pressure and draft; (3) air pressure and flow; (4) feedwater pressure and flow (including low water); (5) flue-gas flow and composition; and (6) proper ignition and burner-flame control.

Modes or the manner in which a control acts and reacts to restore a variable on a boiler may be classified into on-off or two-position control, positioning control, and metering control. As described in Chap. 10, *on-off controls* are usually found on smaller firetube and watertube heating boilers, and are not efficient because of the loss of heat during the off-firing cycle.

The combustion control generally used on process boilers is a *positioning control system,* because it is more flexible. (See Fig. 11.19 for combustion controls used for different fuels.) Steam pressure is the measured variable, and a master pressure controller responds to changes in header pressure and (by means of power units or actuators) positions the forced-draft damper to control airflow and the fuel valve to regulate fuel supply. An independent controller, positioning the uptake damper, maintains furnace draft within the desired limits.

Although positioning-type control systems are an improvement over the on-off type, airflow and fuel supply are at their theoretically correct ratio at only one setting. This is usually the point at which they are calibrated on installation. Positioning control also assumes that a given output signal from the master controller always produces the same changes in the flow of combustion air, in stoker speed, or fuel-valve setting. But stoker speed might be affected by line-voltage variations, and airflow by boiler slagging or barometric conditions. Thus manual adjustment is still necessary, not only on load changes but also to counteract these longer-term effects.

A *metering control* measures the fuel flow and airflow, then modifies the valve and damper positions to maintain these measured flows rather than implied ones. Thus it holds an optimum air-fuel ratio over a wide load range without manual intervention. Especially valuable is its inherent compensation for such variables as boiler cleanliness, voltage swings in electric actuators, lost motion in mechanical or pneumatic devices, and changes in fuel quality.

Computer combustion and load control is growing rapidly with the growth of sensors and electronic transmission of data to a computer for calculating the optimum control setting in order to obtain desired design results. In a direct *digital control* system, a variable is measured through a sensor with the readings transmitted electronically into a computer. The computer may receive many such readings from

Type of fuel	Type of control systems	Boiler capacity range (Mbh)	Precision of pressure or temperature control—tolerance limits (in percent of set point values)	Precision of fuel/air ratio control (%O_2)
Coal	On/off	400–3,000	±6	Not applicable
	High/low/off	2,000–5,000	±5	Not applicable
	Modulating positioning	3,000–62,000	±3	±1.0 for firing rates of or above 33% of maximum; ±2.0 for firing rates below 33% of maximum
	Semimetering	35,000–72,000	±3	±0.9 for firing rates of or above 33% of maximum; ±1.8 for firing rates below 33% of maximum
	Full metering (steam flow/air flow basis)	38,000–80,000	±3	±0.8 for firing rates of or above 33% of maximum; ±1.6 for firing rates below 33% of maximum
	Full metering (steam flow/air flow basis) with O_2 compensating	68,000 and above	±3	±0.6 for firing rates of or above 33% of maximum; ±1.2 for firing rates below 33% of maximum
Oil	On/off	400–3,000	±6	Not applicable
	High/low/off	2,000–5,000	±5	Not applicable
	Modulating positioning	3,000–66,000	±3	±0.5 for firing rates of or above 33% of maximum; ±1.0 for firing rates below 33% of maximum
	Semimetering	36,000–66,000	±3	±0.4 for firing rates of or above 33% of maximum; ±0.8 for firing rates below 33% of maximum
	Full metering (fuel flow/air flow basis)	36,000 and above	±3	±0.3 for firing rates of or above 33% of maximum; ±0.6 for firing rates below 33% of maximum
	Full metering (fuel flow/air basis) with O_2 compensation	69,000 and above	±3	±0.2 for firing rates of or above 33% of maximum; ±0.4 for firing rate below 33% of maximum
Gas	On/off	400–3,000	±6	Not applicable
	High/low/off	2,000–5,000	±5	Not applicable
	Modulating positioning	3,000–66,000	±3	±0.4 for firing rates of or above 33% of maximum; ±0.8 for firing rates below 33% of maximum
	Semimetering	36,000–66,000	±3	±0.3 for firing rates of or above 33% of maximum; ±0.6 for firing rates below 33% of maximum
	Full-metering (fuel flow/air flow basis)	36,000 and above	±3	±0.2 for firing rates of or above 33% of maximum; ±0.4 for firing rates below 33% of maximum

Figure 11.19 Types of combustion controls used for different fuels and boiler capacities.

sensors, calculate the desired set points, and feed this information to a controller for action in a very short time. The controller is thus responding to a very narrow range of desired results, which improves efficiency. See Fig. 11.20.

A *distributed control system* divides a large control system into multiple smaller control systems, each having a certain control to perform. The multiple systems communicate, as programmed, to a main

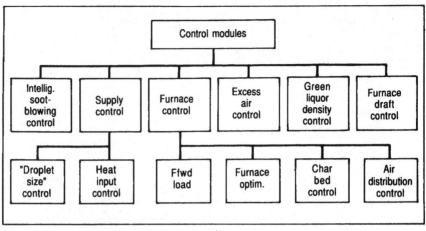

(a)

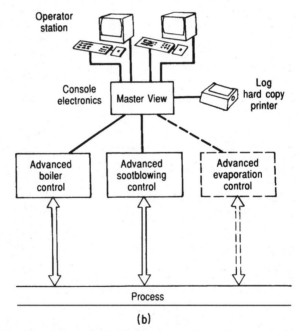

(b)

Figure 11.20 Distributed control. (*a*) Separate control modules are used for the different sections of an operating recovery boiler. (*b*) The control room features readouts on the different variables for comparison to set points. (*Courtesy TAPPI magazine.*)

computer, which in turn is programmed to review each station's results, calculate overall objective set points, and feed back to the individual controller corrections that may be needed on the system to attain overall set points or efficiency.

Figure 11.20a shows a distributed control system regulating different sections of a recovery boiler system. The results are displayed in an operator station as shown in Fig. 11.20b. From this display, the operator can make adjustments on all the variable set points that have been provided, such as viscosity of the black liquor, heat input to the boiler, and some of the other variables shown in the control module. This particular system even has a trend display. Note the availability of a printer on any desired displayed readings.

Note also that one operator may be in charge of this fairly large and complicated boiler system. The driving force behind improved instrumentation and better, more accurate system management is the need to cut operating costs by more automation, and improved efficiency to save on fuel costs.

Superheat control.   Large high-pressure boilers generate superheat and reheat steam. A steam temperature of 1000°F is common, and units have been installed for 1050 and 1100°F. Because these high temperatures are limited only by metallurgy, steam temperatures must be held to close limits for safety as well as economy. Six basic methods are used for controlling the temperatures of superheated steam leaving the boiler:

1. Bypass damper control with a single bypass damper or series-and-shunt damper arrangement for bypassing flue gas around the superheater as required.

2. Spray-type desuperheater control where water is sprayed directly into the steam with a spray-water control valve for temperature regulation. (See Fig. 10.2b.)

3. Attemperator control where a controlled portion of the steam passes through a submerged tubular desuperheater and a control valve in the steam line to the desuperheater or attemperator is used.

4. Condenser control with desuperheating condenser-tube bundles located in the superheater inlet header and water-control valve or valves to regulate a portion of the feedwater flow through the condenser as required.

5. Tilting-burner control where the tilt angle of the burners is adjusted to change the furnace heat absorption and resultant steam temperature.

6. Flue-gas-recirculation control where a portion of the flue gas is recirculated into the furnace by means of an auxiliary fan with a damper control to change the mass flow through the superheater and the heat absorption in the furnace, as required to maintain steam temperature.

Safety Controls

Safety controls generally are those that limit energy input and thus shut down the equipment when unsafe conditions develop. They are: (1) pressure-limit or temperature-limit switches; (2) low-water fuel cutoffs; (3) flame-failure safeguard systems; (4) automatic ignition controls; (5) oil and gas fuel-shutoff-valve controls; (6) air and fuel pressure interlock controls; and (7) feedwater regulating controls.

The safety valve (or relief valve) is the most important safety device. While not considered a control in the usual sense, it is the last measure against a serious explosion.

Safety controls guard against the following hazards:

1. Overpressure, leading to explosions from the water side or steam side.

2. Overheating of metal parts, possibly also leading to explosion in a fired boiler (mainly because of low water or poor circulation).

3. Fire-side explosions (furnace explosions) due to uncontrolled combustible mixtures on the firing side.

These types of accidents are considered major and may lead to loss of life and serious property damage. Other potential sources of accidents are cracking; bulging from local overheating because of scale; deforming, such as tubes bowing; thinning of vital pressure parts, which can lead to cracking or localized rupture; and expansion and contraction failures, causing cracking or rupturing of metal parts.

Manufacturers and state laws are trying to prevent, with safety control equipment, the three major types of accidents of overpressure, dry firing, and furnace explosions. While the other types of failures are controlled somewhat by automatic controls, prevention is mostly by legal inspection requirements and by proper operation and maintenance practices expected from the owner-user of a boiler. Included are good feedwater treatment and testing of controls at periodic intervals, including safety relief valves.

ANSI/ASME CSD-1, titled "Controls and Safety Devices for Automatically Fired Boilers," is a standard being adopted by many jurisdictions as a requirement, similar to the ASME Sections I and VIII Boiler Code. This safety standard applies to nonresident boilers and fuel inputs to 12,500,000 Btu/hr.

Among the safety controls recommended are the following:

1. For hp and lp steam boilers
   a. Two (2) low-water fuel cutoffs with manual reset features.
   b. Operating pressure cutout switch.

c. Upper limit pressure cutout switch with manual reset features.

d. Flame failure safety controls for oil- and gas-firing units.

e. NFPA recommended fuel train for oil- and gas-firing. See Fig. 11.21.

f. Properly sized and set safety valve set at or below the AWP of the boiler.

g. No shutoff valve between the boiler and the upper limit pressure switch.

h. All wiring and electric controls must meet the National Electric Code for the expected operating environment, but as a minimum must be dripproof. Electrical appliances must have UL or similar approval organization's stamp.

2. For hw boilers

a. One low-water fuel cutoff switch, with manual reset features.

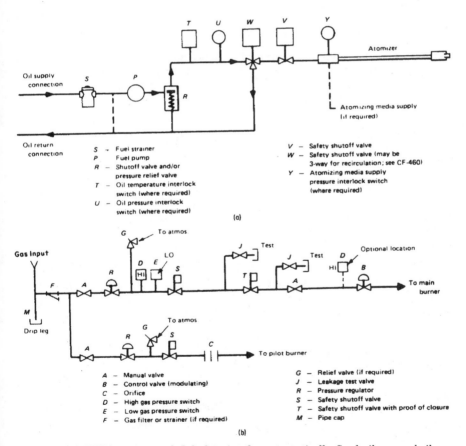

Figure 11.21 NFPA recommended fuel trains for automatically fired oil or gas boilers with input to 12,500,000 Btu/hr. (a) Oil-fired; (b) gas-fired.

b. Operating temperature cutout switch.
c. Upper limit temperature cutout switch with manual reset.
d. Flame failure safety controls for oil- and gas-fired boilers.
e. NFPA recommended fuel train for oil and gas firing.
f. Properly sized and set safety relief valve set at or below AWP.
g. All wiring and electrical appliances must meet the conditions detailed under (h) for steam boilers.

Periodic testing. This standard is for automatically fired boilers below the 12,500,000 Btu/hr input, but the standard recognizes the need for testing the controls according to the schedule in Table 11.1 per item listed. The testing is to be performed by the following, depending on the skill and experience needed: (1) the operator checking the boiler's operation; (2) a service technician; (3) by a manufacturer's representative.

TABLE 11.1   Periodic Testing Schedule for Automatically Fired Boilers

Item	Frequency	Item	Frequency
Gauges, monitors, and indicators	Daily	Operating control	Annually
Instrument and equipment settings	Daily	Low draft, fan, air pressure, and damper position interlocks	Monthly
Firing rate control	Weekly	Atomizing air-steam interlock	Annually
	Semiannually		
	Annually	High and low gas pressure interlocks	Monthly
Flue, vent, stack, or outlet dampers	Monthly	High and low oil pressure interlocks	Monthly
Igniter	Weekly	High and low oil temperature interlocks	Monthly
Fuel valves pilot and main	Weekly		
Pilot and main gas or main oil	Annually	Fuel valve interlock switch	Annually
		Purge switch	Annually
Combustion safety controls		Burner position interlock	Annually
Flame failure	Weekly	Rotary cup interlock	Annually
Flame signal strength	Weekly	Low fire start interlock	Annually
Pilot turn-down tests	As required/ annually	Automatic changeover control (dual fuel)	At least annually
Refractory hold in	As required/ annually	Safety valves	As required
Low-water fuel cut-off and alarm	Daily/weekly	Inspect burner components	Semiannually
High limit safety control	Annually		

Burner Flame Safeguard Systems

A flame safeguard system is an arrangement of flame detection systems, interlocks, and relays which will sense the presence of a proper flame in a furnace and cause fuel to be shut off to the furnace if a hazardous (improper flame or combustion) condition develops. Modern combustion controls are closely interlocked with flame safeguard systems and also pressure-limit switches, low-water fuel cutoffs, and other safety controls that will stop the energy input to a boiler when a dangerous condition develops. Thus it becomes obvious that a modern flame safeguard system performs actually two functions: It senses the presence of a good flame or proper combustion and programs the operation of a burner system so that motors, blowers, ignition, and fuel valves are energized only when they are needed, and then in proper sequence.

A furnace explosion is the ignition and almost instantaneous combustion of explosive or highly inflammable gas, vapor, or dust accumulated in a boiler setting. Often it is of greater expansive force than the boiler setting can withstand. In minor explosions, called puffs, flarebacks, or blowbacks, flames may blow suddenly for a distance of many feet from all firing and observation doors. Thus anyone in the flame path may be seriously or fatally burned. Such minor explosions indicate dangerous conditions, even if no real damage is done. Heavier explosions may shatter gas baffles, bulge setting walls, loosen refractory, blow brick tops of boiler settings through roofs, blow the sidewalls out from under the boiler, break connecting piping, and even demolish boiler housings. The main causes of firebox explosions are:

1. Flame failure resulting from liquids or inert gases entering the boiler fuel system.

2. Insufficient purge before the first burner is lit.

3. Human error.

4. Faulty automatic fuel-regulating controls.

5. Fuel-shutoff-valve leakage.

6. Unbalanced fuel-air ratio.

7. Faulty fuel supply systems.

8. Loss of furnace draft.

9. Faulty pilot igniters.

There have been several incidents of *furnace implosions,* resulting in considerable damage to utility boilers. These furnace implosions have occurred in balanced-draft, fossil-fired boilers, where sufficient force was developed to exceed the structural strength of the boiler furnace. The implosions have occurred on boilers fired with oil or coal with the reported incidents divided approximately equally between the two fuels. No incidents have been reported on natural-gas-fired units, because few gas-fired boilers are balanced-draft. All the reported incidents occurred on balanced-draft boilers with induced-draft fans with high-head capability, operating, at least momentarily, under low-flow conditions. All the implosions have occurred without main fuel firing, some have occurred prior to light-off of main fuel, and others have occurred following a main fuel trip.

Two conditions have caused furnace implosions, both involving operation of high-head induced-draft fans under low-flow conditions:

1. A malfunction of the equipment regulating the boiler gas flow path, including air supply and gas removal, resulting in the furnace being exposed to the full induced-draft-fan head capability; usually occurs at the low-flow, high-head range of a centrifugal fan, and often combined with an interruption of the forced-draft airflow path.

2. Rapid decay of furnace gas temperature and pressure following a rapid reduction in fuel input or a main fuel trip, similar to a flame-out condition.

Flame-monitoring devices. Many flame-monitoring devices are based on the following physical principles of a flame (see Fig. 11.22a).

1. A flame produces an ionized zone, meaning that it can conduct a current through it. Conductivity-flame rod detectors use the principle of a conducting flame for flame-detection monitoring.
2. A flame can rectify an alternating current. This is done by making one electrode across a flame larger than the other, thus making electrons flow through a flame much more readily in one direction than in the opposite direction.
3. Radiation of light is a known phenomenon of any fire. A flame radiates energy in the form of waves which produce heat and light. Three types of radiation from a flame are:
   a. Visible light that can be seen by the human eye. The wavelengths of visible radiation extend only from 0.4 to 0.8 micrometer (μm) (formerly microns). When cadmium is exposed to visible light, it emits electrons with the strength of the visible light. Thus, if a cadmium phototube is designed in an appropriate electronic circuit, electricity will flow through the circuit when the cadmium is exposed to sufficient light. This electricity can be used to trigger relay circuits for flame detection.

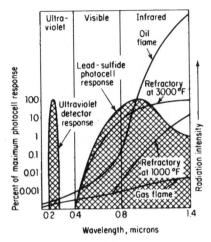

(a)

COMPARISON OF FLAME SAFEGUARDS

Principle of Flame Detection	Rectification		Infrared	Visible Light	Ultravision
Type of Detector	Rectifying Flame Rod	Rectifying Phototube	Lead Sulfide Photocell	Cadmium Sulfide Photocell	Ultravision Detector Tube
Advantages					
Same detector for gas or oil flame			X		X
Can pinpoint flame in three dimensions	X				
Viewing angle can be orificed to pinpoint flame in two dimensions		X	X	X	X
Not affected by hot refractory	X				X
Checks own components prior to each start		X	X	X	
Can use ordinary TW plastic-covered wire for general application, no shielding needed		X	X	X	
No installation problem because of size			X	X	
Disadvantages					
Difficult to sight at best ignition point			X		
Exposure to hot refractory may reduce sensitivity to flame flicker and require orificing			X		
Flame rod subject to rapid deterioration and warpage under high temperatures	X				
Not sensitive to extremely hard premixed gas flow			X		
Temperature limit too low for some applications	X	X	X	X	X
Shimmering of hot gases in front of hot refractory may simulate flame			X		
Hot refractory background may cause flame simulation		X			
Electric ignition spark may simulate flame					X

(b)

Figure 11.22   (a) The properties of a flame are used by detectors to scan a flame. (b) Comparison of flame detectors.

b. Infrared radiation covers most of the useful band of wavelengths and also covers most of the radiation strength. Infrared detectors are suitable for gas and oil flames. Because hot refractories also radiate infrared, scanners must avoid hot refractories. Lead sulfide cells are used in photocells to sense infrared radiation. Unlike the cadmium phototube, it does not emit electrons, but has the property of having its electric resistance reduced while exposed to infrared radiation. The greater the strength of the radiation, the lower the resistance of the lead sulfide. If it is connected to a designed electronic circuit, this principle is used for flame detection purposes.

c. Ultraviolet radiation is the latest flame detector based on the phenomenon of sensing the strength of ultraviolet radiation in a flame. It is insensitive to visual and infrared radiation and is not affected by hot refractories, since these usually do not give off appreciable ultraviolet radiation. When radiation from a flame passes through the typical quartz viewing window of one of these detectors into the flame-sensing tube, the tube becomes electrically conductive. The strength of the detector signal, or current passed through the sensing tube, depends on the kind of fuel, size and temperature of the flame, and distance between the flame detector and the flame. Figure 11.22*a* shows some typical wavelengths in a flame and response percentage of total wavelengths of typical flame pickup devices.

Detector types. The type of flame detector depends on the fuel used, the type of burner, and the size and arrangement of the boiler. Flame detectors vary from those used on small domestic boilers to those on large boilers. Types of detectors and the task each does are:

1. Stack switch, heat sensing.
2. Rectifying flame rod, heat sensing.
3. Rectifying phototube, visible-light sensing.
4. Lead sulfide photocell, infrared-light sensing.
5. Cadmium sulfide photocell, visible-light sensing.
6. Ultraviolet flame-detector tube, ultraviolet-light sensing.

While each of the flame-sensing devices can offer a substantial amount of protection if properly installed, they are all subject to certain limitations that must be allowed for. For example, the lead sulfide cell and the photoelectric cell are subject to the following limitations (there is some variation between each type):

1. *Discrimination between burners:* With more than four burners in one firebox, it becomes difficult to locate the sensing cell where it

will not be actuated by the flame of an adjacent burner when the burner on which it is mounted has been extinguished.

2. *Ambient temperature of sensing cell:*   High ambient temperatures of the sensing cells (which are easily obtained at the locations where they must be mounted) can result in erratic signals, false signals, and a short life. Air and water cooling have been used to prevent or minimize this limitation.

The flame rod is subject to: (1) short service life at high flame temperatures; (2) fouling of insulators, causing short circuits and shutdowns; (3) difficulty in providing sufficient rod areas to ground the flame; and (4) limited generally to pilots and small burners. Figure 11.22*b* makes some comparisons of the different flame safeguard systems.

The rectified impedance system (Fig. 11.23) operates on the principle that either a flame or a photocell sighted at a flame is capable of conducting, as well as rectifying, an alternating current. This alternating current is applied to either a flame electrode inserted in the flame or a photocell sighted at the flame. The resultant rectified current, which can be produced only when a flame is present, is in turn detected by the relay.

NFPA Code Requirements

The prevention of furnace explosion is a requirement detailed in the National Fire Protection Association (NFPA) codes as well as by fire insurance engineering groups, such as the Industrial Risk Insurers and the Factory Mutual groups. Codes of the NFPA are useful in detailing requirements to prevent furnace explosions. For example, the NFPA Code for gas firing and Code for oil firing stipulate that register-starting procedures must be programmed. The open-register start-up procedure for gas firing is outlined briefly thus:

1. Set all, or most, burner-air registers in the normal firing position. Then purge the furnace and boiler setting, using not less than 25 percent of full-load airflow for 5 min.

2. Throughout the start-up period, maintain the same register settings and the same total airflow used for purge.

3. Set fuel header pressure at a value which will provide a burner fuel flow compatible with the burner airflow.

4. Light burners one at a time as increased heat input is required, keeping the burner fuel-header pressure and register settings at their initial settings. As each new burner is lit off, close the burner

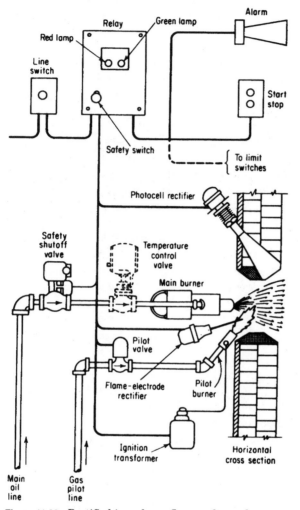

Figure 11.23 Rectified-impedance flame safeguard system operates on the principle that a flame is conducting.

register to light-off position. Since the furnace is air-rich, additional burners may be cut in with no increase in airflow until the fuel flow approaches 25 percent (or whatever airflow rate was used during the purge).

Another requirement found in fire codes, as well as in the ANSI/ASME CSD-1 Safety Code entitled "Controls and Safety Devices for Automatically Fired Boilers," is the double shutoff-valve systems on fuel trains. This generally requires that the main gas-burner train have two manual-reset safety shutoff valves. This arrangement ensures burner cutoff even if one valve stays in an open stuck position.

State laws are being expanded to include the hazard of furnace explosion in state boiler inspection programs. For example, one state has the following requirements for automatic-heating boilers in its inspection code:

1. *Gas-fired boilers:*

   a. Pilot has to be proved, whether manual or automatic, before permitting the main gas valve to open, either manually or automatically, by completing an electric circuit.

   b. A timed trial for the ignition period is established based on the input rating of the burner. For instance, for input rating of 400,000 to 5,000,000 Btu/hr per combustion chamber, the trial for the ignition period for the pilot of automatically fired boilers cannot exceed 15 sec. And the main-burner trial for ignition also cannot exceed 15 sec.

   c. The burner flame-failure controls must shut off the fuel within a stipulated time, again depending on the fuel input of the burner. For a burner rated with an input of 400,000 Btu/hr or more, the electric circuit to the main fuel valve must be automatically deenergized within 4 sec after flame failure. And the deenergized valve must automatically close within the next 5 sec.

2. *Oil-fired boilers:* Similar provisions have been adopted, with requirements on response time for controls to shut off the burner based on fuel input in gallons per hour, instead of Btu per hour. The flame must be continuously supervised by the controls.

Questions and Answers

1 What three conditions determine whether the chemical reactions required for burning will take place?

ANSWER: The three conditions generally needed are: (1) proper fuel-air mixture; (2) intimate mixing of fuel and air in the combustion zone so that all fuel particles have oxygen; and (3) maintenance of an ignition temperature.

2 A flue-gas analysis showed the following percentages by volume: CO_2, 16 percent; oxygen, 5 percent. What was the excess air based on these readings?

ANSWER:

$$\text{Excess air} = \frac{\text{Oxygen by volume}}{CO_2 \text{ by volume}} = \frac{5}{16} = 31.25\%$$

3 What air temperature must be provided a burner for the following conditions? Stack temperature, 625°F; coal burned at 12,000 lb/hr with a heating value of 15,000 Btu/lb; flue gas formed, 13.8 lb/hr fuel burned; boiler efficiency, 80 percent?

ANSWER: Heat lost up the stack is 12,000 $(1 - 0.8)(15,000) = 36$ million Btu/hr. The weight of the flue gas going up the stack is $12,000(13.8) = 165,000$ lb/hr. The heat lost equals $WCp(T_2 - T_1)$. Substituting and solving for T_1 with $Cp = 0.25$, we get

$$36,000,000 = 165,000(0.25)(625 - T_1)$$

Thus $T_1 = 245°F$ and preheated air must be used.

4   What fuel is considered to have two heating values and why?

ANSWER: Fuels containing hydrogen have a higher and lower heating value. The higher heating value is determined in a calorimeter. From this, the heat lost by the water vapor that is formed in burning hydrogen must be deducted to obtain the lower heating value. In most engineering calculations, the higher heating value is used.

5   A boiler burning fuel oil with a heating value of 18,500 Btu/lb has an 82 percent efficiency as determined by test. Fuel-oil consumption averages 1150 lb/hr. How much in Btu's is lost?

ANSWER:

$$\text{Heat released} = 1150(18,500) = 21,275,000 \text{ Btu/hr}$$

$$\text{Heat loss} = 21,275,000(1 - 0.82) = 3,829,500 \text{ Btu/hr}$$

So about 3829 lb/hr of steam is lost.

6   In a pulverized-coal burning boiler, what is primary, secondary, and tertiary air?

ANSWER: Primary air for burners is the air mixed with the fuel at the burner. The fuel is intimately entrained in primary air, resulting in a rich air-fuel mixture, and is about one-third of the total needed for complete combustion. Secondary air is added around the burner or through furnace openings to provide additional air for combustion. Normally for most fuels, this is about two-thirds of the total needed; however, for large pulverized-coal burning boilers, additional air, or tertiary air, is introduced through other openings in the furnace in order to control the desired shape of the flame after it leaves the burner nozzles and provide the final air for complete combustion in the furnace volume.

7   What is the harmful effect of sulfur in a fuel?

ANSWER: Sulfur burns to sulfur dioxide which, when mixed with water or water vapor, forms sulfurous acid that is corrosive to tubes, breechings, and economizer sections on larger boilers. The dew point (temperature at which water vapor condenses) of the flue gas has to be watched (especially with high-sulfur-content fuel) because the gas becomes cooler and cooler while going through a furnace, so as to prevent the combination of water vapor with sulfur dioxide. Recent air pollution laws are tough on high-sulfur-content fuels, because sulfur dioxide is known to pollute the air. Percentages of permissible sulfur content in fuels are slowly being lowered from the previous 4 percent maximum to 1 percent in the near future.

Sulfur in oil-fired boilers leads to fireside corrosion of tubes, especially in those boilers cycling on and off. In cooling on the off cycle, the sulfur dioxide combines with water vapor and with water from leaks to attack the tubes by means of the resultant acid formed.

8 Convert 3 in. of draft into pounds per square inch or into ounces per square inch.

ANSWER: Since water at normal temperature weighs 62.4 lb/ft³, dividing this by the number of cubic inches in a cubic foot gives

$$\frac{62.4}{1728} = 0.036 \text{ psi/in.}^3 \text{ water}$$

Thus for 3 in., pressure = 3 × 0.036 = 0.108 psi. And since there is 16 oz in a pound, the ounces per square inch = 0.108 × 16 = 1.73-oz pressure per square inch.

9 What is meant by balanced draft?

ANSWER: A boiler using both forced-draft and induced-draft fans can be regulated and balanced in the amount of air and flue gas handled, so that the furnace pressure is almost atmospheric. This results in better control of air leakage from the furnace and thus control of the fuel-air ratio in the furnace.

10 What are the usual percentages of excess air in burning the various common fuels in boilers?

ANSWER: For coal, usually 50 percent excess air is used. With oil, gas, or pulverized coal, excess air is 10 to 30 percent.

11 To what capacity of boiler do federal clean-air laws apply? What are some of the requirements on permissible limits?

ANSWER: The Environmental Protection Agency (EPA) has established minimum standards for fossil-fuel-burning boilers with capacities of 250 million Btu/hr of input or larger. However, in their general authority, severe violations below this capacity could also be cited for exceeding the following requirements: (1) no particulate emission in excess of 0.10 lb per 1 million Btu input; (2) no opacity exhibit greater than 20 percent, except up to 40 percent opacity for 2 min is permitted in any hour; and (3) sulfur dioxide emission limited to 1.2 lb for every 1 million Btu fired.

12 What is meant by a dry-bottom furnace compared to a wet-bottom furnace?

ANSWER: On pulverized-coal burning boilers, the furnace design depends on the type of coal to be burnt; however, designers can design furnaces to burn coal of any fusion range. If in burning the coal, the resulting ash is removed in a dry state, fly ash goes up the stack, and the remaining ash goes to the bottom of the furnace, the term "dry-bottom furnace" is used. For low-fusion coals, the ash resulting from burning coal may be in liquid form and flow to the bottom of the furnace to be removed into hoppers after being solidified by cooling. This type of furnace is termed a "wet-bottom furnace."

13   Explain suspension, fuel-bed, and overfire burning in a furnace.

ANSWER: *Suspension firing* involves ejecting gas, liquid, or pulverized fuels from a burner. Since gas is ready for quick mixing with air, gas burners need to proportion only the volume of air and gas to ensure thorough mixing. But, oil or coal burners, in addition to proportioning the fuel and air, must mix them. This means converting the oil from a liquid to gaseous hydrocarbons. For coal, it involves distilling off the volatile matter (gaseous hydrocarbons, plus hydrogen and carbon monoxide). This is done in the instant after the fuel enters the furnace, while in suspension.

In *fuel-bed firing,* the fuel is thrown, pushed, or dropped onto a grate inside the furnace. Air flows upward through the grate and fuel bed; the green coal is heated, volatile matter distills off, and coke is left on the grate. The volatile matter of the coal and the carbon monoxide from the coke burn over the fuel bed with the air that has come up through the fuel bed. Secondary air is usually admitted over the grate. From 40 to 60 percent of the coal's heat is in gas that is liberated over the fuel bed.

In *overfire burning,* about one-half of the burning takes place above the fuel bed with the rest taking place away from the fuel bed. Separate streams of fuel in gas form and air, called stratification, prevent good mixing of gas and air; therefore, excess air is supplied so that gas and air are mixed to complete the combustion.

14   How is scrap metal with the coal going to a pulverizer prevented from entering the pulverizer?

ANSWER: Scrap metal could produce sparks and cause a fire in the pulverizer; therefore, magnetic separators are installed in the feed to the pulverizer in order to remove any scrap metal.

15   What are caking coal and free-burning coal?

ANSWER: A caking coal is one which fuses at the surface when burning to form a more or less heavy crust. The term *coking* is also used. A free-burning coal does not form a crust and is friable (easily crumbles) throughout the combustion process.

16   What are some mineral impurities of coal?

ANSWER: One is ash, which is the incombustible mineral matter left behind when coal burns completely. The amount and character of the ash constitute the biggest single factor in fuel-bed and furnace problems such as clinkering and slagging. An increase in ash content usually means an increase in the carbon carried to waste or imperfect combustion. Next are the incombustible gases such as carbon dioxide and nitrogen. When the volatile matter distills off, a solid fuel is left, consisting mainly of carbon but containing some hydrogen, oxygen, sulfur, and nitrogen that are not driven off with the gases. Sulfur in coal burns, but is undesirable. Besides causing clinkering and slagging, it corrodes air heaters, economizers, breachings, and stacks. It also causes spontaneous combustion in stored coal.

17 Why must soot blowers be provided on coal-burning boilers?

ANSWER: Soot blowers must be provided for cleaning ash from the furnace walls and from the convection-heating surfaces. Hoppers are provided at the bottom of the furnace and at other strategic points throughout the boiler to remove the ash collected at these points if the unit is large.

18 Why should coal be sampled and analyzed in a power plant?

ANSWER: Because coal is not homogeneous with a variable composition. Sampling a given shipment may require taking up to 100 increments in order to obtain a good cross section for laboratory analysis of the coal properties.

19 Which stoker system is the most widely used in industrial plants?

ANSWER: The two most widely used are the traveling-grate and spreader stokers.

20 Name three types of pulverizers.

ANSWER: The three most commonly used pulverizers are: (1) the pressurized, airswept table and roller mill; (2) ball mills using a rotating cylinder with steel balls in it to grind the coal; and (3) tube or rod mills using rods in a rotating cylinder.

21 How is pulverized coal burned?

ANSWER: Pulverized coal is first ground to the fineness of flour. Then it is made to flow through pipes and ejected into the furnace in a manner similar to fuel oil. The plant must have equipment for drying and pulverizing the coal, transporting it to the furnace in an airstream, and injecting it with air needed for combustion. Proper air-fuel ratios must be maintained to avoid furnace gas explosions from unburned fuel suddenly igniting.

22 How is pulverized coal burned by the suspension method?

ANSWER: In suspension burning, because the finely ground coal particles are injected into the furnace and exposed to radiant heat, the temperature rises and the volatile matter distills off. Enough primary air enters at the burner to mix with the coal particles and to burn the gas resulting from separate distillation of each particle. The volatiles, mostly hydrocarbons, ignite more easily than the carbon remaining in the particle and heat the particle to incandescence. Secondary air enters around the burner, flows past the hot carbon particles, and burns with them in a flame several feet long.

23   Is more solid fuel burned in stokers or by suspension firing?

ANSWER: Today, there is more fuel-bed firing than suspension firing, and most solid fuels are burned in stokers. Stoker operation involves (1) pushing, dropping, or throwing coal on a grate within a high-temperature region of the furnace; (2) distilling off part of the coal as a combustible gas so that it can burn above the fuel bed; (3) exposing the remaining red-hot solid coke so that it can be scrubbed by the air coming up through the fuel bed until it is burned; and (4) removing the ash from the furnace zone.

24   Name the various types of stokers.

ANSWER: Stokers are divided into the following two broad classes, depending on how the raw coal reaches the fuel bed: (1) overfeeds, in which the fuels come from above, and (2) underfeeds, where the fuel comes from below. Overfeed designs make up two groups: (1) spreader stokers and (2) chain-grate and traveling-grate stokers.

25   How does a spreader stoker operate?

ANSWER: In a spreader stoker, the raw coal is whirled into suspension within the furnace by paddles or wheels or by air or steam jets. The fines burn in suspension as they travel across the furnace, while the larger pieces fall to the grate to form a fuel bed. The fuel bed is usually thin, but it varies with the steam demand and the class of coal. There is seldom more than a few minutes supply of coal on the grate. A layer of ash under the fuel bed, together with the flow of air up through the grate, helps to keep the metal parts at a safe temperature. Air is preheated to 300 to 350°F.

26   How are the grindability qualities of a coal measured for pulverizing?

ANSWER: The ASTM (American Society for Testing and Materials) approves the following two methods: (1) ball-mill and (2) Hardgrove. The ball-mill method measures the energy needed to pulverize different coals by finding the number of ball-mill revolutions needed to grind a sample so that 80 percent passes a 200-mesh sieve (74 μm). The ball-mill grindability index, in percent, is found by dividing the number of revolutions into 50,000.

In the Hardgrove test, a prepared sample receives a set amount of grinding energy in a miniature pulverizer. The results are measured by weighing the coal that passes a 200-mesh sieve. To find the Hardgrove grindability value, multiply the weight of the coal passing the sieve by 6.93 and add 13 to the product.

27   Explain the term *heating value of fuel.*

ANSWER: The heat liberated by the complete and rapid burning of a fuel per unit weight or volume of the fuel is the heating, or calorific, value of the fuel.

For solid and liquid fuels, this is usually expressed in Btu per pound. For gaseous fuels, it is expressed in Btu per cubic foot at a standard temperature and pressure, usually atmospheric pressure at 68°F.

28 What properties of coal are determined in a "proximate analysis"?

ANSWER: The proximate analysis shows the percentage of (1) moisture, (2) ash, (3) volatile matter, and (4) fixed carbon. These percentages add up to 100. You may also determine the total amount of sulfur as a separate percentage, the ash-fusion temperature, and the heating value.

29 How would you report a coal analysis?

ANSWER: There are five ways to report a coal analysis, but only the first three are used in power-plant work: (1) as received, (2) air-dried, (3) moisture-free, (4) moisture- and ash-free, and (5) moisture- and mineral-free. These analyses report the condition of the coal as it is delivered to the laboratory. They are important because they give the condition of the coal as it is shipped or as it is fired—the values needed for practical work.

30 What is coal ash?

ANSWER: Like moisture, coal ash is an impurity that will not burn. It must be removed from the furnace and plant; this increases shipping and handling costs. Because ash causes clinkering and slagging, it is regarded as the no. 1 fuel-bed and furnace problem.

Ash and slag are impurities that do not burn to a gas and usually trouble coal-fired boilers. The solid particles at high velocity are carried through the boiler with gas in suspension. The general term *fly ash* is used for this slag and ash. Ash and slag can be very abrasive to tube sections if flow distribution is concentrated in the convection passages of a boiler.

31 What is a slag-tap furnace?

ANSWER: In a slag-tap furnace, part of the ash is removed from the combustion zone by directing the burner flame downward over a pool of molten slag. The slag pool is kept hot enough to remain molten so that the slag flows continuously over a water-cooled weir, or ring, into a quenching pit where it is broken up by water.

32 What causes unburned coal in ashpits?

ANSWER: Unburned coal in ashpits results from feeding more fresh fuel than can be oxidized by the air moving through the fuel bed. Common causes of this are: running the stoker too fast, shaking the grates too much, defective grate bars, letting the coal fall through the grates, or carelessness in cleaning the grates.

33 How do you define a windbox?

ANSWER: A windbox is a plenum, or enclosed pressurized air duct, from which air for combustion, such as primary air, is supplied to a stoker or to gas or oil burners.

34 Explain the term "tuyères."

ANSWER: These are components of underfeed stokers for admitting air to the coal moving through the retorts.

35 What is meant by fluidization?

ANSWER: Fluidization is the condition that results when air is blown through a grid at a rate sufficient to overcome the static weight of the solids atop the grid to a point where the mass of these solids, referred to as a bed, behaves like a boiling liquid. Hence the term *fluidization.*

36 What are some of the advantages of fluidized-bed combustion?

ANSWER: Elimination of fine pulverizing, ability to inhibit molten-slag formation by holding combustion temperatures below the ash fusion point, and removal of sulfur by chemical interaction with limestone in the fluidized bed.

37 Name two methods of collecting fly ash.

ANSWER: Fly ash, depending on its size and characteristics, is collected in either an electrostatic precipitator or a baghouse. Collection efficiency greater than 99.5 percent may be realized. The baghouse is now becoming competitive with the electrostatic precipitator as a result of the development of fiber glass material.

38 What factors must be considered in modifying an oil- or gas-burning boiler to coal burning?

ANSWER: For a boiler that was not originally designed for coal burning, the following must be considered: (1) the furnace bottom may have to be modified so that furnace ash can be collected; (2) pulverizers, stokers, coal feeders, and bunkers and associated piping must be added; (3) soot blowers may have to be added; (4) superheaters and air heaters may have to be modified to prevent plugging by fly ash through the convection passes; and (5) windboxes will have to be modified to accommodate new fuel burners for coal burning.

39 What are the chief combustibles in fuel oil?

ANSWER: Fuel oils come from petroleum, with the chief combustible ingredients being carbon and hydrogen. The specific gravities of fuel oils vary and are influenced by origin of the petroleum and by temperature. At 60°F, specific gravities can vary from 0.84 to 0.96.

Since hydrogen has a much higher heating value and lower atomic weight than the other principal elements in fuel oil, the proportions of carbon and hydrogen affect both specific gravity and heating value. Because of this, specific gravity forms a reliable guide to an oil's heating value. Specific gravity in degrees API is found by dividing specific gravity with respect to water (at 60°F) into 141.5 and subtracting 131.5 from the answer.

40   What is a potential source of oil getting into the water side of an oil-fired boiler?

ANSWER: A tube failure inside a steam-heated oil heater can cause a boiler to be contaminated with oil. If the oil pressure is higher than the steam pressure, oil will be forced into the steam side of the oil heater and then travel to the water side of the boiler. To avoid this, place a check valve on the steam line to the oil heater so reverse flow cannot take place. Where condensation is returned to the boiler, it should be piped to a condensate receiver equipped with a gauge glass. Then oil, if any, can be seen in the glass. Double-shell-type oil heaters are often used for extra safety.

41   What are the pour point, flash point, and fire point as applied to oil burning?

ANSWER: Liquid petroleum products can be increasingly viscous as their temperature declines. The pour point is the temperature at which an oil ceases to flow, or pour. The flash point is the temperature at which a sample of oil will flash, or catch on fire, when it comes in contact with an open flame. It generally indicates at what temperature the oil is safe to handle. Distillate fuels flash at about 150°F. The fire point is the lowest temperature at which an oil will burn continuously once it has been lighted.

42   Name four methods of atomizing oil for burning.

ANSWER: Oil burners atomize fuel oil into minute droplets which then readily vaporize by the following methods: (1) under high pressure using air or steam; (2) by discharging the oil through orifices or nozzles at high pressure; (3) by passing the oil over a spinning cylindrical cup in the burner mouth to give a centrifugal force to the oil droplets; and (4) by an acoustic-type burner that has a vibrating resonator located just ahead of the point at which the oil leaves the burner nozzle.

43   What difficulties can be expected with water in fuel oil?

ANSWER: Difficulties encountered with water in fuel oil are: (1) flame outage; (2) erratic or sputtering firing; (3) waste of fuel from slow or faulty ignition; (4) loss of heat by vaporizing the water in the oil; and (5) corrosion in lines and storage tanks from the water (this may lead to burner-tip plugging).

44   What are some reasons for incomplete combustion in oil burning?

ANSWER: Incomplete combustion may be caused by incorrect fuel-air ratio, improper atomization of fuel, oil too viscous or at wrong viscosity, nozzles partially plugged that interfere with maintaining correct spray pattern, water in the fuel, poor draft, and improper damper setting on forced-draft fan.

45   What is the effect of vanadium on boiler tubes in fuel oil?

ANSWER: Fuels high in vanadium have a high tendency to form a fluid plastic-type ash at relatively low furnace temperature of 1100°F. The deposits formed on the tubes act as an insulator, reduce heat transfer, and can cause tubes to blister from localized overheating. Vanadium also combines with iron in the molten state and removes it in layers, often called high-temperature corrosion. Removal of these hard vanadium deposits requires physical chipping of the fireside scale.

46 Determine the wattage needed to heat 40 gal/hr of fuel oil from 120 to 200°F.

ANSWER: Good approximation is obtained by using the equation

Watts needed = 1.25 gal/hr $(T_2 - T_1)$

$$= 1.25(40)(200 - 120) = 4000 \text{ W or 4 kW}$$

47 What maintenance is stressed on oil burners?

ANSWER: Make sure that the burner gets uniformly free-flowing oil, clear of sediment that clogs burner nozzles. This means avoiding sludge buildup in storage tanks and keeping strainers in good condition. The preheat temperature must be right for fuel and burner type and must be uniform. Watch for wear caused by abrasion of ash in fuel and for carbon buildup. In rotary-cup burners, worn rims cause poor atomization. If cups are not properly protected after being turned off, carbon forms on the rim. When the burner is shut down, always take out the cup and insert a flame shield. Worn or carbonized mechanical-atomizing nozzles give trouble. Always replace worn nozzles and keep them clean.

48 What are the properties of commercial grades of fuel oil?

ANSWER: Commercial fuel oil is numbered from grade 1 to grade 6. Its common characteristics are: (1) viscosity, Seconds Saybolt Universal (SSU), or Seconds Saybolt Furol (SSF); (2) specific gravity, in degrees API and Bé; (3) flash point; (4) coefficient of expansion; (5) fire point; (6) vanadium content of ash; (7) heating value; (8) sulfur content; (9) moisture and sediment, in percentage by volume, which should be kept below 1 percent; and (10) pour point.

49 What rules should be followed in fuel oil burning?

ANSWER: There are nine rules to remember when burning fuel oil. (1) Atomize the oil completely to produce a fine uniform spray. (2) Mix the air and fuel thoroughly. (3) Introduce enough air for combustion, but limit the excess air to a maximum of 20 percent. (4) Maintain a clean, steady supply of oil to each burner. (5) Keep the orifices and sprayer plates clean and in good condition. (6) Maintain the proper relation between the sprayer-plate nut and the diffuser hub. (7) Maintain the proper relation between the diffuser and the burner tile. (8) Keep the refractory throat tile in good repair. (9) Use the proper size of sprayer plate (for mechanical type) and the proper number of burners to accommodate the anticipated steaming load.

50 How will moisture or sediment in a fuel affect combustion?

ANSWER: If allowed to enter the atomizer, the moisture or sediment in a fuel will cause sputtering, possibly extinguishing the flame, reducing the flame temperature, or lengthening the flame. Sediment is very annoying because it clogs strainers and sprayer plates. If there is a lot of sediment, frequent cleaning of the strainer and atomizer is necessary. If water, sediment, or both are present to any great extent, they can be separated from the oil by heating. This is done while the oil is in the settling tanks or, occasionally, by passing the oil through a centrifuge. Strainers are always used.

51 What could be wrong if a furnace shows evidence of pulsating?

ANSWER: Panting or pulsating is caused by firing with a considerable air deficiency through the registers. You can usually prevent this by raising the air supply or by cutting down the fuel supply, but it's hard to do if the setting is leaking badly. Too high an oil temperature also causes panting, and the resulting vibration makes loose dampers change their position.

52 How are electronic controls considered superior to pneumatic controls in managing the combustion process with load?

ANSWER: Because of their increased use, costs are beginning to be less than for mechanical and pneumatic controls. Their chief attractions are in the fine control that is possible provided sensors are available with similar accuracy, and the fact that solid-state electronic devices have a good reliability record. Their speed of response has improved efficiency of boiler fuel consumption.

The development of reliable sensors to monitor pressure, temperature, flows, and similar power-plant variables has permitted rapid transmission of these measurable quantities to electric actuators, such as motor-controlled or solenoid-actuated valves and dampers in order to quickly and accurately regulate pressure, temperature, flow, and similar variables within established set points.

Electronic signals from sensors can also be quickly displayed on data loggers and CRT screens for operator review, analysis, and adjustment (if needed) of the variables being controlled. This assists the operator in diagnosing troubles on equipment in the power cycle.

53 How are microprocessors being applied to boiler plant operation?

ANSWER: They are being applied on a segmented basis instead of on a total plant computer basis, for example, for fine air-fuel ratio control. Energy is lost in combustion when the proportions of fuel and air are not in the correct chemical or stoichiometric ratio. Sensors have been developed, such as optical-opacity monitors, zirconium oxide fuel cells that sense oxygen, and infrared carbon monoxide analyzers.

These sensors can be employed with small or microprocessor computers which can quickly calculate the chemical equation for the proper airfuel ratio and feed this information to appropriate actuators (valve, damper, speed control on fans, etc.) in order to obtain the desired air-fuel ratio. They are considered very efficient in load swings. Many of the developing microprocessors have built-in calibration, self-diagnostic circuits, and integral alarms.

The biggest advantage of microprocessor units is in their ability to control variables within narrow set points so that there is very little drift in the controlled parameters. This can result in fuel savings by avoiding fuel-consumption swings that occur under normal mechanical or pneumatic controllers.

Benefits cited in retrofitting existing systems with modern electronic controls are:

1. Existing controls may be obsolete, with parts difficult to obtain.

2. Some existing controls may not be up to present safety standards.

This applies especially to flame-safeguard systems.

3. Additional maintenance may be required on old, worn controls.

4. Efficiency is increased in maintaining boilers on the line within closely defined fuel-control limits.

54 What is meant by mode of control?

ANSWER: Mode of control means the manner in which the automatic controller acts and reacts to restore a variable quantity on a boiler, such as pressure, flow, or temperature, to a designed control or desired value. The three controller systems used to control a boiler are pneumatic, electric, and electronic.

55 Why are manual-reset controls useful on pressure or temperature high-limit controls?

ANSWER: The manual-reset mechanism on the high-limit control calls attention if the operating control (not high-limit) has malfunctioned, thus prohibiting further boiler operation until corrected. At times, this malfunction may be due to fused contacts, a leaking gas valve, a shorted wire, etc. Thus the boiler should not be operated until this is corrected.

56 What conditions are necessary to cause a furnace explosion?

ANSWER: Usually three: (1) accumulation of unburned fuel; (2) air and fuel in an explosive mixture; and (3) a source of ignition, such as hot furnace walls, improper ignition timing, faulty torch, and dangerous light-off procedures on manually started boiler combustion systems.

57 What is the purpose of a flame-sensing device?

ANSWER: Any flame-sensing device is an integral part of the combustion control system. The prime purpose of a flame-sensing device is to ensure that flame or burning conditions are safe and that during light-off proper sequences are followed in order to obtain a safe ignition. In the event of flame failure during operation, the flame-sensing device must be capable of sending a signal so that the fuel is cut off, preferably an alarm initiated, and the furnace and gas passes purged with air before any attempt is made to relight the burners.

58 What does a flame safeguard system prevent?

ANSWER: Furnace explosions (combustion explosions). These are caused by the sudden ignition of accumulated fuel and air in the fire side of the boiler. This can also lead to devastating property damage and loss of life. The flame safeguard system used on different boilers firing different types of fuel is designed to sense, and sometimes anticipate, this accumulation of unburned fuel and air in the fire side of the boiler. It safely shuts off the firing equipment, with purging of the furnace usually following in a time sequence so as to drive the unburned fuel-air mixture out of the furnace.

59 How is the efficiency of a boiler affected by the addition of automatic controls?

ANSWER: Through precise regulation of fuel, air, feedwater flow, and similar variables, tests have shown that the efficiency of a boiler is generally improved with the addition of automatic combustion controls. Uniform steam pressure is maintained, brickwork lasts longer, and better regulation is possible up to maximum load. The operator's skill is now directed to maintaining the controls in good condition and to detecting signs of malfunction before the control can become seriously impaired.

60 What limit devices are generally recommended on a combustion-control safety system?

ANSWER: The safety system should include the following limit devices incorporated into limit circuits so that the unit will shut down automatically if any indicates an unsafe condition: (1) high steam pressure; (2) forced-draft-fan interlock; (3) low fire on start; (4) high fuel pressure; (5) low oil temperature; (6) low water; (7) loss of flame; and (8) additional limit devices recommended by the following property groups—Factory Mutual, Industrial Risk Insurers, and the National Fire Protection codes.

61 What precautions must be observed in adjusting a burner on a two or more burner system?

ANSWER: When an operator makes any change to the fuel flow or airflow to one of the burners, he or she should also check the effect of these changes on the other burner's flame stability.

62 What precautions should an operator take to prevent furnace explosions?

ANSWER: Seven basic precautions are:

1. Check the operation of the boiler periodically.
2. If a burner goes out accidentally, shut off the igniter and fuel supply and thoroughly scavenge the furnaces and gas passes before again attempting ignition. Always determine and remedy the cause of the stoppage.
3. Keep burners and all allied equipment clean.

4. On boilers using both forced- and induced-draft fans, test the interlock periodically.
5. Do not attempt to secure excessively high CO_2 by using too rich a fuel-air ratio or by an inadequate secondary air supply.
6. Keep the temperatures and pressures for preheated air, drying air, fuel oil, etc. at the right levels.
7. Never allow an unstable flame condition to continue uncorrected.

63 What is a major burning consideration in refuse-fired boilers?

ANSWER: High firing temperatures and sufficient residence time in the combustion zone is required in order to obtain complete combustion in the furnace, and not in the convection section of the boiler, and also to make sure air quality standards on emission are being met while burning a variable-quality fuel.

64 What was the purpose of a fusible plug, and why are they not evident in gas- and oil-fired boilers today?

ANSWER: The purpose of fusible plugs was to give warning of a low-water condition on hand- and grate-fired coal-burning boilers. The hollow core had a low-melting-point metal, which melted during a low-water condition. The escaping steam filled the combustion chamber, and made a loud noise so that an operator could take corrective action on the developing low-water condition. With fuels fired in suspension, such as oil or gas, it can be dangerous to smother a flame in the combustion chamber without first shutting the fuel off. A furnace explosion could result from an accumulation of fuel in the furnace. The fusible plug has been replaced by the low-water fuel cutoff for this reason.

65 What can influence excess air determinations?

ANSWER: Excess air is used to obtain as much complete combustion of the fuel and as near the theoretical amount of air needed as shown by chemical reaction equations. By measuring the oxygen percentage in the flue gas on a volume basis, it is possible to determine the amount of excess air that is being used in the furnace of the boiler. This reading, however, can be influenced by the following:

1. Leaking furnace settings and boiler enclosures.
2. In negative pressure boilers, air not used in combustion leaking through leaks beyond the combustion zone.
3. In positive pressure boilers, strong draft in this mode of operation increasing the amount of air going through the boiler with the flue gas.

12

Boiler Auxiliaries
and External Water
Treatment Equipment

The extent of boiler plant auxiliaries is determined by the type and size of the plant (heating, process, or power generation), the type of fuel used, and environmental regulation requirements on discharges into the air or ground. The most common type of auxiliary equipment that an operator may encounter are pumps, fans, precipitators, baghouse filters, feedwater heaters, evaporators, deaerators, water softeners, and similar water treatment equipment.

Pumps

Feedwater pumps in general use may be divided into general classes: *reciprocating, rotary,* and *centrifugal* types. The reciprocating type makes use of a water cylinder and a plunger directly mounted on a common rod from a direct-connected steam cylinder. One or two water (and steam) cylinders in parallel, known as *simplex* and *duplex* pumps, respectively, are the most common types of reciprocating feed pump (Fig. 12.1*a*). Triplex and quadruplex feed pumps often have each plunger rod connected by cranks to a mechanically driven crankshaft.

Pump data and terms. The following data apply to pumps using water.

1 psi gauge = 2.31 ft of water

1 ft of head = 0.434 psi

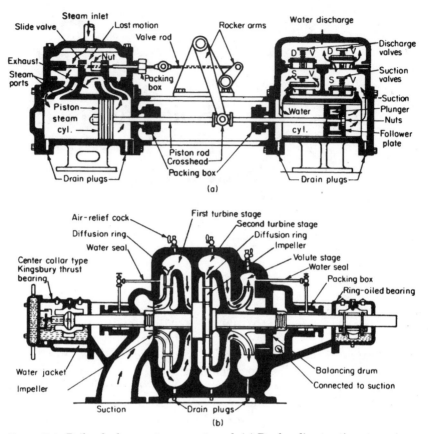

Figure 12.1 Boiler feed-pump types most used. (*a*) Duplex direct-acting steam type;
(*b*) multistage centrifugal.

1 ft^3 of water = 62.4 lb in weight

1 ft^3 of water = 7.48 U.S. gal = 6.24 imperial gal

1 U.S. gal = 8.33 lb in weight

1 imperial gal = 10 lb in weight

1 U.S. gal = 231 in.^{3}

1 imperial gallon = 277 in.^{3}

1 hp = 33,000 ft•lb/min of work

Head. Liquid pressure in pump applications is considered equivalent to a column of liquid of a height sufficient because of the weight
of the column to produce this pressure. For example, for water at

atmospheric temperature, 2.31 ft of water = 1 lb/in.^{2} pressure. *Static head* is the height in feet of the fluid above a designated gauge point. *Pressure head* is the static head *plus* the gauge pressure expressed in feet plus the friction head (if the fluid is flowing). *Velocity head* is the vertical height or feet required to produce a certain speed of flow, expressed by the equation

$$\text{Velocity head, } h_v = \frac{v^2}{2g}$$

where $g = 32.2$ ft/s^2, or the acceleration of a free falling body at sea level. *Pump operating head* is the difference between the pressure and suction heads.

Except for water velocities well above average, or for large volumes at low heads, the velocity head is often left out in calculations.

Friction head is the feet of liquid required to overcome the resistance to fluid flow in pipes and fittings.

Velocity. Velocity of flow per the velocity head equation is expressed in feet per second and is important in flow calculations. In order to determine the flow past a given point, use $Q = Av$, where $A =$ cross-sectional area in square feet of the pipe or fluid conduit, $v =$ velocity of flow in feet per second and $Q =$ ft^3/s. Velocity heads are calculated from pressure heads on both sides of a venturi nozzle or orifice plate installed in the pipe per hydraulic standards. Some approximation is possible for a circular pipe with an inside diameter d, in:

$$v = \frac{0.4085 \, (\text{gpm})}{d^2}$$

Example A test on a 3-in. inside diameter pipe showed that the flow velocity was 68.08 ft/s as measured by an orifice plate. What is the gpm flow?

$$68.08 = \frac{0.4085 \, (\text{gpm})}{3^2}$$

and solving for gpm:

$$\text{gpm} = \frac{68.08(9)}{0.4085} = 1500 \text{ gpm}$$

Work in pumping. The work required of a pump is influenced by the amount of head the liquid will be raised, the force required to pump it into a higher-pressure system, and that required to overcome friction. This work is called the hydraulic horsepower, or theoretical pump horsepower, expressed as follows:

$$\text{Theoretical hp} = \frac{\text{gpm}(H)(s)(\text{lb/gal})}{33,000}$$

where gpm = flow rate
   H = total head of liquid, ft
   s = specific gravity of fluid   For water = 1

For water, which weighs 8.33 lb/gal at ordinary temperature, the equation becomes

$$\text{Theoretical hp} = \frac{\text{gpm}(8.33)H}{33,000} = \frac{\text{gpm}(H)}{3962}$$

The brake horsepower (bhp) is the above theoretical horsepower divided by the pump efficiency E, or

$$\text{bhp} = \frac{\text{gpm}(H)}{3962(E)} \qquad \text{Note that } E = \frac{\text{Theoretical hp}}{\text{bhp}}$$

Example   If a pump has an efficiency of 70 percent and is delivering 1500 gpm of water from ground level against a total head of 1000 ft, what size motor will be required? Substituting in the bhp equation,

$$\text{bhp} = \frac{1500(1000)}{3962(0.70)} = 541 \text{ hp}$$

Viscosity.   Viscosity is a term used to indicate the internal friction of a fluid. In fluid mechanics, dynamic viscosity is expressed as

$$\frac{\text{Shearing stress of fluid}}{\text{Rate of shearing strain}}$$

Another term used in fluid mechanics is *kinematic viscosity*, which is the dynamic (also called absolute viscosity) divided by the density of the fluid. Kinematic viscosity can be expressed as ft²/s or cm²/s, which is called a *stoke*. As can be noted, viscosity is important in pump design and in some fluid flows, such as oils in pipes, in order to determine internal friction to flow. Viscosity varies considerably from one fluid to the other and decreases with rising temperatures. This is why viscous liquids are heated when they are pumped from one point to another. Viscous liquids require more horsepower to pump and reduce the pump efficiency and capacity because of this internal resistance of some fluids.

Suction.   Total suction lift is the reading of the gauge at suction flange of pump, which is converted to feet minus the velocity head in feet at that point. Total suction head is the same as lift except the velocity head is *added*.

From experience, it is known that pumps have suction limitations even though in theory they should lift a liquid to the feet height represented by atmospheric pressure, or 14.7 atmospheric pressure = 2.31 × 14.7 = 34 ft of lift possible. Factors which reduce the possible lift are internal friction to flow, vapor pressure of the fluid, pump speed, capacity, and internal pump design. Vibration and possibly cavitation occur when a pump is trying to operate with a suction lift that it cannot handle.

Reciprocating pumps

Reciprocating pumps are positive displacement pumps and can be used to obtain very high pressures by staging the cylinders or by using more than one pump. In staging, the discharge pressure from one cylinder is the suction pressure for the next cylinder and this can be carried out with each cylinder boosting the pressure to the desired result.

Steam-type reciprocating pumps.   Steam-type reciprocating pumps are classified as *direct acting* if a steam cylinder is in line with the pumping cylinder or steam *power driven* if the steam engine has a crank, flywheel, and crosshead. The term *simplex* means it has one water cylinder. The term *double acting* means it pumps water from both ends of the piston or plunger of the pump. See Fig. 12.1 for pump arrangements.

A *duplex pump* of the steam direct-acting type has two water cylinders whose operation is coordinated to obtain the final desired pressure. A duplex pump can also be driven by a steam engine with a crank, flywheel, and crosshead arrangement. *Triplex pumps* have three water cylinders in parallel, can be driven by steam, and are either direct acting, steam power driven, or have a motor drive.

Reciprocating pumps are also driven by electric motors, diesels, and gas, and steam turbines either directly through one shaft or coupling or through gearing.

Reciprocating pumps must have *packing* to prevent water from leaking past the piston or plunger and also where any rod comes out of the cylinder, called gland packing. Material for packing depends on the fluid being handled, temperatures, pressures, and pump material. The manufacturer's maintenance and installation instructions should be followed when the packing requires replacement to avoid excessive leakage. Strainers should be installed on the suction side of reciprocating pumps in order to prevent foreign substances from damaging the valves or cylinders. *Foot valves* are used for pumps under suction pressure to prevent backflow on the suction line. *Aspirators* are used to drain air out of suction lines to prevent the pump from becoming air bound. It is a form of priming the pump by making sure only water flows to the pump from suction lines.

Water pressure calculation. A common question about reciprocating pumps of the direct-acting steam type is how low-pressure steam can produce high discharge water pressure on the pump.

Example   A simplex pump has a 10-in.-diameter steam cylinder operating at 100 psi, with a single-acting water pump with a diameter of 5 in. If efficiency is neglected, approximately what water pressure would be generated if a final 20 percent must be considered for pump slip?

The easiest way to solve this is to compare the force on the steam piston and equate it to the force on the water piston. Assume the cylinder diameter is the same as piston diameter. Then

Steam piston area $\times$ steam pressure = water piston area $\times$ water pressure

$$\text{Steam piston area} = \frac{\pi(\text{diameter})^2}{4} = \frac{\pi(10)^2}{4} = 78.54 \text{ in.}^2$$

$$\text{Water piston area} = 0.7854(5)^2 = 19.64 \text{ in.}^2 \quad \left(\text{Note: } \frac{\pi}{4} = 0.7854\right)$$

Equating forces by multiplying each area by corresponding pressure,

$$100(78.54) = \text{water pressure}(19.64)$$

$$\text{Water pressure} = \frac{7854}{19.64} = 400 \text{ psi}$$

From this must be deducted 20 percent of 400 for pump slip, giving a final water pressure of $400 - 80 = 320$-psi water pressure.

Horsepower calculation. Another calculation that is quite often needed is the *indicated horsepower* (ihp) that is developed by the steam cylinder of steam-type reciprocating pumps. Reference is made to the foot•pounds of work per minute performed by the piston in terms of the following:

$$A = \text{piston area being acted on by the steam} = 0.7854(d)^2$$

where d is the steam piston diameter, inches.

$$P = \text{mean effective pressure of the steam}$$

This is usually determined by a steam engine indicator card in order to get the average pressure during a stroke of the engine, in psi.

$$L = \text{length of stroke, ft}$$
$$N = \text{number of strokes per minute}$$

$$\text{ihp} = \frac{PLAN}{33,000}$$

Example What is the indicated horsepower for the steam cylinder in the previous example if the stroke is 20 in., the mean effective steam pressure is 90 psi, and the number of strokes per minute is 60? Using the above equation and substituting the known data,

$$\text{ihp} = \frac{90(20/12)(78.54)(60)}{33,000} = 21.42 \text{ hp}$$

Reciprocating steam pump sizes are always given with the steam cylinder diameter in inches listed first, followed by the water piston size and then the stroke in inches.

Relief valves for pumps. Since reciprocating pumps are positive displacement types, it is possible to build up high pressure if a discharge valve is closed; therefore, it is essential to have a relief valve on the discharge line *before* any closing valve to prevent a possible overpressure condition developing, such as blowing off the head of the pump.

Centrifugal pumps

This type of pump has as its main component a casing in which an impeller rotates (see Fig. 12.2). The fluid to be pumped is directed through the inlet pipe to the center of the pump, called the eye. The impeller throws the liquid out radially through pump passageways, and this develops pressure by converting the kinetic energy. *Volutes* convert the velocity energy to pressure energy. In *diffusor-type* or turbine pumps, guide vanes are placed between the impeller and casing chamber, but the transformation of velocity energy to pressure follows the volute casing design (see Fig. 12.2).

Centrifugal pumps may be of the single- or double-suction type. In the double-suction type, the fluid enters from both sides of the pump.

Multistage centrifugal pumps (see Fig. 12.1*b*) are used for service pressures not attainable from single-stage pumps and are found in such service as water supply, fire, boiler-feed, and charge pumps in the refinery and petrochemical industry. Multistage pumps can be of the volute or diffusor type. Volute-type pumps usually have single-suction impellers with half the impeller inlets facing one direction and half in the opposite direction in order to balance the thrust forces. In the diffusor-type pump, impeller inlets generally face one direction with the thrust force neutralized by a differential pressure arrangement, or *balancing piston* or drum. The unbalanced pressure differentials across each impeller create an axial thrust toward the suction end, which is counteracted by a balancing piston located near the end of the last impeller at the outboard end of the rotor.

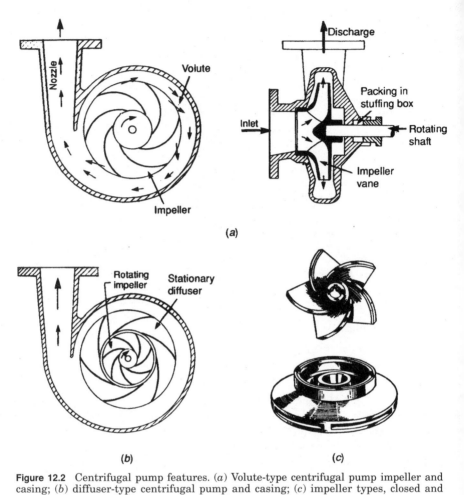

Figure 12.2 Centrifugal pump features. (*a*) Volute-type centrifugal pump impeller and casing; (*b*) diffuser-type centrifugal pump and casing; (*c*) impeller types, closed and open.

Kingsbury thrust bearings at the outboard bearing are also used to take any thrust fluctuation caused by abnormal operations.

The centrifugal pump is not considered a positive displacement pump, as is the reciprocating pump. For example, if the discharge valve on a centrifugal pump is closed completely, the pressure will only rise to a limited pump value with the rotating impeller churning the fluid and the work on the fluid being converted to heat. There will not be a rise in the head on the discharge end of the pump. In contrast, if the discharge valve on a reciprocating pump is closed, pressure continues to build up unless a pressure control stops the pump or a relief valve opens or the overload protection on the driver stops the pump. If none of the above work and there is unlimited horsepower drive available, something would have to burst from overpressure.

All centrifugal pumps are designed to operate on liquids. Whenever they are used on mixtures of liquid and vapor or air, shortened rotating-element life can be expected. If the liquid is high-temperature or boiler feedwater with vapor (steam) present, rapid destruction of the casing can also occur. This casing damage is commonly called *wire drawing* and is identified by wormlike holes in the casing at the parting, which allow liquid to bypass behind the diaphragms or casing wearing rings.

Cavitation. Cavitation can occur on pumps when the fluid pressure equals the vapor pressure at the existing temperature, and as a result vapor bubbles alternately form and collapse. The fast formation of bubbles causes the liquid at high velocity to fill the void with impact force on the internal parts of the pump. These surges into the voids are equivalent to explosions on small areas, but the forces produced on the internal parts can exceed the tensile strength in that part of the pump where they are occurring. This causes particles to be knocked off, and rapid pitting and erosion take place, even to the extent that pieces break off internally, producing serious damage to the pump.

To prevent cavitation, most pump manufacturers stamp their pumps with a *net positive suction head,* or *NPSH,* which should not be exceeded in order to avoid cavitation damage. The fluid being pumped and the corresponding temperature need to be known since vapor pressure has an effect on the NPSH. In power-plant applications, feedwater pumps have been ruined due to cavitation damage because the feedwater temperature control went astray, and this resulted in the water "flashing," which affected the permissible NPSH. Many pumps, especially those designed to handle fluids above atmospheric temperatures, have a bypass valve from a stage of the pump back to the suction side. This bypass is activated if the NPSH allowed on the pump is approached in operation.

Whenever wire drawing or cavitation is detected, an immediate check of the entire suction system must be made to eliminate the source of vapor. Vapor may be present in high-temperature water for several reasons. The NPSH available may be inadequate, resulting in partial or serious *cavitation* at the first-stage impeller and formation of some free vapor. The pump may be required to operate with no flow, resulting in a rapid temperature rise within the pump above the flash point of the liquid, unless a proper bypass line with orifice is connected and is open. (This can also cause seizure of the rotating element.) The submergence over the entrance into the suction line may be inadequate, resulting in vortex formation and entrainment of vapor or air.

When a pump becomes vapor-bound or loses its prime, a multistage pump becomes unbalanced and exerts a maximum thrust load on the thrust bearing. This frequently results in bearing failure; if it is not detected immediately, it may ruin the entire rotating element because of the metal-to-metal contact when the rotor shifts, with probable seizure in at least one place of the pump.

Some other *operating problems* on centrifugal pumps:

1. Low water flow can be caused by improper speed, plugged suction strainer, airbound pump, open air vent valves used to prime the pump, worn wearing rings, damaged impeller.

2. Vibration can be caused by misalignment, bearing wear, impeller unbalance due to wear, and corrosion on pump parts.

3. Progressive shaft thinning and cracking can be caused by improperly installed shaft packing or chemical attack on the material from the fluid handled that may cause stress-corrosion cracking to occur.

Barrel-type pumps.   Barrel-type centrifugal pumps do not have their casing split horizontally, but consist of a double-case cylinder with access to the pump internals being made through removable end heads. These barrel-type pumps are used for high-pressure boiler feed of up to 6000 psi and 600°F water temperature. For the common 2600-psi service, the pump runs at 3600 r/min and has 12 stages or more.

Centrifugal pump *performance curves* are used to show the various relationships of head, power input, and efficiency at various speeds. Figure 12.3a shows the performance characteristics of a centrifugal pump with water as the fluid and operating at constant speed. Note how the head developed drops off after or near rated capacity. Efficiency rises to 91.7 percent and then drops. Figure 12.3b shows the pump performance curves at *different* constant speed tests. The analysis of centrifugal pump behavior can be quite complex when one considers the complicated, turbulent flow inside the pump.

Centrifugal pump laws

Centrifugal pump laws have been established for approximate relations, which are helpful in indicating performance trends. The derivation of these centrifugal pump laws is treated in texts on fluid mechanics. For a pump operating at *constant speed,*

Total head varies directly as the pump impeller diameter *squared,* D^2

Capacity of the pump varies as the pump impeller *cubed,* D^3

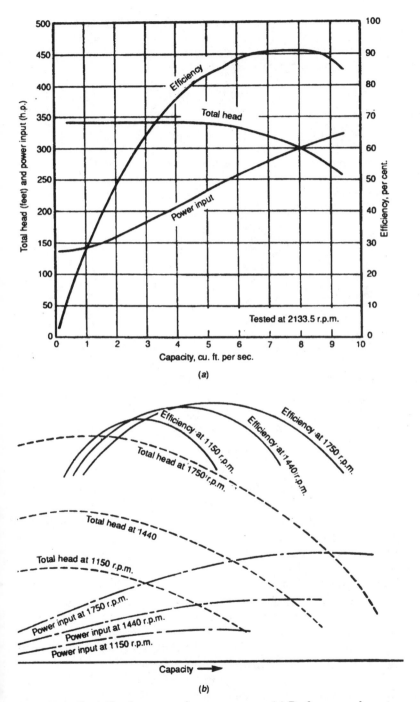

Figure 12.3 Centrifugal pump performance curves. (*a*) Performance characteristics of a single-stage, single-suction centrifugal pump operating at *constant
speed*; (*b*) performance characteristics of a centrifugal pump at *different speeds.*

Fluid power developed by the pump varies by D^5 with D = pump impeller diameter

For a centrifugal pump operating at *different speeds,*

Pump capacity varies directly as the speed, N

Total head varies directly as N^2

Fluid power developed by the pump varies directly as N^3

Example of Pump Laws   A centrifugal pump operating at 1150 r/min produced a total head of 37.6 ft and a capacity of 800 gpm. Using pump laws, what would be the head and capacity at 1750 r/min? This involves the relationships of a pump at different speeds. By proportion, to find head,

$$\frac{37.6}{(1150)^2} = \frac{H}{(1750)^2}$$

$$H = \frac{37.6(1750)^2}{(1150)^2}$$

$$= 87.1 \text{ ft}$$

To find capacity,

$$\frac{800}{1150} = \frac{C}{1750}$$

$$C = \frac{800(1750)}{1150}$$

$$= 1217.4 \text{ gpm}$$

Steam injectors as pumps.   *Injectors,* or *inspirators,* are used commonly for feeding water to small boilers or as an auxiliary means of mechanical feed for medium-sized boilers. They were quite common in railroad locomotive practice. They make use of an elongated nozzle or *Venturi tube,* so that steam may feed water back against its own pressure.

Steam enters one end of the Venturi tube in a jet. The vacuum produced around this entering jet draws the feedwater fed to the jet chamber into the steam flow. As the steam-and-water mixture passes through the reduced area of the throat of the tube, a very high velocity of flow is produced. The weight of the water content in this steam-and-water mixture attains sufficient momentum to open the feed-pipe check valve against boiler pressure, with water being thus fed to the boiler.

Steam-type injectors with properly designed nozzles and tubes can develop *water pressure* up to 50 psi above the steam pressure used, but the incoming water temperature must be less than 150°F to prevent flashing in the throat of the nozzle. The added benefit of a steam injector is that it also heats the feedwater to the boiler; however, its pumping efficiency is in the 5 percent range because of the low expansion of the steam. The main advantages of the steam-water injector are simplicity, compact construction, and no moving parts.

Figure 12.4 shows a double-tube injector, which is designed to handle water hotter than the single-tube injector previously described. Two nozzles are used, one for lifting the water and the other to force the water into the boiler. A lever handle is used to operate the main steam valve and the so-called forcer valve. In operation, the lever handle is pulled just enough to open the main steam valve, which then begins the lifting part of the cycle. Once lift is established, the lever is pulled further to open the forcer steam valve and to close the overflow valve as in the single-tube design. Steam from the forcing nozzle then forces the water from the lifting tube into the forcing tube, where it is discharged at a pressure sufficient to lift the boiler check valve and enter the boiler water.

Injector troubles can include the following: feedwater too hot; suction lift too high; leaky boiler check valve, which would allow steam to blow back into the injector discharge line; leaks in suction pipe destroying vacuum so that water cannot be lifted; steam supply pressure too low for nozzle design, thus impairing lift; steam too saturated, thus not allowing steam to condense properly per design of the nozzle; and obstructions, scale, dirt, and rust plugging in suction lines, tubes, and nozzles.

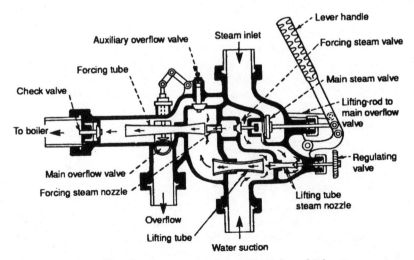

Figure 12.4 A double-tube steam-water injector can force higher temperature water into a boiler than a single-tube steam-water injector.

The manufacturer's maintenance instructions should be followed so that this emergency-type pump is available for use when needed.

Reserve high-pressure boiler feed pump. Boilers having more than 500 ft^2 of heating surface should have two means of feed. However, the Code allows one means of feed under these conditions: Boilers fired by gaseous, liquid, or solid fuel *in suspension* may be equipped with a single feedwater system, provided means are furnished for the immediate shutoff of heat input if the water feed is interrupted. If the boiler furnace and fuel systems retain sufficient stored heat to cause damage to the boiler if the feedwater supply is interrupted, two means of feed are still needed.

For boilers firing solid fuel *not in suspension,* one means of feed must be steam-operated. The source of feed must be such as to supply water to the boiler at a pressure at least 6 percent higher than any safety-valve setting.

Stoker-fired boilers, or boilers with refractories or beds of hot fuel, require special attention to the availability of feedwater in the event of electric power interruption to motor-driven boiler feed pumps or electric controls. It is recommended that a spare steam-driven boiler feed pump or injectors of sufficient capacity be maintained to provide feed to the boiler in the event of electric power failure involving the power supply to the boiler. Another alternative is to have in-house electric power generation that could supply the boiler feed pump motors, preferably on a separate emergency circuit. Where there is an inherently large amount of heat stored in a furnace, there is the risk of overheating, and even melting tubes, if electric-driven pumps cannot be operated due to an electrical failure.

Mechanical-drive turbines. Mechanical-drive steam turbines are used in boiler plants to drive boiler feed pumps and induced and forced draft fans, as well as for emergency electric generator drive. It is also common to employ dual drive, with a steam turbine at one end of a boiler feed pump or fan and an electric motor on the other end, so that if power fails when the unit is driven by the motor, the steam turbine is automatically placed into service. It is also common to use the turbine drive in winter to make use of the exhaust steam to improve the plant's heat balance.

Sizes of steam turbine drives can vary depending on the size of the plant. The smaller turbines have been standardized as respects horsepower, steam pressure inlet, exhaust pressure, and speed, with governors available for constant- and variable-speed operation. Figure 12.5*a* shows a single-stage unit equipped with an oil relay governor, forced-feed lubrication, and also hand-valve-controlled nozzles for adding or reducing steam flow with load changes. Hand valves are also used to

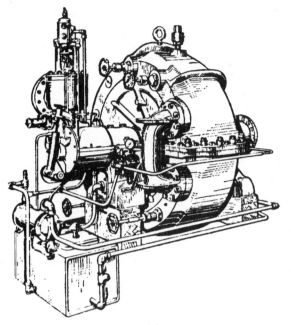

(a)

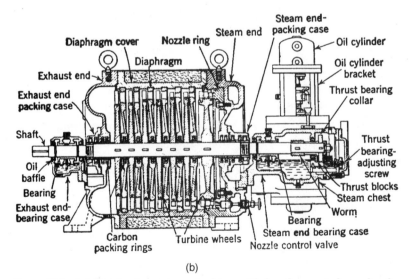

(b)

Figure 12.5 (*a*) Single-stage steam turbine with hand-operated nozzles for restricting or adding additional steam to the turbine beyond governor capability. (*b*) Multistage impulse turbine has seven pressure stages.

provide good starting power when beginning a plant operation at lower pressure, to operate at low speed beyond the range of the governor, and also for extra load capability during periods of high demand. As the power need becomes great, multistage steam turbines are used as is shown in Fig. 12.5*b*. This unit has seven pressure stages.

The selection of a steam turbine for mechanical drive requires a careful plant heat balance study to determine if the exhaust steam from the turbine can be used efficiently in the plant for feedwater or process heating. If exhaust steam has to be vented to the atmosphere, approximately 1000 Btu/lb are wasted.

Operation and maintenance of steam turbines should follow good mechanical machinery practices, including following the turbine manufacturer's instructions. Among these are:

1. Make sure all water or condensate is removed from the casing, steam chest and steam line before opening the throttle valve to the turbine. This will avoid water striking blades and nozzles that were designed to handle the lighter steam. It will also avoid blades and nozzles being corroded or possibly cracked because of wet steam.

2. Always check the oil reservoir level and the operation of any oil pumps to make sure the bearings are properly lubricated. Check oil reservoirs for water that may come from a leaking oil cooler or condensate from leaking seals entering the bearing oil.

3. Warm up the steam turbine by the manufacturer's recommended ramp rate to prevent excessive thermal stresses.

4. Check the emergency overspeed trip by using the hand trip before starting and annually by an actual overspeed test per the manufacturer's instructions.

5. Keep track of operating performance, such as vibration, shaft seal leakage, first stage and exhaust pressure, and steam consumption. Abnormal operating conditions may require further investigation as to cause and may involve opening the turbine to check on blade and nozzle wear, bearing wear, and similar internal parts that may require repair or replacement.

It is also important to have a relief valve on the exhaust end of a back pressure steam turbine before any exhaust shutoff valve, to prevent the turbine casing's being exposed to overpressure if the exhaust line valve is closed by mistake when starting a steam turbine.

ID and FD fans

Induced-draft fans pull the products of combustion in a boiler and direct them to the chimney for atmospheric discharge. The *forced-*

draft fan takes air from the atmosphere and delivers it through air ducts to air preheaters and burners and even directly into the furnace, depending on the size of the boiler and its arrangement. Forced-draft fans produce some air pressure, and if the boiler's casing is not tight, furnace gases may escape into the boiler room through leaking joints or cracks. In coal-fired plants, the boiler room may start to be covered by unburnt coal and ash. This may be the penalty for operating with a *positive-pressure* furnace. In a *negative-pressure* furnace, the induced-draft fan creates a partial vacuum condition in the furnace, and atmospheric air rushes in through leaking joints or cracks of the furnace casing. This reduces the efficiency of burning by introducing excess air into the furnace. To make the effects of leaks minor, *balanced draft* is used in most large boiler systems; this keeps the furnace slightly vacuum at about 0.1 in. of water vacuum.

Interlocks are used between fans and the combustion or burner equipment in order to avoid boiler combustion problems. Large induced-draft fans on large watertube boilers have caused high negative furnace pressures resulting from failure of interlock systems. The large negative pressures have caused *implosion* to occur in furnaces that included the cave-in of waterwall tubes. Blade erosion on induced-draft fans from particulates in the flue gas requires designs to prevent rapid blade material deterioration. Vibration monitoring is used to detect unbalance due to abrasive wear or abnormal deposits being lodged on the fan. Periodic inspection, cleaning, and balancing must be employed on induced-draft fans to keep them operating reliably.

The centrifugal fan is the most common type of fan, and this is divided into two classes: (1) *steel-plate* as shown in Fig. 12.6a, and (2) *multiblade* as shown in Fig. 12.6b. In the steel-plate fan, the wheels consist of one or two spiders, each having 6 to 12 arms. Each pair of arms has a flat blade extending radially as shown in Fig. 12.6a. The blades may be straight or curved either forward or backward. These types of fans are usually used with engine drives.

The *multiblade* fan is built up of two or more annular rings and has many narrow curved blades riveted or welded in place between the annular rings as illustrated in Fig. 12.6b.

Designation of centrifugal fans includes the following: (1) number of inlets—single or double, (2) width of wheel, (3) diameter of wheel, (4) discharge—horizontal, vertical, top, bottom, or angular, (5) housing—full, in which the fan scroll is completely above the base; seven-eighths, and three-quarters, where the scroll is located below the top of the supporting base, and (6) rotation of clockwise or counterclockwise as noted when looking at the fan from the drive side.

Fans that are in high-temperature and abrasive service need special attention. For example, induced-draft fans on coal-burning boil-

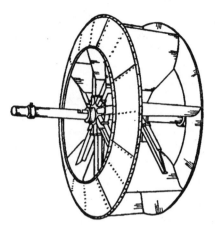

(a)

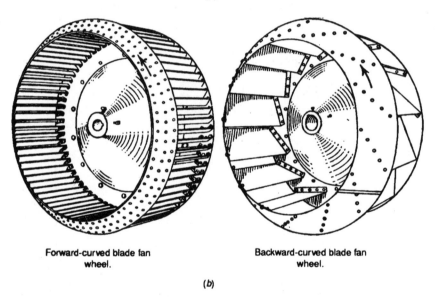

Forward-curved blade fan
wheel.

Backward-curved blade fan
wheel.

(b)

Figure 12.6   Steel-plate and multiblade fan wheels. (*a*) Steel-plate fan wheel is used
for lower speeds; (*b*) multiblade fans may have forward- or backward-curved blades.

ers should have water-cooled bearings because of the high temperature of the exhaust gases that the fan is exposed to. The abrasive
action of fly ash requires more rugged construction and material such
as stellite tips on blades in order to prevent rapid wear of the blades
and perhaps a sudden rupture of a part of the fan.

Volume and pressure control on the gas flow can be manual or automatic. There are several methods of controlling the volume and pres-

sure: (1) varying the speed of the driver, (2) louvre dampers placed at the outlet of the fan with constant speed drive, (3) adjustable vanes placed at the inlet of the fan as shown in Fig. 12.7a, variable-pitch blades on vane axial fans, and (5) bypass or spillover where some of the gas is directed back to the suction side or discharged to the atmosphere.

Centrifugal fan performance laws

Similar to centrifugal pumps, for a given fan size, duct system, and air density, the following relationships exist on performance (see Fig. 12.7b):

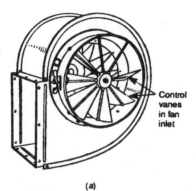

(a)

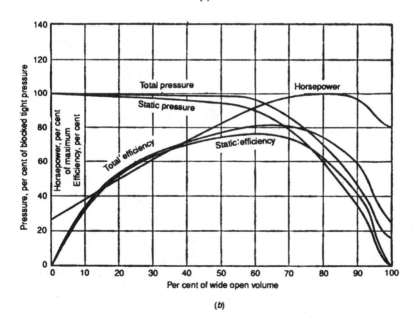

(b)

Figure 12.7   (a) Control is by vanes in fan inlet for this design; (b) typical performance curves for a backward-curved multiblade fan.

1. Capacity varies directly as the speed ratio.
2. The static pressure varies as the speed ratio.
3. The speed and capacity vary as the square root of the static pressure.
4. The horsepower varies as the cube of either the speed or capacity ratio.
5. The horsepower varies as the 3/2 power of the static pressure ratio.
6. The air velocity varies directly as either the speed or the capacity ratio.

Where the static pressure is *constant* at the fan outlet, the following rules apply:

1. The capacity and the horsepower vary as the square of the wheel diameter.
2. The speed varies inversely as the wheel diameter ratio.
3. At constant static pressure, the speed, capacity, and power vary inversely as the square root of the ratio of the air densities.
4. At constant capacity and speed, the horsepower and static pressure vary directly as the ratio of the densities of the air.

Example For conditions 1 and 2 just above, one fan has a wheel diameter of 22.75 in. with a capacity of 7890 cfm at 6½-in. static pressure and a speed of 2064 r/min. Assume each fan operates at the same static pressure of 6½ in. of water. What would be the capacity and speed of a 45.5-in. diameter fan? Per rule 1,

$$\text{Capacity} = 7890 \times \left(\frac{45.5}{22.75} \right)^2 = 31{,}560 \text{ cfm}$$

Per rule 2, the speed is:

$$2064 \times \frac{22.75}{45.5} = 1032 \text{ r/min}$$

Draft

Draft can be natural, such as a chimney in a furnace, or mechanical, where a fan is used to move the air or gas. The movement is caused by a pressure difference between inlet and outlet and is measured in terms of the height in inches of water of a water column that is equivalent to the air pressure acting on the water column. Figure 12.8 shows some types of draft gauges. In U-tube manometers, one side is

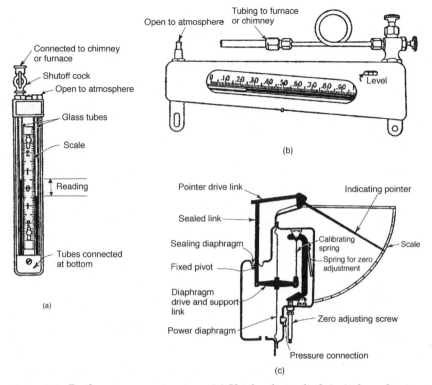

Figure 12.8 Draft measurement gauges. (*a*) U-tube shows draft in inches of water differential readings; (*b*) inclined draft gauge must be kept level to obtain proper reading; (*c*) diaphragm-operated draft gauge.

connected to the duct, chimney, or furnace where the draft is to be measured, while the other end is open to the atmosphere. The difference in the water level is a measure of the pressure difference. For example, to convert 3 in. of draft to pressure, remember that a cubic foot of water weighs 62.4 lb and by dividing this by the cubic inches in a cubic foot, we obtain $62.4/1728 = 0.036$ psi/in.^{3} of water. Thus for a 3-in. draft, the pressure would be $3 \times 0.036 = 0.108$ psi.

Automatic draft control is the automatic regulation of fans and dampers for increasing or decreasing airflow to maintain constant steam pressure on boilers as the load changes, and also to maintain good combustion conditions. Devices to control draft-fan speed and damper position frequently employ diaphragms. Changes of air or steam pressure act upon the sensitive diaphragms to open or close the electrical switches, fluid valves, or steam valves that control fan speed or damper position.

Operation and maintenance of fans. Fans and blowers should be started with minimum flows and gradually brought up to full load to prevent shocking the machine's components and to also avoid seriously overloading the driver. As with all machinery, it should be maintained clean and free of deposits; otherwise possible corrosion attack and imbalance with resultant vibration may occur. Fans and blowers have been torn apart by the destructiveness of prolonged operation with abnormal vibration. For this reason, a daily operational check should include checking the vibration.

Critical fans and blowers, and those located away from operator attention, should have vibration pickups that display this vibration in control rooms. Vibration causes can include misalignment, poor or loose foundation attachments, imbalance due to erosive wear, such as fly ash on induced fans, cracks from sharp corner designs, and stress-corrosion cracking, to name a few. Therefore, fans should be internally inspected annually in order to detect wear and tear on blades and their attachments. This is especially applicable to critically assigned operational fans and those with high monetary repair or replacement costs.

Proper lubrication and the maintenance of oil quality are other important functions of operators who have fans and blowers under their care and control.

Air heaters. Air heaters are applied to boilers firing solid fuels, seldom on boilers firing gas or oil. Air heaters provide hot air to evaporate some of the moisture in solid fuel, which assists in the rapid combustion of the fuel and also extracts heat from exhaust gases, thus saving energy or fuel.

The thermal energy in exhaust gases of large stations is also partly recovered by economizers, which preheat the water going into the boiler. Economizers, however, are used for boilers burning solid, liquid, or gaseous fuels, whether or not an air heater is used or installed. Economizers are simpler heat exchangers than are air heaters.

Air heaters are classified as being (1) recuperative, or (2) regenerative. In the *recuperative* air heater, Fig. 12.9a, heat is transferred from the hot medium side through tubes to the air to be heated on the other side. In the *regenerative air heater,* Fig. 12.9b, heat from the flue gas heats a heat storage medium, such as baskets, which are then rotated to the airflow side to heat the incoming air by giving up the stored heat in the baskets. Corrosion and plugging can be a problem on air heaters, which manufacturers have tried to solve by the use of corrosion-resistant material and by good spacing of the heat-transfer components. A planned inspection and maintenance program, however, is still required on the heat-transfer apparatus to avoid wear and tear casualties.

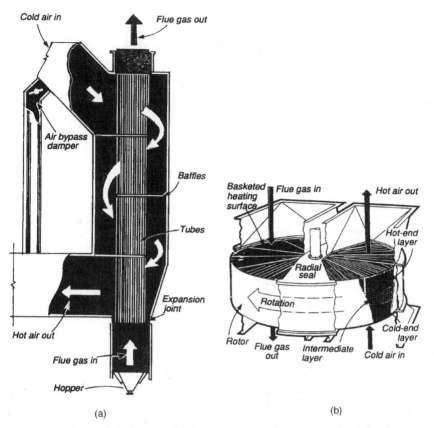

Cold air in

Flue gas out

Air bypass damper

Baffles

Tubes

Expansion joint

Hot air out

Flue gas in

Hopper

(a)

Basketed heating surface

Flue gas in

Hot air out

Hot-end layer

Radial seal

Rotation

Cold-end layer

Rotor

Flue gas out

Intermediate layer

Cold air in

(b)

Figure 12.9 Types of air heaters. (*a*) Recuperative air heater transfers heat through tubes; (*b*) regenerative air heater stores heat in baskets on one side of rotation, and then transfers heat to air on the other side of rotation.

Fire-side deposits and emissions

Soot blowers. Steam or air is used to blow soot from heating surfaces (Fig. 12.10*c*). Frequency of soot blowing depends on exhaust gas temperature rise, which tells the boiler operator when to blow tubes. When blowing, be sure to keep the boiler steaming at a reasonably high rate. This avoids the possibility of flameout or the explosion of dead pockets filled with unburned combustibles.

What precautions must be followed to prevent soot blowers from eroding tubes, tube sheets, or baffles?

Steam or air used should be dry in order to avoid the impact of water droplets on the tube surfaces. Soot blowers should be drained of all condensate before being used. Alignment of the nozzles of nonretractable rotary soot blowers must be maintained as designed, otherwise the jets from the nozzles will impinge on the tubes and cause erosion wear that will cause tube failure.

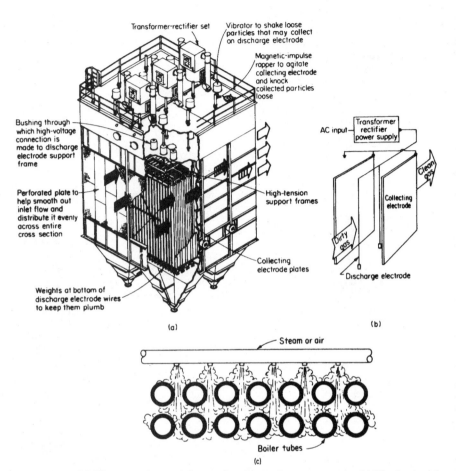

Figure 12.10 (*a*) Electrostatic precipitator showing major components. (*b*) The basics of the electrostatic-precipitator operation: Particles in the dirty gas entering a precipitator are charged by the discharge electrode. The particles then migrate to the collecting electrodes, where they adhere and lose their charge. The particles are discharged from the collecting plates by vibrators, or rappers, and they fall into hoppers below the plates. (*c*) A fixed-position soot blower is used for in-line tube arrangements.

Particulate discharge is a major problem in burning coal and other solid fuels. Fluidized-bed boilers are used to limit sulfur dioxide discharge (acid rain). Scrubbers are used to wash the particulates before venting to the stack. Fabric filters are also used to control fine solid particulate discharge by the installation of baghouses in the flue-gas stream from coal-burning boilers.

Coal-burning and other solid-fuel-burning boilers require auxiliary equipment to remove fly ash and other particulates being emitted to the surrounding atmosphere. The equipment commonly used includes the following:

1. Baghouse employing fabric filters now usually made of fiberglass that can withstand flue-gas temperatures of 275 to 550°F.

2. Scrubbers that wash particulate emissions out of the flue gas and form a sludge that is disposed of in landfills.

3. Electrostatic precipitators (Fig. 12.10a).

These precipitators produce an electric charge between two electrodes through which flue gas is passed. The particles in the flue gas become charged and are attracted to the positively charged and grounded collecting electrode. The particles so collected are discharged into hoppers by rapping the collecting electrode (Fig. 12.10b).

Fabric filters. Emission standards have risen on fossil-fuel-fired utility power plants so that particulate collection equipment has reached 99 percent efficiency. Fabric filters are being used with electrostatic precipitators. Fabric filters require special design considerations on emission flow so that all fabric bags are collecting a proportionate amount, and on periodic cleaning of the bags, usually achieved by air reverse flow. (See Fig. 12.11.)

In the unit shown, dirty gas flows from inside the bags to the outside with the airflow periodically reversed through the fabric filter, which causes significant bag motion that releases the collected dust. Another design, *called a shake/deflate* unit, also collects dust on the inside of the bags, but, to clean the bags, the top end is shaken by a drive linkage designed for the proper frequency and amplitude of shake to be imposed for cleaning the bags. It is usually an off-line cleaning method.

For industrial boiler applications, fabric filters called a *pulse-jet* system are quite often used. The dust-laden flue gas is directed into the bag and cleaning is accomplished with a high-pressure burst of air into the open end of the bag, which is timed so that a valve opens to direct a pulse to the open bag. Pulse-jet units can be cleaned while on-line or off-line.

Fabric selection is a most important item in fabric filters, with utility units being mostly of the woven fiberglass type. Also used in pulse-jet units are woven and felted fiberglass and teflon fluorocarbon fiber. Fabric material is a heavy-wear item because of the abrasive dust exposure and generally requires periodic replacement. Pressure drop across a fabric filter is an important indicator of fabric-filter performance per the manufacturer's guidelines. Too high a pressure drop may require cleaning the fabric manually or require longer periods off-line for cleaning.

Boiler water auxiliaries

Boiler water auxiliaries can be subdivided into three classifications:

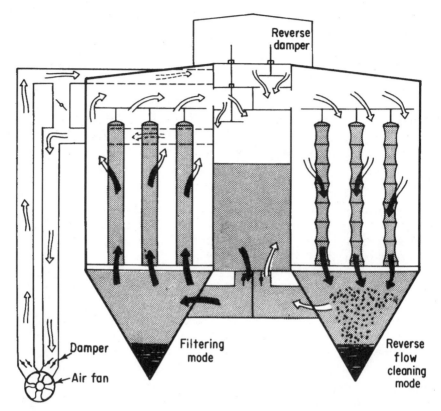

Figure 12.11 By means of duct dampers and automatic timers, flue gas flow to the filter bags is diverted to filtering mode and then cleaning mode by reversing airflow through the filter bags.

1. Equipment that preheats feedwater, but is outside the flow of the products of combustion.

2. Equipment that combines some external feedwater heating with some elimination of undesirable gases or impurities in the makeup water, such as evaporators and deaerators.

3. Equipment that externally treats boiler feedwater to eliminate undesirable scale-producing impurities such as water softeners and demineralizers.

Closed feedwater heaters. Closed feedwater heaters are extensively used in utility plants to preheat feedwater in stages by extracting steam from steam turbines in what is described as regenerative heating. However, industrial boiler plants also use feedwater heating, but sizes and arrangement of the closed feedwater heater will vary, as is shown in Fig. 12.12.

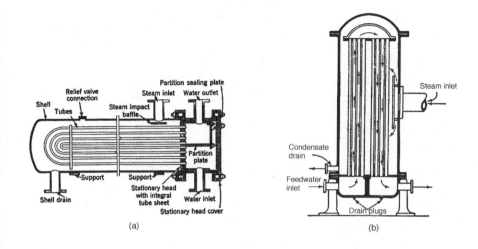

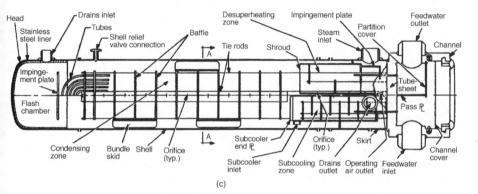

Figure 12.12 Closed boiler feedwater heater types. (*a*) U-tube horizontal closed feedwater heater; (*b*) vertical shell and straight-tube closed feedwater heater; (*c*) three-zone utility size high-pressure closed feedwater heater.

Heat transfer to the water from the steam is dependent on the square feet of heating surfaces, the velocity of the waterflow through the heater, and the mean temperature difference between the water and the steam. Sufficient time must be provided to obtain the desired temperature rise in the feedwater.

Feedwater heaters are used to bring feedwater nearer to the temperature of the boiler water. Each 10°F rise in feedwater temperature increases the overall boiler efficiency about 1 percent, owing to savings in fuel that would have been required to heat the boiler water an equal amount. An added advantage is that temperature stresses in the boiler may be avoided by feeding water at higher temperatures.

The utility heater used shown in Fig. 12.12*c* has three zones of heat transfer: (1) *desuperheating zone* in which superheated extraction steam from a turbine maximizes the feedwater outlet temperature by transferring some sensible heat from the steam to the feedwater; (2) *condensing zone* where the steam from the desuperheating zone is condensed to further heat the feedwater; (3) *subcooling zone* where the drains from the condensing zone recover additional heat from the shell-side drains before they leave the closed feedwater heater.

Venting is important on these heaters, especially in the condensing zone, in order to remove noncondensable gases that may be released by chemicals in the feedwater and from air inleakage. Accumulation of noncondensable gases can produce corrosion problems in the heater and acid attacks in dead pockets of the heater. They also lower the heat-transfer rate in the heater.

In industrial plants, the closed feedwater heater is located after the feedwater pump; thus the water in the feedwater heater is at a higher pressure than the boiler water, making it possible to raise the water temperature higher before flashing may occur. This also prevents suction pressure problems on the feedwater pump.

A *tube failure* on a high-pressure heater using turbine extraction steam can affect not only the heater, but also possibly the steam turbine by high-pressure water flowing into the turbine through the extraction steam line. The resultant thermal stresses inside the turbine can do severe damage and produce a lengthy outage in a generating plant.

Feedwater heaters have water level controls, and also high-water-level alarms to warn operators of water flooding a heater. These must be tested at frequent intervals to prevent deposits or gas pockets in the level connections giving false or improper readings. Extraction lines should be equipped with check valves to prevent water backflow, and these also require periodic testing. The numerous cases of water induction incidents into steam turbines, mostly from leaking tubes on heaters, prompted the ASME to publish a recommended guideline to prevent water induction, which operators with steam turbine generators and feedwater heaters using extraction steam should obtain as reference material for inspection and testing guidelines.

Most tube failures can be traced to vibration of the tubes with variable load, which produces erosion wear on intermediate supports, corrosion from venting problems of noncondensable gases or air leakage, and stress-corrosion cracking at high-stressed areas of the tubes.

Many nondestructive testing methods are now available to check feedwater heaters, usually on a five-year schedule after the initial operating period of a heater. Eddy-current testing is the most used and is usually performed by firms specializing in tube testing.

Open feedwater heaters. Feedwater heaters of this type are *direct-contact* heaters because they use the heat of the steam to heat water as they are mixed. See Fig. 12.13*a*. The open heater operates at low pressure, from atmospheric to 30 psig with the water and steam at the same pressure. The unit shown in Fig. 12.13*a* features a steam and oil separator at the inlet. Water trickles down from the top inlet over trays, which breaks the water up to obtain better mixing with the steam and thus obtain better heat transfer. The heated water passes through a coke filter before entering the boiler feed pump suction.

The direct-contact heater has two definite subdivisions, namely, the standard *open heater* and the *deaerating heater.* The open heater was originally designed to utilize exhaust steam for feedwater heating and is essentially a low-pressure heater. It is always located on the suction side of the feed pump, and the heater must be at a sufficient elevation above the pump suction to prevent steam binding. (When hot water is subjected to vacuum, it flashes into steam. Thus, a pump handling hot water must have its suction fed under positive pressure, or no water will flow to the pump. See Fig. 12.14. A steam-bound pump will race and so may be damaged.) The required elevation depends on the maximum water temperature.

The principle of the open heater is to pass cold makeup water from the top down over a series of metal trays. Low-pressure steam enters between these trays, condensing and mixing with the water.

Important functions performed by the open heater in addition to raising the water temperature are:

1. Depositing solids that cause "temporary" hardness in the water.

2. Removing a considerable proportion of free oxygen by bringing the water to the boiling point and venting the gases to the atmosphere.

Step 1 may reduce scale formation in the boiler; step 2 helps to reduce corrosion and pitting, which are accelerated by free oxygen.

Deaerating heater. The *deaerating heater* (Fig. 12.13*b* and *c*) is a development of the open heater and increases its oxygen-removal function by operating at temperatures corresponding to pressures above atmospheric. Although for this reason it is no longer an "open" heater, it is, nevertheless, still a direct-contact heater. It is used with excellent results in moderate- to large-sized plants where a sufficient volume of low-pressure steam (5 to 50 psi) is available for the heating process.

Oxygen and noncondensable gases are vented with steam through a vent condenser on top of the heater. Here the steam condenses and the condensate returns to the system, with the oxygen and other noncondensable gases being vented through a vacuum pump to the atmosphere.

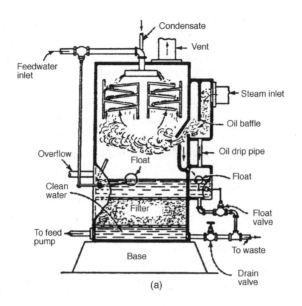

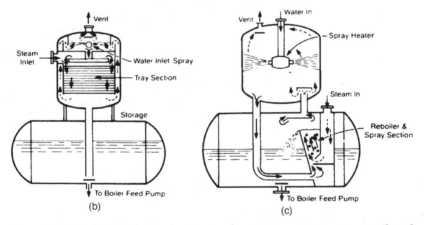

Figure 12.13 (*a*) Open feedwater heater uses low-pressure steam, removes oil, and vents noncondensable gases. (*b*) Tray-type deaerator; (*c*) spray-type deaerator.

Deaerators are used to reduce oxygen and other dissolved gases in a steam plant. (See Fig. 12.13*b* and *c*.) The steam used in the deaerator raises the feedwater temperature, and this lowers the solubility of the oxygen in the water, leaving the water to be vented as a gas.

Recent *cracking incidents* in the United States on deaerators have resulted in an awareness of the need to periodically inspect the welds on these vessels, and also to avoid sudden pressure changes on con-

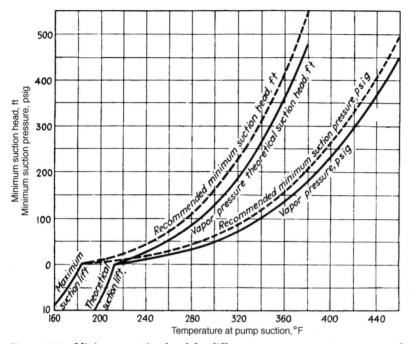

Figure 12.14 Minimum suction head for different water temperatures at eye of boiler feed pumps, as recommended by Hydraulic Institute.

nected piping, which can occur from trips on turbogenerators. Usually, extraction lines from a steam turbine supply steam to these vessels. There have been water-hammer incidents when valves to or from the heater closed too fast during tripping sequence. Other incidents involved corrosion fatigue cracks in weld areas.

The Heat Exchange Institute, 1300 Summer Ave., Cleveland, OH 44115, has revised its standards on deaerating heaters, which now include the following on new heater designs:

1. Corrosion allowance increased from ⅟₁₆ in. to ⅛ in.

2. Prohibits the use of SA-515 carbon steel on new construction.

3. Requires smooth grinding of internal weld seams to avoid stress concentration and the forming of corrosion pits.

4. Requires stress-relieving of storage tank welds to eliminate residual weld stresses.

5. Requires full x-ray examination of all shell and head weld seams.

6. Requires wet, fluorescent, magnetic-particle inspection of nozzle-to-shell welds.

The reinspection of older deaerators has stressed the use of the wet, fluorescent, magnetic-particle inspection of all internal weld seams, plus nozzle connections to supplement any internal visual inspections. It is also recommended that NDE be performed by qualified personnel in accordance with SNT-TC-1 requirements. Any crack repairs should be approved by a qualified jurisdictional Code inspector. Inspections have been made more frequent since incidents of crack failures received industry attention. Recommended intervals are:

1. One year after installation on new units.
2. No cracks found and no repairs required: three-year interval.
3. Units with repaired cracks: one year or less after repair.

Evaporators. The *evaporator* shown in Fig. 12.15a is a still in which raw (impure) water is evaporated into steam. This steam is condensed into pure condensate for feedwater. The condensate is often high in oxygen content, and it is customary to include with the installation a deaerating heater for oxygen removal.

The evaporator consists of a shell into which the raw water is fed to maintain a constant level. Tube coils through which steam at 10- to 150-psi pressure passes are submerged in the water. The condensate in the steam coils is trapped back into the feedwater system. The vapor or evaporated steam from the raw water passes through an evaporator condenser where the steam is condensed for use as makeup in the feedwater supply.

Most of the scale-forming impurities are left in the water in the evaporator shell. As the concentration of the raw water builds up, it should be reduced by blowing down the evaporator and refilling.

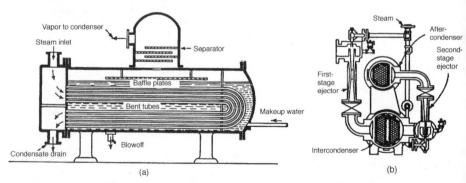

(a)   (b)

Figure 12.15 (*a*) U-tube evaporator prepares makeup water by evaporating and condensing the water with some dissolved gases removed by heat, but solids remaining in the evaporator require periodic removal, either mechanically or chemically. (*b*) A two-stage injector used to remove air from a steam condenser.

With a number of units, evaporators may be installed in series (multiple effect). Usually, four effects in series (quadruple effect) are sufficient to produce pure water from raw water as impure as is possible to use and yet be productive of maximum practical efficiency.

Steam supply to evaporators usually is extracted from steam turbines. The evaporator is more practical in moderate- to large-sized plants using a small percentage of makeup.

Evaporators are classified by the method of vaporization used as:

1. *Flash type.* Hot water is pumped or injected into a chamber under vacuum, where the water flashes into steam.

2. *Film type.* Water in a thin film is passed over steam-filled tubes.

3. *Submerged type.* Steam-filled tubes are submerged in the water to be evaporated. (See Fig. 12.15*a*.)

Evaporators are most often used in power plants that condense the steam from the turbogenerators and feed it back to the boiler, thus requiring minimal makeup water. In plants with large water feed makeup requirements, the use of evaporators would require large sizes, and the operating costs would be too great; therefore, chemical water treatment methods are used to prepare the makeup water.

Air ejectors. In generating plants with shell and tube condensers connected to steam turbogenerators, air ejectors are used to remove air from the condenser. As Fig. 12.15*b* indicates, an air ejector is a steam nozzle that discharges a high-velocity jet of steam at about 3500 ft/sec. The steam flows across a suction chamber and through a venturi-shaped compression tube. The air or gases to be evacuated enter the ejector suction where they are entrained by a jet of steam and then discharged through the throat of the ejector. The velocity of the kinetic energy is converted into pressure in the throat of the ejector and compresses the mixture to a lower vacuum or a higher absolute pressure. Two-stage ejectors have a compression ratio of about 8:1 (the ratio between the discharge pressure and the suction pressure). The ejector discharges to either a small condenser or a feed-water heater where the steam is condensed, and the air and gases are vented to the atmosphere. The two jets, shown in Fig. 12.15*b*, are in series with intercondensers between the stages. These intercondensers condense the steam and cool the air-vapor mixture. Aftercondensers are used for the same purpose.

Raw water preparation

Many plants use raw water from rivers or lakes for boiler makeup water. This raw water is externally treated by sedimentation, filtration, softening, and removal of dissolved gases.

Sedimentation.   Sedimentation allows the solids to settle out of the water by dropping to the bottom of a basin or impounding reservoir. It can be assisted by the use of coagulants such as alum or aluminum sulfate, ferrous sulfate, ferric chloride, sodium aluminate, and magnesium oxide. The raw water that is available will determine the coagulant to use. In addition, with many waters it is necessary to add an alkali, such as lime or soda ash, to bring the water to the best or required pH value.

Natural sedimentation combines (1) the mixing of chemicals to aid in the suspended solids clinging to the coagulants, and (2) the small particles are then brought together, *called flocculation,* by gentle mixing to form larger particles that settle out faster. This is performed by baffles, or by mechanical mixers. *Pressure filters* are also used to remove smaller amounts of suspended solids, such as makeup water for boilers, as shown in Fig. 12.16. The advantage of mechanical and pressure settling or filtering is that the raw water needs to be retained for less time to remove the suspended solids.

Color in some waters is removed chemically by aluminum sulfate and chlorinated copperas. These compounds react with color in water to form a precipitate that settles with the sludge in the sedimentation process.

Filtration.   Filtration differs from sedimentation in that smaller and lighter particles of suspended and coagulated matter remain after sedimentation and now must be removed by filtration (see Fig. 12.16). Most common filters use suitably graded beds of sand or anthracite

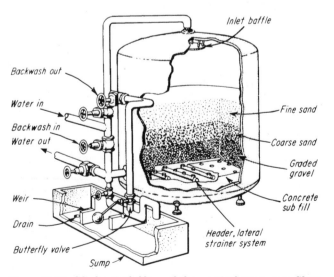

Figure 12.16   Mechanical filter of the vertical type uses filter media to remove finer solids from raw water.

coal. When the fine particles enter a filter, they settle out in the top few inches of the bed and, with time, build up on the surface. If this is not corrected, it starts to restrict flow.

Backwash is used to remove the particles from the filter bed. Water is passed upward through the bed at a rate of 4 to 7 times the filter rate, and the accumulated suspended particles are washed out of the bed and sent to waste.

Activated carbon filters are used to *remove odors* and improve water taste. Their construction is similar to that of the sediment filter and usually is under pressure. However, the carbon is not a strainer, but an *absorber* of the odorous substance, and must be periodically replaced.

Water softening. Water softening methods are continually being improved with the study of water chemistry. Operators may be exposed to the following types of water softening equipment:

1. Water softening by *precipitating out of solution* the compounds that cause water hardness. The earliest used was the *lime softening* method followed by the *zeolite softening method*. Today, this is called the *carbonate treatment* method, because calcium carbonate is precipitated out of the water. This compound is insoluble and settles out of solution. It must be removed by bottom blowdown.

Water containing appreciable amounts of calcium and magnesium compounds in solution is called *hard water*; the name was derived from the fact that when using soap with hard water, it is difficult to obtain a lather. Hard water is especially objectionable for use in boilers because the calcium and magnesium salts deposit on the tubes, forming a stonelike layer on the inner walls of the tubes, commonly called scale. This scale acts as an insulator, preventing proper heat transfer between the fire and water sides of the tubes. This adds to fuel consumption. Severe scaling can cause tubes to become overheated and rupture, which can be dangerous in a boiler plant operation.

There are two types of hardness. One is called *temporary hardness,* or temporary hard water, which contains large quantities of calcium bicarbonate, $Ca(HCO_3)_2$. This water may be softened by boiling it, with the calcium carbonate, $CaCO_3$, precipitating out of solution, and carbon dioxide gas being released. In industrial plants, water is softened by the addition of a sufficient amount of lime to precipitate out the calcium carbonate. This soft sludge must be "blown out" of the boiler to prevent buildup as a precipitate in drums or headers.

Water that contains calcium and magnesium sulfates is not softened by boiling and is referred to as having *permanent hardness*. It can be softened by the addition of sodium carbonate (NA_2CO_3). Calcium and magnesium carbonate precipitate out.

The *cold-lime and soda softening process* treats raw water with lime, calcium hydroxide, and soda, or sodium carbonate, to partially reduce hardness. The water usually then requires further internal treatment in boilers (Fig. 12.17a).

The *hot-lime and soda softening process* operates at 212°F and over and uses steam as a heat source. The heat causes a faster chemical reaction to take place. Calcium hydroxide, lime and soda, or sodium carbonate are also used in the hot lime method of softening (Fig. 12.17b).

Above about 250 psi and high temperature, sodium carbonate which precipitates out of water with the lime treatment can decompose into caustic soda, (NaOH), and liberate carbon dioxide gas, (CO_2), both of which can be detrimental to boiler metals. Another factor is the large amount of blowdown required to eliminate the precipitated sodium carbonate. The carbonate treatment was replaced by the phosphate treatment, especially for high-pressure boilers.

2. The *zeolite softening process* uses a sandlike substance called zeolite, which can be of synthetic or natural derivation. This substance is also arranged inside a tank as a filter bed (see Fig. 12.18). Zeolite has the remarkable property of base exchange. When the hard water passes through the bed of zeolite, the calcium and magnesium compounds pass into the zeolite and are replaced by sodium from the zeolite. The calcium bicarbonate becomes sodium bicarbonate, and magnesium sulfate becomes sodium sulfate. These sodium compounds do not form scale; thus, this ionic exchange makes the water soft by freeing it from the hardness compounds. Eventually, the zeolite loses its sodium concentration because the sodium is combining with the calcium and magnesium compounds, and this causes the zeolite to lose its exchange power.

Zeolite regeneration involves irrigating the zeolite with a strong brine solution (sodium chloride). A reverse reaction causes the sodium from the brine to replace the calcium and magnesium in the zeolite. Figure 12.18 shows a zeolite softener that uses automatic controls. In the softening cycle, water flows downward through the zeolite bed. As the water-softening capacity of the zeolite declines to a set point, automatic valves cut off the downward flow of water and *backwash* upward to loosen the material and also remove deposited dirt. In the third step, a measured quantity of common salt brine is admitted at the top of the bed. After a timed interval, a stream of rinsing water is introduced to remove excess salt and to clean the zeolite, after which the bed is ready for another softening cycle.

In a boiler water application, zeolite-treated water shows zero hardness by a soap test, but the water now has soluble sodium salts in solution. Blowdown is necessary to limit the concentration of these

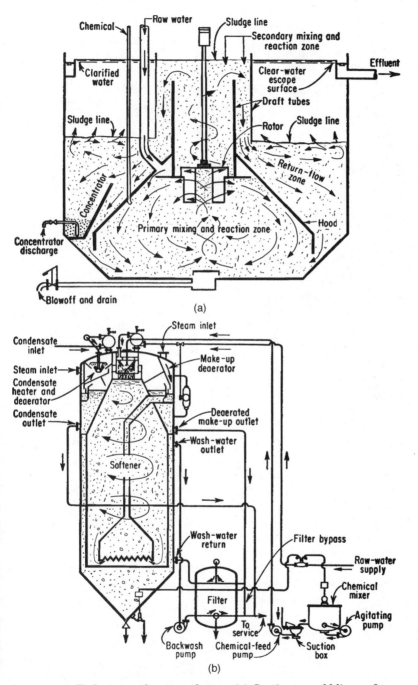

(a)

(b)

Figure 12.17   Early types of water softeners. (a) Continuous cold-lime softener;
(b) hot-lime softener also incorporates a deaerating section for condensate
return flow, for removal of noncondensable gases. (Courtesy The Permutit Co.)

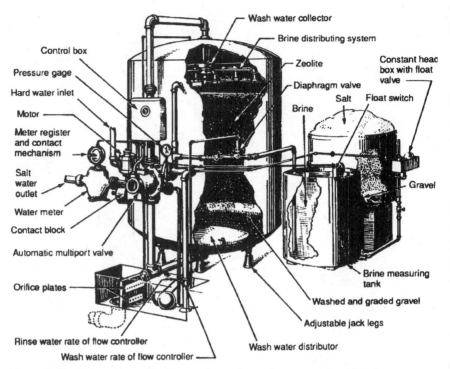

Figure 12.18 Automatic zeolite-type water softener is common in older, low-pressure steam plants.

salts so that priming and foaming do not occur in steam boilers. A high percentage of sodium carbonate may cause steel embrittlement under certain conditions; thus, zeolite softeners are better suited to handle magnesium sulfate hardness. To remove the carbonate hardness, the hot-process lime treatment is used ahead of the zeolite softener.

The amount of salt required for regeneration depends on the grain of hardness. The usual figure cited is ¼ to ½ lb of salt per 1000 gal of water per *grain hardness*. For example, regeneration of 50 ft^3 of a high-capacity synthetic zeolite requires 300 lb of salt. If the water has a 10-grain hardness and ¼ lb of salt is required per 1000 gallons, how many gallons of water can now be softened?

$$\text{Gallons softened} = \frac{4 \times 300 \times 1000}{10} = 120,000 \text{ gal}$$

Hot-process chemical softening and zeolite softening can be combined into a *hot-process–hot-zeolite treatment* system that will deliver, especially to high pressure-boilers, a hot feedwater with zero hardness.

3. *Ion exchange* is the term applied to the exchange of calcium and magnesium minerals for sodium minerals, because of the minerals being ionic in nature in a solution or having an electrical charge. The ions are further classified as being either positively or negatively charged, with the positive ions called cations, and the negative ions called anions. Positive cations in the ionic form of calcium, magnesium, iron, and manganese cause water hardness. By using ionic exchange, these hardness ions are removed to soften water and thus reduce scale formation in boilers.

The ion-exchange method, using new and more versatile ion exchangers, has replaced the original and synthetic-type zeolite materials.

Ion exchange in water treatment is based on the principle that impurities which dissolve in water dissociate to form positively and negatively charged particles known as ions. These impurities, or compounds, are called electrolytes. The positive ions are named cations because they migrate to the negative electrode (cathode) in an electrolytic cell. Negative particles are then anions since they are attracted to the anode. These ions exist throughout the solution and act almost independently. For example, magnesium sulfate ($MgSO_4$) dissociates in solution to form positive magnesium ions and negative sulfate ions.

Ion-exchange material has the ability to exchange one ion for another, hold it temporarily in chemical combination, and give it up to a strong regenerating solution. The chart in Fig. 12.19*a* lists the

Ion - exchange materials	Regeneration chemical
Cation exchangers:	
Sodium cycle:	
Natural greensand	NaCl
Synthetic gel	NaCl
Sulfonated coal	NaCl
Styrene resin	NaCl
Hydrogen cycle:	
Sulfonated coal	H_2SO_4
Styrene resin	H_2SO_4
	H_2SO_4
	HCl
Anion exchangers:	
Weakly basic (aliphatic amine)	Na_2CO_3
Weakly basic (Phenolic)	Na_2CO_3
Weakly basic (Styrene)	NaOH
Strongly basic (Type I)	NaOH
Strongly basic (Type II)	NaOH

(a)

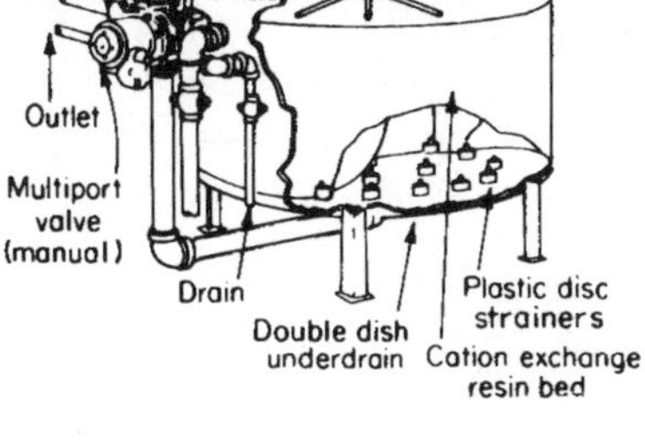

(b)

Figure 12.19 (*a*) Ion-exchange material, and regeneration materials used; (*b*) ion-exchanger of the cation type.

ion exchangers and the regeneration chemicals commonly used in water treatment.

Figure 12.19*b* shows a water softener of the cation-exchanger type. Water softening using the ion-exchange process is accomplished by means of passing hard water through a bed of synthetic resin. The hardness-forming calcium and magnesium ions in the water are removed by exchanging them for the nonhardness-forming sodium ions which are attached to the resin. When all the sodium in the resin has been used up, the resin bed no longer has the ability to soften the water and must be regenerated. This is done by passing an excess quantity of sodium chloride brine through the resin bed to drive off the calcium and magnesium and replace these elements with sodium. The brine is then rinsed out of the bed with water before being placed back into operation on the softening cycle.

4. *Demineralizers* remove dissolved matter from boiler pretreated water by contacting the water with ion-exchange resins. These are spherical beads made of insoluble acids and bases that are made from cross-linked polymer chains. These resins remove dissolved solids by an ionic exchange that leave harmless ions in the water. Resins can be regenerated for further use by backwashing, but eventually must be replaced with a new bed. See Fig. 12.20. Demineralization of water in certain industries requires the water to be completely free from

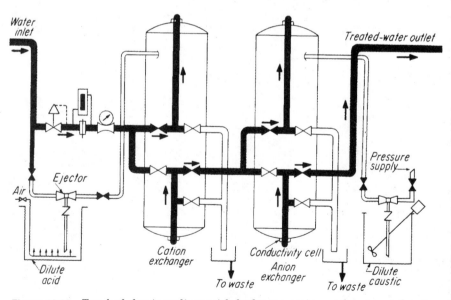

Figure 12.20 Two-bed demineralizer with hydrogen cation exchanger and anion exchanger that may exchange acid radicals to form water (weak basic anion), or strong basic anion that also removes acids, silica, and carbon dioxide. (*Courtesy Power magazine.*)

mineral salts. This also applies to central power station's boiler water. Distillation is one method, but it is costly. Ion exchange is a two-step method, the first being the hydrogen cycle, the *cation* exchange, followed by the second step, the *anion* exchange. The anion exchangers are further divided into weakly and strongly basic anion exchangers. The weakly basic unit will not remove weak acids, such as carbonic or silicic, and therefore the treated water may contain silica and carbon dioxide. Weak base resins are regenerated by alkali, such as ammonia, caustic, or soda ash.

The strongly basic anion exchanger can remove both strong and weak acids, producing water that is free of silica and carbon dioxide, but it is more expensive to operate. Regeneration is with caustic.

Demineralizers are used in boiler plants that operate over 1000 psi. Demineralizers resemble the ion-exchange process. The cation exchange is operated on the hydrogen cycle where the metal ions are replaced with hydrogen ions by using the appropriate resin in the bed. The anion exchanger operates on the hydroxide cycle using specially prepared resins saturated with hydroxide ions. The salt anions, such as bicarbonate, carbonate, sulfate, and chloride, are replaced by hydroxide ions. The final effluent consists essentially of hydrogen ions and hydroxide ions or pure water.

In *mixed-bed demineralizers,* the two types of resins are mixed together in a single tank. Regeneration in a mixed bed can be performed because the two resins can be hydraulically separated into different beds.

Condensate polishing is used to purify returned condensate, and demineralizers are used in power plants to remove corrosion products and ionized solids that come from connected piping, turbines, heaters, or condensers. This improves turbine-generator efficiency, protects the steam loop from the effects of condenser leakage, and avoids having harmful deposit or corrosion products in the boilers.

In power-plant applications of condensate polishing, flows are very large in comparison to raw water makeup, because all the steam going to a turbogenerator is condensed and returned to the boiler. The main impurities to be removed are metal oxides, termed "crud," silica, various types of scale, and fragments of resins and filter media. Resins used in the ion-exchanger, or demineralizer, consist of gel and macroporous resins. A typical regeneration system has a cation-separation regeneration vessel, an anion regeneration vessel, and a resin mix-and-hold tank.

Membrane demineralizing is another method receiving attention for removing impurities from boiler water. The membrane treatment consists of passing pressurized fluids through a semipermeable membrane, usually polymeric. In conventional particle filtration, called macrofiltration, the entire influent stream passes through the filter media, leaving particulates behind. This is a developing field concentrating on the type of membrane, the membrane material's properties, membrane and flow arrangement, and on the purity of the resulting filtered water that can be expected from water coming from various sources. Pumping pressures to drive the water through the membrane can range from 150 to 400 psi for nonbrackish water to as much as 1000 psi for seawater desalination.

Membrane technology is being combined with demineralizers, especially in nuclear generating plants, to reduce total organic carbon in treated water from 2000 to 3000 ppb down to 2 to 20 ppb. The membrane treatment uses less energy than the evaporator method. Its increased use in the production of high-purity water for makeup water will be driven by this economic factor: less cost to produce the desired water quality.

The methods described for conditioning boiler water are essentially external treatment methods. The next chapter is devoted to water chemistry, water testing, and internal treatment.

Questions and Answers

1 A boiler feed pump has a discharge pressure of 200 psi gauge and delivers water weighing 62.3 lb/ft^3 at 1250 gal/min at rated load. If the mechanical efficiency of the pump is 85 percent and suction head is neglected, calculate the hydraulic or water horsepower required and the motor horsepower needed to drive the pump.

ANSWER:

$$\text{Head developed} = \frac{200 \times 144}{62.3} = 462.3 \text{ ft}$$

$$\text{Weight of 1 gal of water} = \frac{62.3 \times 231}{1728} = 8.216 \text{ lb/gal}$$

$$\text{Weight of water handled per minute} = 1250 \times 8.216 = 10{,}270 \text{ lb/min}$$

$$\text{Hydraulic or water horsepower} = \frac{10{,}270 \times 462.3}{33{,}000} = 143.9 \text{ hp}$$

$$\text{Motor horsepower} = \frac{143.9}{0.85} = 169.3 \text{ hp}$$

2 What medium is used in permanently installed soot blowers?

ANSWER: The cleaning medium used can be dry superheated steam and compressed air depending on the design of the boiler and the sootblower equipment. Sootblowing is best performed with negative pressure in the furnace in order to avoid the soot coming into the boiler room through leaking settings.

3 What is the difference between a plunger and piston-type pump?

ANSWER: See Fig. 12.21. In the plunger pump, the plunger does not have piston rings or packing inside the cylinder, but instead it is kept tight against water leakage by passing through outside the cylinder-packed packing box as shown in Fig. 12.21. In contrast, the piston pump has the piston entirely within the cylinder and has rings or packing fitted into grooves of the piston (circumferentially), and these are held in place by a follower plate as shown in Fig. 12.21. Note that one advantage of a plunger pump is that the packing can be replaced without having to open the cylinder as with a piston pump.

4 What is the function of a foot valve?

ANSWER: See Fig. 12.21. The function of the foot valve is twofold: (1) to act as a check valve in two directions, when water is lifted in a suction pipe to a water pump and to prevent water from flowing back out of the suction pipe to the pump. (2) As Fig. 12.21 shows, a strainer is also incorporated into the foot valve construction to stop foreign objects from entering the pump suction. Another advantage of a foot valve is that it keeps the pump suction line above the valve full of water, making it easier for the pump to start up with water on the suction side.

5 What is the water cylinder diameter of a steam-type reciprocating pump that has the following name plate: 10 × 18 × 12 in.?

ANSWER: The water cylinder is 18 in. The steam cylinder size is always listed first, being 10 in., while the stroke of 12 in. is listed last.

6 A simplex water pump of 8 × 4 × 10 in. operates at 125-psi steam pressure. What theoretical water pressure can be developed?

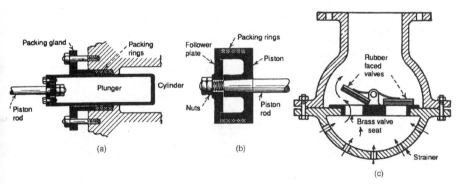

Figure 12.21 (*a*) Plunger pump has outside packing box; (*b*) piston pump has packing rings around piston; (*c*) foot valve has strainer, and acts as check valve on suction side of pump.

ANSWER: Equate steam piston area times steam pressure to water piston area times water pressure and solve for water pressure as follows with $\pi/4 = 0.7854$:

$$0.7854(8)^2(125) = 0.7854(4)^2 P_w$$

$$P_w = \frac{64(125)}{16} = 500 \text{ psi}$$

7   How many gallons of water will a pump, size 6 × 4 × 6 in., double-acting, duplex, pump at 80 strokes per minute if overall water pump efficiency is 85 percent?

ANSWER: One gallon = 231 in.^{3} Duplex = two water cylinders pumping. Double-acting = pumping at both sides of piston. Therefore, volume in in.^{3} swept by water side of pump is

$$\text{volume} = 0.7854(4)^2 \times 6 \times 2 \times 2 \times 80 \times 0.85 = 20,508 \text{ in.}^3$$

$$\text{gallons} = 20,508/231 = 89 \text{ gpm}$$

8   What does the term *valve deck* mean?

ANSWER: A valve deck is a plate that contains either the suction or discharge valves of a direct-acting pump.

9   What do *air bound* and *steam bound* mean?

ANSWER: Air bound indicates air is leaking into the suction side of the pump, incapacitating the ability of the pump to draw water on the suction side of the pump. Steam bound indicates the water is too hot for the suction of the pump, and the water flashes into steam with reduction of pressure. Air binding usually requires venting the pump and then priming it with water to get the pumping action started. Steam flashing can cause pump seizure, vibration, and cavitation damage. Check the upper permissible temperature limits for the water as established by the pump manufacturer and do not exceed these limits.

10   What can cause knocking on a steam-type reciprocating pump?

ANSWER: There are several items that can cause these conditions: (1) Condensate in steam lines or steam cylinder. Drain both, especially when starting, to avoid a head being blown off or other internal damage to the steam side of the pump. (2) Make sure there is no air on the water side of the pump. Use vent valves and, if required, prime the pump. (3) Make sure no valves have been closed on the suction or discharge side of the pump and that water is flowing properly with no obstructions. (4) Check the operation of the valves to make sure no lost motion exists. (5) Check all linkages, piston rods, bearings, and similar mechanical parts for excessive wear.

11   What are pump slip and volumetric efficiency?

ANSWER: Pump slip is the *difference* between the displacement of volume of a reciprocating pump, which is calculated as the theoretical discharge volume, and the actual discharge of the pump. Slip is expressed as a percentage of the theoretical displacement, and these have values ranging from 3 to 15 percent. The volumetric efficiency is the ratio of the volume of water actually delivered to the theoretical displacement of the pump.

12   A centrifugal pump is operating with a 60-psi suction pressure and a 200-psi discharge pressure. What would the total head on this pump be if friction and velocity head totaled 25 ft?

ANSWER: Net increase in pressure performed by the pump is $200 - 60 = 140$ psi. Converting to feet, $140/0.434 = 323$ ft and total head $= 323$ plus $25 = 348$ ft.

13   A steam condensate pump on a 50,000-kW turbogenerator has a friction and external head of 60 ft and suction of 29.5 in. The turbine has a steam rate of 12 lb/kWh. The velocity head for the condenser pump is only $\frac{1}{2}$ ft. What are the gallons of flow per minute through the pump and the theoretical horsepower required of the pump?

ANSWER: It is necessary to convert to workable units for use in pump equations.

$$\text{Converting steam rate to gpm} = \frac{50,000(12)}{60(8.33)} = 1200 \text{ gpm flow}$$

For the vacuum head, the vacuum on condensers is measured in inches of mercury, which weighs 0.49 lb/in.^{3} Therefore, 29.5 in. $\times 0.49 = 14.455$-psi negative pressure. Converting to feet of water $= 14.455 \times 2.31 = 33.4$ ft of suction head.

$$\text{Total head} = 60 + 33.4 + 0.5 = 93.9 \text{ ft}$$

$$\text{ihp} = \frac{93.9(1200)(8.33)}{33,000} = 28.4 \text{ hp}$$

14   A reciprocating pump has the following nameplate information: $12 \times 6 \times 14$ in. What is the steam cylinder size, and if the pump is operating at 90 strokes per minute, what are the piston speed and total piston travel in ft/min?

ANSWER: The first number shown is the steam cylinder diameter: 12 in. The stroke is 14 in.; therefore, the speed in feet per minute is $14/12 \times 90/2 = 52.5$ ft/min, and the distance traveled is $52.5 \times 2 = 105$ ft/min.

15   A centrifugal pump designed for variable speed is rated for a top speed of 3500 rpm. At 2000 rpm, the pump generates a total head of 600 psi. What total head would be developed at 3000 rpm?

ANSWER: Using the centrifugal pump laws, total head varies as speed squared, or

$$600 \text{ psi} = 1386 \text{ ft head, then}$$

$$\text{New head} = \frac{1386(3000)^2}{(2000)^2} = 3118.5 \text{ ft} = 1353 \text{ psi}$$

16 How does an injector force water against boiler pressure when it uses steam at the same pressure?

ANSWER: Because of the restricted area of the nozzle, high velocity steam carries the drops of water so that their momentum causes flow into the boiler.

17 What is the purpose of a forced-draft and an induced-draft fan?

ANSWER: The forced-draft fan supplies air to the combustion space. The induced-draft fan draws gases and products of combustion and delivers them to the stack breeching.

18 What arrangement is desired for safety between the induced- and the forced-draft fans?

ANSWER: An interlock so that the forced-draft fan cannot be operated with the induced-draft fan shut down; otherwise, fire might be blown out of the doors and observation ports.

19 What are some causes of fan or blower vibrations?

ANSWER: Vibration may be caused by: (1) bearings that are worn or poorly lubricated; (2) a defective thrust bearing; (3) excessive deposits on blades; (4) unevenly eroded blades; (5) rotating parts rubbing on or binding with stationary parts; (6) unit misaligned with gear set and driver; (7) bent shaft or some other bent component; (8) unit that is loose from foundation; (9) one side running hotter than other side, causing misalignment; and (10) cracks in rotating element.

20 How is draft measured?

ANSWER: Air and gas under flow conditions are measured in inches of water by instruments such as manometers, where one side of a tube is connected to the duct or furnace in which the draft is to be measured and the other is connected to the atmosphere or inlet duct. The difference in the water level in the two columns indicates the inches of water of draft, which is a measure of the difference in pressure.

21 Air flows in a duct 24 × 30 in. with an average velocity of 20 ft/s when its static pressure is 0.75 in. of water. The air is at 50°F and weighs 0.07788 lb/ft³, while the water temperature in the gauge is at 60°F and weighs 62.34 lb/ft³. What is the *total pressure* of the air in the duct and the volume of flow in cubic feet per minute?

ANSWER: Use the equation

$$V = \sqrt{\frac{2gDh_v}{12d}}$$

where $V = 20$
$$g = 32.17$$
$$D = 62.34$$
$$d = 0.07788$$

and solve for h_v. Substitute

$$20 = \sqrt{\frac{2(32.17)\ 62.34 h_v}{12 \times 0.07788}}$$

and by solving, we obtain

$$20 = 65.51\ h_v$$

thus, $h_v = 0.305$ in. of water. The *total pressure* is the sum of the static and velocity, or

$$0.75 + 0.305 = 1.055 \text{ in. of water.}$$

And the flow Q = duct cross-section area × average velocity of flow, or

$$Q = \frac{24 \times 30}{144} \times 20 = 100 \text{ ft}^3/\text{s} \qquad \text{or 6000 ft}^3/\text{min.}$$

22 What are some conditions that may prevent a steam turbine from developing full power?

ANSWER: The following may contribute to low power output: The machine is overloaded, the initial steam pressure and temperature are not up to design conditions, the exhaust pressure is too high, the governor is set too low, the steam strainer is clogged, turbine nozzles are clogged with deposits, and there is internal wear on nozzles and blades.

23 Why is it necessary to open casing drains and drains on the steam line going to the turbine when a turbine is to be started?

ANSWER: To avoid slugging nozzles and blades inside the turbine with condensate on start-up; this can break these components from impact. The blades were designed to handle steam, not water.

24 What is a regenerative cycle?

ANSWER: In the regenerative cycle, feedwater is passed through a series of feedwater heaters and is heated by steam extracted from stages of a steam turbine. This raises the feedwater to near the temperature of boiler water, thus increasing the thermal efficiency of the cycle.

25 What are some common troubles in surface-condenser operation?

ANSWER: The greatest headache to the operator is loss of vacuum caused by air leaking into the surface condenser through the joints or packing glands. Another trouble spot is cooling water leaking into the steam space through the ends of the tubes or through tiny holes in the tubes. The tubes may also become plugged with mud, shells, debris, slime, or algae, thus cutting down on the cooling-water supply, or the tubes may get coated with lube oil from the reciproacting machinery. Corrosion and dezincification of the tube metal are common surface-condenser troubles. Corrosion may be uniform, or it may occur in small holes or pits. Dezincification changes the nature of the metal and causes it to become brittle and weak.

26 Where would you look for a fault if the air ejector didn't raise enough vacuum?

ANSWER: In this case, the trouble is usually in the nozzle. You will probably find that (1) the nozzle is eroded, (2) the strainer protecting the nozzle is clogged, or (3) the steam pressure to the nozzle is too low.

27 What are the main causes of turbine vibration?

ANSWER: Turbine vibration is caused by (1) unbalanced parts; (2) poor alignment of parts; (3) loose parts; (4) rubbing parts; (5) lubrication troubles; (6) steam troubles; (7) foundation troubles; and (8) cracked or excessively worn parts.

28 What is the most prevalent source of water induction into a steam turbogenerator?

ANSWER: Leaking water tubes in feedwater heaters, which have steam on the shell side supplied from turbine extraction lines. The water at higher pressure can flow back into the turbine because the extraction steam is at a lower pressure. Check valves are needed on the steam extraction line to prevent the backflow of water into the turbine.

29 What is meant by the water rate of a turbine?

ANSWER: It is the amount of water (steam) used by the turbine in pounds per horsepower per hour or kilowatts per hour.

30 What are three types of condensers?

ANSWER: (1) Surface (shell-and-tube), (2) jet, and (3) barometric.

31 Why is there a relief valve on a turbine casing?

ANSWER: The turbine casing is fitted with spring-loaded relief valves to prevent damage by excessive steam pressure at the low-pressure end if the exhaust valve is closed accidentally. Some casings on smaller turbines are fitted with a sentinel valve which serves only to warn the operator of overpressure on the exhaust end. A spring-loaded relief valve is needed to relieve high pressure.

32 What is the difference in principle between an open and a closed feedwater heater?

ANSWER: The open heater brings low-pressure steam in direct contact with the water and operates at or slightly above atmospheric pressure. The closed feedwater heater consists of shell and tubes with indirect contact between steam and water and may operate at high pressure.

33 What is the main difference in the purpose and function of a deaerating and an open feedwater heater?

ANSWER: The open heater reduces oxygen content by heating the feedwater to about 212°F, venting the contents at atmospheric pressure. The deaerating heater removes practically all oxygen by heating the feedwater with steam of about 30 psi or higher. Its shell is vented at pressure, through a vent condenser and vacuum pump.

34 What factors determine the collection efficiency of a precipitator?

ANSWER: Time of flue-gas retention, amount of electric power supplied to the precipitator discharge system, the size of the fly-ash particles, and the resistance of the dust to flow.

35 Name three types of synthetic fibers used in fabric-type filters.

ANSWER: The limiting factor of fabric filters or collectors has been the rapid wear, or inability to withstand temperature and corrosiveness of flue gases. The development of synthetic fibers such as polyesters, acrylics, and fiberglass has greatly expanded the application of fabric filters.

36 What is meant by low-temperature corrosion in a boiler?

ANSWER: Low-temperature corrosion occurs when the flue gas contacts surfaces that are at a temperature below the dew point of corrosive constituents in the gas. These low temperatures are found at the water inlet of economizers if the feedwater temperature is too low and at the cold end of an air heater. They may also occur in scotch marine boilers subject to on-off operation and furnace purging cycles. Low temperatures cause sulfur and other corrosive gases to form sulfurous and sulfuric acid that attacks boiler metal.

37 List the impurities that produce hard and soft scale and corrosion in boilers.

ANSWER: Hard scale is caused by calcium sulfate, calcium silicate, magnesium silicate, and silica. Soft scale is caused by calcium bicarbonate, calcium carbonate, calcium hydroxide, magnesium bicarbonate, iron carbonate, and iron oxide. Corrosion is caused by oxygen, carbon dioxide, magnesium chloride, hydrogen sulfite, magnesium sulfate, calcium chloride, magnesium nitrate, calcium nitrate, sodium chloride, and certain oils and organic matter.

38 How can a dissolved impurity leave a solution and become a solid?

ANSWER:

1. By a temperature increase, reducing the solubility of the solid in the water.
2. By exceeding the saturation point of the dissolved impurity in the water. Water can hold only a limited amount of the dissolved impurity, so the concentration is important.
3. By chemical changes of the impurity with heat, causing it to break down and form insoluble substances.

39 What is the chief chemical compound in temporary hard water?

ANSWER: Calcium bicarbonate, $Ca(HCO_3)$.

40 What two methods are used to soften water with temporary hardness?

ANSWER: Temporary hard water may be softened by *boiling* the water, with the calcium carbonate precipitating out as soft sludge and carbon dioxide being liberated. Temporary hardness can be eliminated by *adding lime* in the right quantity, with calcium carbonate being precipitated and water being a by-product of the chemical reaction.

41 What chemical compounds in water cause permanent hardness?

ANSWER: Calcium sulfate and magnesium sulfate.

42 How is hard water softened?

ANSWER: Permanent hard water can be softened by the correct addition of sodium carbonate or by passing the water through a zeolite softener.

43 What is clarification?

ANSWER: Clarification is the removal of suspended matter and/or color from water supplies. The suspended matter may consist of large particles which settle out readily. In these cases clarification equipment merely involves the use of settling basins and/or filters. Most often, however, suspended matter in water consists of particles so small that they do not settle out and even pass through filters. The removal of these finely divided or colloidal substances therefore requires the use of coagulants.

44 What is coagulation?

ANSWER: Coagulation is the clumping together of finely divided and colloidal impurities in water into masses which will settle rapidly and/or can be filtered out of the water. Colloidal particles have large surface areas which keep them in suspension; in addition, the particles have negative electric charges which cause them to repel each other and resist adhering together. Coagulation, therefore, involves neutralizing the negative charges and providing a nucleus for the suspended particles to adhere to.

45 What various types of coagulants are used?

ANSWER: The most common coagulants are iron and aluminum salts such as ferric sulfate, ferric chloride, aluminum sulfate (alum), and sodium aluminate. Ferric and alumina ions each have three positive charges; therefore their effectiveness is related to their ability to react with the negatively charged colloidal particles. With proper use these coagulants form a floc in the water which serves as a kind of net for collecting suspended matter. In recent years synthetic materials called polyelectrolytes have been developed for coagulation purposes. These consist of long, chainlike molecules with positive charges. In some cases organic polymers and special types of clay are used in the coagulation process to serve as coagulant aids. This aids coagulation in making the floc heavier, causing it to settle out more rapidly.

46   What is chemical precipitation?

ANSWER: In precipitation processes, the chemicals added react with dissolved minerals in the water to produce a relatively insoluble reaction product. Precipitation methods are used in reducing dissolved hardness, alkalinity, and in some cases silica. The most common example of chemical precipitation in water treatment is lime-soda softening.

47   What is ion exchange?

ANSWER: When minerals dissolve in water, they form electrically charged particles called ions. Calcium carbonate, for example, forms a calcium ion with plus charges (a cation) and a carbonate ion with negative charges (an anion).

Certain natural and synthetic materials have the ability to remove mineral ions from water in exchange for others. For example, in passing water through a simple cation-exchange softener, all the calcium and magnesium ions are removed and replaced with sodium ions. Ion-exchange materials usually are provided in the form of small beads or crystals which compose a bed several feet deep through which the water is passed.

48   What is demineralization?

ANSWER: This involves passing water through both cation- and anion-exchange materials. The cation-exchange process is operated on the hydrogen cycle. That is, hydrogen is substituted for all the cations. The anion exchanger operates on the hydroxide cycle which replaces hydroxide for all the anions. The final effluent from this process consists essentially of hydrogen ions and hydroxide ions, or water.

The demineralization process may take any of several forms. In the mixed-bed process, the anion- and cation-exchange materials are intimately mixed in one unit. Multibed arrangements may consist of various combinations of cation-exchange beds, weak- and strong-based anion-exchange beds, and degasifiers.

49   How does lime react in the softening process?

ANSWER: Hydrated lime (calcium hydroxide) reacts with soluble calcium and magnesium bicarbonates to form insoluble precipitates. This is shown by the following equations:

$$Ca(OH)_2 + Ca(HCO_3)_2 \rightarrow 2CaCO_3 + 2H_2O$$

Lime   Calcium   Calcium   Water
              bicarbonate   carbonate

$$2Ca(OH)_2 + Mg(HCO_3)_2 \rightarrow Mg(OH)_2 + 2CaCO_3 + 2H_2O$$

Lime   Magnesium   Magnesium   Calcium   Water
              bicarbonate   hydroxide   carbonate

Most of the calcium carbonate and magnesium hydroxide comes out of solution as a sludge and can be removed by settling and filtration. Lime, therefore, can be used to reduce hardness present in the bicarbonate form (temporary hardness) as well as decrease the amount of bicarbonate alkalinity in a water. Lime reacts with magnesium sulfate and chloride and precipitates magnesium hydroxide, but in this process soluble calcium sulfate and chloride are formed. Lime is not effective in removing calcium sulfates and chlorides.

50 How does soda ash react in the softening process?

ANSWER: Soda ash is used primarily to reduce nonbicarbonate hardness (also called sulfate hardness or permanent hardness). It reacts as follows:

$$Na_2CO_3 + CaSO_4 \rightarrow CaCO_3 + Na_2SO_4$$

Soda ash   Calcium   Calcium   Sodium
             sulfate   carbonate   sulfate

$$Na_2CO_3 + CaCl_2 \rightarrow CaCO_3 + 2NaCl$$

Soda ash   Calcium   Calcium   Sodium
             chloride   carbonate   chloride

The calcium carbonate formed by the reaction tends to come out of solution as a sludge. The sodium sulfate and chloride formed are highly soluble and non-scale-forming.

51 What is the purpose of blowdown, and distinguish the difference between surface blowdown and intermittent or bottom blowdown?

ANSWER: Blowdown is for the purpose of removing precipitated solids, dirt, mud, sludge, and other undesirable materials from the boiler water or a vessel that treats feedwater externally from the boiler. Surface blowdown is used on larger boilers to "skim" sludge or scale off the top of the boiler water. Continuously removing a small stream of water from the top level of water in the boiler keeps the concentration of impurities within the boiler at a constant level.

Bottom blowdown at intermittent time periods is still required to remove sludge and deposits from the bottom of shells, drums, or mud drums. Bottom blowoff is performed manually by the boiler operator, and the frequency will depend on the boiler system in place, such as type, size, pressure, and steam usage. It may also be influenced by the type of water treatment being used. For industrial boilers, blowdown is usually on a 4 to 8 hour basis during boiler idleness or low steaming periods.

52   What is turbidity as applied to boiler water?

ANSWER: Turbidity is the term used to describe sediments in water that are coarse particles that quickly settle out of standing water. It is expressed in ppm. The ppm is reduced to acceptable levels by filtration or by natural sedimentation, such as in lakes or ponds. Potable water, for example, typically requires a maximum turbidity of 10 ppm.

53   How is water treatment determined?

ANSWER: This requires trained water treatment specialists, especially since water quality for industrial, power generation, and process use today requires establishing stringent limits on concentration, acidity, oxygen level, and similar water chemistry criteria developed with advancing technology. Water treatment specialists start with water testing, determine treatment methods to obtain desired results, establish testing controls to note if treatment is effective for the desired quality of water, make adjustments in treatment as testing shows a need, or if conditions change in the water supply, or as a result of process problems. With regard to boiler water treatment, internal inspection is a vital part of checking on the effectiveness of the water treatment.

54   How is membrane technology being applied to boiler water treatment?

ANSWER: Membrane technology is being applied as an adjunct to ion exchangers. This permits longer operational runs. Membrane treatment consists of passing pressurized water with impurities through very fine thin film composites, such as cellulose acetate. When the influent stream passes through this filter media, it leaves particulate matter behind as the purer water passes through. For plants with ion-exchange demineralizers, addition of membrane units upstream is considered cost effective for processing feedwater with total dissolved solids down to about 300 ppm. See Fig. 12.22.

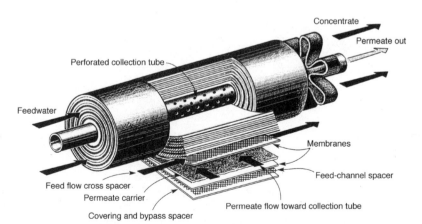

Figure 12.22   A parallel flow, spiral-wound composite membrane has permeate (treated water) come out of a central tube. This arrangement is called cross-flow, as treated water percolates to center tube at right angle to effluent flow. (*Courtesy Power magazine.*)

55   What is reverse osmosis?

ANSWER: This term is often applied to the membrane process of water demineralizing. The physical phenomena of *osmosis* reveals that if pure water is separated from a solution with impurities by a partition such as a membrane with no pressure applied, then the pure water will slowly flow through the partition into the solution containing the impurities.

In *reverse osmosis,* pressure is applied on the solution side with the impurities so that the flow of purer water through the membrane is reversed, hence the term *reverse osmosis.*

13

Boiler Water Problems and Treatment

Standard water testing and treatment specification is more and more being performed by water treatment specialty companies working with boiler plant operators. There is a complementary role for both in solving boiler water effect problems so that the boiler can be operated safely, efficiently, and continuously as needed.

Boiler water problems can cause scale, corrosion, priming and foaming, carryover, stress-corrosion, and embrittlement in the boiler or connected steam-using machinery, such as steam turbogenerators. Many steam turbine blade failures can be traced to steam conditions. As a result, there has been a continuous tightening of permissible concentrations of impurities in steam going to the steam turbine, and this has caused a general improvement in water chemistry as applied to boilers. There are also technological changes being made as research develops new and better chemicals for eliminating the objectionable concentrations of impurities in boiler water. This means operators must be alert to the various methods that are now or will become available in solving boiler water problems.

Basic Water Chemical Characteristics

Definitions applicable to water chemistry:

1. *Acid:* A compound that yields hydrogen ions, such as sulfuric acid.

2. *Alkali:* A substance or salt that will neutralize an acid.

3. *Base:* A compound which can react with acids to form a salt and in water solutions yields hydroxyl ions.

4. *Colloidal:*   A gelatinlike substance which appears to be in a dissolved state but is actually in suspension. It is made of very tiny particles.

5. *Concentration:*   This is applicable to solutions and expresses the ratio of the dissolved substance to the amount of water, usually expressed in a weight ratio.

6. *Corrosion:*   The chemical action of a metal being combined to form an oxide of the metal by the action of oxygen, an acid, or alkali on the metal.

7. *Grain:*   A unit of weight as used in water treatment, with 7000 gr equaling 1 lb.

8. *Hardness:*   A measurement of the amounts of calcium and magnesium compounds in solution in water because the chemical action of these compounds forms an insoluble product, or scale, in the water.

9. *Hydrogen ion:*   An ion formed from hydrogen with the symbol H^-, which forms an acid when combining with certain impurities in water.

10. *Hydroxyl ion:*   This has the symbol OH^- and forms a base compound when combined with some impurities in water.

11. *Ionization of water:*   The breaking up of a molecule of water into hydrogen ions and hydroxyl ions with the reaction proceeding at a higher rate at higher water temperature.

12. *Oxide:*   The chemical combination of oxygen with a metal to form the metal oxide, such as iron oxide, commonly called rust.

13. *Parts per million, or ppm,* is a measure of the amount of impurities on a weight basis to one million pounds of water. Also used: *grains per gallon* = 17.1 ppm; *milligrams per liter* or mg/L.

14. *pH:*   A measurement of the hydrogen-ion concentration in order to measure the relative acidity or alkalinity of a solution. It is the logarithm to the base 10 of the reciprocal of the hydrogen-ion concentration, with 7 being neutral. Below 7, the solution is acidic; above 7, it is base.

15. *Precipitate:*   The solid substance that is separated from a solution by chemical reaction in the solution.

16. *Reagent:*   A substance of a known strength which is used for the detection and measurement of another unknown substance.

17. *Salts:*   Those substances which in solution yield ions other than hydrogen or hydroxyl. They are also the product of an acid and

base combining or an acid and certain metals chemically reacting.

18. *Scale:*   An adherent deposit on metal surfaces in a boiler, which is caused primarily by impurities precipitating out of the water and cementing on the metal as temperatures rise in the boiler.

19. *Soluble:*   The ability of a substance to go into solution by dissolving.

20. *Condensate:*   Steam that is condensed in the steam loop and is returned to the boiler system.

21. *Makeup water:*   Water that must be replenished in the boiler system as a result of leakage, blowdown, and steam process use.

22. *Feedwater:*   The combination of condensate and makeup water that is supplied to the boiler for evaporation.

23. *Blowdown:*   The bleeding of a portion of the water in the boiler in order to remove suspended solids.

24. *Condensate polishing:*   The purification of returned condensate by passing it through demineralizers.

The treatment of boiler water is a problem requiring periodic testing of the water and proportioning the treatment according to the varying conditions. There are a number of reputable laboratories prepared to equip small or large plants with suitable test kits and to supply or advise the proper treatment indicated by the tests.

Chemical tests.   Minimum chemical tests usually prescribed for high-pressure boilers are dependent on ratio of makeup water to condensate making up the boiler feedwater as well as with the treatment used or specified by water-treatment specialists. For information purposes, here are some tests and their purpose:

1. *Acidity or alkalinity test:*   This is used for controlling corrosion and also scale by using the values obtained in calculating the amount of alkali to be added to an acidic raw water, or the quantity of lime and soda that may be needed in a lime-soda water softener.

2. *Testing for hardness, calcium, and magnesium:*   A measurement of calcium and magnesium is a measure of the hardness of raw and softened waters and feedwaters. Hardness produces scale in a boiler, therefore the values obtained for calcium and magnesium can be used to determine the quantity of lime and soda ash that needs to be added to the boiler water, and thus control scale formation.

3. *Testing for hydroxide:* The amount of hydroxide in boiler water is determined in order to control corrosion, embrittlement, carryover, and indirectly, scale control. Hydroxide must be kept at a low enough level so that no carryover takes place by foaming, and to prevent stress concentration points attack of the steel, and to avoid embrittlement of steel in dead pockets of circulation. Hydroxide concentrations are also used to convert hardness which would form hard scale, to sludge that can be blown out of the boiler water.

4. *Testing for phosphate:* Phosphate concentration is controlled in order to produce soluble scale that can be blown out of the boiler. Phosphate concentration is also maintained so that a relationship exists between phosphate and pH or alkalinity in the boiler water so that no free hydroxide is present, and thus avoid embrittlement.

5. *Testing for sulfite:* Sulfite concentration, if slightly excess, will combine with the dissolved oxygen in the water, and thus prevent corrosion. Sulfite treatment is not recommended for use in boilers with drum pressures above 1600 psig, because the chemical reactions become harmful at the higher pressure.

6. *Testing for iron:* This test is to determine if the condensate return has excess iron oxide, or rust from connected piping and steam-using machinery. The term *solid particle erosion* has come into use, because most of the iron is in particle form, and not dissolved in the water. Membrane filters are used to approximate the concentration in the water.

7. *Testing for copper:* Similar effect as iron, but the source is usually heat exchangers or pumping equipment with copper parts. Repairs that replace the copper can reduce the source of this contaminant.

8. *Electrical conductivity test:* This test determines the amount of solids present in the water, and is used for blowdown control. If used on condensate returns, it can be used to detect raw water leakage from condensers and heat exchangers into the condensate, thus instituting corrective actions before serious damage occurs.

There are five possible steps needed in water treatment depending on the supply, pressure, extent of makeup, and similar conditions: (1) pretreatment of raw-water supply; (2) treatment of makeup water going to the boiler; (3) internal treatment of the water in the boiler; (4) treatment of the condensate being returned to the boiler; and (5) blowdown control to remove precipitated sludge from the boiler.

Analyzing a water sample is the process of finding out how much of the various impurities and other chemical substances is present in the water. The results are usually expressed in parts per million

REPORT OF WATER ANALYSIS

	Parts per million	Equivalents per million
Silica as SiO_2	5	
Iron as Fe_2O_3	1.2	
Calcium as Ca	62	
Magnesium as Mg	31	
Sodium and potassium as Na	38	
Bicarbonate as HCO_3	250	
Carbonate as CO_3	0	
Hydroxide as OH	0	
Chloride as Cl	11	
Sulfate as SO_4	138	
Nitrate as NO_3	0	
Carbon dioxide as CO_2	10	
Turbidity	5	
Physical characteristics of sample	Clear when drawn	

Date _____

Source _____

Date analyzed _____

Total dissolved mineral solids _____ ppm

Organic matter none ppm

Suspended solids 5 ppm

Chloroform, extractable (oil, etc.) none ppm

pH 7.7

Phenolphtholein alkalinity as $CaCO_3$ 0 ppm

Methyl orange alkalinity as $CaCO_3$ 205 ppm

Hydroxide alkalinity as $CaCO_3$ 0 ppm

Hardness as $CaCO_3$ 282 ppm

Specific conductance _____ micromhos

Figure 13.1 Water treatment requires periodic water analysis of boiler water, with a listing of impurity data so that treatment can be determined or adjusted.

(ppm) and tabulated as shown in Fig. 13.1. Parts per million is a measure of proportion by weight, such as one pound in a million pounds. Grains per gallon is another way of expressing the amount of a substance present. One grain per gallon equals 17.1 ppm.

No matter what the chemical characteristics of the impurity may be, four states are possible:

1. If the impurity is a soluble solid, it appears in a dissolved state or in *solution* with the water.

2. If the solid is not soluble in water, it is not in solution, but in a state of *suspension.*

3. Those impurities of a gaseous nature that are partially soluble are in an *absorbed* state in the water.

4. *Colloidal* solutions have particles in suspension between those in a dissolved state and those in suspension. Colloids are defined as being smaller than 0.2 micrometers (μm) and larger than 0.001 μm (1 micron = 1 μm = 0.001 mm = 10^{-6} meters). Particles smaller than 0.001 microns are considered in solution.

Chemical Elements and Reaction Equations

Many water chemistry problems involve the following commonly found chemical elements in water chemistry.

Element and symbol	Atomic weight	Valence
Aluminum, Al	27.0	3
Calcium, Ca	40.1	2
Carbon, C	12.0	4
Chlorine, Cl	35.5	Varies
Hydrogen, H	1.0	1
Iron, Fe	55.8	ferrous 2
		ferric 3
Magnesium, Mg	24.3	2
Nitrogen, N	14.0	Varies
Oxygen, O	16.0	2
Phosphorus, P	31.0	Varies
Potassium, K	39.1	1
Silicon, Si	28.1	4
Sodium, Na	23.0	1
Sulfur, S	32.1	Varies

Valence is defined by chemists as the number of an element that represents the capacity of its atomic weight to combine with, or displace, the atomic weights of other elements. It is further defined as the valence of an element as being the number of atoms of hydrogen or of chlorine that the atom of the element can combine with or displace. Hydrogen is taken as the standard with a valence of one. Acids behave as if composed of two radicals, the hydrogen and the radical such as SO_4, NO_3, PO_4 for sulfuric, nitric, and phosphoric acid, respectively. The formula for sulfuric acid is H_2SO_4; therefore, the radical SO_4 has a valence of two since it takes two hydrogen atoms to form with the radical. For nitric acid, HNO_3, the radical NO_3 has a valence of one since it combines with one atom of hydrogen to form HNO_3. The formula for phosphoric acid is H_3PO_4; therefore, the PO_4 radical has a valence of three. Calcium has a valence of two and combines with the radical CO_3 to form scale, $CaCO_3$; therefore, the radical CO_3 has a valence of two. The best way to remember the valence of elements is to remember the formula of a compound containing that element. For example, oxygen has a valence of two. What is the valence for silicon in the compound SiO_2? Since the total valence of the two oxygen atoms is $2 \times 2 = 4$, then that is the valence of silicon.

Many elements exhibit more than one valence. Thus, iron can form ferrous chloride, $FeCl_2$, and would have a valence of two since chlorine has a valence of one. For ferric chloride, $FeCl_3$, the valence of iron would be three.

Valence is used in the calculation of ppm of ions in terms of calcium carbonate in a water sample undergoing analysis. Chemists use this to calculate the amounts of reacting chemicals in a given equation by using *equivalent weights*. Calculating the concentration of a given ion in terms of its calcium carbonate equivalent is done by comparing equivalent weights, where equivalent weight = atomic weight of element or compound divided by the valence of that element or compound.

Chemical Reactions

Chemical reactions of chemical compounds with each other are shown by chemical equations where the elements on one side of the equation must equal the number for that element on the other side of the equation. The equation for reducing the calcium and magnesium bicarbonate, "temporary hardness," to calcium carbonate sludge is expressed as follows in a chemical reaction equation, with lime being used to precipitate out the insoluble calcium carbonate.

$$Ca(OH) + Ca(HCO_3)_2 = 2CaCO_3 + 2H_2O$$

Lime   + calcium   = calcium + water
             bicarbonate   carbonate

Note there are six Ca elements on the left side of the equation and six on the right side. The elements must always balance out on both sides. The equation for converting magnesium bicarbonate is

$$2Ca(OH)_2 + Mg(HCO_3)_2 = Mg(OH)_2 + 2CaCO_3 + 2H_2O$$

Lime   + magnesium = magnesium + calcium   + water
            bicarbonate   hydroxide   carbonate

In changing the permanent hardness of calcium sulfate and calcium chloride by soda ash, the following chemical reaction takes place.

$$Na_2CO_3 + CaSO_4 = CaCO_3 + Na_2SO_4$$

Soda ash + calcium = calcium + sodium sulfate
            sulfate   carbonate

For the calcium chloride reaction,

$$Na_2CO_3 + CaCl_2 = CaCO_3 + 2NaCl$$

Soda ash + calcium = calcium + sodium chloride
chloride   carbonate

In both equations, the number of elements on each side balance. Some chemists use the symbol ↓ for sludge precipitating like the calcium carbonate coming out of solution, and the symbol ↑ for gases that may come out of solution.

Another way water treatment chemists use chemical equations is by applying the total of atomic weights per compound in the reaction. Let us assign the atomic weights in the equation for converting calcium chloride by the use of soda ash, and the equation can be written as follows:

$$Na_2CO_3 + CaCl_2 = CaCO_3 + 2NaCl$$
$$106 \quad + 111.1 = 100.1 \quad + 117$$

The numbers represent the sum of the atomic weights of the elements in that compound. Chemists use the atomic weights to calculate the weights of reactions by proportioning. For example, let us assume a check showed 150 lb of soda ash, Na_2CO_3, were used to treat water with calcium chloride, $CaCl_2$. How much of the other compounds would be formed or used?

1. For $CaCl_2$, $106/111.1 = 150/CaCl_2$ and solving, $CaCl_2 = 157.2$ lb
2. For $CaCO_3$, $106/100.1 = 150/CaCO_3$ and solving, $CaCO_3 = 141.7$ lb
3. For $2NaCl$, $106/117 = 150/(2NaCl)$ and solving, $2NaCl = 165.6$ lb

The weights on one side of the equation equal the weights on the other side of the equation.

Calcium and magnesium bicarbonates shown in the chemical equations produce a *temporary* hardness. The term "temporary" is used because carbonate hardness may be partially removed as water temperature climbs. This can be shown by the following equation:

$$Ca(HCO_3)_2 + Heat = CaCO_3 + CO_2\uparrow + H_2O$$
$$\text{Gas}$$

Water with temporary hardness when cold prevents soap from lathering, but after being heated, it allows the soap to lather.

In *high-pressure boilers,* soda ash reacts differently.

It has been shown that soda ash, Na_2CO_3 assists in removing hardness from water containing calcium sulfate and calcium chloride; however, above about 250 psi boiler pressure, Na_2CO_3 decomposes to caustic soda with the liberation of carbon dioxide as shown by the following equation:

$$Na_2CO_3 + H_2O = 2NaOH + CO_2$$

The carbon dioxide may be absorbed by the water to form a weak carbonic acid, H_2CO_3, as shown by the following equation:

$$CO_2 + H_2O = H_2CO_3$$

A chemical chain reaction can follow per the following reaction equations.

The acid breaks down to the following ions:

$$H_2CO_3 = H_2^+ + CO_3^-$$

The hydrogen ions can act with carbonate ions from any calcium or magnesium carbonates to form the bicarbonate form as shown for calcium bicarbonate by the equation

$$H^+ + CO_3^{--} = HCO_3^-$$

The end result is the formation of calcium bicarbonate again as follows:

$$CaCO_3 + CO_2 + H_2O = Ca(HCO_3)_2$$

Phosphate treatment. This results in "temporary hardness" again. The action of carbon dioxide causing this temporary hardness when calcium and magnesium carbonate are present has restricted the use of soda ash treatment to lower pressures. This treatment was replaced by the phosphate treatment with an established excess of the chemical used in order to have present at all times a sufficient amount to ensure the completion of the chemical reaction. For illustrative purposes, the reaction equation for the most objectionable scale former, calcium sulfate with trisodium phosphate is

$$3CaSO_4 + 2Na_3PO_4 = Ca_3(PO_4)_2 + 3Na_2SO_4$$

phosphate soft sludge   soluble salt

The use of phosphates for internal chemical treatment of boilers has been modified as boiler pressures increased; a discussion of this first requires a review of acidity and alkalinity.

Acids, bases, and pH value. Water treatment chemists refer to the pH value of a sample of water being tested. This term is used to denote the presence of acids or bases in water or solutions. For example, hydrochloric acid, HCl, also called muriatic acid, forms electrically charged atoms, called ionization, in water to form H^+ and Cl^- ions. In the same manner, a base or alkali, such as sodium hydroxide, NaOH, called caustic soda, will form Na^+ and OH^- ions with the hydroxide ion, OH^-, the alkaline reaction agent.

Chemists consider the reaction of a base and acid as a neutralizing action by the action of $H^+ + OH^- = H_2O$ forming neutral water as follows:

$$NaOH \quad + HCl \quad\quad = NaCl + H_2O$$

Caustic + muriatic = table + water
soda   acid   salt

Both the acid power and the alkaline power have been eliminated and a salt and water formed that is neutral. When there are impurities in water, this balancing out of acid and base may be disturbed as may be the concentration of each in the solution. Thus the water or solution may be acidic or alkaline. Chemists use the term "pH value" as a reference for acidity or alkalinity. It actually expresses the concentration of H^+, or hydrogen ions, present and also the amount of OH^-, or hydroxide ions present, because for water at any one temperature, the concentration of hydrogen ions times the concentration of hydroxide ions equal a constant. This means a change in the concentration of one requires a change in the concentration of the other. Chemists have determined that the concentration of pure water (neutral) is 0.0000001 grams per liter $= 1/10^7$ grams per liter. To avoid the use of large decimal fractions, the method was adopted that 7 is a neutral water, below 7 it is acidic, and above 7 it is alkaline.

Solubility. In water chemistry, the majority of impurities are found to be in the dissolved state in boiler water or in solution. However, temperature has an effect on solubility so that in some cases a slight change in the solution temperature has the immediate effect of causing a dissolved substance to become insoluble and settle out as suspended matter.

Some water treatment chemists classify a substance that becomes more soluble with *an increase* in temperature as having a *positive solubility* ratio, while those substances that become *less soluble* with an increase in temperature are classified as having a *negative solubility* ratio. This can be shown as follows:

Substance	Formula	Effect on solution, temperature increase
Sodium chloride	NaCl	Positive solubility, increases
Calcium carbonate	$CaCO_3$	Positive solubility, increases
Calcium sulfate	$CaSO_4$	Negative solubility, decreases
Calcium hydroxide	$Ca(OH)_2$	Negative solubility, decreases
Magnesium hydroxide	$Mg(OH)_2$	Positive solubility, increases

The degree of solubility is also important in determining when a certain concentration of the impurity in the dissolved state may drop out of solution as a solid because the water has become saturated with the impurity and can no longer dissolve the impurity.

Those impurities having negative solubility characteristics are the chief producers of scale deposits in boilers. This is because as the temperature increases in the boiler with pressure rise, these impurities become insoluble and precipitate out to start scaling heat transfer surfaces of the boiler. The other problem with impurities is that even if they remain dissolved through wide ranges of temperature and concentrations, they may produce other detrimental effects, such as corrosion, foaming, and priming.

Scale. Water treatment specialists classify impurities by the effect they produce on the water side of boilers. *Scale* is defined as an adherent deposit on the heat-transfer surface on the water side of boilers caused by the following impurities settling or being baked on heating surfaces:

Hard scale	Soft scale	Scale or corrosion former
Calcium sulfate	Calcium bicarbonate	Calcium nitrate
Calcium silicate	Calcium carbonate	Calcium chloride
Magnesium silicate	Calcium hydroxide	Magnesium chloride
Silica	Magnesium bicarbonate	Magnesium sulfate
	Magnesium carbonate	Magnesium nitrate
	Magnesium hydroxide	Alumina
	Calcium phosphate	Sodium silicate
	Iron carbonate	
	Iron oxide	

Substances with various effects on heat-transfer surfaces are oils and greases and suspended matter.

The third group of impurities may not cause scale by themselves, but may be bound to other scale formers, thus appearing in the composition of the scale. Most are more active in producing corrosion than scale.

Scale and the effects of scale.   Chemically, before any dissolved impurity can produce scale, it must leave the solution and become a solid in the following ways:

1. By the reduction of solubility with increase in the temperature of the water with those impurities classified as having negative solubility.
2. By exceeding the saturation point so that the water can no longer hold the dissolved impurity in this state, and it precipitates out of the solution.
3. By chemical changes with heat to form insoluble substances, such as the bicarbonates of calcium and magnesium.

Thus, scale may involve several chemical changes that can occur in boiler water.

There are two definite objections to scale on boiler heating surfaces: (1) Scale is a very efficient *nonconductor* of heat, the degree of nonconduction varying somewhat with its density. Its presence in appreciable thickness means less heat absorption by the boiler water, with consequent loss of boiler efficiency. (2) Because scale is a poor heat conductor, the heating surfaces thus insulated from boiler water on one side and exposed to hot gases on the other may soon reach a dangerously high temperature. Serious damage—rupture of tubes (Fig. 13.2) and even boiler shells—has resulted.

Scale formation often increases with the rate of evaporation. Thus, scale deposits will often be heavier where the gas temperatures are highest. As an example, quite often a tube failure in a high-pressure watertube boiler can be traced to the high heat absorption zone where steam bubbles form. The water envelope surrounding the bubble now contains the impurity from the steam bubble *and* the impurities it already had in solution. The area beneath the steam bubble of the tube is momentarily dry, and this causes the tube temperature to rise. For negative solubility substances, the solubility of that compound is now at a low level, and this causes the compound to scale out onto the tube surfaces, because the solution has reached its saturation point at that tube temperature. The deposits can build up and act as an insulator so that poor heat transfer results. This can cause the tube metal to overheat underneath the scale and cause tube ruptures as shown in Fig. 13.2. The average scale has a thermal conductivity about the same as firebrick, or about $\frac{1}{48}$ that of steel.

Scale is usually more serious in a watertube boiler than in a firetube boiler. A coating of scale $\frac{1}{16}$ in. thick on water tubes exposed to radiant heat may cause tube failure, whereas much heavier deposits of scale on fire tubes cause loss in efficiency but may not be dangerous. The

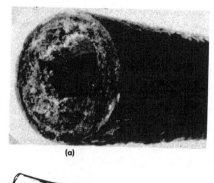

(a)

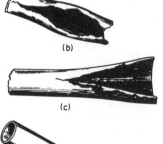

(b)

(c)

(d)

Figure 13.2 Tube scale and ruptures. (a) Excessive phosphate deposits; (b) deposits and chemical-attack-ruptured tube; (c) long-term overheating rupture; (d) short-term overheating damage, shown by knife-edges of fractured surface.

reason for this difference is that fire tubes absorb their heat by convection of gases, and not by radiant exposure.

Scale in firetube boilers also can affect tubes, but generally the tubes sag from overheating. However, furnaces of scotch marine boilers that absorb radiant heat have bulged severely from scale overheating as have bottom shells of HRT boilers. Scale formation can also block the proper operation of controls and even safety valves by plugging connections or causing the valve or control to stick. A primary cause of safety-valve failures is the buildup of deposits around the seat, which cause the valve to stick in the closed position. A regular program of valve testing will prevent this type of failure as will improvement in water and steam purity.

Heavy scale deposits are usually an indication of neglect, for scale can be prevented in most cases by proper treatment of the water. Where scale has formed to an appreciable thickness, it should be removed, and once a clean boiler is attained, proper steps should be taken to prevent its recurrence.

Identifying scale. Scales of different consistency may appear in a boiler and may require further chemical analysis for identifying the impurity. The scales described here are to show the characteristics of the different scales that may be found in boilers.

1. *Calcium sulfate* causes a very hard scale that adheres tenaciously to heating surfaces. This scale is considered the most objectionable because of its extreme hardness, difficulty of removal, and low heat conductivity.

2. *Calcium carbonate* is a soft, muddy-type scale, usually chalky in appearance, and is easily removed by water washing.

3. *Magnesium carbonate* forms a soft, sludgy scale similar to calcium carbonate.

4. *Silica* does not form scale alone but imparts a glassy structure to the calcium sulfate deposits, which produces a very hard scale, very brittle and practically insoluble in acid. Silica, in high pressure boilers of the central electric generating type, volatizes and travels with the steam to the steam turbogenerator to deposit out as a hard scale, resembling porcelain on turbine internal parts.

5. *Calcium and magnesium silicate* both tend to produce scales of dense crystalline structure, very adherent to heat-transfer surfaces and with low-heat transfer characteristics.

6. *Calcium and magnesium hydroxide* produce soft deposits, which can adhere or cement to other substances.

7. *Iron carbonate* is often found in other scales and is an undesirable substance because it adds a corrosive nature to the scale in which it appears.

8. *Calcium and magnesium phosphate* are by-products of phosphate water treatments and appear as a soft mud that is easily removed by blowdown.

9. *Magnesium sulfate* is not too common in scale, but where it appears alone, the scale is comparatively soft. However, in combination with calcium carbonate or calcium sulfate, a very hard flinty incrustation may result.

Preventing scale formation. The following strategies are generally employed in preventing scale formation:

1. Prevent the use of *hard water* in the boiler. This hardness property is primarily from the presence of the salts of calcium and magnesium. In water analysis, hardness conditions of water may be reported as carbonate, noncarbonate, and total hardness. Units of hardness are expressed as equivalent ppm of calcium carbonate. (See Fig. 13.1.)

Water treatment chemists report hardness as though all of it were due to calcium carbonate alone. This is done by using equivalent atomic weights of the elements involved in the chemical equation. As an example, consult the readings in the Report of Water Analysis, Fig. 13.1. This shows magnesium as 31 ppm and calcium as 62 ppm. The total hardness as calcium carbonate is obtained as follows. The atomic weight to the nearest tenths of the elements involved are calcium (Ca) = 40.1, magnesium (Mg) = 24.3, carbon (C) = 12.0, and oxygen (O) = 16.0. Thus calcium carbonate $(CaCO_3) = 40.1 + 12 + 3(16) = 100.1$.

Calculating the concentration of a given ion in terms of its calcium carbonate equivalent is done by comparing the equivalent weights of the two. The equivalent weight of an element is equal to atomic weight/valence. This term refers to the amount of an element combining with a unit weight of hydrogen. Since most hardness is due to calcium and magnesium mineral matter, the water analysis report in Fig. 13.1 refers to the total hardness as calcium carbonate in terms of the calcium and magnesium hardness. This hardness is, respectively, 62 ppm for calcium and 31 ppm for magnesium. To convert to total hardness as calcium carbonate, the equivalent weight for calcium is 40.1/2 = 20.05, for magnesium it is 24.3/2 = 12.15. For calcium carbonate it is 100.1/2 = 50.05. From this the ratio of 50.05/20.05 = 2.50 is the multiplier for converting ppm of calcium to equivalent ppm as CO_3; for magnesium it is 50.05/12.15 = 4.1. The total hardness as CO_3 as shown in Fig. 13.1 is then

$$(62 \times 2.50 + 31 \times 4.1) = 282 \text{ ppm for hardness as } CO_3.$$

2. Scale is also minimized by keeping concentrations of impurities within acceptable limits using treatment and blowdown to control the concentration. See Fig. 13.3 for the recommendations of an ASME Research Committee in Thermal Power Systems that reflects the desirable concentrations of impurities for different pressures on boiler systems.

3. Scale is also minimized by following prescribed external and internal chemical treatment.

4. Bottom blowdown must be integrated with chemical treatment that produces sludge that must then be removed from the boiler water.

Scale prevention—external treatment. As detailed in Chap. 12, pretreatment of water may be necessary because of the variance in supply, makeup water requirements, type of process or application of the boiler system, pressure and capacity, and similar conditions. Thus, external treatment for scale prevention may include reduction of suspended solids by filters, hardness reduction by lime softening, zeolite softening, ion exchange equipment, demineralizers, evaporators and

	BOILER WATER				FEEDWATER[3]			
Boiler Pressure ppm[1]	Total Solids ppm[1]	Total Alk ppm as CaCO₃[1]	Silica ppm as SiO₂[2]		Hardness ppm as CaCo₃	Iron ppm as Fe	Copper ppm as Cu	Oxygen ppm as O₂
0- 300	3500	700	75-50		0-1 Max.	0.10	0.05	0.007
301- 450	3000	600	50-40		0-1 Max.	0.10	0.05	0.007
451- 600	2500	500	45-35		0-1 Max.	0.10	0.05	0.007
601- 750	2000	400	35-25		0-1 Max.	0.05	0.03	0.007
75:- 900	1500	300	20-8		0-1 Max.	0.05	0.03	0.007
901-1000	1250	250	10-5		0-1 Max.	0.05	0.03	0.007
1001-1500	1000	200	5-2		0	0.01	0.005	0.007
1501-2000	750	150	3-0.8		0	0.01	0.005	0.007
2001-2500	500[4]	100[4]	0.4-0.2		0	0.01	0.005	0.007
2501-3000	500[4]	100[4]	0.2-0.1		0	0.01	0.005	0.007

Feedwater organics should be zero and pH in the range of 8.0 to 9.5[3]
References and Notes:
1. American Boiler Manufacturers Assoc. 1958 Manual
2. Above 600 psig silica level selected to produce 0.02 ppm SiO₂ steam

3. Babcock & Wilcox Publications (a) Water Treatment for Industrial Boilers, BR-884, 8-68 and (b) J. A. Lux, Boiler Water Quality Control in High Pressure Steam Power Plants, 9/62
4. J. A. Lux, 3(b), recommends levels as low as 15 ppm TDS above 2000 psig.

Figure 13.3 ASME Research Committee recommended boiler and feedwater impurity limits for drum-type boilers at different operating pressures.

deaerators for oxygen and gas removal. Membrane technology or reverse osmosis is also being applied in external treatment systems. See Fig. 13.4 for external treatment's possible results.

Scale prevention—internal treatment. The selection of chemicals for scale prevention in internal treatment is directed to the control of mineral impurities that slip through the pretreatment program. This is especially applicable as boiler pressures rise, such as for electric power generation. Pretreatment, whether by chemical precipitation

	Average analysis of treated water				
Method of treatment	Hardness ppm as CaCO₃	Alkalinity ppm as CaCO₃	CO₂ in steam (potential)	Dissolved solids	Silica
Cold lime-soda	30–85	40–100	Medium-high	Reduced	Reduced
Hot lime-soda	17–25	35–50	Medium-low	Reduced	Reduced
Hot lime-soda phosphate	1–3	35–50	Medium-low	Reduced	Reduced
Hot lime-zeolite	0–2	20–25	Low	Reduced	Reduced
Sodium-cation exchanger	0–2	Unchanged	Low to high	Unchanged	Unchanged
Anion dealkalizer	0–2	15–35	Low	Unchanged	Unchanged
Split-stream dealkalizer	0–2	10–30	Low	Reduced	Unchanged
Demineralizer	0–2	0–2	0–5 ppm	0–5 ppm	Below 0.15 ppm
Evaporator	0–2	0–2	0–5 ppm	0–5 ppm	Below 0.15 ppm

Figure 13.4 Possible results for different water conditions from the use of various external water treatment methods.

or ion exchange, will reduce, but may not eliminate, the problem-causing scaling tendencies of impurities in boiler water.

Combined-phosphate treatment was developed from the use of phosphate treatment to eliminate the calcium and magnesium sulfate scales, with a sludge developing that can be removed by proper blow-down, especially for boilers operating below 600 psi. The sludge is called hydroxyapatite. The chemical reaction equation water treatment specialists use is the following:

$$10Ca \quad + 2OH \quad + 6PO_4 \quad = Ca_{10}(OH)_2(PO_4)_6$$
Calcium + hydrate + phosphate = hydroxyapatite
hardness   alkalinity

Another reaction is that of magnesium hardness in the presence of adequate silica forming magnesium sludge, called *serpentine* as shown by the following chemical equation:

$$Mg \quad + 2SiO_2 + 2OH \quad = MgO + 2(SiO_2 \cdot H_2O)$$
Magnesium + silica + hydrate = serpentine
hardness   alkalinity

The treatment with only phosphates produces orthophosphates residual between 30 and 60 ppm as PO_4, with hydrate alkalinity levels of 200–400 ppm (as OH).

There are several undesirable features of the conventional phosphate treatment. The reaction adds to the suspended solids content of boiler water, which is not desirable as boiler pressure is near or above 1000 psi. The hydrate alkalinity levels are also considered too high for this boiler pressure, because of the danger of caustic corrosion. The usual limits that water treatment chemists prescribe for alkalinity limits are:

Boiler pressure, psi	Nondeionized water alkalinity limit, ppm	Deionized water alkalinity limit, ppm
0–300	700	350
301–450	600	300
451–600	500	250
601–750	400	200
751–900	300	150
901–1000	200	100
Above 1000	—	—

As boiler pressures rise, there is more chance of caustic corrosion occurring due to the metal being stressed more by the increase in pressure and temperatures. As can be noted above, alkalinity limits decrease with rise in boiler pressure for this reason.

Coordinated-phosphate/pH control. This system of water treatment was developed for the higher pressure boilers in order to avoid caustic corrosion. It requires the maintenance of a fixed relationship between boiler-water pH and phosphate concentration. Figure 13.5a shows such a curve for a utility boiler and represents the relationship of trisodium phosphate, Na_3PO_4, to pH. The trisodium phosphate has a sodium to phosphate ratio of 3. If either the pH or the phosphate concentration changes, this ratio of sodium to phosphate changes. The program was based on the principle that an increase in free-hydroxide concentration would be prevented by a shift in the ionic equilibrium in the direction favoring reformation of the Na_3PO_4. According with the information in Fig. 13.5a, operators were instructed on drum-type boilers to maintain pH and PO_4 concentration below or to the right of the curve, since it was considered devoid of free hydroxide. The shaded area shows the normal operating range for this plant. The curve represents a Na to PO_4 ratio of 3. The water treatment specialist recommended a pH range of 9.6 to 10.0 ppm and a simultaneous phosphate range of 5 to 10 ppm, but at all times staying below the curve shown in Fig. 13.5a. Both pH and phosphate require daily checks for concentration in order to maintain the low hydroxide level.

Hideout and congruent-phosphate/pH control. Hideout is caused by the precipitation of sodium phosphate salts, usually caused by long-term full load operation on utility-type drum boilers. Hideout causes a buildup of phosphate in "dead" pockets of the water circulation and consequently reduces the phosphate concentration in the other parts of the water loop. This increases the pH and alkalinity level of the boiler water. Laboratories worked on this problem of phosphate hideout and found evidence that sodium hydroxide could be produced as a result of hideout from solutions of trisodium orthophosphate at or above a sodium-to-phosphate ratio of 2.8. The term *congruent control* was then applied, which is a reference to the congruent composition in which the liquid and solid phases are the same. Guidelines were established to maintain a sodium-to-phosphate ratio above 2.6 but below 2.8 with PO_4 concentrations between 1 and 6 ppm. See Fig. 13.5b.

In the coordinated phosphate controls, the pH is regulated by introducing sodium, per water treatment specialists' advice, with the phosphate. The desired ratio is maintained by controlling the proportions of trisodium, disodium, and monosodium forms of phosphate in the boiler water. Caustic may be used to raise alkalinity and pH and blowdown to reduce them. Typical utility boilers employing congruent control operate above 1200 psi.

Chelants are chemicals that combine with hardness salts before they form boiler sludge, and this is another method to prevent scal-

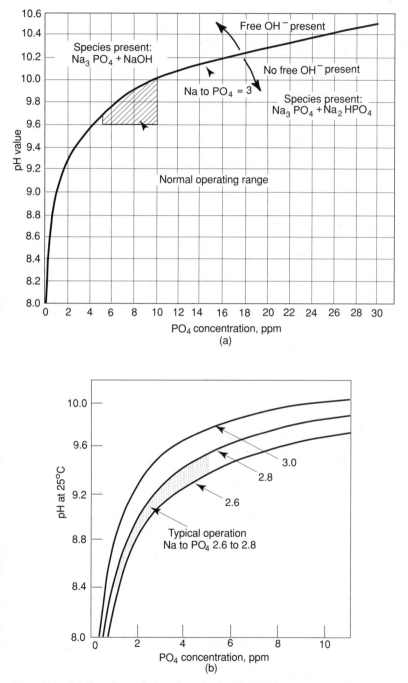

Figure 13.5 (a) Coordinated-phosphate and pH control maintains a fixed relation between boiler water pH and phosphate concentration. The Na to PO_4 ratio is maintained at 3. (b) In congruent-phosphate and pH control, the Na to PO_4 ratio is maintained between 2.6 and 2.8.

ing. However, feedwater must have low hardness, of less than 2 ppm, for this treatment to be economical. This makes its use limited to softened or demineralized makeup water. The two commonly used chelating agents are EDTA (ethylenediaminetetraacetic acid), and NTA (nitrilotriacetic acid). Both agents form stable salts with calcium and magnesium. However, it requires 10 ppm of EDTA and 5 ppm of NTA to tie up each 1 ppm of hardness. This makes the treatment expensive, so it is mostly applied to low hardness pretreated feedwater. Another problem is that EDTA starts to break down at 300 psi, and at about 1500 psi it loses its ability to chelate. NTA loses its chelating ability at about 900 psi. Thermal degradation makes it impractical to monitor the free chelant residuals in boiler water, which makes control of dosages difficult. Chelants themselves can cause corrosion in the boiler if overfeed of the chelant occurs for any lengthy period of time.

Below 400 psi, another group of chelating-type chemicals are used, termed *organic phophonates* with the abbreviated terms AMP (aminomethylene phosphonate), and HEDP (hydroxethylidene), which have the property of inhibiting calcium carbonate scale. All solubilizing agents show capability similar to EDTA and NTA for keeping calcium and iron in solution, but are not as strong, and therefore do not pose as great a threat as chelants for causing corrosion attack. The phosphonates can be used more economically than the chelants at feedwater levels of 50 ppm and higher.

Polymer or sludge conditioning.   In industrial boilers, the carbonate-cycle of scale control involves the intentional precipitation of calcium hardness salts as calcium carbonate with polymer additions to provide a conditioned sludge. Anionic polymers are widely used in industrial boilers where the polymer molecules wrap themselves around suspended boiler sludge. This introduces to the sludge a degree of dispersancy, or fluidity, which permits easier removal of sludge by bottom blowdown. There are various synthetic polymers sold by water treatment companies. For example, Nalco Chemical Co. uses the name Transport-plus for its polymers. It is applied to boilers up to 1500 psi, and the term "transport" is used to indicate that it can transport virtually 100 percent of feedwater impurities, including hardness, silica and iron impurities, through the boiler system because the polymer makes the sludge more fluid for eventual blowdown control. The anionic polymers inhibit growth of the crystal-lattice structure of scale. This process also weakens the scale as the polymer becomes absorbed in the scale structure, and smaller particles of scale are formed as a result.

Control of the feed rate and polymer depends on the test methods used to check on concentrations. Overtreatment is still a threat;

therefore, some water treatment specialists use blends of polyacrylates and polymethacrylates as the major components of a blended polymer treatment program.

Oil deposits

Oil in boilers is a dangerous condition. Oil is an excellent heat insulator, and its presence on heating surfaces exposed to high temperatures may cause serious overheating and damage to the boiler.

A common cause of this condition was the use of reciprocating steam-equipment exhaust containing cylinder oil for condensate return to the boiler feed system. Also, fuel-oil heating equipment may leak oil into the steam system and cause this difficulty if the condensate is returned to the boiler. A minimum amount of high-grade properly compounded cylinder oil should be used for lubrication of steam engines and pumps where condensate is returned, and an efficient type of oil separator should be used in the exhaust system. Oil may also enter the feed through its presence in such raw-water supplies as rivers and streams contaminated by mill, marine, or trade wastes.

Scale and oil removal. Water-side *scale removal* is accomplished by one of three methods: mechanical removal, water treatment, and acid cleaning. Mechanical removal of scale is effected while the boiler is idle and empty. The accessible parts of shells, drums, heads, and braces are chipped with a dull chisel or scaling hammer, care being taken not to score the metal. Scale may be ground off the internal surfaces of water tubes with a tube turbine. Water is generally used to wash out ground "scale sludge" while the tube turbine is in operation. Care should be exercised not to operate a tube turbine too long in one place or to force it unduly, for damage to the tube may thus result.

Extreme care should be taken in removing existing scale in a boiler by treatment of the water. If the scale is removed too quickly, it may drop down in large quantities, with serious damage to the boiler being the result because of restricted circulation and overheating. In a watertube boiler, ruptured tubes may be the consequence; in a firetube boiler, bulges and even rupture in the shell have followed.

Scale deposits on external surfaces of fire tubes may be vibrated loose with a tube rattler or by shaking a long, heavy bar in each tube. Extreme care should be taken after such mechanical treatment to see that *all* loosened scale is removed from the boiler before closing it up for operation. Many cases of serious damage have resulted from loose scale accumulations left in boilers. See Fig. 13.6.

Acid cleaning of boilers is often used to remove metallic oxide. The solvents used for acid cleaning are varied. Some use hydrochloric acid;

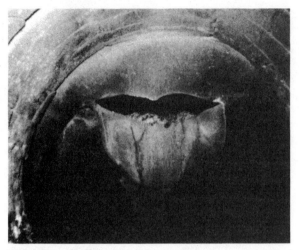

Figure 13.6 Collapsed scotch marine boiler's corrugated furnace caused by scale and dry firing. (*Courtesy Royal Insurance Co.*)

others, phosphoric acid. The usual procedure is to fill the boiler until the solution overflows at the air vent (acid is added outside the boiler). The solution is *allowed* to soak the boiler from 4 to 6 hr, followed by refilling with a neutralizing agent. If hydrochloric acid is used for soaking, a weak solution of phosphoric acid is used. After draining, fresh warm water is used for flushing; then the boiler is immediately filled with an alkaline solution and boiled again for several hours. This solution is drained; the boiler is flushed again and then refilled with normal service water, with proper feedwater treatment started immediately.

A precaution to be observed in acid cleaning of boilers equipped with a superheater and other such bent tubes is to make sure all traces of acid are thoroughly cleaned out of U bends. This is critical in the neutralizing and flushing stage after the tubes are soaked with an acid solution. Compressed air may have to be used to force the solution out of dead pockets. If this is not done, the acid solution may not be completely cleared, and thinning of tubes will result. Acid cleaning of older riveted boilers can be dangerous because the acid may settle under the butt straps or lapped plates and eat up these holding elements. Thus riveted boilers are usually not acid-cleaned.

Areas subject to high local stress and repetitive application of stress may be affected by acid treatment. Tubes that have been repeatedly rolled and acid-cleaned may develop tube-roll leakage. Then with further rolling impossible, new tubes may be required. But the prevention of scale in a boiler is still the best method of keeping a boiler clean.

To guard against the consequences of superheater chemical contamination, one of the most important practices prior to proceeding with a chemical cleaning operation is to carefully review the entire piping arrangement and procedures. All the cognizant personnel involved in the chemical cleaning must be familiar with the cleaning method and flow paths to be used in filling, draining, adding chemicals, backfilling the superheater, etc. All possible paths (such as drain lines from the superheater connected with lines or manifolds being used for filling or draining the boiler) should be examined closely to ensure that cleaning solutions being fed to or drained from the boiler do not have a possible flow path to the superheater. The lines used to fill the superheater with water during the boiler cleaning should preferably not be connected with boiler chemical fill or drain lines. If flow paths exist between the superheater and the boiler fill or drain lines or manifolds, positive means of isolation must be provided. Double valves with a telltale connection between them are the minimum recommendation.

Caustic soda and soda ash are the old standbys for cleaning oil from the water side of boilers. One pound of each chemical is added for every 1000 lb of water required to fill the unit. After the boiler is filled with water, drum vents open, and a light fire is started and maintained until the vents issue steam. After the vents are closed, pressure is built up to 25 psig and held while boiling for 24 hr. Some operators blow down to half a gauge glass after about 4 hr of boiling. But 24 hr after boiling, the solution is dumped, and the unit is refilled with fresh hot water with the vents open. After this, flushing water is dumped and a thorough internal inspection is made. If necessary, a hose is used for spraying hot water for final cleaning.

Water treatment specialists also recommend the following if small amounts of oil entering the boiler water is a continual problem: (1) For free oil in the water going to the boiler, use the flotation method to separate out the oil from the water, as described in previous chapter; (2) For oil that is emulsified, use special chemicals prescribed by the treatment company to break up the emulsion, then filter the water with special leaf-type filters, assisted by diatomaceous earth.

Corrosion and Its Effects

Corrosion is the second cause of boiler water problems after scale formation. While there are many causes of corrosion in boiler water, including chemical reactions from wrong dosages of chemicals, the causes of corrosion can be grouped together by their attack on the boiler metal. Corrosion in boilers is the deterioration of metal by chemical reaction. The metal is dissolved or eaten away. The corrosive effect can seriously weaken the metal so that unexpected failure of a pressure-containing part of the boiler can occur. This section will review the principal causes of corrosion as:

1. The relative acidity of the boiler water.
2. The presence of dissolved oxygen in the boiler water.
3. Electrolytic action.

Sources of acid. Acidic conditions in boiler water usually result from chemical reactions and not from direct entrance of an acid from condensate or feedwater. Acid-forming substances are (1) carbon dioxide gas; (2) magnesium chloride; (3) magnesium sulfate; (4) sodium chloride; (5) certain oils.

The most important of these is carbon dioxide gas, which is the most soluble of all gases in water and has a tendency to form carbonic acid, H_2CO_3. The chemical breakdown of the bicarbonate salts under heat always liberates carbon dioxide gas, as shown by the following:

$$Ca(HCO_3)_2 + Heat = CaCO_3 + H_2O + CO_2$$

The carbon dioxide continues the chemical reactions by combining with water as follows:

$$CO_2 + H_2O = H_2CO_3$$

The carbonic acid can combine with iron or boiler metal as follows:

$$Fe + H_2CO_3 = FeCO_3 + H_2$$

The ferrous carbonate may combine with water to continue the chemical reaction as follows:

$$FeCO_3 + 2H_2O = Fe(OH)_2 + H_2CO_3$$

Note that carbonic acid was formed again, and this can continue to combine with boiler iron, thus weakening the metal. It also explains why a small amount of acid can continue the corrosion process. See Fig. 13.7a. Magnesium chloride, a component of seawater, can form hydrochloric acid by the following chemical reaction:

$$MgCl_2 + 2H_2O = Mg(OH)_2 + 2HCl$$

Magnesium chloride can also be formed by the chemical reaction of sodium chloride salt with magnesium sulfate salt. Both are present in seawater. The reaction is:

$$2NaCl + MgSO_4 = Na_2SO_4 + MgCl_2$$

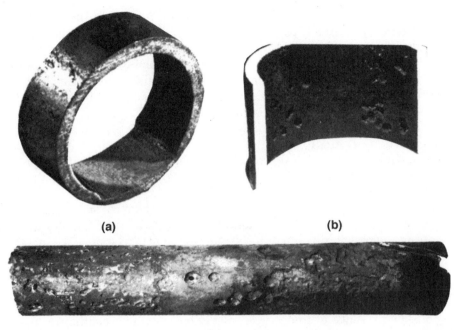

(a)   (b)

(c)

Figure 13.7 Types of boiler metal attacks from water impurities. (*a*) Carbonic acid attack on watertube boiler; (*b*) pitting attack from oxygen in watertube boiler; (*c*) pitting attack on outside surfaces of firetube boiler tube.

The magnesium chloride will react with boiler water to form hydrochloric acid, HCl, as shown in the previous equation. Hydrochloric acid reacts with iron or boiler metal as follows:

$$Fe + 2HCl = FeCl_2 + H_2$$

The iron chloride, $FeCl_2$, can combine with water as follows:

$$FeCl_2 + 2H_2O = Fe(OH)_2 + 2HCl$$

Again, an acid was formed to continue the attack on boiler metal. The entire corrosive action of these acids is due to the behavior of the hydrogen ions they contain. The iron atoms become ions in water solutions. Iron accepts only positively charged ions, which hydrogen has. The hydrogen gives up its charge to the iron, and thus assists it in leaving the boiler metal to combine with the acid radical, such as CO_3.

Acid corrosion detection, besides visual inspection for corrosion, depends on the various tests that are available when such acidic conditions are present so that the operator, with assistance from water treatment specialists, can take action before serious damage may result. The most often-used tests are the pH and alkalinity tests.

The prevention of acid corrosion depends on keeping the hydrogen ion concentration as low as possible and within certain limits. These limits have been set from experience, so that acid corrosion can be prevented by maintaining a boiler water value above 9.6 on the pH scale, with the usual operating range being between 10.0 and 10.5 pH. Since neutral water has a pH of 7.0, it is necessary to add compounds containing a base or alkaline solution to raise the pH value to the desired level by careful, calculated dosages. The advice of water treatment specialists must be followed as well as the methods of conducting acidity and alkaline tests per the usual instructions provided by chemical suppliers. For example, there are instructions for alkalinity that apply if the indicator solution is phenolphthalein or if the indicator is methyl orange.

Oxygen corrosion

Oxygen causes the corrosion of boiler metal in two ways:

1. The presence of free oxygen in the boiler water results in a pitting attack on the boiler metal as shown in Fig. 13.7b and 13.7c if it is of a localized area. Oxygen will also unite with the boiler metal in a general way to produce iron oxide (rust). Free oxygen can be produced as the temperature of the boiler water rises, and the oxygen is forced out of the solution. The oxygen then attaches itself in the form of a bubble or gas to a heating surface of the boiler to start the chemical reaction between oxygen and iron. The solubility of oxygen in water varies with the temperature of the water solution; it is generally assumed that oxygen comes out of solution usually above 750°F.

2. The second type of oxygen attack is as a catalyst, where it does not unite with the metal directly, but assists other corrosive elements in the boiler water to react with the metal or speeds up the reaction. This action is related to the exchange of metals by the natural ionization of water into H^+ ions and OH^- ions. Any free oxygen present in the boiler water tends to unite with the hydrogen, which generally is plated out on the metal surfaces. The H^+ ions give up their electrical charge to the iron atoms, allowing them to leave the boiler plate and combine chemically with the hydroxyl ion, OH^-. This produces ferrous hydroxide in the boiler water. Any free oxygen present in the boiler water tends to unite with the hydrogen in the film adhering to the boiler metal, and this forms water. This permits more hydrogen to settle on the boiler metal and combine with the boiler metal to form ferrous hydroxide. Thus, the amount of oxygen present determines the rate of hydroxyl reaction with boiler metal to form ferrous hydrox-

ide. The more oxygen present, the more rapidly is the hydrogen film destroyed on the metal, and the greater will be the iron going into solution to form ferrous hydroxide.

Oxygen-type corrosion is prevented (1) externally by the use of deaerators and (2) chemically by the use of oxygen-scavenging chemicals. Chemicals commonly used for this purpose are sodium sulfite, hydrazine, and catalyzed hydrazine.

Sodium sulfite reacts with oxygen to form sodium sulfate as shown by the chemical equation

$$2Na_2SO_3 + O_2 \quad = 2Na_2SO_4$$

Sodium   + oxygen = sodium sulfate
sulfite

It is generally required to use eight parts of sodium sulfite for each part of dissolved oxygen. Sodium sulfite for the scavenging of oxygen is applied to boilers to about 1800 psi. It is not used above this pressure because thermal decomposition of the sodium sulfite chemical produces acidic gases, SO_2 and H_2S.

Hydrazine reacts with equal parts of oxygen to produce inert nitrogen and water as shown by the equation

$$N_2H_4 \quad + O_2 \quad = N_2 \quad + H_2O$$

Hydrazine + oxygen = nitrogen + water

This chemical reaction produces products that are volatile or neutral, and thus does not increase the dissolved-solids concentration in boiler water. Hydrazine is effective at low levels of application. Since it is highly volatile itself, it may decompose thermally to ammonia and nitrogen as follows:

$$3N_2H_4 + heat = 4NH_3 \quad + N_2$$

= ammonia + nitrogen

The evolution of ammonia may restrict hydrazine application to preventing an increase in pH above the desired control point. Plants without condensate polishing systems sometimes use hydrazine in condensate based on pH control criteria.

The sluggish reaction of hydrazine with oxygen at low temperatures has resulted in high-pressure plants using this chemical only to scavenge oxygen. This has led to the development of organically catalyzed hydrazine, so that low- and medium-pressure boiler plants may also use it for oxygen scavenging. These catalysts also speed up the reactivity. One of the hydrazine catalysts is hydroquinone.

Hydrazine is considered carcinogenic; therefore, protective clothing is required in handling the chemical. Many plants use automatic feed systems to avoid human contact. This concern has also led to the development of substitute scavengers, such as carbohydrazide, erythorbic acid, methylethyl ketoxime, and DEHA (diethylhydroxylamine). Operators must carefully review their application and their effects on the boiler conditions on a case-by-case basis.

In higher-pressure boilers it is also necessary to control boiler-water *silica concentration* by automatic analysis. Silica is the firmest, toughest, and most difficult to remove of all the dissolved minerals. This removal is critical, for silica is apt to carry over with the steam. Silica's glasslike deposits inhibit heat transfer, resulting in tube burnout. When deposited on turbine blades, silica will reduce efficiency, which often results in rotor unbalance, necessitating costly and premature shutdowns.

Figure 13.3 provides some boiler and feedwater limits that are recommended by various authorities. The ASME has published, through its Committee on Water in Thermal Power Systems, a guide on water quality for industrial boilers entitled "Consensus on Operating Practices for the Control of Feedwater and Boiler Water Quality in Modern Industrial Boilers." This paper should be used as a guide for establishing water-quality limits for the boiler size involved.

After-boiler corrosion and treatment. *Condensate returns* are a way of saving fuel; however, condensate also produces water problems. The condensate-return system may have corrosion products from steam and condensate piping, which can form highly insulating deposits on boiler surfaces. Improved steam and return-line corrosion control must be used to limit this sludge. The most common attackers of the condensate systems are dissolved oxygen and carbon dioxide that find their way into the steam system.

The condensate-return system is a very revealing sample point to monitor total boiler water-treatment performance. The amount of contaminants found, and their nature, will often point out malfunctions and suggest corrective action in the rest of the system. Permissible contaminant levels would depend on the nature of the troublesome constituent, boiler design, and operating pressures. Dissolved oxygen and carbon dioxide, for example, can be directly or indirectly responsible for many system failures and are often monitored in condensate.

Carbon dioxide in condensate return systems has received increased attention because it can form carbonic acid and attack the steel metal of a boiler system. Carbon dioxide is evolved from the breakdown of feedwater bicarbonate and carbonate alkalinity.

Monitoring and control of hardness, conductivity, and specific contaminants (such as iron and copper) will also permit maximum condensate reuse.

Nuclear boiler systems deserve further considerations, for condensate return is usually considered a radioactive substance and requires special monitoring.

Condensate polishing is the term used for the removal of solid traces as well as dissolved solids that are found in condensate systems. In the utility field, the two condensate polishing systems used are the externally applied precoat filter and the mixed-bed demineralizer in order to eliminate the solids. Condensate systems also can be chemically treated to control corrosion damage caused by water, carbon dioxide, and oxygen. The treatment chemicals include neutralizing amines, filming amines, hydrazine, and sometimes ammonia.

Corrosion in steam lines and return lines is caused primarily from the reaction of CO_2 with water to form carbonic acid. The CO_2 may come from condenser or other steam-using equipment that permits air to leak into the system at low or vacuum pressure conditions. It was also shown that bicarbonate alkalinity can produce CO_2. Buildup of acidic conditions will reduce the condensate pH value, and this results in metal loss in piping and general corrosion. Oxygen present in the condensate will accelerate the corrosion.

Two methods are used to combat after-boiler corrosion attack:

1. *Neutralizing amines* such as cyclohexylamine, morpholine, and diethanolamine are used to neutralize the condensate pH. In utility boilers, a *volatile treatment* of hydrazine and ammonia is often applied. The main benefit of hydrazine/ammonia is that it constitutes an all-volatile approach by minimizing the introduction of organic constituents into the system, and provides a lower cost of condensate-pH control. However, an excessive use of ammonia could lead to an elevation of condensate pH, with resultant increase in corrosion. The disadvantage of neutralizing amines for industrial boilers is the possible poor control of chemical injection.

2. *Filming amines* are used to establish a continuous protective film over the surfaces of after-boiler piping systems. This method prevents contact of any potential corrosive steam/condensate constituents with the metal of the piping system. Filming amines are quite often supplemented by the use of neutralizing amines to protect the metal from any discontinuity in the protective film. This is especially applicable to boiler systems operating above 200 psi with 75 percent or more condensate returns.

Priming, foaming, and carryover. *Priming, foaming,* and *carryover* are factors usually controllable by the operating engineer. *Priming* is the lifting of boiler water by the steam flow. The water may be lifted as a spray or in a small body; as it enters the steam line, its weight and velocity may cause severe damage to equipment. Ruptured steam-line fittings or wrecked turbines or engines have resulted from "slugs" of water. Unless priming is induced by faulty boiler design (which is not uncommon), it is caused by carrying too high a water level for the demands for steam flow. The water level in the drum should be kept several inches lower than normal if the steam flow fluctuates very much, for a sudden rush of steam sometimes tends to pick up water from the surface directly below the nozzle.

Operators can avoid priming by:

1. Not forcing or overfiring any boiler connected to steam-using machinery, such as engines or turbines.

2. Maintaining a constant and steady water level in the boiler and avoiding rapid water level fluctuations.

3. Following good surface and bottom blowdown procedures to eliminate mud and sludge from internal treatment.

4. Avoiding sudden openings of throttle valves to steam-using machinery.

Scaling of superheater tubes may occur from deposits resulting from carrying over impurities with water slugs.

Foaming is more a chemical than a mechanical problem. High surface tension of the boiler water causes many of the steam bubbles to be encased by a water film. These film-encased bubbles rise and pass out in the steam flow. The cause of high surface tension is usually a high concentration of solids in the boiler water. Organic matter, too, may produce this trouble. Periodic checks on boiler-water concentration and control of blowdown to hold the concentration within allowable limits will prevent foaming. The density of the boiler water is a measure of its concentration. Specially calibrated hydrometers are available at low cost for direct reading of this condition, as are conductivity meters.

The general effect of foaming is a reduction in steam quality by increase in the moisture content of the steam. Foaming promotes priming and carryover. Foaming may be corrected by (1) using surface blowdown more frequently until the surface foam has been removed, (2) correcting the salt content; remember foaming can be caused by a high dissolved salt content in the boiler water as well as excessive alkalinity and similar boiler water problems. If the concentrations are found to be high, use the bottom blowdown and feed fresh water to dilute the concentration to acceptable levels. Trace the system for any leakage in condensate returns or condensers, and also check the prescribed dosage of chemical compound treatment to make sure over- or under-treatment is not occurring.

Antifoaming agents may also suit some plants. As dissolved solids and alkalinity increase in boiler water, foaming and resultant carryover also increase. Chemical antifoaming agents typically function by becoming increasingly insoluble as water temperature increases. Somewhere in this higher temperature range, the agents break up the foam, disrupting its cohesive nature and reducing the potential for carryover. Silicones, polyglycols, and polyamides are used up to about 800–900 psi boiler operating pressure. Above this pressure range, they become undesirable, because their addition increases the suspended-solids content of boiler water.

Carryover from steam boilers. Clean steam plays an important part in economical power plant operation. When the system is contaminated with water, mineral solids, or other impurities, numerous troubles develop and costs automatically increase. Foreign matter entrained in otherwise clean steam leaving a boiler drum is commonly termed *carryover*. It can be eliminated or minimized by determining its cause and then applying the right correction. The magnitude of the losses occasioned by carryover is not generally realized. Fuel consumption, equipment maintenance costs, and plant safety are all affected.

There are many cases where carryover still persists, even though all ordinary preventive measures are used. To reduce entrained solids to an absolute minimum, steam washers or steam separators are installed. Washers use the relatively pure incoming feedwater to wash outgoing steam. Separators remove entrained water and solids by impinging the steam against baffles or suddenly changing its direction of flow so that foreign particles are thrown out by centrifugal force. Separators or purifiers are also valuable as insurance against any unexpected slugs of water which might damage power plant equipment.

Total dissolved solids when maintained at too high a concentration in the boiler water may cause excessive carryover, which can result in mechanical damage and deposits in turbines and feedwater heaters. The measurement of dissolved solids is by conductivity. For most boiler waters 1 micromho = 0.9 ppm dissolved solids. For condensate, the average is 1 micromho = 0.5 ppm dissolved solids. Dissolved solids control is by blowdown.

Acid attack, or low-pH operation, causes the magnetic iron oxide film on metal surfaces to disappear, and the metal itself is attacked. The loss of metals is in the form of smooth, rolling contours, usually called *gouging* (see Fig. 13.7*a*). A secondary-type attack could be hydrogen damage or embrittlement, as hydrogen is released during the acidic attack on the metal, once the oxide coating is lost.

Blowdown. *Blowdown* is an integral part of the proper functioning of a boiler water treatment program and usually requires continuous monitoring for positive control. It is through blowdown that most of the dirt, mud, sludge, and other undesirable materials are removed from the boiler drum.

In most systems, surface blowdown is accomplished continuously, and the optimum blowdown interval is such that sludge or scale on heating surfaces is minimized. At the same time, the loss of heat and chemical additives is also kept to a minimum.

In the larger, more critical boilers, continuous surface blowdown is usually combined with a regular bottom blowdown. In many high-pressure boilers, it is desirable to minimize boiler blowdown to reduce heat and water losses.

Blowdown analysis is complicated by sample conditioning considerations (a sample cooler is usually required), but the control parameter is generally conductivity. Typically, a conductivity meter limit is maintained within a control range, and blowdown is activated by a certain deviation from that range. Other blowdown monitoring parameters include pH, silica, hydrazine, and phosphate.

Intermittent or bottom blowdown is taken from the bottom of the mud drum, waterwall headers, or lowest point in the circulation system. The blowoff valve is opened manually to remove accumulated sludge, about every 4 to 8 hr, or when the boiler is idle or on a low-steaming rate. But hot water is wasted, and control of concentrations is irregular and requires operator trial-and-error to establish quantity and time of blowdown.

Continuous *surface* blowdown automatically keeps the boiler water within desired limits. Continuously removing a small stream of boiler water keeps the concentration relatively constant. Savings by transferring heat in the blowdown to incoming makeup often pay for the investment.

Percentage blowdown calculations. There are many operating engineers who use cycles of concentration in calculating percentage blowdown. The author prefers equating flows as shown in Fig. 13.8 to obtain flows of the items listed in the figure. Marine engineers at one time used the term "concentration factor." For example, if the feedwater shows 0.5 grains of impurities, and the boiler water shows 15 grains per gallon, the concentration factor is $15/0.5 = 30$. From this, the general equation for percent blowdown was used, namely

$$\text{Percent blowdown required} = \frac{\text{Feedwater concentration}}{\text{Concentration factor}}$$

Amount of blowdown is calculated as a percentage of makeup water flow into the boiler in order to keep the concentration of impurities in the boiler at an acceptable level, or at acceptable ppm as shown in Fig. 13.3. If no blowdown was performed, the solids concentration in the makeup water would *add* to the existing solids concentration as steam is evaporated. Blowdown removes the concentrated impurities to acceptable boiler water concentrations, in modern practice, expressed in ppm.

The best way to demonstrate the *balance of flow method* in calculating required percentage blowdown is to use an example. See Fig. 13.8a.

Example   The stated conditions are: A 200-psi boiler generates 50,000 lb/hr steam with 100 percent makeup. An analysis shows boiler water is maintained at 3500 ppm total solids, while makeup is at 200 ppm total solids at 60°F.

1. Find percentage of blowdown required to maintain 3500 ppm total solids in the boiler water.
2. Find blowoff and feedwater flow.
3. See Fig. 13.8b. Find blowdown percentage, blowdown, condensate return, and makeup flow for 90 percent condensate return.

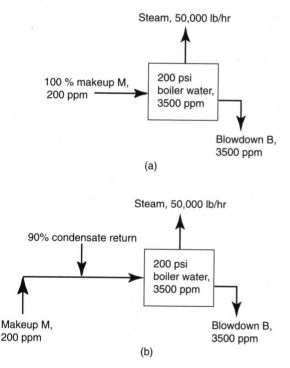

(a)

(b)

Figure 13.8   Layout diagram of boiler steam flow, makeup and blowdown assists in calculating percentage blowdown and the flows involved. (a) Makeup is 100 percent. (b) In this arrangement, makeup is 10 percent.

solutions

1. Let M = makeup flow, lb/hr and B = blowdown flow, lb/hr. To hold solids at a given level in the boiler, amount entering with makeup must equal that removed by blowdown, or

$$200M = 3500B$$

from this

$$B = \frac{200}{3500}\,M = 0.057M = 5.7\% \text{ of } M$$

2. At 100 percent makeup, feedwater = makeup = M flow

Feedwater flow = 50,000 + Blowdown

$$M = 50,000 + 0.057M$$

$$0.943M = 50,000$$

$$M = 53,022 \text{ lb/hr, and } B = 3022 \text{ lb/hr}$$

3. At 10 percent makeup (see Fig. 13.8b),

Feedwater flow = 50,000 + Blowdown

$$M + 0.9(50,000) = 50,000 + 0.057M$$

Solving for M,

$$0.943M = 5000$$

$$M = 5302 \text{ lb/hr and}$$

$$B = 0.057(5302) = 302 \text{ lb/hr}$$

Note. The usual practice is to assume condensate has low solids, and ignore it.

Quite often, blowdown and makeup calculations also involve the cost of chemicals used. The following problem illustrates the method.

Problem A 500 hp boiler operating at 150 psi has an average load of 80 percent of its rating with 10 percent condensate return and 10 percent blowdown. The water treatment chemical supplier quotes the following for the chemicals deemed necessary for treatment: One pint per 1000 gallons of makeup at a cost for the chemicals of $2.80 per gallon of chemicals needed. Find (1) the amount of chemicals that will be needed for a 30-day period and (2) the cost of these chemicals for the 30-day period.

solutions (1) 34.5 lb/hr = one boiler hp, rating of boiler is 500(34.5) = 17,250 lb/hr, and average load is 0.8(17,250) = 13,800 lb/hr. Let M = makeup flow, lb/hr.

Feedwater flow = boiler flow + blowdown flow

$$M + 0.1(13,800) = 13,800 + 0.1M$$

$$0.9M = 12,420 \text{ lb/hr}$$

$$M = 13,800 \text{ lb/hr}$$

$$\text{Gallons of makeup needed for 30 days} = \frac{13,800}{8.33} \times 24 \times 30 = 1,192,797 \text{ gal}$$

$$\text{Pints of chemicals needed} = \frac{1,192,797}{1000} = 1192.8 \text{ pt}$$

(2) Cost of chemicals is

$$\frac{1192.8}{16} \times \$2.80 = \$208.74$$

Water treatment instrumentation. As boiler operation continues to become more automated, sensor development for continuous monitoring of feedwater has eliminated many water treatment tests previously accomplished, for example, by titration of grab samples of water, which basically involved manual analysis of the water samples. Utilities strive to continuously monitor operating water conditions, and the Electric Power Research Institute has published guidelines. These are listed in Table 13.1. However, grab samples will still be performed by water treatment specialists and water treatment chemists in order to check on the proper operation of available instruments and for calibration of such instruments.

Conductivity measurements are the most often used criteria to determine dissolved contaminants in boiler water. This is because water containing ionized contaminants, or electrolytes, exhibits con-

TABLE 13.1   Feedwater Monitoring Guidelines

Item to be monitored	Recommended frequency*	
	Optimum	Acceptable
Conductivity	Continuous	Continuous
TOC	Continuous	Semiweekly
Dissolved SiO_2	Continuous	Daily
Total SiO_2	Daily	Semiweekly
Na^+, Cl^-	Continuous	Continuous
F^-, $SO_4^=$, K^+	Semiweekly	Weekly
Ca^{++}, Mg^{++}	Semiweekly	Weekly
Fe, Cu	Semiweekly	Weekly
Dissolved O_2†	Continuous	Continuous
Suspended solids	Continuous	Semiweekly
Bacteria‡	Semiweekly	Weekly
pH§	As needed	As needed

*Dictated by availability of on-line analyzers or meters; continuous monitoring instrumentation may not be available for all parameters.
†For the system with a vacuum degasifier.
‡Critical for systems with RO or UF membranes; useful diagnostic for other systems.
§For diagnostics.
SOURCE: From Electric Power Research Institute guidelines.

ductivity of electricity, which is dependent on its nature and concentration. Instruments have been developed that measure conductivity by immersing two electrodes in the solution under test with a low ac voltage applied across them. The resultant current between the electrodes is measured by an ac bridge, and the available commercial instruments amplify the current signal and convert it to conductivity values. See Fig. 13.9a. These instruments have gained in accuracy by the use of titanium conductivity cells and microprocessors. There is automatic electronic adjustment for cell fouling.

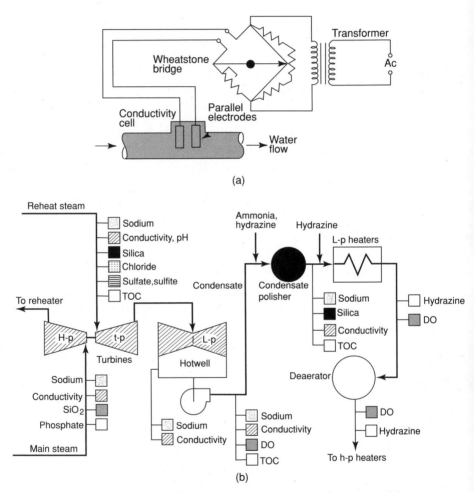

Figure 13.9 (a) Conductivity meter has electrodes in water to be measured for amperes between electrodes, which is converted by Wheatstone bridge into instrument conversion to ionized contaminant content in the water. (b) Typical utility tracking of impurities at different points of steam and water flow, primarily to protect the steam turbine from contaminant damage. This is made possible by modern continuous-reading instrumentation.

pH monitoring depends on the electrochemical potential at the surface of a pH-sensitive glass-electrode membrane. A reference electrode is incorporated in commercial instruments to complete the electric circuit between the measuring and reference electrodes.

Sodium analyzers are important instruments in utilities with steam turbogenerators in order to limit the potential for stress-corrosion failures in the steam turbines due to the presence of sodium. The operation of sodium analyzers is based on the elevation of sample pH and the use of a sodium-sensitive ion-selective electrode to measure the sodium concentration. The electrodes generally used, with silver/silver chloride internal elements in a potassium chloride solution, can measure sodium down to 0.1 ppb.

Organics are an important source of contamination or impurities entering the boiler water. Ion chromatography separates chemical compounds, plus a mass spectrometer identifies impurities by comparison with the spectral response of known compounds. The process has been improved by the development of alternative instruments, based on the measurement of the total organic carbon (TOC) content of the water being sampled. Further details are available from the suppliers of these instruments to measure organics.

Figure 13.9*b* shows the scope of utility measurement of water chemistry items because the criteria is the protection of the steam turbine from contaminants more than the boiler. This is because boiler-water specifications may not guarantee acceptable turbine steam, especially those operating at high pressure. As can be seen, monitoring and analysis by specialists are done at many points in the water-steam cycle in order to limit impurities to a few parts per billion. Continuous on-line analysis is performed of pH, O_2, conductivity, cation conductivity, sodium, and chlorine in the steam to check on corrosion and deposition. This is combined with liquid ion chromatography for sodium, chlorine, ammonia, potassium, fluoride, and sulfate in the water. Grab samples are used by specialists to supplement any continuous monitoring.

Computer use to collect and analyze data is increasing; computers display the data graphically so that further diagnostic analysis can be made.

Other instruments available for water treatment applications include:

1. Oil trace measurement: directing a beam of light on the surface of a water sample. A photodetector generates an alarm if the reflection is caused by an oil film.

2. Borescopes for visual inspections of pipe and tubes to detect visible corrosion, gouging, and deposits.

3. Dissolved oxygen detectors using probes, typically gold or silver/gold or silver cathodes and platinum or lead anodes, separated by an oxygen permeable membrane. Dissolved oxygen in the sample or water diffuses through the membrane, and this can create a proportional current or voltage signal for instrument application.

Operators should review with their chemical treatment company or specialist what additional instrumentation might assist them in maintaining the best water chemistry conditions for their boiler plant.

Questions and Answers

1 What type of impurities may raw water have?

ANSWER: Water may have the following types of impurities:

1. Suspended matter, such as sand, mud, and organic wastes.
2. Dissolved or in-solution solids that precipitate out of the water at higher temperatures, such as limestone (calcium carbonate), dolomite (magnesium carbonate), gypsum (calcium sulfate), epsom salt (magnesium sulfate), sand (silicate), common salt (sodium chloride), Glauber salt (hydrated sodium sulfate), and traces of iron, manganese, fluorides, and aluminum, depending on the source of water.
3. Dissolved gases, such as oxygen, carbon dioxide, hydrogen sulfide, and even methane.
4. Fine particles called colloids in suspension and having a size between particles in a dissolved state and those in suspension.

2 How are grains per gallon and ppm related?

ANSWER: One grain per U.S. gallon = 17.1 ppm. This makes 1 ppm = 0.058 gr/gal.

3 How is hardness expressed?

ANSWER: In U.S. practice, it is expressed as parts of calcium carbonate per ppm of water or grains of calcium carbonate per gallon of water.

4 How is acidity or alkalinity expressed?

ANSWER: This is expressed by the pH scale, which ranges from 1 to 14. A pH of 7 is considered neutral; anything below this value is acidic, and anything above this value is alkaline. The pH values are related to the laws of dissociation of water into OH and hydrogen ions. It is an exponential function with 10^{-7} being the neutral point. This means that a pH of 5 is 10 times more acidic than a pH of 6.

5 In neutralization, what chemicals are used to raise the pH?

ANSWER: Low pH indicates an acidic condition, which can be corrected by the addition of base substances, such as lime, soda ash, or caustic.

6 What are the four major boiler-water control tests used by stationary engineers? What water condition does each test determine?

ANSWER: See Fig. 13.10. The four major tests are for hardness, alkalinity, chlorides, and pH. (1) Hardness affects scale. (2) Alkalinity indicates required amounts of treatment chemicals (caustic soda or soda ash). (3) Chlorides control the concentration of solids and check on the surface condenser for leaks (especially where sea water is used, as in marine or tidal power plants). (4) pH (hydrogen ion) is a type of alkalinity or acidity to check to note if proper alkalinity control is maintained.

7 What compounds cannot be removed by boiling, making the water have *permanent hardness*?

ANSWER: Water containing calcium and magnesium sulfates cannot be softened by boiling; thus they have permanent hardness, which is solved by chemical treatment with the addition of sodium carbonate as noted by the following equations:

$$CaSO_4 + Na_2CO_3 = CaCO_3 + Na_2SO_4$$

$$MgSO_4 + Na_2CO_3 = MgCO_3 + Na_2SO_4$$

The carbonates precipitate out of solution.

8 Using the above equations, if 13 lb sodium carbonate is used to treat calcium sulfate in boiler water, what would be the other weights in the chemical equation when using the atomic weight method?

ANSWER: The compounds in the calcium sulfate equation have the following atomic weights: $CaSO_4 = 136.2$; $Na_2CO_3 = 106$; $CaCO_3 = 100.1$; $Na_2SO_4 = 126$. Then, by proportion,

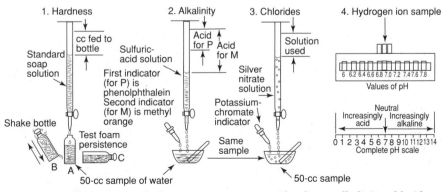

Figure 13.10 Four common boiler room water tests are hardness, alkalinity, chlorides, and acidity or pH.

$$CaSO_4 = \frac{13(136.2)}{106} = 16.7 \text{ lb}$$

$$CaCO_3 = \frac{13(100.1)}{106} = 12.3 \text{ lb}$$

$$Na_2SO_4 = \frac{13(126.1)}{106} = 15.5 \text{ lb}$$

9 What is the purpose of an electrical conductivity test of boiler water?

ANSWER: To measure the extent to which dissolved substances are concentrated in the boiler water. This test then helps in controlling carryover of dissolved solids, which condense in lines or equipment such as turbine blades, into the steam system.

10 What are some ways to reduce the oxygen content of boiler water?

ANSWER: Mechanical deaerator removal and the use of sodium sulfite or hydrazine.

11 Why is hydrazine used for oxygen control on high-pressure boilers?

ANSWER: Hydrazine (N_2H_4) is the oxygen scavenger most often used for high-pressure boilers. Its advantages over sulfite, which is used in most low-pressure units, are that it adds no dissolved solids to the boiler. Hydrazine scavenges oxygen according to the following reaction:

$$N_2H_4 + O_2 \rightarrow N_2 + 2H_2O$$

A major disadvantage to the use of sulfite in high-pressure boilers is that it may break down to corrosive H_2S and SO_2. The following reactions are involved:

$$Na_2SO_3 + H_2O \text{ heat} \rightarrow 2NaOH + SO_2$$

$$4Na_2SO_3 + 2H_2O + \text{heat} \rightarrow 3Na_2SO_4 + NaOH + H_2S$$

Another method used is *volatile treatment*. This system is based on the use of hydrazine and neutralizing amines or ammonia. Major advantages claimed of volatile treatment are that it adds no solids to the boiler and affords good pre-boiler protection. Boiler pH is controlled between 8.5 and 9.0, and the hydrazine residual in the feedwater is controlled between 0.01 and 0.10 ppm.

12 Why are chelants used in water treatment?

ANSWER: For boilers operating under 1000 psi, chelants have been used to control metal oxide deposits in boiler systems. The chelants react with metal ions to form soluble compounds which can then be flushed from the boiler water. The degree of combining is dependent on the system environment and the reactivity of the chelant to the metal ion that may be encountered. The chelant agents' reaction is considered to be applicable for soluble metal ions and not for insoluble metals. For the latter, dispersant polymer agents have been developed. The polymer is adsorbed onto the iron or insoluble metal oxide, thus controlling insoluble oxide deposits by altering their charge characteristics and preventing agglomeration. When used this way, the polymers are called *dispersants*.

13 What are major treatment problems with heating boilers?

ANSWER: Corrosion and pitting. Scale is not a problem because the same feedwater is used continuously, and the initial treatment usually lasts throughout the heating season.

14 What are some boiler deposit characteristics?

ANSWER: All of the following deposits can cause boiler parts to overheat if the scaling is too thick. Some features of deposits are (1) *calcium carbonates* are granular and porous. If dropped in an acid, bubbles of carbon dioxide are released. (2) *Sulfate deposits* are hard, dense, and brittle but do not pulverize easily. (3) *Iron deposits* are dark colored and magnetic and, if placed in a hot acid, become soluble, forming a dark-brown solution. (4) *Silica deposits* are very hard, resembling porcelain, and are brittle but difficult to pulverize. They are not soluble in hydrochloric acid. (5) *Phosphate deposits* arise from internal boiler treatment and are soft brown or gray. They can be removed by normal cleaning methods (washed out).

15 A 150-psi boiler, average load of 15,000 lb/hr, is found by test to have 1500 ppm solids, and the makeup water has 100 ppm solids. What is the percentage blowdown required to maintain the 1500 ppm solids, and what flow is required for makeup and blowdown, assuming no condensate return?

ANSWER: Let M = Makeup and B = Blowdown in lb/hr. Then,

$$B = \frac{100}{1500} M, \text{ or } B = 0.067M, \text{ or } 6.7 \text{ percent of makeup.}$$

Equating in and out flows of water to the boiler, with M = Feedwater flow at 100 percent makeup

$$M = 15,000 + 0.067M$$

and solving for M

$$M = 16,077 \text{ lb/hr} \quad \text{and} \quad B = 1077 \text{ lb/hr}$$

16 How does the flow method compare with the concentration method in determining blowdown percentages and flows?

ANSWER: For the same concentrations as given in Problem 15, some stationary and marine engineers would state that the feedwater can be concentrated $1500/100 = 15$ times. This means that for every 1000 lb of water fed into the boiler, $1000/15 = 66.7$ lb must be blown down to keep the dissolved solids from exceeding 1500 ppm. Thus the percentage of blowdown is $66.7/1000 =$

6.7%, the same as for the previous problem. With a feedwater flow of 16,077 lb/hr, blowdown would still be 0.067 × 16,077 = 1077 lb/hr, or the same answer.

17 What is the purpose of blowdown?

ANSWER: Blowdown reduces any suspended solids or solids in solution. When the water blown out of the boiler is replaced by lower-solids feedwater, the boiler water is diluted to acceptable limits provided the amount of blowdown and its frequency is properly controlled.

18 How is the oxygen level controlled?

ANSWER: By external treatment in a deaerator and by chemical scavenging of oxygen by sodium sulfite and by hydrazine.

19 How is pH controlled?

ANSWER: When the makeup water is treated by the lime-soda or lime-soda-zeolite method, the effluent may be within the range of 8.0 to 9.5 pH, which is recommended by most boiler manufacturers. However, if excessive sodium bicarbonate is formed, this decomposes to form caustic soda, or excess alkalinity. The ABMA recommends that the alkalinity content should not exceed 20 percent of the total solids of the boiler water. *Split-stream* softening is used to reduce the alkalinity content.

20 What is galvanic action?

ANSWER: It is an electrolytic flow between dissimilar metals in a boiler electrolyte, such as water, resulting in localized deterioration of the metal.

21 What is the purpose of detergent cleaning vs. chemical cleaning of boiler internal surfaces?

ANSWER: Detergent cleaning is conducted with an alkaline solution to remove primarily oil and grease, while chemical cleaning uses a solvent solution to remove primarily mill scale and corrosion products.

22 How is corrosion caused on the internal surface of boilers?

ANSWER: Dissolved gases such as oxygen and carbon dioxide in the feedwater are the main cause of corrosion on the waterside of boilers. These gases can be removed by mechanical deaeration of the condensate or feedwater. Chemical scavengers are used for internal treatment to eliminate these gases such as sodium sulfite and hydrazine.

23 What is priming?

ANSWER: Priming is caused by the sudden or rapid carryover of boiler water into the steam space that usually occurs by: (1) a sudden change in demand for steam, (2) surging or spouting in the drum. The danger of priming is that it causes slugs of water to pass out with the steam, and these can cause impacts to produce water hammer in piping, drop in superheat temperature, and damage to turbine blades, or other machinery using the steam from the boiler.

24 What causes foaming?

ANSWER: This is when bubbles or froth form on the surface of the boiler water and are carried outside the boiler with the steam. It is caused by impurities such as alkalies, oils, fats, greases, and some organic matter being present as impurities in the water. Oil contamination can come from the oil used on steam-using equipment, such as pumps, compressors, and turbines, getting into the condensate returns. It is believed that soaps are formed by the oil reacting with water alkalies.

25 What are acids and bases and the pH relation of each?

ANSWER: Acids are those ions in water that form hydrogen ions, H^+. For example, hydrochloric acid ionizes in water to form H^+ and Cl^-. An alkali forms hydroxide ions in water, OH^-. Also, sodium hydroxide (NaOH) ionizes in water to NA^+ and OH^-. When a base and an acid are put in a solution, a salt is formed. For example, $NaOH + HCl = NaCl + H_2O$, with NaCl being the salt formed by combining caustic soda with hydrochloric acid. The base OH^- neutralizes the acid H^+. The pH value was derived from the hydrogen-ion concentration in *pure water,* which is

$$0.0000001 = \frac{1}{10,000,000} = \frac{1}{10^7}$$

To avoid the use of large decimal fractions, the exponential 7 in the above equation is used to denote neutral or pure water, without acid or alkaline properties. In this way it is easier to remember that 7 is neutral; values below 7 are acidic solutions, and above 7 they are alkaline solutions. The exponential relationship is brought out this way: If the hydrogen ion in pure water is increased 10 times, we have

$$10 \times 0.0000001 = 0.000001 = \frac{1}{10^6}, \text{ and the pH is 6}$$

26 Why is sodium objectionable in boiler water?

ANSWER: Sodium is objectionable in boiler water because it can combine with hydroxide ions that may be present in the water. The chemical combination can produce the caustic compound sodium hydroxide. Highly stressed areas of the boiler piping or steam turbine can be attacked by this compound and cause stress-corrosion cracks to occur.

27 Describe the visible effects of the destructive action in boilers known as (1) pitting; (2) grooving; and (3) corrosion.

ANSWER:

1. Pitting takes the form of a local action by which small circular areas on shell plates, heads, or tubes are attacked. Pustules having a brown oxide covering form over the pits and when active are filled with black oxide of iron.

2. Grooving commonly occurs in the knuckles of flange or at rivet seams. It is generally caused by a breathing action and may result in cracks. Grooving is a combination of mechanical action and corrosion, each aiding the other.

3. Corrosion is oxidation or rusting of the metal, and it may be produced by acids in feedwater, exposure to air, or other oxidizing agents. Corrosion takes the form of rust and is usually general or local wasting away of the metal. Pitting attacks the inner surface of boiler from the waterline downward. Grooving occurs mainly in knuckles of flanges or near rivet sections on braced or stayed surfaces. Corrosion may occur anywhere on a boiler where the surface of metal is exposed to an oxidizing agent.

28   What impurity in water requires critical attention on very high-pressure boilers?

ANSWER: Silica in high-pressure water volatilizes and passes over with steam to the turbine. As the steam expands and drops in pressure and temperature, the silica condenses and deposits on the turbine blades, cutting the machine's efficiency. Vaporous carryover involving the volatilization of matter (hot gases) increases most rapidly around 2000 psi. Above this range, mechanical carryover predominates, especially above the critical pressure (3206.2 psi), for there is no longer a liquid-to-steam phase. Steam quality on these units is no longer measured by a calorimeter in percentage of moisture but in parts per billion by the use of flame photometers.

29   Are there many water treatment methods available?

ANSWER: Yes, but they all come under three broad classifications: *mechanical, heat,* or *chemical treatment.* Mechanical treatment includes filtration and boiler blowdown. Heat treatment includes makeup distillation and steam purification. But distillation is limited to boilers with small amounts of makeup. Steam purifiers are used where the process demands very dry steam. Chemical treatment, both internal and external, is most widely used. External treatment adjusts raw-water analysis before the water enters the boiler. Internal treatment adjusts the boiler-water analysis by feeding chemicals directly to the boiler.

30   How is condensate corrosion in return lines prevented?

ANSWER: Condensate return line corrosion is usually from carbonic acid formed when carbon dioxide gas combines with water. A neutralizing agent such as the amines morpholine and cyclohexylamine may be used. Cyclohexylamine is more alkaline than morpholine and is also more volatile.

If a deaerator is used in feedwater treatment, it may vent to the atmosphere. With no deaerator, cyclohexylamine stays closer to the carbon dioxide in the steam, and there is very little lost in boiler blowdown.

31   What method is used to test boiler water for phosphate?

ANSWER: The usual method is by color comparison as directed by a water treatment specialist. Special testing chemicals are used that act on the phosphate in the water to form a blue solution. The shade of blue is compared to standards usually supplied by water treatment chemical suppliers. The shade of blue determines the concentration of phosphate in the water sample under test.

32   A water analysis report shows 40 ppm for calcium and 26 ppm for magnesium in a boiler water sample. What is the total hardness as calcium carbonate?

ANSWER: The equivalent weight for calcium is atomic weight/valence, or $40.1/2 = 20.05$, and for magnesium it is $24.3/2 = 12.15$. For calcium carbonate it is $100.1/2 = 50.5$. The equivalent weights are now compared to calcium carbonate, or for calcium it is $50.5/20.5 = 2.46$, and for magnesium it is $50.5/12.15 = 4.1$. These numbers are used to multiply the ppm of calcium and magnesium to obtain the ppm of hardness as CO_3, or $(40 \times 2.50)+(26 \times 4.1) = 206.6$ ppm hardness as calcium carbonate.

33   Where is the preferred method of feeding chemicals for internal treatment of boiler water?

ANSWER: Most chemicals are fed through solution tanks or proportioning pumps. Chemicals are usually fed after any deaeration before the entrance of phosphates, chelates, or caustic solutions into the boiler drum. It is preferred that chemicals react with the boiler water at feed entrance before they enter the steam-generating loop of the boiler. It is also preferred that chemicals used to prevent oxygen attack, such as sulfites and hydrazine, and those introduced for preventing scale and acid corrosion, such as caustics and organics, should be fed continuously into the water going into the boiler.

34   What is used to prevent suspended solids in boiler water from depositing on heat-transfer surfaces?

ANSWER: Organic sludge conditioners are used to prevent suspended solids from depositing on tubes. The usual sludge conditioners employed are:

1. Synthetic polymers for all types of sludge.
2. Tannins for conditioning high-hardness makeup waters.
3. Starches for conditioning high-silical feedwaters, and oil contamination.
4. Lignins for conditioning phosphate-type sludges.

35   What would be the cost for chemicals for a 30-day period to treat a 100-hp boiler, operating at 200 psi, with 3500 ppm dissolved solids, and being supplied with 150 ppm dissolved solids makeup at 100 percent makeup? Chemicals cost \$3.50/gal, with one quart of the chemicals treating 2000 gallons of makeup water.

ANSWER: Let M = makeup water flow, in pounds per hour, then

$$\text{Blowdown} = \frac{150}{3500} \, M$$

Flow in = flow out and output for 100 hp = 3450 lb/hr.

$$M = 3450 + \frac{150}{3500} \, M$$

$$0.957 \, M = 3450$$

$$M = 3605 \text{ lb/hr} \qquad \text{or} \qquad \frac{3605}{8.33} = 432.8 \text{ gal/hr of makeup}$$

$$\text{Amount of chemicals needed} = \frac{432.8}{2000} = 0.216 \text{ qt/hr}$$

$$\text{Number of gallons needed for 30 days} = \frac{0.216(24)(30)}{4} = 38.9 \text{ gal}$$

$$\text{Cost for 30 days} = 38.9 \times 3.50 = \$136.15$$

36   On what basis is the choice made between neutralizing and filming amines to combat corrosion from condensate contaminants?

ANSWER: Water treatment specialists recommend neutralizing amines for systems with low makeup water flow, low alkalinity, and good oxygen level control. Filming amines are recommended for systems with high makeup and alkalinity and high air in-leakage on the condensate. Filming amines are also recommended for boiler systems that do not operate on a regular basis.

37   What problems does oil in boiler water cause?

ANSWER: Oil contamination can cause: (1) coating of heat-transfer surfaces permitting other suspended solids to stick to that surface, which can produce overheating damage to the heat-transfer component; (2) foaming and boiler water carryover that can damage steam-using machinery, such as turbines.

38   What danger do boilers out of service present?

ANSWER: Corrosion can be caused by the exposure of wet metal to oxygen in the air. The *wet method* of boiler lay-up should be followed. Store the boiler full of water and add chemicals such as sulfite, organics, and caustic per your water treatment specialist's advice. Nitrogen gas blankets are also used to keep air intrusion out of the boiler water. On high-pressure boilers with superheaters, use demineralized water or pure condensate treated with hydrazine and a neutralizing amine to protect the superheater.

39 What is the danger in nitrogen blanketing of idle boilers?

ANSWER: Before entering any drum for inspection, make sure all nitrogen gas is removed so there is no danger of suffocation. Follow confined space entry rules and have the atmosphere in the drum checked for the proper level of oxygen before entering.

40 When is the *dry storage method* used?

ANSWER: This method is usually employed for longer periods of boiler outage. It consists of draining the boiler and cleaning and drying the internal surfaces. A moisture-absorbing chemical such as hydrated lime or silica gel is placed in trays inside the critical areas of the boiler, and the boiler is closed tightly to prevent in-leakage of air. If the outage is long, periodic replacement of the moisture-absorbing chemicals is required.

41 List the main reasons for internal chemical water treatment.

ANSWER: These usually apply to high-pressure boilers.

1. To eliminate alkalinity from the feedwater (makeup and condensate)
2. To adjust alkalinity to prescribed levels in order to prevent acid or oxygen corrosion from chemical compound reactions
3. To treat any suspended matter with sludge conditioners in order to produce a sludge nonadherent to boiler metal
4. To prevent foaming by adding antifoam protection so that a reasonable concentration of dissolved and suspended solids will not cause foaming and carryover.

14

In-Service Problems, Inspection, Maintenance, and Repairs

In-service problems can always develop with time as deposits of impurities slowly build up on heat-transfer surfaces, corrosion eats metal away, erosion reduces tube thickness, cycling causes incipient cracks to appear, control connections become plugged, creep affects the strength of the metal to resist stress, and similar operating problems occur due to wear and tear on equipment. There are many operating problems that affect output and efficiency. Modern boilers can have many operating controls, as indicated in Fig. 14.1. Quite often these controls and associated gauges and instruments need adjustment to maintain normal operating efficiencies or output. It is here that the skill and knowledge of the operator may be very important in making proper adjustments without affecting other control functions, which may also need adjustment. For example, a low-water safety device may operate to shut off a burner. The operator checks his gauge glass, and sees the water level is at the proper half-full glass level. He then assumes the low-water fuel cutoff is defective and blocks the switch into a closed position in order to keep the boiler operating. Unknown to him, the gauge glass valves to the boiler were closed during a gauge glass replacement and never reopened by maintenance personnel. The level in the gauge glass was thus not a true indication of the water level. The operator became aware of this only after the induced fan housing on the boiler became cherry red as the tubes slowly melted inside the boiler.

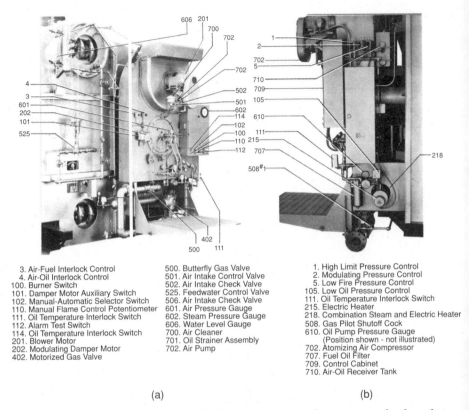

3. Air-Fuel Interlock Control
4. Air-Oil Interlock Control
100. Burner Switch
101. Damper Motor Auxiliary Switch
102. Manual-Automatic Selector Switch
110. Manual Flame Control Potentiometer
111. Oil Temperature Interlock Switch
112. Alarm Test Switch
114. Oil Temperature Interlock Switch
201. Blower Motor
202. Modulating Damper Motor
402. Motorized Gas Valve

500. Butterfly Gas Valve
501. Air Intake Control Valve
502. Air Intake Check Valve
525. Feedwater Control Valve
506. Air Intake Check Valve
601. Air Pressure Gauge
602. Steam Pressure Gauge
606. Water Level Gauge
700. Air Cleaner
701. Oil Strainer Assembly
702. Air Pump

1. High Limit Pressure Control
2. Modulating Pressure Control
5. Low Fire Pressure Control
105. Low Oil Pressure Control
111. Oil Temperature Interlock Switch
215. Electric Heater
218. Combination Steam and Electric Heater
508. Gas Pilot Shutoff Cock
610. Oil Pump Pressure Gauge
    (Position shown - not illustrated)
702. Atomizing Air Compressor
707. Fuel Oil Filter
709. Control Cabinet
710. Air-Oil Receiver Tank

(a)   (b)

Figure 14.1   An industrial watertube boiler has many controls, gauges, and valves that may need an operator's attention. (*a*) Watertube boiler burner front; (*b*) control cabinet front. (*Courtesy Cleaver Brooks Co.*)

Overheating damage due to low water is still the most prevalent cause of boiler outages on low- and medium-pressure industrial boilers. Utility boilers have tube failures as the leading cause of outages.

Effect of Low Water

Low water in a boiler may lead to anything from leakage to an explosion, depending a great deal on the type of boiler, the rate of firing, and just how low the water gets. If the boiler is a type with a crown sheet over the firebox or combustion chamber, such as a locomotive or scotch marine type, a rupture of the crown is almost inevitable if the water drops below the level of the crown sheet, for the bared metal soon attains such a temperature (red heat) that its tensile strength drops to a dangerously low point. A rupturing crown sheet often is exceedingly violent, and many serious explosions have resulted.

In firetube boilers, the first result of the water level dropping below the safe level when a hot fire is carried may be leakage at the rear ends of the upper rows of tubes. As the water recedes from tubes exposed to high-temperature gases, the expansion of the tubes is so great that their rolled-in seat is broken. Leakage may appear from the rear ends of each succeeding row of tubes as the water level drops further, until distortion of the shell plates and heads, with leakage at the seams, usually occurs. An explosion due to low water is uncommon in this type of boiler because of the many points of leakage to give warning. However, the tubes may collapse.

The effects of low water on water tubes are similar to those on fire tubes. The tubes expand as the water leaves them, and they break their expanded seats, leakage being caused. Excessively low water may result in tube rupture and tube melting. Figure 14.2 shows the tubes that are left after a severe dry firing. The boiler was equipped with two low-water fuel cutoffs, a feedwater regulator, and a low-water alarm. When these devices were periodically tested was not logged. Figure 14.3 shows a low-water fuel cutoff full of sludge deposits because of failure to periodically blow or flush the sludge out of the chamber. Persistent management insistence of logging actual testing of these vital safety devices would prevent a majority of low-water failures.

Figure 14.2   Melted tubes due to low water in watertube boiler equipped with two LWCOs and a low-water alarm. (*Courtesy Factory Mutual Engineering*)

Figure 14.3 Low-water fuel cutoff is plugged with sludge from lack of testing. (*Courtesy Factory Mutual Engineering*)

Improvements in the operation and care of a boiler to minimize the possibility of developing a low-water condition depend on frequent checking and testing of feed, condensate return, pumps, and similar components of a boiler water loop system that is supposed to keep the boiler supplied with water. Regulators and associated alarms with low-water cutoffs are the next and usually last defense against low water, unless an operator takes corrective actions. The best way to test the low-water cutout (LWCO) is by duplicating an actual low-water condition. Slowly drain the boiler (through the blowdown line) while under pressure. If a heating boiler does not have proper drains for doing so, be sure to correct this condition. Many operators drain only the float chamber of the cutout for this test. But the float-chamber drain is only for blowing sediment out of the float chamber. Usually the float will drop when this drain is opened because of the sudden rush of water from the float chamber.

Tests often show that draining the float chamber will indicate that the cutoff performs satisfactorily, but when proper testing is done by draining the boiler, the cutoff fails to work. Daily checking of the LWCO by draining the float chamber (or electrode chamber) is good practice. But at the beginning of each season, duplicate an actual low-water condition.

To test a boiler for whether the lower gauge glass is obstructed even though the gauge glass is half full of water, open the try cocks on the water column. If all show steam, it means the bottom connection is obstructed, permitting steam from the top connection to condense in the gauge glass. The boiler should be shut down immediately and inspected for possible dry-firing damage. Naturally, the bottom connection of the water column and gauge glass should be cleaned of all obstructions before the boiler is returned to service.

If an unusually high feedwater pressure is necessary to maintain the water in the boiler, check the feedwater valves and lines to make sure that a valve has not broken off its seat or that there is not some obstruction in the line itself. Some methods of feedwater treatment have been known to deposit chemicals inside the feedwater line, making it impossible to get water into the boiler. Also look for leaks due to cracked or corroded piping of the feedwater (or condensate line on heating boilers) especially if it is buried anywhere in the system.

If water is not visible in the gauge glass because of failure of the feedwater supply, immediately do the following:

1. Shut off the fuel to the burners and secure the burners.

2. Check the water level by trying the try cocks and water-column drain. If definite low water is indicated below the gauge-glass level, close the main steam valve and feedwater valve.

3. If the boiler is equipped with one, open the superheater drain.

4. Continue operating forced-draft and induced-draft fans until the boiler cools gradually.

5. Let the pressure reduce gradually and when the furnace area is sufficiently cooled, check for leaking tubes and other signs of overheating damage. On firetube boilers, look for cracked or warped tube sheets, broken and leaking stay bolts in the water legs. On scotch marine boilers, check for cracked or leaking furnace-to-tube sheet welds. On cast-iron boilers, look for cracked sections. On steel boilers, check for leaking joints on longitudinal or circumferential welds or riveted joints.

6. If no leakage is evident, give the boiler a hydrostatic test of 1½ times the allowable working pressure. Then again check for leakage at all critical parts of the boiler. If leakage is observed during the initial check or during the hydrostatic test, notify the authorized boiler inspector immediately so she or he can inspect the boiler and advise on permissible repairs.

Tube Failures in Generating Plants

The top 10 failure mechanisms for tubes in generating plants are high-temperature creep, fly-ash erosion, short-term overheating due to flow disturbances, sootblower erosion, welding defects, fire-side corrosion, corrosion fatigue, falling slag, thermal fatigue, and vibration fatigue.

Boiler tube failures are the largest single source of fossil plant availability loss per a recent study of the North American Electric Reliability Council. Coal-fired units of 200 MW or larger have experienced an average of 3.3 percent plant availability loss from tube failures.

By closer tracking of deterioration by NDT, and good recordkeeping on trends, tube failures can be reduced by analyzing the causes and then taking corrective actions *before* a failure can be expected. This involves a cooperative assessment effort and scheduling appropriate operations, maintenance, and inspection activities.

Grate-fired boilers need special attention in order to prevent low-water damage during a power failure. A steam-driven boiler feed pump, or similar back-up pump is required in the event a motor-driven boiler feed pump cannot be operated because of a power failure. The effects of a power outage on the dumping of grated burning contents must also be considered, as many grate dumping systems are electrical-motor driven. An emergency gas engine should be incorporated into the grate dumping system. The effect of a power failure extends to the functioning of electrically operated controls. A grate with burning fuel on it can cause overheating damage to tubes if no water can be supplied to the boiler.

It is also essential to make sure all fuel-feeding devices on waste fuel boilers actually stop the flow of feed into the boiler during any emergency operating condition, such as low water. A remote disconnect switch to the feeding device is a good safety investment. For example, making sure all black liquor to a recovery boiler stops flowing to the burners in the event of a tube failure.

Poor circulation due to design is difficult to detect, and quite often tube failures from this cause are attributed to water chemistry problems. If water chemistry is not properly controlled, a buildup of internal scale can cause overheating tube damage in the affected area. However, mass flow velocities in tubes that do not match fire-side heat mass flows can also cause tube overheating. These low-water velocities can also cause film boiling inside the tubes, which acts as an insulator and can cause above-normal tube metal temperatures. Local overheating of tubes can result. If the circulation problem is not solved, repeated tube failures may result. Correction of poor circulation usually requires the manufacturer of the boiler to review the design. This usually requires an adjustment of the "pumping" head to the tubes to ensure adequate circulation.

Corrosion

Corrosion damage can be a slow process of deterioration and is usually controllable by water treatment and internal inspections. The most frequently encountered corrosion is that due to: (1) dissolved oxygen; (2) caustic attack which generally causes gouging out of metal; and, in some boiler treatment programs, (3) chelate corrosion which also destroys the thin film of magnetic iron oxide which protects metal surfaces against corrosion. Corrosion is further classified into *general*; *localized* as in pitting; *crevice* corrosion that occurs as a corrosive liquid settles into a crevice; and *galvanic* corrosion. In galvanic corrosion, when two dissimilar metals exposed to each other in a conductive environment come in contact, if there is a potential between them, a current flows between them. The less resistant metal, called anodic, loses metal, while the more resistant metal, called cathodic, has the less resistant metal deposit portions of its lost metal at the juncture point.

Another form of corrosion is called *leaching,* or dealloying, and in some cases dezincification, such as for brass material. Gray cast iron can be dealloyed; the ferrite is converted to iron oxide in a slightly acidic water. *Fretting corrosion* occurs when two metals rub each other and thus destroy the protective film, exposing their surface to new oxidation.

Internal corrosion. Internal corrosion is an electrochemical deterioration of the boiler surfaces, usually at or below the water line. See Fig. 14.4. The pH value of the water is a measure of its alkalinity or

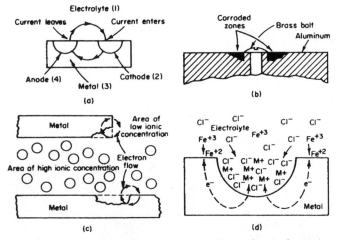

Figure 14.4 Types of corrosion attacks. (*a*) Electrochemical reaction theory favors ionic current flowing in electrolyte corrosion. (*b*) Galvanic corrosion attacks less noble of two metals in contact with each other. (*c*) Crevice corrosion occurs at localized structural fault. (*d*) Chemical pitting is caused by high chloride or oxygen concentration.

acidity and usually has a direct bearing on the corrosive properties. All water contains alkaline (hydroxyl, OH), ions and hydrogen (H) ions. The product of these concentrations is always approximately 10^{-14}. The pH value of the water is the log of the reciprocal of the H ion value.

If the water is neutral, the OH ion concentration is 10^{-7}; therefore, the H ion must also be 10^{-7}. Then the pH is 7. Waters with an H ion concentration greater than 10^{-7} are acid. Hence, a pH below 7 indicates acidity; over 7 designates an alkaline condition.

The detrimental effects of corrosion depend on its rate of penetration. Corrosion affecting large areas of boiler plate is not likely to penetrate as rapidly as localized corrosion in small areas. The former condition sometimes is difficult to see, and it may progress unnoticed to a dangerous extent.

Localized corrosion may be in the form of pitting or *grooving* (Fig. 14.5). Pitting is caused by repeated breaks at the same spot in the protective oxide film. It is affected by the type of surface, especially if mill scale or such surface irregularities are present. The pits may be as small as a pinhead or as large as a half dollar.

Effects of corrosion.   The strength of a shell or drum is reduced as corrosion or the formation of closely spaced pits progresses. The allowable pressure quite often must be reduced from the weakening effects of a plate corroding away.

Example   A firetube boiler, 60-in. ID with a longitudinal efficiency of 100 percent had an original thickness of 0.5 in. Pitting was evident during an inspection, and thickness tests indicated an average loss of ³⁄₁₆ in. of shell thickness. The boiler was originally designed for 250 psi with a S.F. of 4, and the safety valve is set for this pressure. The material for the shell has a tensile strength of 70,000 lb/in.² Must the pressure be reduced because of thinning?

Use the shell equation illustrated in Chap. 9.

<table>
<tr><td>(a)</td><td>(b)</td></tr>
</table>

Figure 14.5   Two types of corrosive attacks on tubes. (*a*) Pitted tube; (*b*) caustic gouged tube.

$$P = \frac{SEt(t - C)}{R + (1 - y)(t - c)}$$

where $S = 70,000/4 = 17,500$
$E = 1.0 \quad t = 0.5 - 0.1875 = 0.3125$ in.
$C = 0$
$R = 30$
$y = 0.4$

Substituting,

$$P = \frac{17,500(1)(0.3125)}{30 + 0.6(0.1875)}$$

$$= 181.2 \text{ psi}$$

The pressure must be reduced, and new safety valves are needed.

Grooving is a form of deterioration of boiler plate by a combination of localized corrosion and stress concentration. See Fig. 14.6. It is found usually in areas adjacent and parallel to riveted seams or in the flanges of dished heads. The groove is usually ⅛ to ½ in. wide and may be several inches to several feet in length. Since the reduction in thickness occurs in a part that is subjected to stress concentration, grooving may be very serious. If it occurs to any extent in the seams of an unstayed boiler shell or drum, no repairs are possible. The allowable pressure must be reduced considerably, or the boiler must be permanently removed from service. In all such cases, the advice of an authorized inspector should be followed.

Cracking Problems

Crack initiation and propagation is a matter of concern in all inspections of power plant components. Crack growth has three components.

1. An incubation period such as cold working, cyclic stressing, chemical attack on the grain structure of the material.

2. Crack propagation from an initiating point of metal separation due to items mentioned in (1).

3. Cracking up to the *critical size* where a component is now too weak to carry the load imposed in it, and final failure to destruction occurs rather rapidly.

Fatigue and creep can initiate cracks. *Fatigue* usually occurs at a stress concentration point such as a corner of a part, or a pitted surface, and usually originates at the surface. *Creep* usually causes cavi-

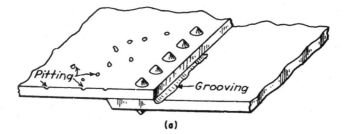

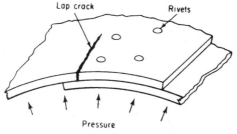

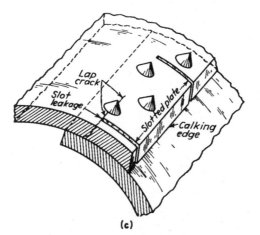

Figure 14.6   Problems with old riveted boilers. (a) Metal deterioration from grooving; (b) lap crack from joint not being a true
circle, causing stress concentration at lap; (c) slotting was used
to determine if there existed a "through" lap crack.

ties to form at grain boundaries, either within the material or at the
surface, and then the cavities coalesce to form a crack. Creep is a
long-term damage phenomena as the material loses its ability to
stretch with load.

Exposure of metals to temperatures that are only slightly above
their design temperatures can cause a dramatic reduction in creep

life. A temperature increase of 50°F (approximately 5 percent) can reduce the expected creep life from 40 years to 4 to 8 years. Further, operating only 10 to 20°F over the metal's design temperature for long periods of time can cause a life reduction of 10 to 20 years. This consideration is especially important in boiler tube and header assemblies, where metal temperature can vary across the width of the boiler and can be excessively high in one area, and also on high-temperature utility steam piping.

Superheater header cracking has been experienced by some utilities on boilers subject to cycling service. This type of cracking is classified as being due to a combination of *fatigue and creep*. See Fig. 14.7. High temperatures and repetitive stress with cycling accelerates the cracking as units age. Many of these cracks were found during life-extension studies of a boiler. The firm B & W also alerted their customers to check headers manufactured of 1¼ Cr-½Mo-Si (SA 213-T11 and SA 335-P11) that were operating at 975°F and above, and were manufactured between mid-1951 and mid-1965. The ASME code was changed in 1965, which reduced the allowable stress for these materials as more service data became available. The tube hole to tube hole header cracks shown in Fig. 14.7 represent creep-fatigue cracks. As the age of high temperature steam-generating units increases, inspection for ligament cracking on superheater headers is required. Experienced boiler engineers from the manufacturer stress several measures to determine the remaining life of units subject to long-term, time-dependent failure mechanisms such as creep-fatigue: measuring creep swelling of the suspected boiler component and surface replication.

Corrosion fatigue usually causes transgranular cracking of metals that are cycled under stress in a corrosive environment. *Stress corrosion* cracking occurs where a metal is in contact with a corrosive medium while under heavy stress. It requires the presence of highly stressed metal and a corrosive medium. Normal stresses are then magnified above normal.

Caustic cracking (caustic embrittlement) is a serious type of boiler metal failure characterized by continuous, mostly intergranular cracks. The following conditions appear to be necessary for this type of cracking to occur: (1) the metal must be stressed; (2) the boiler water must contain caustic sodium hydroxide; (3) at least a trace of silica must be present in the boiler water; and (4) some mechanisms, such as a slight leak, must be present to allow the boiler water to concentrate on the stressed metal. Caustic cracking was a particular problem in older boilers with riveted drums because of stresses and crevices in the areas of rivets and seams. While this type of cracking has become less frequent since the advent of welded drum boilers,

(a)

(b)

Figure 14.7   (a) Creep-fatigue superheater header cracks. (b) Superheater longitudinal weld seam crack due to creep. (*Courtesy Babcock & Wilcox Co.*)

rolled tube ends are still vulnerable areas of attack. The possibility of caustic cracking should be considered in establishing any inspection program.

Any abrupt section in a boiler component under stress causes a stress concentration that multiplies the normal expected stress at that section. If the boiler is cycled, fatigue cracks may appear. Sharp corners require alertness during inspection to detect possible incipient cracking.

Cycling of boilers is very evident in those units subject to peaking service. This is also prevalent in heat recovery steam generators (HRSG) that are used in combined-cycle cogeneration plants; combined with fast-starting gas turbines, they are often used as peaking units. The cycling service causes repetitive stresses on rolled or welded tube joints from severe thermal swings as a unit is brought on and off the operational cycle. Eventually the following components of the boiler may develop fatigue cracking from thermal cycling: economizer inlet headers, furnace wall tube joints, and tube to header welds. Most older boilers were designed for baseload service. Cyclic operation causes differential expansion and contraction between headers and tubes and other pressure-containing parts. Cyclic service thus produces heavy strains as on-off cooling occurs, and eventually cracks appear due to fatigue or repetitive stresses and strains. Economizer inlet headers are also very susceptible to thermal cycling cracks as cold water enters a warm header during adjustments in drum water level.

A common operator procedure is not to open superheater drains on any shutdown of peaking units, but to open the drains during startups. However, to avoid condensate buildup in the superheater during cyclic shutdowns, the superheater drains should be opened to avoid a quenching effect that produces large thermal stresses from contraction.

As is evident, the *method of operation* is also a consideration when making inspections for possible boiler problems.

Lap crack was a very pronounced concern on old riveted boilers. Probably the most dangerous of all fatigue cracks was the lap crack developing unseen between rivet holes of the longitudinal lap-riveted seam of lap boiler shells. This defect was induced by the fact that a lap-seam boiler is not rolled into a true circle, and a bending stress is concentrated at the offset of the lap by the breathing action of the shell (see Fig. 14.6b). Many serious explosions have resulted from such cracks. Usually, leakage from the seam was a warning of a lap crack. If any leakage or suspicion of this defect existed, the plate was slotted (Fig. 14.6c). If any leakage came from this slot, a crack was usually indicated. It was often necessary to remove several rivets in suspected regions so that the inside of the rivet holes could be examined. No repairs were permitted on lap cracks. The boiler was condemned immediately.

Hydrogen-induced damage usually occurs in the following identifiable boiler components:

1. Furnace tube failures usually are located in the high-heat input zones of the furnace. The hydrogen damage is traceable to the inside surface of tubes exposed to high-temperature furnace gases. It usually occurs beneath relatively dense mineral deposits, termed "underdeposit corrosion." It may also occur in areas of fluid disturbance, such as tube bends. The hydrogen produced in the corrosive reaction between the fluid and tube metal causes the hydrogen to diffuse into the metal, decarburizing the tube metal on the inside and causing microcracks to appear in the tube metal. These microcracks grow until a piece of the tube is blown out as shown in Fig. 14.8a.

2. Figure 14.8b illustrates the role of diffusible hydrogen in causing a delayed crack in a weld. Dissolved hydrogen entering into the weld may be the result of the welding process, such as from moisture being present. The use of low-hydrogen classified electrodes is now the acceptable method to avoid hydrogen infusion. Electrodes must also be dry prior to any welding operation. Recent developments include the manufacture of electrodes with coatings that may resist moisture pickup. However, welders may open a sealed container of very dry electrodes and use them slowly one by one. This will expose the elec-

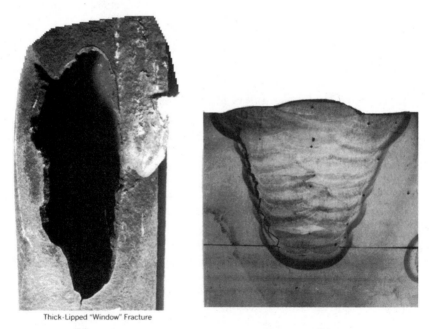

Thick-Lipped "Window" Fracture

(a)   (b)

Figure 14.8 Hydrogen-induced cracking failures. (a) Furnance tube failure due to hydrogen infusion. (*Courtesy Babcock & Wilcox Co.*) (b) Heat-affected zone cracking due to hydrogen infusion. (*Courtesy Welding Journal of AWS.*)

trodes in the container to moisture in humid atmosphere conditions, which may cause hydrogen to diffuse into the weld and cause hydrogen-induced cracking. Electrodes should be heated at the work site to avoid moisture contamination.

Weld cracks. Modern construction of boilers and pressure vessels has been possible with the advancement of welding knowledge. However, many variables must be considered in obtaining a proven, good welded joint. It is necessary to consider the following variable factors: weldability of the base metal, shape of the joint, welding process to be used, procedure to be followed in performing the weld, size and type of electrode, current (or temperature of weld) to be applied, preheat and postheat to be used, NDT to be applied to check the joint, and the use of properly qualified welding procedures and welders. Final acceptance considerations may also involve a hydrostatic test.

Defects in welding techniques or procedures can produce weld cracks (Fig. 14.9). Small cracks can usually be veed out and rewelded. Methods of NDT such as dye penetrant are used to determine if the end of a crack has been reached in the veeing-out process. Preheat and postheat treatment may be required on the weld repair following ASME Section IX rules. The repair must usually be approved and then witnessed by an authorized inspector.

Defects in welds considered unacceptable are cracks, areas with incomplete fusion or lack of penetration, and slag inclusions of ¼ in. for plate thickness up to ¾ in. with the maximum slag-inclusion length being ¾ in. for plate thickness over 2¼ in. The ASME publishes porosity charts which detail the number and size of permitted porosity in a given length per weld thickness. Welds can fail from the appli-

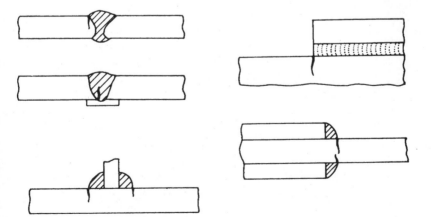

Figure 14.9   Areas of welded joints to check for cracks.

cation of repeated stress, especially where a discontinuity may be present. Figure 14.9 illustrates typical areas of welds that should be suspect during inspection for possible fatigue crack initiation. Heat-affected-zone cracking has been the subject of much research. Brittle fractures in the heat-affected zone may be initiated from the presence of hydrogen, reheat cracks, fatigue cracks, or lack of root or sidewall fusion. Postweld heat treatment will generally improve HAZ toughness of most grades of carbon steels and this will help in preventing brittle failures in the HAZ of a weld. As a quality-control device in ensuring a sound weld, NDT inspections of welds have been materially beneficial in reducing weld crack failures.

Fire-Side Problems

External corrosion or deterioration of boiler surfaces on the fire side may be a continuous process. It is a chemical combination of the metal, known as oxidation or rust. Normally, this action would not progress appreciably in the life of a boiler. However, most boiler surfaces are coated with soot or ash on the fire side. The sulfur content of the soot combines with any moisture to form a sulfurous acid which is highly corrosive. Hence, a minor leak may cause a serious defect to develop within a few years; even though there is no leak, the boiler may "sweat" when idle in humid weather, and such moisture in combination with the soot will cause trouble.

Continued leaks from any source should not be tolerated whether they are from a roof, valve packing, gaskets, piping, or other sources. Water dripping onto a boiler will cause damage. Leaky soot blowers are a frequent source of external corrosion of water tubes. The soot-blower valves should be kept tight, and the piping drained of condensate before blowing soot.

Handhole- and manhole-gasket leakage frequently causes damage to the flange or surrounding plate by external corrosion. The dry sheet of boilers with internal manholes or handholes should have a ¾-in. drain hole in the bottom so that any leakage from a manhole gasket, soot blower, or tube will drip through the hole and give an indication of the leak. Otherwise, water might collect on the bottom of the dry sheet and cause serious damage.

Erosion is closely allied with external corrosion in its effect but it is purely a mechanical action, a wearing of external surfaces by abrasion. The gas-entrance ends of tubes in firetube boilers may become thin after 10 to 20 yr because of the scouring action of soot particles entering the tubes at high velocity. This effect may be due also to the fact that internal corrosion is more rapid where the high-temperature zone causes a higher evaporation rate.

Erosion by improperly adjusted soot blowers is not uncommon. In a few weeks of use, a hole may be worn through several tubes by one faulty jet of a soot blower. The action resembles sandblasting. Erosion as a result of flame scrubbing probably does not have the opportunity to become serious before the damage done by localized overheating makes the condition evident.

When large quantities of ash are produced in coal-fired boilers, slagging and fouling problems may occur in the furnace and convection sections of the boiler, especially if soot blowers are not operated sufficiently to remove the ash. Plugging of convection passes can cause the fly ash to become sticky, thus accelerating further slagging. Metal under slagging may be attacked in the presence of moisture, leading to fire-side deterioration. Soot blowers and other boiler auxiliaries (Figs. 14.10a and 14.10b) require preventive maintenance to avoid boiler problems.

Cold-end corrosion frequently occurs on economizer tubes, or any boiler with low exhaust temperatures sweeping heat-transfer surfaces. The term *dew-point corrosion* is also used. It occurs primarily in solid fuel burning boilers and in oil-burning boilers with some sulfur in the fuel and also compounds of vanadium and sodium. As the exhaust gases are cooled below about 325°F, condensation of water in the flue gas can occur. Sulfuric acid formed by the combining of sulfur in the fuel with moisture also condenses at about 280°F. Thus, acid and moisture will attack any heat-transfer surface at the back end of the boiler. This includes HRSG used in oil-fired gas turbine combined-cycle plants. Modern practice is to use more expensive corrosion resistant materials, such as economizers, in the cold end of the boiler. Figure 14.11a shows the damage to an inlet economizer of a HRSG from operating below the dewpoint temperature.

Power plants *burning municipal wastes* experience tube wastage from high concentrations of chlorides, alkali metals, and heavy metals in the fuel being fired. Figure 14.11b shows the effect to a water-wall tube that suffered corrosion-induced damage. Most refuse burn-

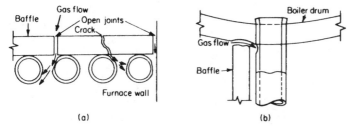

(a)   (b)

Figure 14.10 Plugging of gas passages and faulty baffles can cause tube failure from erosion. (*a*) Tube being eroded by scrubbing gas flow; (*b*) boiler drum being eroded by concentrated gas flow scrubbing through faulty baffle.

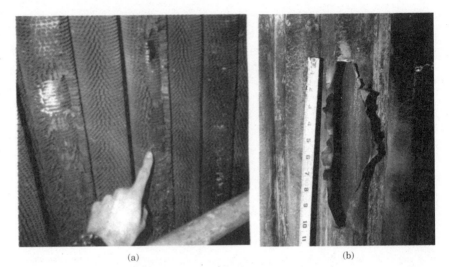

(a)   (b)

Figure 14.11 (*a*) Economizer tube of a HRSG being attacked by corrosion due to feed-water temperature being a few degrees below the dew point of flue gas. (*b*) Refuse-derived fuel burning boiler has waterwall tubes attacked by corrosion-induced reaction from waste fuel. (*Courtesy Power magazine.*)

ing plants have to resort to alloy 625 weld overlays to arrest the tube wastage. This alloy has high nickel content that provides more resistance to chloride attack than carbon or stainless steels. Newer installations may also use composite tubes with an alloy 825 on the outer side of the tube. This has about 42 percent nickel content.

Fly ash corrosion is accelerated in the dew-point temperature operating range and when a boiler is idle. This occurs in fossil-fuel firing boilers. Fly ash can accumulate in the flue gas path, and this acts like a sponge to collect moisture and acid, causing tube corrosion especially as the boiler cools during shutdown periods.

Minimizing fly-ash erosion. Tube failures due to fly-ash erosion represent the second largest cause of availability loss in fossil plants. The most important variables affecting fly-ash erosion are particle velocity, angle of impingement, and coal and fly-ash quality. The effect of impact velocity is particularly important because the rate of erosion loss is usually found to be proportional to the velocity raised to the power of between 2 and 4. Thus, slight increases in flue-gas velocity have a substantial effect.

The *remedial measures* most commonly used consist of the temporary, sacrificial type. Pad welding and arbitrary placing of shields are widely used, with replacement of tubes as a final resort. These measures do not eliminate the failure cause and thus inevitably result in repeated failures and expensive maintenance and ultimately to decreased availability.

High-flow flue gas control. Another method to control fly-ash erosion is by installing screens and baffles in regions of high flue gas velocity in order to control the impact velocity of the flue gas on the tubes.

The technique developed for the proper placement of the screens and baffles is referred to as *cold-air velocity testing.* This involves the use of detailed and local airflow measurements in the boiler during an outage with the fans at some nominal output. Once high-flow regions are determined, diffusion screens and baffles are designed and installed. Effectiveness is checked with another cold-air velocity test and the unit is returned to service for practical testing of particle velocity control to minimize fly-ash erosion. This method of minimizing fly-ash erosion by controlling particle velocity was developed in Canadian fossil-fuel-burning plants.

Solid particle erosion. When oxide scale exfoliates from the inside of boiler, superheater, and reheater tubes, and also steam leads in piping, these solid particles are carried by the high-velocity steam into steam turbines, where the particles erode steam nozzles and valves, buckets, and blades as they impact like sandblasting on the turbine components.

This erosion or wear eventually changes the shape of blades, nozzles, and valve seats, reducing the efficiency of steam flow through the turbine, and also potentially causing severe enough wear so that parts can break inside the turbine, resulting in serious damage within the machine and causing expensive repairs.

Oxide-side exfoliation usually results from up and down temperature swings that cause tubes to expand and contract, thus loosening the scale.

Solid particle erosion reduction is possible by:

1. Avoiding too many temperature swings in a boiler (baseload operation).

2. Chemical cleaning of tubes whenever oxide scale is noted.

3. Chromating the inner surfaces of tubes to prevent the formation of oxides in a recent development.

4. Improving water-treatment control to prevent metal oxidizing.

5. Use of condenser bypass systems during boiler startup and shutdowns which lessens temperature swings in superheaters and reheaters, and thus prevents scale from flaking off due to rapid temperature changes.

6. Coating critical steam path turbine parts with coatings, such as chromium carbide to resist the impact of the eroded scale coming from boiler tubes and leads.

Flame impingement. *Flame impingement* is a source of damage to boilers and refractory. If the flame impinges directly on the boiler shell (Fig. 14.12a) excessive evaporation will be caused on the water surface over that point. The high temperatures may cause damage through local scale formation or corrosion which otherwise would be dormant, or the temperature may be high enough to cause serious damage by overheating the plate.

Direct impingement of flame on tubes of watertube boilers may cause steam pockets. That is, evaporation and resultant circulation upward in a tube may be more rapid than the rate at which cooler

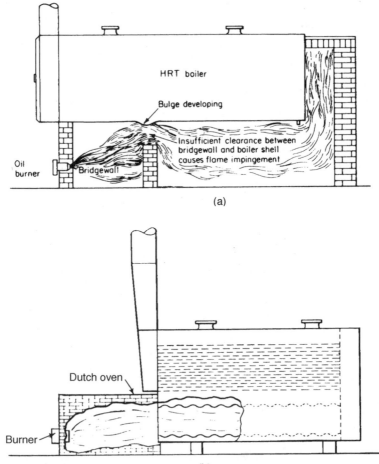

HRT boiler

Bulge developing

Insufficient clearance between bridgewall and boiler shell causes flame impingement

Oil burner

Bridgewall

(a)

Dutch oven

Burner

(b)

Figure 14.12 (a) Flame impingement can cause severe local overheating damage; (b) Dutch oven increases furnace time for fuel burning.

water can be supplied from its lower end. A steam pocket, serious overheating, and failure of the tube usually result. Water tubes are also susceptible to the same results of flame impingement as those mentioned with reference to firetube boilers.

A reduction in thickness, due to erosion by particles of burning carbon and of fly ash, will result if flame impingement continues. Flame scrubbing of refractory greatly shortens its life, owing to erosion and overheating.

Combustion problems. Combustion processes and combustion problems are many and involve the burning of different fuels in proportioned furnaces so that proper air-fuel ratios and good ignition with a stable flame are maintained.

Oil fires sometimes pulsate or flutter to the extent that the entire boiler setting may vibrate. This effect may be traced usually to a pulsating oil-burner pressure resulting from a reciprocating oil pump. Use of an air-cushion chamber usually solves this problem.

Burners should never be lighted by the heat of the refractory or from the flame of a burner. Use the igniter. Furnace-gas explosions may result if these igniting precautions are neglected.

Ignition stability is important to safe burning of fuels in suspension. Coal of low volatile (gas) content is sometimes unstable when pulverized and when operation is at low loads. Oil-burner instability can be traced usually to a clogged oil system or to improper oil temperature.

One of the most common causes of furnace explosion is the momentary failure of ignition during regular operation. During the pause, unburned fuel enters the furnace, and highly combustible gases, distilled by the heat of the firebox, fill the boiler setting. These gases may penetrate a crevice in the refractory or ash where red heat exists, and a blast results. Furnace explosions may be caused also by accumulations of unburned combustible igniting spontaneously. These explosions may cause serious damage.

Modern, properly installed combustion flame safeguard systems will help in eliminating many, if not most, furnace explosions resulting from poor burning. See Chap. 11. Main-flame and pilot scanning cells with electronic relays that can shut down a burner system within 2 to 4 sec after loss of flame are now commonly specified. Programmed purging prior to light-off and pilot proving are now required. Fuel trains must provide redundant safety shutoff valving.

The inherently limited combustion volume of internally fired boilers and the requirements of fuels such as wood-waste products sometimes demand additional volume. This is often produced by a *Dutch oven,* which is actually an external, primary furnace that leaves the boiler furnace as a secondary combustion chamber (Fig. 14.12*b*).

In the paper industry, *recovery boilers* present an additional hazard besides fire-side fuel explosions. This is the problem of *smelt-water reactions* taking place in the furnace. The best prevention is to maintain pressure-part integrity in order to avoid water entering the furnace in any form. The concentration of the black-liquor fuel must also be monitored so that too dilute a solution does not enter the furnace and cause a smelt-water reaction to take place. Use of water to wash down tubes or deposits in the presence of a smelt bed in the furnace must be avoided. The Black Liquor Recovery Boiler Advisory Committee, which consists of users, manufacturers, and insurance representatives, stresses the need to remove as rapidly as possible any water source that could enter the furnace. Emergency procedures are recommended under these conditions, one of which is to drain the boiler as rapidly as possible and in accordance with manufacturers' recommendations to a level 8 ft above the low point of the furnace floor.

Combustion air for small boilers needs attention. To ensure proper draft-air opening for a packaged boiler installed in a closed machine room there should be fixed opening for fresh air, an average area of 2 ft^2 for each 100 boiler hp. Opening windows is not the answer because they are often closed in cold weather. Then the boiler starves for air. For each boiler horsepower, about 10 ft^3/min of air is needed.

Some *safety precautions* to be observed in operating and maintaining an automatic burner system are the following:

1. Always close off all manual fuel valves before working on a burner or disconnect the wiring to automatic fuel values or do both.

2. Never stand in *front* of a burner or boiler during start-up.

3. Never manually push in relays unless the manufacturer's instructions so advise.

4. Never permanently block in relays with rubber bands, sticks, or other devices.

5. Never change the safety-switch timing of a flame supervisory control. If the system is locking out, correct the cause, *not* the symptom.

6. Never install jumper wires or bypass any safety interlock switches.

7. Before starting a burner, visually inspect every combustion chamber to make sure there is no accumulation of combustible.

8. Regard every system lockout as a safety lockout until proved otherwise by competent personnel.

Piping problems. *Piping* evaluations constitute an important part of plant design and maintenance. Many industrial plants replace the boilers after 20 yr or so with new boilers designed and operated at higher pressure. Often, the original boilers operated at below 125 psi, and many of the pipe fittings and valves throughout the mill were of cast iron, 125-lb standard. In installing new boilers for higher pressure, caution should be exercised to replace all low-pressure pipe fittings with fittings designed for the new operating pressure. Whenever a new system is started up, not only should the boiler be boiled out, but this process should be applied to the complete system. Every piece of pipe has foreign matter deposited inside. Unless it is cleaned out properly by personnel familiar with boiling out and flushing piping so as to avoid undue strains on piping, trouble may develop. Elbows, bends, and other dead pockets in piping are dangerous spots and must be thoroughly cleaned.

High-pressure *steam pipe failures* is a utility concern because of the unexpected ruptures of hot reheat pipes in several steam-generating stations during the last few years. The failures occurred in a welded longitudinal seam weld that were over 20 years old, with the pipe material being a Cr-M0 steel. A periodic inspection program for creep, crack detection, thickness reduction, and similar steam-pipe evaluation methods are being used to estimate the suitability of the steam pipe for further service, and whether repairs may be required during the outage. Many utilities are replacing the pipe with modern seamless pipe to avoid such a disastrous failure as a pipe explosion.

Fracture mechanics and life-prediction programs are also being used to monitor the future life of older steam piping.

Each failure in the welded piping from utility generating plants has resulted in extensive studies of the failed welds (Fig. 14.13), with several explanations offered as to cause, namely, creep, toe of weld crack initiation, porosity of weld, heat-affected zone crack initiation, nonmetallic inclusions, and similar weld crack initiation criteria. Angle-beam ultrasonic testing with a high degree of sensitivity is the primary inspection method used to detect this type of piping cracking, following Electric Power Research Institute guidelines.

Pipe size and supports. The size of pipe should be adequate for conservative flow velocities, or else excessive friction head will result. The maximum recommended velocity for steam flow is 5000 ft/min for heating service (up to 15 psi), 10,000 ft/min for high-pressure saturated steam, and 14,000 ft/min for high-pressure superheated steam. The Crane Co. has found velocities of up to 20,000 ft/min reasonable for high-pressure superheated steam in large pipes.

Example In order to select the proper size of pipe, the following formula may be used:

$$D = 12 \left(\sqrt{\frac{C}{0.7854F}} \right)$$

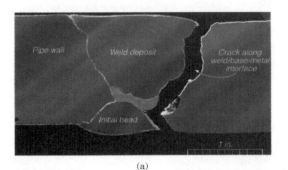

(a)

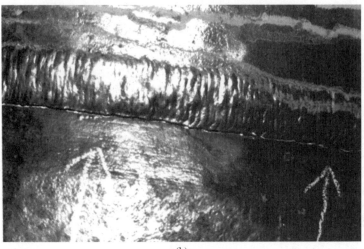

(b)

Figure 14.13 Utility steam reheat pipe failures. (*a*) Longitudinal weld cracking failure on utility reheat steam line. (*Courtesy* Power *magazine*) (*b*) Circumferential welding crack in utility reheat steam line. (*Courtesy Babcock & Wilcox Co.*)

where C = steam flow, ft³/min
$\quad\quad\quad D$ = required diameter of pipe, in.
$\quad\quad\quad F$ = permissible steam-flow velocity, ft/min

To select a proper pipe size for 12,000 lb/hr of saturated steam at 300 psi, first find the density for the pressure from the steam tables. For 300-lb pressure, it is 0.67 lb/ft³. Therefore,

$$\frac{12,000}{0.67} = 17,910 \text{ ft}^3/\text{hr or } 298 \text{ ft}^3/\text{min}$$

Substituting in the foregoing formula, we have

$$D = 12\left(\sqrt{\frac{298}{0.7854 \times 10,000}}\right) = 2.3 \text{ in.}$$

It would be customary to go to the nearest *larger* standard pipe size, and so 2½-in. pipe would probably be used in this case.

The density of steam rises with the pressure; thus, the *volume* for a given weight of steam is less at higher pressures, and smaller-sized piping may be used.

Pipe supports should be designed to carry the weight of the piping *full of water.* They should be spring-mounted for heavy service in order to provide proper support during vertical motion of the pipe resulting from expansion. Hanger-type supports should be adjustable by means of a turnbuckle or nut threaded to the top of the hanger rod so that compensation may be made for settling of the supporting structure.

Expansion of pipe anchored at each end may set up severe stresses. It is good practice to provide an expansion joint of the slip, bellows, or loop type for long horizontal runs of steam piping. The length of expansion is calculated by:

$$(t_1 - t_2) \times L \times 0.0000065 = \text{expansion, in.}$$

where t_1 = steam temperature
t_2 = room temperature
L = horizontal length of pipe section, in.

A Holly loop is a piping arrangement used to return condensate from steamline separators to the boiler (Fig. 14.14a).

In operation, the flow of the condensate-and-steam mixture from the steam separator to the surge and separation tank is established by causing a slight pressure drop in the surge tank. This is accomplished by having the vent valve open slightly in a small line from the top of the surge tank to a heater or hot well. The surge tank is located at an elevation sufficient to give static head pressure so that the condensate will return against boiler pressure by gravity.

Use of a Hartford loop is confined usually to heating boilers and eliminates the need of a check valve to prevent condensate returns from backing up the return line in case the steam valve is closed (Fig. 14.14b).

Safety-valve problems. *Safety-valve care* in the operating schedule should include frequent periodic tests of safety valves. In boilers of moderate pressure, the valve should be lifted by its lever at least once each week of operation, and the pressure should be raised to the popping point to test the safety valves at least once each year of operation. If the safety valve does not blow at its set pressure, the lever should be tried at that pressure, for the spindle may be stuck slightly in its bushing. If the valve is not freed or if it does not blow at its set

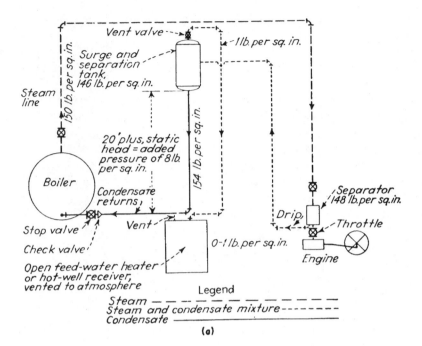

Legend
Steam — — — — — — — — —
Steam and condensate mixture - - - - - - - -
Condensate ————————

(a)

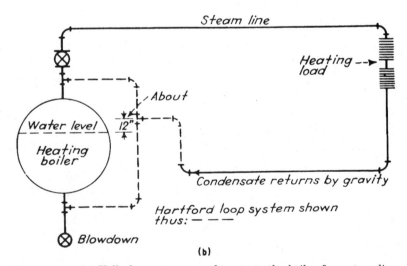

(b)

Figure 14.14 (*a*) Holly loop returns condensate to the boiler from steamline separators. (*b*) Hartford loop is used to return condensate in gravity return system.

pressure on several subsequent trials, the boiler should be shut down at once for safety-valve overhaul.

Owing to the cutting action of high-pressure steam, weekly tests of safety valves set at over 500 psi may not be advisable. The manufacturer and local boiler authorities should be consulted on this point, but it is seldom wise to allow more than a month to pass between tests except on superheater safety valves.

If the safety valve blows at a lower pressure than normal, according to the pressure gauge, the gauge should be tested. If the gauge is correct, the gauge piping should be blown out to be sure that it is clear. If the safety valve then blows at too low a pressure, it may be adjusted by a competent person, taking extreme care that plenty of clearance is left between the spring coils when the valve is wide open. It is always best to check the boiler pressure at two different points on the boiler to eliminate the possibility of incorrect pressure indication. Any considerable deviation from normal operation of safety valves should receive immediate competent attention.

Safety-valve escape pipes may become clogged with sediment owing to leaking safety valves. The periodic test should disclose this condition. Leaking safety valves should be reground, *never* tightened by adjustment to stop leakage.

Outdoor installation.   Package boilers or economizers may be located *outdoors* between the boiler house and stack. In freezing temperatures, the economizer or boiler should be drained when out of service. When it is in use, a bypass around the blowdown valves should be provided, and this bypass should be opened slightly to prevent freezing of the blowdown line.

Also, all exposed pipes out of the main circulatory system should be heavily insulated. Soot accumulations should be removed periodically for they may impede heat transfer and draft. Large quantities of soot in the base of economizers may be washed out with a water hose.

Auxiliaries need maintenance attention too. If neglected, serious explosions can result. Deaerating heaters have failed violently from cracks around welds that were not stress relieved during manufacture. The Code permitted this at the time on unfired pressure vessels. See Chap. 12 on auxiliaries for the required inspections on deaerating heaters.

A recent serious explosion involved a vented condensate tank. (See Fig. 14.15.) The investigation indicated that the float for the water feed control inside the condensate tank became loose and floated to the vent opening, thus making the condensate tank a pressure vessel. With no relief valve on the tank, pressure increased until a violent failure occurred. A relief valve set to the permissible pressure for the condensate tank would have relieved the high pressure. Most safety

Figure 14.15 Condensate tank explosion due to plugged vent seriously damaged entire boiler installation. (*Courtesy National Board Bulletin.*)

engineers also recommend a cage or similar barrier around the vent opening that can prevent a loose float from blocking the vent opening, and thus prevent a pressure buildup in the tank.

Boiler lay-up considerations.   Controlling corrosion and eventual deterioration of boiler equipment is as important during shutdown as it is during operating periods. Corrosion can be initiated by oxygen, water, and low pH. If either water or oxygen is kept out of the system, the other will do no damage.

Cast-iron and steel heating boilers are susceptible to corrosion damage during summer outages. Soot, if not cleaned, will form sulfuric acid in damp or sweating basements. Water, if not treated, will corrode the inside of the boiler. Burning trash intermittently in a boiler can cause dry firing if the water level is not watched and can also cause acid-type attacks on the fireside when the boiler is idle. Heating boilers should be flushed and cleaned on the water side. The fire side should be cleaned of all soot deposits. Then lay up the boiler either *wet* or *dry*. The big problem with the dry method is keeping the insides dry. Air blast with independent outside hot air after draining. When it is dry, place shallow pans of quicklime inside; then close all openings tightly. Place trays between tubes and one in each steam drum (watertube type) and bottom of shell (firetube type). Open the boiler every 30 days and if the quicklime is saturated with water, replace it (or whatever material is used).

The *wet-storage* method is best when freezing is not a problem and if the unit will not be needed for at least a month. After it is prepared for storage, fill the boiler to water level with deaerated water. If no deaerated water is used, open a top vent. Then build a light fire to boil the water for 8 hr so dissolved gases are driven to the atmosphere. Use 1½ lb of sodium sulfite for each 1000 gal of water stored in the boiler to protect against oxygen. The concentration should be about 75 ppm. Use caustic soda to obtain alkalinity of 375 ppm. Keep the water temperature as low as possible and test the water weekly for the smaller packaged boilers usually found in industrial plants.

Maintenance and inspection programs. Most boiler problems can be minimized by established maintenance and inspection programs, which with automatic operation, is often not understood. Safe boiler operation requires management commitment to safety, efficiency, and continuity of dependable operation, and this requires management support of the operators and maintenance personnel in correcting plant problems as they appear. Many automatic devices have been developed to make boiler control easier, safer, more efficient. But all this automatic equipment demands more head work to replace the strong back required of yesterday's engineer. The automatic equipment must be understood and maintained. If it should fail, the operating engineer must be capable of picking up manual control of many operations on a split second's notice.

Boiler accidents can occur. Here are some of the causes:

1. More unattended automatic operation of boilers, with complete reliance on automatic controls for overpressure and fire-side damage prevention. Although controls can malfunction in many ways, their installation can lead to a false sense of security.
2. Failure to test safety relief valves on a consistent, regular basis.
3. Failure to maintain boiler and auxiliaries properly. The latter includes reserve boiler feed and low-water fuel cutoff. Maintenance is often neglected on water treatment, cleaning, and checking of controls.
4. As automatic boilers become more complex in control arrangement, tampering with controls or blocking the safety controls may lead to a failure.
5. The higher firing rates with suspended fuel on today's more compact boilers can quickly lead to dry firing or to improper fuel-air ratios that trigger fireside explosions, again if safety controls do not work fast enough.

Figure 14.16 Boiler explosion due to low water, causing rupture of pressure parts. (*Courtesy Factory Mutual Engineering.*)

The results of a boiler explosion should not be minimized. Even a small boiler may cause terrific damage. See Fig. 14.16. Small boilers, indeed, must often be considered more hazardous than those in large plants, owing to their frequent lack of competent attendants. However, close adherence to ASME Code standards for construction and care of power boilers and a high grade of operation and inspection will practically eliminate the possibility of accidents of this nature.

Cost of forced outages. Boiler plant operation is generally rated by efficiency of operation, costs, reliability, and what is sometimes taken for granted, safe operation. As boiler size and capacity increase, the possibility of a forced outage, especially one resulting from a tube failure or explosion, takes on greater significance. The length of outage and the cost of repairs are proportionate to the size of the boiler. Thus every effort must be made to prevent failure of pressure parts by adequate inspection and maintenance. Visual inspections are still required to get as close to the parts of the boiler, both internal and external, as is practicable.

On older-type installations this was possible because a greater percentage of the surfaces of the pressure parts was accessible because of design. And more openings were provided in the setting. But with newer boilers this is not often possible. Thus inspection of large boilers should also include a review of instrumented readings so as to pinpoint areas of trouble. For example, an increase in pressure drop across a bank of tubes may indicate tube fouling in that bank, and is quicker than looking at the tubes.

In larger plants output data are compared to input data, and this can also help in pinpointing possible areas of the boiler, or its controls, that may require attention. For example, a loss in efficiency may require a review of the internal conditions of heat-transfer surfaces, or perhaps combustion controls. The causes can be many, and it requires experienced personnel to find the reason, by reviewing operating data, for any output drop.

Maintenance programs. Operators have certain general, daily and weekly or monthly routines to follow, and quite often these are dictated by the boiler manufacturer's instruction for the particular boiler that is involved. Broad guidelines are:

1. *General maintenance* includes observing for steam, water and fuel leaks, and repairing the leaks when they are noticed. Tightness of connections that contain pressure parts or fuel lines, valves and similar connections including control apparatus should be part of the routine. The proper functioning of gauges, controls, and instruments must be observed at all times.

2. *Daily maintenance* includes checking the operation of firing equipment and associated auxiliaries such as burners, nozzles, pumps, fuel storage tanks, flame failure devices, and general burner operation as respects air-fuel ratios, stack temperatures, smoke and similar items involving proper combustion. Water level maintenance includes checking the water column and gauge glass connections by blowing down the column to make sure level readings are correct. Follow the recommended blowdown schedule per water test results. Check samples of boiler water and feedwater per the advice or guidelines of the water treatment program being followed for the boiler. Clean fire sides by using soot blowers if so equipped.

3. In many plants *weekly* or *monthly* readings are taken to check on boiler performance, such as amount of fuel burned, heating value, and this is compared to Btu of output for the period.

Log readings and notes of items needing attention during the annual inspection is another feature of maintenance.

Types of maintenance. Many plants operate on a *breakdown maintenance* basis, which means equipment is allowed to operate to failure before it is repaired or replaced. This type of maintenance requires little planning, but it does produce inefficient use of workers, who are paid overtime pay in emergencies, and causes excessive unplanned downtime in a service or production. *Engineered preventive mainte-nance* combines predictive analysis and testing techniques to deter-mine the frequency of overhaul or part inspection and repair or replacement in order to maximize operating time and to eliminate unnecessary overhaul work based only on frequency schedules. Some inspection work is dictated by jurisdictional requirements, and these are usually used as a starting base for preventive maintenance work. As pressures and temperatures rise and the effect of an accident or the expense of an outage becomes apparent, the legal inspections must be supplemented by plant-engineered preventive maintenance programs.

With the advent of on-line diagnostic monitoring, *predictive mainte-nance* is now an accepted method of controlling maintenance costs by monitoring critical machine parameters, and process variables such as cyclic service, and then comparing these results with past readings or initial baseline readings. Changes affecting efficiency or output can thus be recognized, and maintenance can be performed to bring the results back to the baseline readings. This diagnostic approach is also used to monitor mechanical conditions. For example, keeping track of tube thickness due to erosion and replacing the affected tubes when minimum thickness is reached.

The problem of maintenance should be approached with the view of how much it will cost *not* to carry on an active maintenance program. The unexpected expenses will be sure to be far greater in frequency and extent when such a program is abandoned or deferred than when it is in effect.

Large plants use a card-index system and even have established a computer maintenance scheduling program. One card is made out for each piece of equipment and has on it identification information and a space for entering records of tests and remarks. A signaling system can be used sometimes to indicate the date on which equipment is due for tests, overhauls, and inspection. Even the smallest plant can create a system that will give this information in an alphabetical card-index file, or on computer tape to be reviewed daily.

Preventive maintenance in an industrial boiler plant has been influenced by legal requirements and is performed to protect person-nel and to prevent equipment damage that could lead to costly repairs and loss of productive capacity. In fact, preventive maintenance directed *specifically* to maintaining boiler efficiency has been the exception rather than the rule. But rising fuel costs have placed increasing emphasis on conscientious maintenance, which is neces-sary for preserving high efficiencies. These preventive maintenance practices are often justified easily on an economic basis.

Efficiency-related boiler maintenance is that directed at correcting any condition which increases the amount of fuel required to generate a given quantity of steam. Thus, at a specified boiler load, any condition which leads to an increase in: (1) flue-gas temperature; (2) flue-gas flow; (3) combustibles content of ash or flue gas; (4) convection or radiation losses from the boiler exterior, ductwork, or piping; or (5) blowdown rates is considered an efficiency-related maintenance item. Generally, attention to such items also can forestall more serious consequences, which could cause damage to equipment or injury to personnel.

Use of logs. A log in which each operator can keep hourly readings and remarks of any unusual occurrence is of value. The value is enhanced by recording notes on readings and observations from the more remote parts of the plant that may be visited but seldom during the normal course of duties. An operator, unless utterly incompetent, hesitates to walk past equipment to take a reading without observing its condition.

The central "clearinghouse" for daily study and writing up of these logs depends on the size of the plant. In any case, the reports should receive careful daily attention and then be systematically filed for reference at any time. Most important readings, such as fuel consumption, steam evaporation, and gas temperatures, should be broken down into efficiencies and equipment factors for recording in graphical form. A graph showing daily, weekly, or monthly progress in these results should be posted where everyone responsible for the operation may see them. A gradual depreciating operation is shown in this manner where otherwise a loss might be unnoticed.

Individual boilers should have a log in order to record the operation and performance of controls and safety devices, as shown in Fig. 14.17. The minimum tests and checks shown will assist operators in determining whether the boiler and its controls are in good operating condition. The tests must be performed at established frequencies, and if any malfunctions are found, they should be corrected immediately. A boiler with a defective safety valve, improper water-level control, or nonoperating low-water fuel cutoff should never be operated unattended until these vital devices are in good working order.

Log sheets are a forced reminder to check certain components of a boiler to prevent trouble from developing later and to note if proper operation is taking place. A log sheet should record all important operating data, such as pressures and temperatures, and it should also record procedures such as the time at which the soot blowers and water columns are blown and when blowdown valves were operated. A continuous record of operating data and important procedures carried out is then at hand when needed. It is also important to log when testing safety appurtenances.

Any irregular operation or event should be recorded in a separate book, with a description of the irregularity and the corrective mea-

To Test

WATER COLUMN AND GAGE GLASS—open drain valve quickly and flush water from glass and column. When drain is closed, water level should recover promptly.

LOW WATER FUEL CUT-OFF AND WATER LEVEL CONTROL—drain float chamber when firing equipment is operating. Proper operation of the control should shut off the firing equipment and start feed pump. If controls are of probe or other type that require lowering of water level in boiler to test. DO NOT lower water level to point below bottom of gage glass.

HIGH-PRESSURE POWER BOILER LOG
READINGS TAKEN EACH 8 HR. WATCH

BOILER NO. _____
WEEK BEGINNING _____

CHECK OR TEST OR RECORD EACH 8 HR WATCH	MON.			TUES.			WED.			THUR.			FRI.			SAT.			SUN.		
	8 to 4	4 to 12	12 to 8	8 to 4	4 to 12	12 to 8	8 to 4	4 to 12	12 to 8	8 to 4	4 to 12	12 to 8	8 to 4	4 to 12	12 to 8	8 to 4	4 to 12	12 to 8	8 to 4	4 to 12	12 to 8
WATER LEVEL-PROPER																					
STEAM PRESSURE, PSI																					
FEED PUMP PRESSURE, PSI																					
FEED WATER TEMPERATURE, °F																					
CONDENSATE TEMPERATURE, °F																					
FLUE GAS TEMPERATURE, °F																					
LOW WATER CUT-OFF-TESTED																					
WATER LEVEL CONTROL-TESTED																					
WATER GAGE GLASS-CLEAN																					
FEED PUMP IN GOOD REPAIR																					
CONDENSATE TANK & FLOAT TESTED																					
BURNER OPERATION-NORMAL																					
FUEL SUPPLY-ADEQUATE																					
FLAME FAILURE SAFEGUARD																					
WATER TREATMENT-TESTED																					
BOILER BLOWDOWN-PERFORMED																					
MONTHLY-SAFETY VALVE TESTED																					
OPERATOR'S INITIAL																					
REMARKS																					

Figure 14.17 High-pressure boiler log permits recording and checking of controls and safety devices. (*Courtesy Royal Insurance Co.*)

sures taken. In this book all orders should be written and initialed by the operator.

Firemen or shift engineers reporting for duty should read the notations made by the previous watches. Then they will know what the past operation has been, what orders have been issued for future operation, and what trouble spots to watch for. They should then initial those items for which they are responsible so as to indicate that they are familiar with the situation. Log sheets are supplied by some insurance companies for small low-pressure plants and industrial high-pressure boilers. Many plants design their own log sheets to record important data pertaining to their specific plant details and layout.

Good housekeeping is another maintenance item that needs attention. *Cleanliness* and leak-free operation must exist in the boiler plant. Otherwise increased operating costs and maintenance, accidents, and lost time due to illness may be expected in a dirty plant. But the operating personnel cannot be expected to maintain cleanliness if not all reasonable steps to prevent dirt have been taken. Nothing will break down the aims of those cleaning the plant quicker than to have clouds of soot and dust pour from cracks in a boiler setting just after they have cleaned the plant. Savings gained by long deferment of needed repairs are not economy. All steps possible to prevent dirt should be taken, and then a full effort can be expected of those responsible for maintaining cleanliness.

Testing. Test programs in the boiler plant are essential if operating efficiencies are to be maintained at a high level. Even though a full complement of recording and indicating instruments is installed to permit most economical operation, these instruments should receive periodic calibration and check by actual test.

There are some plant functions for which instruments have not yet been developed to give direct readings and for which analytical methods of securing the desired information are necessary. Tests on fuel and water are examples of the methods employed in checking on this group of functions.

In the small plant, it is sufficient usually to test each lot of fuel received. But large plants storing coal outdoors find this system impractical, and for them a periodic test of the fuel as fired is necessary. Similarly, a small plant may find it sufficient to test the boiler water once a week, but a large plant operating under widely varying conditions finds it advisable to make this test at least once a day.

Testing and calibration of pressure gauges and meters should be carried out on routine schedule. The date of the test and the initials of the person making it should be printed on a slip and pasted on the instrument or otherwise recorded for future reference.

Computer logging and analysis of data are now being practiced on large boiler systems. This does not eliminate the need for periodic review of developed data in order to maintain the efficiency of the equipment, especially when a conflict may arise between production needs and reliable operation of the equipment.

Testing of safety devices is an extremely important consideration in maintaining a safe plant. This includes safety valves, pressure cutout switches, flame failure devices, interlocks on fans, burners, and feedwater pumps as well as low-water fuel cutouts.

Low water is a frequent cause of boiler outages, especially on automatically fired packaged boilers. Three major areas must be checked and tested periodically to prevent low-water accidents:

1. *Feedwater system:* It must have the capability to maintain normal water level in the boiler.

2. *Low-water safety controls:* They must sense when the water level in the boiler drops below prescribed limits.

3. *Fuel-cutoff actuators:* They must be able to interrupt fuel flow when receiving a signal from the low-water sensors.

It is necessary to test *float controls* periodically to make sure they move freely. On power boilers operating around the clock, this should be done on every shift; once a week is adequate for heating boilers. Note that float controls often become sluggish or stuck when sludge and sediment build up in the float bowls. Avoid this by blowing down the float bowl daily. Also, dismantle the entire component once each year and clean scale and sediment manually from the operating mechanisms. One further note: Blow down the water column after examining float controls; clean it thoroughly during the annual inspection.

Probe-electrode-type controls require periodic checking in the following areas:

Examine insulators where electrodes penetrate the boiler shell or water column to make sure they are not cracked or bridged with dirt or moisture. This could cause current to flow from the electrode to the shell, destroying the sensing value of the electrode.

Look for broken electric connections to the electrodes and for wiring touching the boiler. These conditions also can limit probe's sensing value.

Make sure electrodes of the proper length are installed on replacement jobs. Otherwise, it may be possible to insert a unit that senses at a point below the safe water level.

Minimize scale, mud, and sediment in a water-column-mounted electrode to stop these impurities from completing the circuit at a false water level.

Annual maintenance and inspections. The annual maintenance and inspection program involves checking the entire boiler system, including auxiliaries. Large boilers may have a manufacturer's service representative present to assist in the inspection and make adjustments in controls or other performance problems noted by operators. The maintenance will include cleaning the internal and external heating surfaces, inspecting for wear-and-tear damage, and making necessary repairs so that the boiler system can be properly returned to service.

There are numerous reasons for making inspections of boiler systems, but the paramount one is to assure safe and reliable service. Boilers become affected by service conditions. Corrosion, cracking, thinning, fouling, grooving, and other metal deterioration continuously occur from service effects.

Each type of boiler has its own peculiar area to be watched, either because of process effects or because of the nature of design and application. Some boilers may have an inherent hazard related to material properties. For example, a cast-iron boiler has a cracking hazard if, in operation, too wide a temperature swing is experienced. Inspection programs are usually patterned on plant needs and the physical conditions and economic factors that can influence output or plant reliability. From a broad safety perspective, an inspection program can consist of the following activities:

1. Establish benchmark conditions when equipment is new or has had an extensive overhaul. Equipment record cards and code construction data will assist in this initial evaluation.

2. Set up an operating inspection program to note if the boiler is operating within rated parameters and within the pressure and temperature ratings. This would include a check on the alarms and shutdown and pressure-relieving devices that provide protection against abnormal conditions.

3. Review with operation and test personnel the efficiency of performance and whether the boiler may require internal cleaning to restore output efficiency. A review of pressures, temperatures, flows, and similar indicators of normal and abnormal performance will assist in determining if an internal inspection may be required.

4. Review past incidents of failure, abnormal condition, or crack discovery in order to develop schedules of inspection for fatigue cracks after a given number of cycles. This will require coordination with the operating personnel and careful record keeping.

5. Perform similar reviews on such items as water treatment, concentrations of mix, etc., where past analysis of failures indicates a need for controlling these items in order to avoid stress-corrosion cracks or similar detrimental conditions.

6. When calculations of wear rate and records indicate that corrosion allowances may have been used up, conduct more frequent internal inspections to track this thinning action.

7. Determine the condition of connected valves during operating inspections in order to note if leakage is occurring through wearable items such as packing and valve seats.

8. Follow manufacturers' instructions; most manufacturers have explicit instructions on what inspections may be required on their equipment.

9. Cooperate with the jurisdictional inspector when mandated inspection is due.

Types of inspections. Inspections can be broadly divided into the following types:

1. Jurisdictional inspections

2. Internal inspections

3. External or operating inspections

4. Nondestructive testing and inspection

Various states and municipalities have laws pertaining to inspection of boilers. The main purpose of these laws is to protect the life and limb of employees and the public as well as to avoid property damage. The legislation sets up standards for design, installation, and inspection. These are usually ASME or NB rules, made legal requirements. Most of the laws provide for periodic inspection of boilers coming within their scope, by state, municipal, or insurance company inspectors. The owner or operator must arrange for these inspections at required, stipulated intervals. For power boilers, the requirement is usually a yearly internal inspection and an external inspection six months later. Low-pressure heating-boiler inspections may be on a yearly or biannual basis, depending on the jurisdiction.

Safety in performing internal inspections

Vessel entry. Vessel entry for internal inspections requires a safety system approach to avoid asphyxiation or contact with harmful residues. A vessel previously filled with harmful contents should be steamed and flushed until its contents have been reduced to a tolerable limit.

No person should be permitted to perform hazardous work, or enter a vessel or *confined space,* until a work permit signed by an authorized person has been issued. The permit should have as a part of it a checklist that indicates the following:

1. The atmosphere inside the vessel or space has been tested and found safe for entry, and all the necessary lines have been blanked

or disconnected, and the lines so identified and listed. Where lines cannot be blanked off because of interference with production, valves to the vessel should be closed and locked. Warning signs that people are working inside should be placed close to the valves. If the valves on a vessel containing hazardous substances fail to shut tight, the vessel *should not be entered* until conditions have been corrected and rendered safe.

2. The vessel has an access hole large enough to allow a person wearing safety harness, lifeline, and emergency respirator to enter and leave easily.

3. All the relevant safety equipment is present, including a list of it and confirmation that it is in working order.

4. A ventilator has been set up and is in working order.

5. Specific instructions have been issued for safely making the repairs or inspection.

The permit is issued to the authorized person of the maintenance crew, who signs it, indicating that it has been read and that the equipment has been accepted for work. A permit should be valid only for the shift on which it was issued. If the job requires more time than that, the permit should be signed back to the issuer and a new one written after the shift change.

If required, checks should be made frequently with an explosimeter or chromatograph, or both, to ensure that it is safe to enter where contents may have been flammable. Vessels should be purged with air both to cool it and to eliminate any oxygen deficiency. A test for oxygen adequacy by means of an oxygen probe, Orsat, or chromatograph analyzer (preferably by two different sampling and testing methods) should be made.

Nitrogen blanketing storage. This method is used to limit air intrusion into boilers that are on a long term reserve status. All low-point drains must be used to make sure that all water has been removed from the boiler. See Fig. 14.18. A nitrogen blanket pressure of 5 psi is generally used.

Internal inspections on a boiler that has been stored with a nitrogen blanket requires special attention on the purging of this gas from drums and headers. An oxygen deficiency monitor should be used inside the drum and headers to make sure that at least 19 percent of oxygen is present inside all areas of the boiler in order to prevent any bodily harm to the person entering the shell, header, or drum.

Jurisdictional inspections. One of the jurisdictional inspections generally required on high-pressure boilers is the internal inspection. The purpose of the internal inspection by plant personnel and an

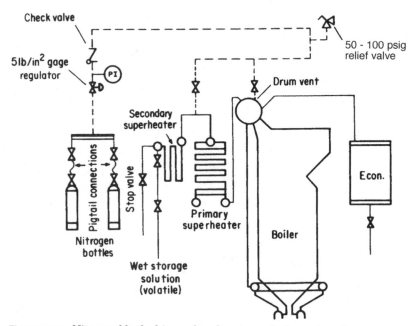

Figure 14.18 Nitrogen blanked is used to dry-store a boiler on long-term reserve status in order to prevent corrosion. (*Courtesy Betz Laboratories, Inc.*)

authorized inspector is to check on the structural soundness of the pressure-containing parts and to note any conditions that can affect its strength to confine the pressure. Wear, deterioration, corrosion, scale, oil, cracks, grooving, thinning, and other such weakening conditions require inspection. Most boilers develop their own areas of trouble spots, depending on design, operating conditions, and maintenance practices. Check all exposed metal surfaces inside the boiler for effectiveness of water treatment and scale solvents, also for oil or other substances that enter with feedwater. Oil or scale on heating surfaces weakens the metal, causing bagging or rupture. Corrosion areas next to a seam are more serious than in a solid plate away from seams. Thinning on a joint is dangerous because the strength of a joint is less than that of a solid sheet.

Check for evidence of grooving and cracks along longitudinal seams of shells and drums. Carefully look for internal grooving in fillets of unstayed heads. Inspect stays and stay bolts for even tension, fastened ends for cracks where stays or stay bolts are punched or drilled for rivets or bolts. Manholes and other openings are subject to corrosion thinning and cracks. See that openings to water column connections, dry pipes, and safety valves are free of obstructions such as mud and scale.

Ligaments between tube holes in heads (of all types of boilers) often crack, then leak and weaken the boiler. Also, on both watertube and firetube boilers, the beading and flaring on tube ends need checking for erosion and corrosion, cracks, and thinning. Welded nozzles and other such openings require inspections for weld washout, cracks, and evidence of deterioration of the joints.

Oil is usually hard to detect, especially if only a very small amount is present. Run the back of your finger along the waterline. If it is stained and the stain cannot be washed off with soap and water, oil is entering with the feedwater and is being distributed throughout the boiler by circulation. Oil, being lighter than water, rises gradually and forms a scum along the water level. The real danger comes from tiny solid particles sticking to the oil before it adheres to the drum sides. Then gradually this weighted oil settles to the heating surfaces, causing tubes to blister or completely fail.

Carefully inspect the plate and tube surfaces that are exposed to the fire. Look for places that might become deformed by bulging or blistering during operation. Solids in the water side of lower generating tubes cause blisters when sludge settles in tubes and water cannot carry away heat. The boiler must be taken out of service until the defective part or parts have been properly repaired. Blistered tubes usually must be cut out and replaced with new ones.

Old *lap-joint* boilers are apt to crack where plates lap in a longitudinal or straight seam. If there is evidence of leakage or trouble at this point, remove the rivets and examine the plate carefully if cracks exist in the seam. Cracks in shell plates are usually dangerous, except fire cracks that run from the edge of the plate into the rivet holes of girth seams. Usually, a limited number of such fire cracks are not very serious.

Test *stay bolts* by tapping one end of each bolt with a hammer. For best results, hold a hammer or heavy tool at the opposite end while tapping. A broken bolt is indicated by a hollow sound.

Tubes in firetube boilers deteriorate faster at the ends toward the fire. Tapping the outer surface with a light hammer shows if there is serious thinness. Tubes of vertical tubular boilers usually are thin at the upper ends when exposed to the products of combustion. Lack of water cooling is the cause. Tubes subject to strong draft often thin from erosion caused by impingement of fuel and ash particles. Improperly used soot blowers will also thin the tubes. A leaky tube spraying hot water on nearby sooty tubes will corrode them seriously from an acid condition. Short tubes or nipples joining drums or headers lodge fuel and ash and then cause corrosion if moisture is present. First clean and then thoroughly examine all such places.

Baffles in watertube boilers often move out of place. Then combustion gas, short-circuiting through baffles, raises the temperature on portions of the boiler, causing trouble. Heat localization from improper or defective burners, or operation causing a blowpipe effect, must be corrected to prevent overheating.

The inspections needed on the *external fitting* of boilers include the following: safety valves are the most important attachments on a boiler. There should be no rust, scale, or foreign matter in casings to hinder free operation. The best way to test the setting and freedom of safety valves is by popping the valve with pressure. If this cannot be done, test by try levers. Inspect the discharge pipe to make sure it is secure. Operators have been killed because a valve discharging into the boiler room fills the space with steam in a few seconds. The opening in the discharge line must not be plugged.

Pressure gauges have to be removed to test by comparing with a standard test gauge. Blow out the pipe leading to the pressure gauge. Make sure water-column connections are free by removing plugs, or the tees. Examine the condition of the water column and gauge-glass attachments.

Examine the *supports* of the boiler structure. Make sure that ash and soot will not bind the boiler structure to produce excessive strains from expansion under operating conditions. Look also for evidence of corrosion from soot on structural supports. Check the blowdown valves to see that they work freely and are packed and that external piping and fittings are not corroded or damaged.

Use of nondestructive testing. Nondestructive testing equipment is being used more in boiler inspection to locate potential areas of failure. Five major nondestructive tests are used: ultrasonics, radiography, magnetic-particle, dye-penetrant, and eddy-current. Ultrasonic equipment is now portable for field use and is extensively used for plate- and tube-thickness checks. These instruments become useful as tracing instruments for determining causes of failure of a repetitive nature. For example, tube failures in waterwalls are a common problem. After one tube failure, adjacent tubes can be checked ultrasonically and thinned tubes replaced prior to failures.

A similar practice is followed on tubes subject to fly-ash erosion. The thickness of the tubes in a suspected area is checked by ultrasonic equipment, and those found thinned are replaced during normal outages. Plate thickness around manhole and handhole openings, water legs, shells, and heads is checked ultrasonically in order to determine allowable pressure.

Flaw detection, such as checking for laminations, cracks, porosity underneath plate surfaces, or welds on inaccessible visual parts, is playing a more important role. Pulse-echo instruments are now available for field testing to do flaw detection. See Fig. 14.19.

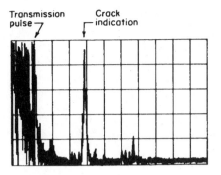

Transmission pulse ⌐ /

Crack ⌐ indication

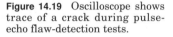

Figure 14.19 Oscilloscope shows trace of a crack during pulse-echo flaw-detection tests.

Radiography, so important in new construction, is extensively used in field testing. Welded repairs on high-pressure boilers are tested by x-ray or other radiographic equipment.

Magnetic-particle inspection finds its chief use in surface crack detection. Its main use is on piping and joints of boilers.

Eddy-current testing is finding its chief use in nonmagnetic-tube searching for defects, such as condensers and heat exchangers connected to a boiler.

Nondestructive testing of nuclear reactors in service of course supplements any traditional visual internal inspection which is limited because of the radiation hazard.

Remaining life. Large older boilers are being restored because it costs less money than buying a new boiler. Life assessment procedures are being used to determine what parts of the boiler need replacement, and whether the unit can be salvaged so that future trouble-free service can be expected. Metallurgical examinations of boiler components are made. These include residual stress measurements, plastic field replication, miniature boat specimen testing of material, thickness testing, as well as other NDT tests. Creep damage to high-temperature components is made by taking plugs and boat samples from headers or drums, and applying a stress to the component samples at high temperature in order to determine if the material was not damaged by creep. It is not uncommon on utility units to replace whole headers, with new tubes installed on site.

Studies have shown that the various stages of creep—i.e., formation of voids, linking, and microcracking—can be detected through the use of replication techniques and conventional NDT methods. The amount of remaining life can then be evaluated based on the condition of the material, operation effects, and other parameters.

The replication technique can detect creep only on the surface. Subsurface examination requires using other nondestructive test methods. Replication of a spot does not positively assure that adjacent material is without creep damage. Creep voids tend to locate preferentially near existing flaws, which are typically subsurface. Thus, supplemental inspection methods are increasingly important in evaluating steam piping and boiler component life. Methods used include: visual, magnetic particle, liquid penetrant, ultrasonic for thickness measurement and flaw detection, eddy current, metallography, and replication.

Replication method. When using replication, a surface is prepared by cleaning and etching. Acetate tape is softened by acetone, and the tape applied to the prepared surface. An image of the surface is imposed on the tape, which is then removed, shadowed to improve contrast, and examined in an optical microscope. The replica can also be examined in a scanning electron microscope if high magnification resolution is required.

The replicas are examined for metallurgical characteristics such as graphitization and for creep voids and cracks in the metal's grain structure. A decision can then be made if the part is still satisfactory for service or needs to be replaced.

Boiler Repairs

Most jurisdictions in the United States have reinspection laws on high-pressure boilers, as detailed in Chap. 1. The possible pressure-part failures that may result on a boiler will determine repairs that may be needed to correct the defect.

National Board R stamp. As more jurisdictions in the United States pass laws on installation of boilers and unfired pressure vessels in conformity to ASME Code standards and also pass requirements for reinspection at periodic intervals, a program of registering qualified repair organizations has evolved which in some respects duplicates the issuance of ASME new-construction stamps to boiler and pressure vessel manufacturers. The aim of registering repair organizations is to promote uniform repair procedures. This involves using Code-approved material, establishing a welding procedure, using welders that have been qualified for the repair or alteration that may be required, and having an inspection agency verify that the repairs or alterations meet Code and NB requirements. The National Board of Boiler and Pressure Vessel Inspectors has developed an R stamp issuance program to any organization, manufacturer, or owner-user group that desires to be certified as a registered repair organization. Most jurisdictions have accepted this program of *nationally* certifying repair organizations, because it promotes uniform repair procedures and safe practices throughout the United States.

The National Board publishes a document entitled *The National Board Inspection Code,* which is used as a guide by inspectors and others to help maintain the integrity of boilers and pressure vessels. It contains rules and guidelines for inspection after installation, repair, alteration, and rerating, thus helping to ensure that this machinery continues to be used safely. When a weld repair or alteration is completed, the organization that carried out the work is required to complete a record-of-weld-repair form that details the type of repair or alteration made by that organization. The record-of-weld-repair form is required to be signed by the organization carrying out the work, verifying that it was carried out in accordance with the *National Board Inspection Code* and the jurisdictional requirements. In addition, this form is to be signed by the authorized inspector who accepted the repair or alteration; the form verifies that the work was carried out in accordance with the *National Board Inspection Code.* The weld-repair form is forwarded to the enforcement authorities for boilers and pressure vessels in a jurisdiction. In the event of a repair or an alteration to a boiler or pressure vessel, the authorized inspector of the jurisdiction should be contacted for guidance in complying with the National Board rules or those existing in a jurisdiction on repairs or alterations (usually similar to the nationally recognized NB rules).

Requirements and guidelines concerning the NB R stamp include the following, but the NB rules should be consulted, as these can change with time:

Authorization to use the stamp bearing the official National Board repair symbol as shown in Fig. 14.20 will be granted by the National Board pursuant to compliance with its rules and requirements.

Repair organizations, manufacturers, contractors, or owner-users that make repairs to boilers and pressure vessels classified under the *National Board Inspection Code* may apply to the National Board of Boiler and Pressure Vessel Inspectors by completing the application forms for the loan of a repair stamp and for authorization for its use. Each repair symbol stamp must be serialized and used only by the repair organization to which it was issued.

The holder of the National Board repair symbol stamp or holder of an applicable ASME code symbol stamp must submit to the authorized inspection agency that accepts the work and to the state or city of the United States or province of Canada (jurisdictional authority) a National Board Form R-1, Report of Welded Repair, in accordance with the *National Board Inspection Code* (see Fig. 14.21).

(a)

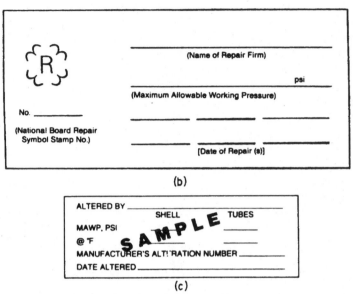

(b)

(c)

Figure 14.20 National Board repair stamps. (a) National Board "R" stamp to show repairs were made to a code vessel; (b) nameplate showing the name of the repair organization and date of repairs; (c) alteration nameplate.

Alterations of a boiler or pressure vessel must be made in accordance with requirements of the *National Board Inspection Code.*

A nameplate may be used, or where permitted, the repair symbol may be impressed directly below the original stamping on the vessel. If a nameplate is used, it must be welded or permanently attached either below or adjacent to the original stamping. Each repair organization, manufacturer, contractor, or owner-user that completes the repairs of a boiler or vessel shall hold a valid Certificate of Authorization for use of the National Board repair symbol stamp when required by the jurisdictional authorities of the states of the United States or provinces of Canada.

Before issuance or renewal of a National Board Repair Certificate of Authorization, all requirements, including a written quality control system, material control, repair design, fabrication, examination, and authorized inspection, must be met. In addition, it is

FORM R-1, REPORT OF WELDED ☐ REPAIR OR ☐ ALTERATION as required by the provisions of the National Board Inspection Code

1. Work performed by _____
   (name of repair or alteration organization)   (P.O. no., job no. etc.)

   _____
   (address)

2. Owner _____
   (name)

   _____
   (address)

3. Location of installation _____
   (name)

   _____
   (address)

4. Unit identification _____ Name of original manufacturer _____
   (boiler, pressure vessel)

5. Identifying nos.: _____
   (mfr's serial no.) (original National Board no.) (jurisdiction no.) (other) (year built)

6. Description of work: _____
   (use back, separate sheet, or sketch if necessary)

   _____
   _____
   _____
   _____

   _____ Pressure test, if applied _____ psi

7. Remarks: Attached are Manufacturers' Partial Data Reports properly identified and signed by Authorized Inspectors for the following items of report: _____

   _____
   _____
   _____
   _____
   _____

   (name of part, item number, mfr's name, and identifying stamp)

CERTIFICATE OF COMPLIANCE

The undersigned certifies that the statements made in this report are correct and that all design, material, construction, and workmanship of this _____ conform to the National Board Inspection Code.
(repair or alteration)

Certificate of Authorization no. _____ to use the _____ symbol expires _____, 19 ___

Date _____ 19 ___ _____ Signed _____
(repair or alteration organization) (authorized representative)

CERTIFICATE OF INSPECTION

The undersigned, holding a valid Commission issued by The National Board of Boiler and Pressure Vessel Inspectors and certificate of competency issued by the state or province of _____ and employed by _____ of _____ has inspected the work described in this data report on _____, 19 ___ and state that to the best of my knowledge and belief this work has been done in accordance with the National Board Inspection Code.

By signing this certificate, neither the undersigned nor my employer makes any warranty, expressed or implied, concerning the work described in this report. Furthermore, neither the undersigned nor my employer shall be liable in any manner for any personal injury, property damage or loss of any kind arising from or connected with this inspection, except such liability as may be provided in a policy of insurance which the undersigned's insurance company may issue upon said object and then only in accordance with the terms of said policy.

Date _____, 19 ___ Signed _____ Commissions _____
(Authorized Inspector) (National Board (incl endorsements) state, prov and no.)

This form may be obtained from the National Board of Boiler and Pressure Vessel Inspectors, 1056 Crupper Ave. Columbus, OH 43229

NB-66
Rev 4

Figure 14.21 The National Board welding repair and alteration form is now required by most jurisdictions with boiler and pressure vessel inspection laws.

required that the applicant successfully demonstrate the implementation of its written quality control system. In areas where there is no National Board member jurisdiction or where a jurisdiction elects not to review a manufacturer's facilities, the review will be performed by a representative of the National Board of Boiler and Pressure Vessel Inspectors.

The quality-control system must describe and explain what documents and procedures the repair organization will use to validate a repair that would restore the boiler or pressure vessel to a safe condition and maintain the validity of the original vessel.

Safety-relief-valve repairs. The National Board now has a registration and stamp-issuance procedure for qualified safety-relief-valve repairers. Many states are adopting this registration procedure as a legal requirement. See Fig. 14.22. When a safety-relief valve is repaired per NB rules, a metal repair nameplate must be affixed to the repaired valve, showing the name of the repair organization, the VR symbol, set pressure, blowdown capacity, and date of repair. The purpose of this accreditation system is to prevent safety-valve tampering on set pressure and capacity, and thus retain the original design identity of the safety-relief valve.

Boiler component repairs. *Tube troubles* in boilers can be caused by many factors, especially on large steam generators that are equipped with economizers, superheaters, and reheaters. Generating tubes fail mostly because of overheating due to the presence of scale or sludge, obstructed circulation, or heat concentration in an area so that the normal water and steam flow cannot move the fireside input heat fast enough. Tubes are also attacked chemically. The problem of chemical attack is in some cases related to overheating, for high temperatures

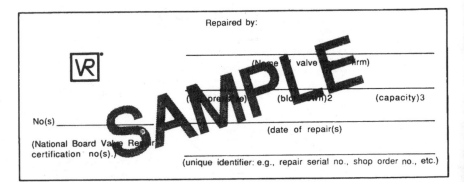

Figure 14.22 National Board stamping for officially repaired safety relief valve.

greatly speed up certain chemical reactions involving iron. Most chemical attack can be prevented, however, by proper boiler water treatment.

External thinning may be caused by the improper placement of soot blowers and by abrasive particles in the flue gas, such as fly-ash impingement. Failure of a fire tube by pitting or corrosion always requires renewal unless the defect is close to one end.

Superheater tubes rarely fail because of corrosion, unless it has occurred during periods of standby. Failures of these tubes are almost always attributed to overheating, which may be caused by the insulating effect of deposits carried over from the boiler water or by starvation.

Deposits having their origin in the boiler water are carried over because of foaming in the steam drum, resulting from excessive concentrations of dissolved solids or alkali or the presence of oil or other organic substances. Foaming caused by high concentration of solids and alkalinity may be minimized by increasing the boiler blowdown or by the addition of an antifoam material to the boiler water. The elimination of organic contaminants may require special pretreating equipment or the insertion of an oil separator in the feed-water line. It should be emphasized that pure hydrocarbon oils do not cause carryover, but additives in them frequently do.

Carryover in the form of slugs of water is another cause of superheater deposition, and this may be remedied by lowering the water level or by obtaining better water-level control. However, heavily swinging loads in excess of those for which the boiler was designed may also contribute to this condition.

Starvation of superheater tubes may result from poor start-up practice or by operating the boiler at undesirably low rating. If, during start-up, the furnace temperature is brought up too rapidly before full boiler pressure is attained, there will be insufficient steam to cool the tubes. In boilers with a radiant-type superheater, this same effect may be noted at low loads.

Creep cracking of tubes can be identified by the signs of extremely fine cracks or strain markings.

Repair of tubes involves considering the extent of the defects. Tube replacement is common. Localized repairs can be made by window patching or cutting out a section of the tube and circumferentially welding in a piece. Preheat and postheat treatment may be needed. An authorized inspector should approve the repair.

Removal of a tube is accomplished by cutting the beading or flaring off of each end. One end is slotted with a ripping chisel, with care being taken not to score the seat. The tube end is then reduced in diameter with a crimping tool (Fig. 14.23c) so that it may be drawn

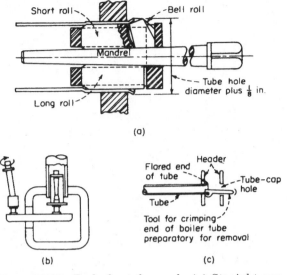

Figure 14.23   Tools for tube work. (*a*) Straight-run expander; (*b*) right-angle tube expander; (*c*) crimping tool for tube removal.

through the hole. An acetylene cutting torch may be used to advantage in some operations. The new tube is then put in place. If a fire tube, it is expanded and beaded. If a water tube, it may be expanded and flared (Fig. 14.23*a* and *b*).

If a *tube hole* should be damaged by scoring, it may be reamed out slightly oversize, and a copper or a soft-iron ferrule used between the tube and the tube hole. The tube is then expanded against this ferrule.

It is necessary sometimes to weld ends of a water tube, as with circulating tubes to a header which is being renewed. The edges of the tube should be ground to a V and a ⅛-in.-thick copper backing sleeve used inside the tube so that the weld metal will fuse into the bottom of the V, attaining full penetration without "icicles" hanging into the tube. The copper is used because the weld metal will not fuse to it; thus it may be removed easily. Steel backing rings as left in piping usually are not desirable in boiler tubes, for even a slight restriction in flow may not be tolerated.

Large-capacity steam generators have very long tube lengths that must be welded together during field erection of the boiler. While the ASME Code does not require circumferential tube welds to be x-rayed, as in shells or drum requirements, many users are specifying x-ray examination of all welds in the furnace-area waterwalls, some high-temperature superheater tube welds, and some tube-to-header welds. This is a form of quality control on the welding performed, but it is also an attempt to prevent an expensive outage from a relatively small weld failure, such as lack of fusion, porosity, or small HAZ crack. On recovery-type boilers, x-raying of welds around the furnace area, in order to prevent water getting into the furnace, is also becoming common practice.

A *bulge* is caused by overheating of shell plates and it affects the entire thickness of the plate, whereas a *blister* is caused by a slag inclusion forming a lamination at the time the plate was rolled in the steel mill. The entire thickness of the plate may not be affected. Usually, the area between the lamination and the fireside cracks and blisters. Bulges and blisters in the boiler shell often require repairs. If a bulge is not down more than 2 percent of its length, it is usually best to leave it alone unless the metal has been burned badly. If a blister has not reduced the thickness so that the percentage of new thickness to original thickness is less than the efficiency of the longitudinal seam, repairs are not usually necessary.

If a bulge is down more than 2 percent of its length, but not over about one-eighth of its length and the plate has not ruptured or burned, it is usually advisable for it to be driven back by experienced boilermakers. A form holding a charcoal fire may be used to heat the entire bulge simultaneously. A short-handled, heavy striking hammer is used around the circumference, and work is gradually toward the center.

Bulges or blisters of more serious extent or where the plate is burned badly are repaired only by patching. Usually the size of the area to be patched may be reduced by driving back the circumference of the affected section, provided that the plate is not burned. A *blister* is actually a separation of the metal from the shell plate, caused by impurities rolled into the shell plate when formed. But only the outside layer will blister from the heat because the remaining thickness is not affected. A blister cannot be driven back but must be cut out and the edges trimmed; if the remaining metal is sound, the pressure on the boiler must be reduced to correspond to the thickness of the remaining sound metal. If the pressure cannot be reduced, the entire blistered section, including sound metal, must be cut out, a flush-welded patch formed and butt-welded in, with localized stress relieving required after welding. If the length of the blister is over 8 in., the weld must be radiographed. After this is shown to be satisfactory, a hydrostatic test of 1½ times the maximum allowable pressure is required to check the repair. All welding must be done by a certified welder, and the repair must be approved by a commissioned boiler inspector.

Corroded areas of tube sheets may be built up by welding where tubes act as stays. But all tubes in such corroded areas must be removed before welding is applied. After welding, the tube hole should be reamed before new tubes are installed.

Stayed surfaces and tube sheets must also conform to the following requirements before welding. Corroded areas in stayed surfaces may be built up by fusion welding, provided the remaining plate has an average thickness of not less than 50 percent of the original thickness and the areas so affected are not corroded sufficiently to impair the safety of the boiler per ASME Code. An authorized inspector must approve and witness this type of repair.

Handhole and manhole openings are commonly oval, with the cover fitted inside the boiler (Fig. 14.24). A woven gasket seals this joint. The gasket softens when heated so the nut against the dog must be tightened more as pressure is brought up on the boiler. Unless it is tightened gradually as pressure builds, the gasket may leak or be blown out, scalding personnel.

Persistent *gasket leakage* results in the boiler plate being thinned as a result of corrosion, which can only lead to expensive repairs. Always correct gasket leakage as soon as possible.

Handhole rings will be required to strengthen edges of handholes deteriorated badly by external corrosion. This corrosion is caused usually by leakage past the gasket. The elliptical ring must be of the same thickness as the sheet, and it should be flush-welded to the opening to sound metal. The thin edges of the opening should be cut back so that a bead may be applied by electric welding to seal the ring to the edge of the hole. Corrosion at the edges of handholes may be prevented by keeping the handhole gasket made up tight against leakage.

The subject of *stay bolts* and *furnace-sheet* repairs in firebox boilers is covered by NB repair rules. Threaded stay bolts sometimes develop leaks, which can be through telltale holes or around the boiler plate.

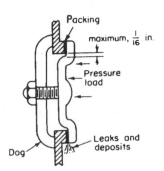

Figure 14.24 Leaks around handholes and manholes can cause metal deterioration, requiring repairs if not corrected.

Repairs are not permitted because the leaking telltale hole indicates the stay bolt is cracked inside the boiler. A new stay bolt should be installed.

Stay-bolt heads cannot be welded to stop leakage around the heads. The heads should be recaulked. If leakage persists, it may be an indication of a corroded sheet on the boiler. The old stay bolt must be removed, the sheet must be examined for corrosion and thinning, and if it is satisfactory, new stay bolts must be installed. If the sheets are corroded more than 50 percent of the original thickness, the defective section must be cut out and a flush-welded patch installed.

Threaded stays may be replaced by welded-in stays, provided that, in the judgment of the qualified boiler inspector, the plate adjacent to the stay bolt has not been materially weakened by deterioration or wasting away. Stress relieving other than thermal may be used as provided in NB rules for welding.

Violent furnace collapse may result, too, if a number of stay bolts in the same location are broken.

Bulged furnace sheets on VT boilers usually require extensive stay-bolt removal and replacement before the furnace sheet can be replaced. Sometimes the cost of repair may exceed the cost of a new boiler.

Braces or *stays,* unlike stay bolts, seldom break, for they are long enough to have considerable flexibility. If a brace or stay does break, repairs should not be attempted. A new brace or stay should be installed at once.

Through-to-head braces below the tubes in firetube boilers sometimes become bowed when the bottom of the shell is overheated owing to scale or oil. The shell may expand nearly 1 in. because of excessive temperatures; but the braces are subjected to the much lower temperature of the boiler water, and their heat expansion may be only half that amount. As the shell expands, it carries the lower part of the heads and brace attachments with it, stretching the braces far beyond their yield point. Thus, when the shell contracts, the braces have been permanently elongated and will bow. The direction of this bowing is often upward (but may be in any direction).

A slight bow need not be repaired. If the braces are bowed badly, they should be taken out and straightened. If they are elongated so much that a full number of threads is not in contact with the inside nut, new braces should be installed.

Hydrostatic tests are made on boilers that have had any repairs, such as tube installation, or patching. They are applied also to determine the location, extent, or existence of defects. The boiler is filled to the top with water at room temperature, but not below 70°F so that if a defect should "let go," the slight expansion force of the water would not result in an explosion. A pressure of up to 1½ times the maximum allowable pressure is applied by either the feed pump or a special test pump.

The safety valves should be gagged or removed and the connections fitted with a blank flange. The safety-valve adjustment should not be screwed down for hydrostatic tests.

If other boilers are on the line, the drain valve between the stop valves should be left open. If water or steam leaks past the stop valves, showing from the open drip, the hydrostatic test should be delayed until the steam line is fitted with a blank flange. No chance should be taken of cold water pressure building up against a stop valve having steam temperature on its other side.

Any areas where defects are suspected, such as welded or riveted seams, should be exposed for the test by removing all brickwork necessary. The test pressure should be maintained long enough to examine all parts for leakage. A hammer test is often applied to suspected sections while under test pressure.

If any leakage or indication of distress develops, an authorized inspector should be consulted in regard to the cause and the procedure to follow. Whenever possible, it is advisable to have the test witnessed by the inspector.

The first thing to do on completing the test is to remove the gags or blank flanges from the safety valves. This action should never be delayed, for if it is, it may be forgotten, and a serious boiler explosion would result from malfunctioning of controls or operator neglect.

The normal procedure after the hydrostatic test and inspection is completed is to release pressure slowly through a small drain valve, then fully open vents and drains. It is important to make sure that boiler sections normally not containing water during operation, such as superheater, reheaters, and circulating tubes above normal water level are thoroughly drained of water after the hydrostatic test. If the gauge glass was removed during the test with cocks shut off, they should be restored to operating conditions. If any temporary handhole or manhole gaskets were used during the test, these should be replaced with regular operating-type service gaskets prior to preparing for boiler operation.

Acoustic leak detection is being used on large boilers to alert operating personnel to the existence of a leak within minutes after the leak starts. This is done by locating multiple acoustic sensors in different parts of the boiler. The acoustic sensors detect the high-frequency emissions from leaking steam. The acoustic signal is transmitted electronically to a control room display; the exact location of the leak is then traced according to which sensor responded to the leak. The systems are designed to filter out normal boiler operating noise. By early detection of the leak, repair plans can be made quickly before more damage is done to adjoining tubes.

Asbestos insulation on pipes and older boilers must be removed with care and regulatory requirements followed during any repairs or alterations. This adds to the cost of repairs and the time required, and many boiler plants are replacing this material for this reason, as well as for maintaining a healthy work environment.

Questions and Answers

1 What is serpentine scale?

ANSWER: This is scale having a dull green color and mottled appearance, with mineral or rock characteristics designated as hydrous magnesium silicate, $MgO + 2(SiO_2 \cdot H_2O)$.

2 Name three scale-forming elements in water.

ANSWER: Calcium (Ca), magnesium (Mg), and silica (SiO_2).

3 What are the two major objections to scale in a boiler?

ANSWER: (1) Scale is a heat insulator and thus may produce overheating of the parts affected. (2) Scale causes a considerable loss of efficiency.

4 Is a tube turbine used to remove scale mechanically from fire tubes? If not, what is used?

ANSWER: No; a tube rattler or vibrator is used. A tube turbine is used on watertube boilers.

5 Of what benefit is a coating of oil on internal surfaces of a boiler? Explain your answer.

ANSWER: None. It is exceedingly dangerous. Oil is a heat insulator and thus may produce dangerous overheating of affected surfaces.

6 What is the most common source of oil in boilers?

ANSWER: Use of contaminated condensate returns from reciprocating steam equipment or process heat exchangers.

7 How might fuel oil get into a boiler when oil burners are used?

ANSWER: By failure of a tube or coil in a fuel-oil steam heating element when the condensate is returned to the feedwater system.

8 What causes most external corrosion?

ANSWER: Sulfur in soot, coal, ash, and moisture.

9 What objection is there to burying piping under the boiler floor?

ANSWER: External corrosion may progress unnoticed to a dangerous point.

10 What is erosion and what is the most common cause of it in a boiler?

ANSWER: It is the wearing of surfaces by abrasion. It is usually caused by impingement of soot and ash particles.

11 What harm may be caused by flame impingement?

ANSWER: Localized overheating and damage to boiler parts if not protected by refractory.

12 What is the external indication of low water and overheating in a boiler other than possible distortion, discoloration, and leakage?

ANSWER: The soot would be burned off of affected surfaces.

13 Which is the more serious, a lap crack or a fire crack? Where may each be found?

ANSWER: A lap crack is more serious. Lap cracks are found in longitudinal lap-riveted or welded seams. Fire cracks may be found in the lap of riveted or welded seams exposed to radiant heat as in old HRT girth seams and firebox-boiler furnace seams.

14 How may lap cracks be repaired?

ANSWER: No repairs are allowed other than an entire new course of boiler plate. Otherwise, the boiler is condemned.

15 How is a broken stay bolt in an empty, idle boiler detected?

ANSWER: When the ends of solid-type stay bolts are struck with a hammer, a broken stay bolt may be detected by a dull sound. Flexible stay bolts are tested by removing the cap and using a heavy screwdriver in the ball slot.

16 What does leakage from the telltale hole in a stay bolt indicate?

ANSWER: The bolt is either broken or cracked halfway through.

17 What causes stay bolts to break, and what location is most susceptible in firebox boilers?

ANSWER: The expansion and contraction (breathing action) in the boilers bend the stay bolts. Fatigue may break them eventually. The upper rows are most susceptible to breakage, for the expansion is greatest there, the lower part of the firebox being stiffened by the riveted or welded connection to the outer sheet.

18 A battery of several boilers is allowed 100-psi pressure, and all safety valves are set at that pressure. All steam gauges are accurate. One boiler shows 100 psi and its safety valve is blowing, while the other boiler gauges registered only 80 psi. What is wrong?

ANSWER: The stop valve on the boiler under 100 psi is closed, the connecting steam pipe is blanked off, or the dry pipe is stopped up to prevent passage of steam. This condition might also be created if the steam pipe connecting the 100-psi boiler were small compared to the size and steaming capacity of the boiler.

19 (a) Where is internal corrosion most likely to occur in a vertical tubular boiler?

  (b) Where is internal corrosion most likely to occur in a locomotive boiler and why?

  (c) Where is external corrosion most likely to occur on these boilers and why?

ANSWER:

  (a) Internal corrosion in vertical tubular boilers is most likely to occur in the water leg at the bottom ring or ogee flange because this is a point of poor circulation. Also it occurs along the waterline, of both shell and tubes, because this is where air or entrained oxygen is liberated in the ebullition of water while the boiler is steaming. Corrosion also occurs around the plate where the supporting stay bolts enter because of grooving.

  (b) For the same reason, internal corrosion is most likely to occur in locomotive boilers around the mud ring, in the water leg, and along the waterline. This type of boiler is also found to corrode badly around the bottom of the barrel, especially toward the front or smokebox end.

  (c) External corrosion of any boiler is most likely to occur at points where there may be leakage or moisture and especially where coupled with sulfur from soot and ashes. In vertical tubular boilers it occurs around the handholes, the lower part of the outside shell (where metal is in contact with ashes and with ogee construction at the bottom of the furnace sheet below the grate line), and the upper tube sheet and tube ends. In the locomotive type, it occurs around handholes, the lower part of furnace sheets at the grate line, the lower part of external water-leg sheets, the throat sheet, and the lower part of the front head at the tube sheet.

20 (a) In a watertube boiler, of longitudinal drum type, the lower tubes in one bank showed signs of overheating. Where would you look for the trouble?

  (b) The nipples connecting the mud drum in this boiler are continually leaking. They have been renewed and rerolled, and they still leak. What would be the cause of this?

ANSWER:

(a) The down flow pipe could be partially filled with scale, thus obstructing free circulation of water.

(b) The mud drum being bottomed or resting on brickwork would prevent movement caused by expansion and contraction.

21 (a) What appliances on a boiler must readily give the true water level? How can this be checked?

(b) What is the object of a water column? Why does the Code stipulate 1 in. as the minimum-size pipe connection?

ANSWER:

(a) The water glass. It can be checked by using the try or gauge cocks.

(b) The water column is used to give enough volume of water to provide a steady level in the glass for a true reading when the glass cannot be connected directly to the boiler head. Smaller pipe could plug up too easily with scale and sediment.

22 On high-pressure boilers, in addition to conductivity of feedwater, what other measurement should be made of steam going to high-pressure turbines?

ANSWER: Stress corrosion cracking can occur if the recommended level of sodium values of the turbine manufacturer is exceeded. The maximum limit is about 25 parts per billion (ppb) for boilers operating over 600 psi. Ion electrode instruments are used to measure the sodium content of steam to the turbine. The possibility of exceeding permissible sodium levels occurs from upsets in demineralizers, condenser leakage, and leakage of sodium from condensate polishers. Turbine manufacturers are specifying even lower sodium levels (5 ppb) as they try to find material that will perform satisfactorily under high stress in a possible hostile environment that can produce stress corrosion cracking with time exposure.

23 Briefly, what causes through-to-head braces below the tubes in an old HRT boiler to bow?

ANSWER: Overheating of the shell bottom causes excessive expansion and permanent elongation of the braces. When the shell cools and contracts, the braces bow.

24 What causes steam binding of a feed pump? How should you recognize it, and what should you do for it?

ANSWER: Excessive temperatures for the head pressure on the suction are the cause. A reciprocating pump would race, short-stroke, or hammer. A centrifugal pump would overheat, vibrate, and operate noisily. Temporary measures might include slowing the pump down, throttling the water discharge, and/or reducing the water temperature by introducing cold water into the suction if necessary. Playing a stream of cold water onto the suction pipe may aid in emergencies. Permanent correction requires greater suction-head pressure or lower temperature.

25   What is a common cause of excessive feed-pump suction temperature?

ANSWER: Steam blowing through into the feedwater storage owing to defective traps.

26   What operating precaution should be exercised with outdoor packaged boilers or economizers in cold weather?

ANSWER: Freezing of the blowdown should be guarded against, either by use of a small line bypassing the blowdown valves or by leaving the blowdown valves open slightly.

27   How are superheater tubes protected against overheating in starting a boiler?

ANSWER: Flooding the tubes with water is recommended for a few types. More commonly, the superheater is vented to the atmosphere by a free blow valve until the boiler goes on the line, and the firing is kept at a low rate until steam appears from this vent, indicating that steam cooling of the tubes has started.

28   What happens if the lower connection to a gauge glass is obstructed?

ANSWER: The water will return to the glass very slowly after blowing it down. If the obstruction is complete, the glass will fill with water slowly, because of steam condensation. An incorrect water level will be shown by the glass.

29   With reference to question 28, how should the water level be checked? Outline the procedure.

ANSWER: By use of the try cocks. Try to remove the obstruction by attempting to blow down the glass with the steam connection closed. If this is not successful, close the connections and remove the glass. Try to run a stiff, bent wire through the lower connection after opening this valve partway, taking care to keep to one side so as not to get scalded. A careful check on the water level by the try cocks should be maintained. If the obstruction cannot be freed, the boiler should be shut down so that the connection may be dismantled for cleaning.

30   How should you know if the upper connection of a gauge glass were obstructed? What should you do if it were?

ANSWER: Water would rise rapidly to the top after blowing the glass down, although try cocks show a lower true level. Proceed as for lower obstruction.

31   How much space should there be in front of a firetube boiler in planning installation?

ANSWER: Sufficient room for replacement of tubes.

32   If, on looking in the furnace, the bottom of the shell of an old HRT boiler is found to be bulged, what should be done?

ANSWER: The boiler should be shut down immediately and then inspected by an authorized boiler inspector. The inspector's recommendations should be followed before the boiler is returned to service.

33   Why were HRT boilers over 72-in. diameter required to be supported by the outside-suspension type of setting?

ANSWER: Because the weight of the larger boilers may be in excess of the safe load on brickwork. Crushing or buckling of the walls might result from the load of a large boiler full of water.

34   How should a boiler be prepared for inspection?

ANSWER: Extinguish the fire, and allow the boiler to cool slowly. When it is cool, open the blowdown and drain, venting the boiler to atmosphere. The soot and ashes should then be blown and swept clear of all tubes, shell plates, heads, seams, and every accessible external surface.

The blowdown valve should be shut if any other boilers feed this line. The manhole, handhole plates, and inspection plugs should then be removed. All loose deposits of sludge or other sediment should be washed out. The blowdown valve should be opened only when it is certain that there is no pressure in the line and no one inside the boiler. Any scale or oil deposits should be left for the water treatment specialist to see for analysis.

A boiler properly prepared for inspection should be cool, clean, and dry.

It is advisable to attach a red tag marked "Person in boiler" to the steam, blowdown, feed, and fuel valves and also to the manhole plate whenever anyone is in the boiler. Also follow OSHA rules for confined space entry.

35   May an owner-user commissioned inspector act as the authorized inspector when repairs or alterations are made to the owner-user's boilers and pressure vessels?

ANSWER: The NB revised rules of 1996 permit the owner-user's authorized inspector to act as such when repair or alterations are made to boilers or pressure vessels at the owner-user's location. If the owner-user has a NB "R" stamp, the owner-user inspector may also act as the third-party inspector to verify that the stamp holder made a Code-approved repair. The owner-user, however, must have an inspection program at the location that is accredited by the NB. The owner-user inspector under NB rules is limited to making authorized inspections only to owner-user equipment. Check with the NB for further details. The owner-user inspector must also have a valid owner-user commission issued by the NB. The owner-user's inspection program must also meet jurisdictional requirements for that location.

36   How does replication determine the condition of a header or piping surface?

ANSWER: The method provides a metallurgical microstructure image of the surface. A thin plastic film is applied to the surface of a metal in areas under high stress. The film reproduces the surface features that can reveal microstructural changes or failure initiation in the metal surface. Trained technicians in this method of NDT can then determine the current stage of any creep damage to the metal and then determine remaining life.

37   What is the disadvantage of having one float control the on-off water level operation, and also, if the float drops further, the low-water fuel cutoff function?

ANSWER: The disadvantage is that two control functions are activated by one float. If the float becomes stuck or inoperative at what appears to be a "safe" water level, both the level control and the backup and low-water fuel cutoff will not function, and a serious dry-firing accident can be the result. It is preferred by the Code that one device control the water level, and the low-water fuel cutoff be separate from this control.

38   In NDT, what is the difference between surface and volumetric examination?

ANSWER: Surface examination, such as with dye penetrant and magnetic particles, are used to detect discontinuities or cracks on the surface of the boiler component or piping under test for defects, while volumetric examination is used to detect discontinuities including cracks, porosity, and inclusions below the surface or in the internal structure of the part under test. The two most commonly used field methods are ultrasonic inspection and radiographic examination.

39   What three functions can a boiler operator contribute to boiler maintenance?

ANSWER:

1. *Maintaining a safe and reliable plant.*   This includes periodic testing to make sure the low-water alarms and cutoffs function properly; proper water levels are maintained to avoid overheating damage; the water treatment is preventing scale and rust from forming dangerous deposits on heat-transfer surfaces; the flame safeguard system is functional so that no accumulated fuel in the furnace can cause a furnace explosion; all auxiliaries are operating properly so that boiler operation is not endangered from faulty draft, faulty feedwater flow, or faulty fuel flow; pressures are within the AWP; and the safety valve is functional.

2. *Maintaining an efficient plant.*   Heat loss up the stack is a significant percentage of loss in efficiency. Some causes of this are under the control of the operator, such as maintaining proper draft and air-fuel

ratios as well as stack temperatures. Above-normal stack temperatures may be due to scale buildup on heat-absorbing surfaces in the boiler. Too much excess air wastes fuel by heating the air that is uncombined with the fuel. Poor draft may prevent all fuel from being burned in the furnace.

3. *Inspection and logging of items that need to be corrected during the next maintenance period.* This includes auxiliaries. During outages, internal inspections will indicate if corrosion or harmful deposits are occurring, and this may require adjustments in operation or feedwater treatment.

40 What is a common defect at the edges of a handhole? Give cause, prevention, and repair.

ANSWER: Eroded and steam-cut bearing surface. The cause is surfaces not properly cleaned, or an improperly installed gasket. The bearing surfaces should be properly cleaned, and proper gaskets installed. Repairs can be made: Build up the corroded area by welding providing that better than 50 percent of the original metal thickness remains or a reinforcing ring may be used. If the area is built up by welding, the weld is then machined and made smooth for the gasket surface.

41 What causes water hammer and how can it be corrected?

ANSWER: Water hammer is the passage of high-velocity slugs of water through a steam pipe. The impact of the slugs on pipe bends or elbows can cause pipe failures. Slugs or droplets of water are caused by pockets in the steam lines that are not adequately drained and by opening steam valves too rapidly. Providing adequate drains at low points in the steam lines will prevent pockets of water. Valves on cold steam lines should be opened slowly.

42 When are double stop valves required on the steam lines connecting a boiler to another boiler's line, and what are the requirements for these valves?

ANSWER: Boilers connected to common headers must have two stop valves with a free-blow drain valve between the valves. The valve next to the boiler should be a nonreturn type, and the second stop valve should be of the outside-screw-and-yoke type.

The free blow drain will tell an operator if a valve is leaking before the operator enters a drum for maintenance and inspection work. This avoids scalding of the operator.

43 Name three causes of bagging or bulging of boiler shells and tubes.

ANSWER: Bagging or bulging is due primarily to overheating a plate or tube to a point where the tensile strength is lowered and plastic flow is induced by the boiler pressure. Contributing factors are deposits of scale or oil or both on the internal surfaces. Plastic flow may also be induced on clean boiler surfaces where severe flame impingement or faulty circulation causes steam blanketing. Bulging can also develop when a tube or plate has been reduced by corrosion or by erosion due to cutting of a soot-blower jet or fly ash until the wall is too thin to withstand the boiler pressure.

44 What areas of a scotch marine boiler experience the greatest number of failures?

ANSWER:

1. Tube leakage at tube to tube-sheet rolled joints.
2. Tube leakage from metal thinning, wasting of tubes, or blowing of tubes.
3. Weld leakage at furnace to tube-sheet welds.
4. Weld leakage at tube sheet to shell welds.
5. Tube-hole and ligament cracks and leakage.
6. Furnace bags, bulges and cracking.
7. Furnace bags, rupture and explosion.

Item 7 can be a very serious type of failure, causing not only extensive property damage but also personal injury and death.

45 A crack 14-in. long is discovered in a 30-in.-diameter corrugated furnace. The condition of the furnace is otherwise satisfactory. Is it permissible to repair the crack by welding?

ANSWER: Yes, provided the welding is done by a qualified welding procedure and welder and the work is locally stress-relieved and approved by an authorized inspector.

46 How should you repair a broken stay bolt?

ANSWER: Repairs are not allowed. A new stay bolt should be installed.

47 What kind of water should be used in conducting hydrostatic tests on nondrainable sections of steam generators, reheaters, or superheaters?

ANSWER: To avoid scaling, pitting, or stress corrosion to develop in the nondrainable sections of a boiler, deaerated distilled water, demineralized water, or polished condensate should be used.

48 How is ash categorized in coal-fired plants?

ANSWER: Ash is categorized as (1) bottom ash, which is collected and withdrawn from the bottom of the furnace, and (2) fly ash, which is the lighter, fine particles removed downstream of the boiler in filters, electrostatic precipitators, mechanical collectors with the balance being discharged through the stack.

49 When monitoring the air around a repair that may involve asbestos removal, what is the permissible exposure limit per OSHA rules?

ANSWER: The permissible exposure limit per latest OSHA rules for an eight-hr time-weighted average for asbestos is 0.1 fibers per cubic centimeter.

50   What are two ways to approach steam piping in a building that may have asbestos insulation?

ANSWER: Per present OSHA guidelines, if the airborne concentration is below the allowable limits: (1) Leave the asbestos covering as is, but keep monitoring the concentration at stipulated intervals, or when dust particles are noted. (2) Encapsulate the existing piping with new material to prevent the asbestos from contaminating the air. (3) Enclose the area, which means sealing off the area where the asbestos concentration exceeds the allowable limits of concentration. The process of enclosing involves installing a barrier such as reinforced polyethylene.

51   What parts of a watertube boiler may suffer damage from on-off cycling service?

ANSWER: Cycling service is experienced on boilers used for peaking service, and which are subject to load reduction, idle periods then restarts, sometimes daily, and those on severe swing loading. On-off cycling causes thermal-induced stresses from expansion and contraction effects, which can cause thermal-fatigue cracking in such components as economizer headers, furnace wall tubes, tube to header welds, and superheater and reheater headers. On-off cycling of 20 per yr is one criteria used to denote a cycling condition.

52   What causes dew-point corrosion?

ANSWER: Fuels that contain sulfur, such as fossil fuels and most waste fuels, produce products of combustion that contain sulfur dioxide and water vapor. At temperatures below about 300°F, these vapor and gases condense and form sulfurous and sulfuric acids, which attack the boiler metals in what is termed acid corrosion. If the temperature of the flue gas can be maintained above the acid dew-point temperature, this attack on boiler components can be minimized. Special alloys can also be used to resist acid dew-point corrosion.

53   How should a hydrostatic test be applied on a boiler installed in battery with the other boilers in operation?

ANSWER: A hydrostatic test is performed with cool water at a pressure of 1½ times the maximum allowable working pressure. The test is performed on new boilers and on boilers in the field to check major repairs or suspected defects.

All areas to be examined should be exposed. The boiler should be filled to the top with water at room temperature but not under 70°F. The safety valves should be gagged or the connections blanked. Both stop valves on the steam lines should be closed and the valved drip between them open. A test gauge should be used to check the pressure. If any leakage from the drip between the stop valves occurs, the test should be stopped and the steam line blanked off. The water pressure should be raised slowly to not over 1½ times the working pressure and held at that value long enough to make a complete examination of the boiler and a hammer test when advisable. Immediately after the test, the safety-valve gags or blank flanges should be removed.

54   If a safety valve sticks open, what should you do?

ANSWER: Reduce the firing rate. Tap the top of the safety-valve spindle sharply with a *light* hammer. After the valve seats, blow all dust from the external surfaces of the safety valve, and pour some kerosene around the spindle bushing. Operate the valve with the lifting lever until it works freely. If the valve will not seat, remove the boiler from service and overhaul the safety valve. Care should be taken to maintain proper water level when the safety valve is blowing.

55   Must all repairs be approved by an authorized inspector?

ANSWER: Yes, if the strength of the vessel has been impaired in any way requiring repairs involving Code enforcement and interpretation. Repairs not affecting the strength of the boiler or of a minor or routine nature may not require approval. But the inspector should be consulted on the problem if there is any doubt about the safety of the boiler. Crack repairs, welding, tube replacement, safety-valve replacement, and similar repairs or changes require approval. The best rule to follow on any structural repairs or changes on a boiler is to immediately contact an authorized boiler inspector.

56   Is a qualified welder permitted to make welded repairs on any part of a boiler?

ANSWER: Not necessarily. That a welder is qualified to make some welds may not mean he or she is qualified for welding: (1) the particular thickness of plate; (2) the type of material; (3) in the position of welding to be used; or (4) according to the welding method required.

57   If the distance between the bases of gauge-glass stuffing boxes is 12 in., how long should you cut a new gauge glass? Explain your answer.

ANSWER: About 11¾ in. to allow for expansion.

58   What are some destructive tests used in qualifying Code welders and welding procedures for boiler installation and repair work?

ANSWER: Destructive tests used are: (1) Tensile test of specimens to establish the strength of the weld. (2) Bend tests to determine the suitability of the welded joint for the intended service and to qualify the procedure and welders. (3) Hardness tests to check on the heat-affected zone of a hardenable steel to determine the potential for cold cracking. Hardness tests are also quite often used to provide rapid estimates of mechanical properties. For example, 500 times the Brinell hardness number gives an approximate value for the ultimate strength of steel in pounds per square inch. (4) Drop-weight tests are also used to determine the temperature above which a dynamic crack will be arrested. (5) Charpy V-notch tests are used to determine the toughness of steel in resisting cracking if a slight notch in the specimen is present. (6) Some applications require determining the transition tempera-ture zone where material changes from ductile to brittle behavior. Load is

applied as in the Charpy V-notch test but to fracture at different temperatures of the material. A plot is made of the foot-pounds needed for fracture
versus the temperature at which it occurs. At cold temperatures, less energy
is needed, and at the transition zone the material starts to become ductile
and fractures at higher foot-pounds as temperature increases to a final complete ductile fracture zone.

59 On what basis are repairs allowed on boilers?

ANSWER: Repairs permitted are based on restoring the affected part or parts
to as near the original strength as possible. They are governed by Code
requirements for new construction, or by NB rules on permissible repairs,
where the state has adopted NB rules for repairs.

60 Is it necessary to repeat the preheat treatment and postweld heat treatment on a power boiler that underwent repairs as a result of defects found
during a hydrostatic test?

ANSWER: The preheating and postweld heat-treating rules of a welded joint
must be repeated or reapplied to all weld repairs made to pressure parts.

61 What assistance should the owner or operating engineer give the jurisdictional boiler inspector during inspections?

ANSWER: It is the responsibility of the owner to prepare the boiler for the
required legal internal inspection. All openings must be removed. All scale
and mud must be removed so metal surfaces are exposed for inspection. Fire
sides must be cleaned of soot so tubes can be inspected for corrosion, thinning, erosion, and evidence of overheating.

The inspector must be given all the help she or he needs. Point out any
known defects. Station someone immediately outside the boiler during the
internal inspection. If the boiler is in battery with others, make sure that all
steam, water, and blowoff valves are locked and cannot be opened. Make provision for the hydrostatic test if the inspector deems it advisable. In general,
assist in every way to make the examination thorough and complete.

Complete all safety recommendations that may be offered so that the
Certificate of Operation issued by the jurisdiction can be quickly processed as
a result of the jurisdictional inspection.

62 What precaution must be observed before closing a new boiler or a boiler
that has been opened for cleaning or repair?

ANSWER: Make sure all tools, pipes, welding rods, rags, and other such items
are removed from drums, headers, furnaces, and tubes. At times, a mirror
and flashlight must be used to check headers that are otherwise not accessible for inspection of foreign material. Bent tubes that cannot be looked at
from end to end (such as superheater tubes) should be thoroughly flushed,
one tube at a time. On new boilers or where work was done in a tube area,
drop rubber balls, and even steel balls, to make sure the tubes are free of
obstruction. Water and air can be used to push through the rubber balls.

63 What precaution must be observed in turbining the tubes?

ANSWER: Turbining tubes can cause local tube wear or nicking if the turbining tool is forced through a tube or held in one position too long.

64 Is it safe to use portable lamps and extension cords inside boiler drums, shells, or headers?

ANSWER: Only if low-voltage lamps, 32 V or less, supplied by transformers or batteries are used to avoid electric shocks in case a lamp or bulb breaks and creates a current flow through the boiler shell. Never use extension cords without proper waterproof fittings. Make all connections outside the boiler. And light bulbs should have explosion-proof guards. Fittings, sockets, and lamp guards must be grounded.

65 A boiler is to be connected to a steam header supplied by two other boilers. What are the stop-valve requirements for the boiler per Code?

ANSWER: The boiler must have two stop valves when connected to a common steam header, with a free drain between the stop valves so the tightness of the valves can be noted. The valve next to the boiler must be an automatic nonreturn valve, and the second valve near to the header must be of the outside-screw-and-yoke type.

66 Why should a steam safety valve not be placed on the steam line connecting a boiler to the plant steam line?

ANSWER: Safety valves should be installed on the boiler proper, because if installed on the steam line, due to flow of steam in the pipe there would be a difference in pressure in the pipe from that in the boiler; thus the pressure directly under the seat of the valve would not be the boiler pressure. The flow of steam in the pipe would also cause the safety valve to chatter and eventually damage the disc and seat. If the boiler has a dry pipe and it became obstructed, the safety valve would not sense boiler pressure, but only pipe pressure.

15

Boiler Plant Training, Performance and Efficiency Monitoring

Boiler operation and maintenance practices are changing as more plants become equipped with modern controls and instrumentation. Boiler plant performance improvement includes technical considerations to improve efficiency with reliability so as to assure continuous safe service is being provided per existing statutes and what is termed good engineering practice. A new maintenance philosophy combines efficiency centered maintenance with reliability maintenance. The reliability concept involves identifying key physical measurements that could cause a breakdown, and then setting limits for when corrective actions must be taken. The term "computer-based maintenance" is often used. This involves setting up a computer program to:

1. Schedule and track performance output so that changes can be noted. If the performance is below set points, preventive maintenance is scheduled with the necessary work orders.

2. Show trending of various material data, such as heat rate, amount of fuel used and output, pressure drops in the flue gas loop, and similar data to show that corrective actions may be required in order to improve performance or to avoid a failure in operation.

3. Track labor and material costs to maintain the plant on a yearly basis per thousand lb of steam produced.

4. Maintain a spare-parts inventory control and ordering system.

Financial considerations in improving boiler plant performance are interrelated to technical plant performance in such areas as reducing fuel costs by improving efficiency and reducing personnel needs by more automation, and also considering outside maintenance for the annual turn-around inspection. The trend is to view each boiler plant by cost-accounting methods and, for those plants that generate both process steam and electric energy, as a profit center, in other words, to consider whether it is cheaper to buy power or generate power.

Employee selection and training. Modern boilers and auxiliary equipment have been developed along such lines of automatic control that they seem to require little attention. Unfortunately, this development has had its effect on some operators with the consequence that they have acquired the idea that the equipment can think for itself. Regardless of how automatic a plant may be, the human element may still be the weakest link and may become responsible for a costly shutdown. Each individual connected with boiler plant operation and maintenance should be trained to understand that she or he has a highly responsible duty to carry out and that those who cannot assume this responsibility capably have no place in a boiler plant. This attitude should be assumed by each person in the organization, from the ash handler to the chief engineer. The success of the organization depends largely on the careful selection of each prospective employee—a real problem in personnel study.

High availability of power plant equipment is the result of correctly installing reliable hardware and associated systems, which are then properly operated and maintained by skilled people. Part of the effort to keep power plant equipment operating efficiently and safely requires setting up inspection programs of the boiler, auxiliaries, firing equipment, and controls.

Operators are being cross-trained in combined-cycle plants to perform minor maintenance as well. Management will view such personnel as being more valuable to their operation. The basic rule to follow is to learn new skills, and you become more valuable.

Training programs. Study programs should be developed. The size of the plant and available facilities will determine the best approach. In large plants this may consist of periodic classes for the operating personnel under the supervision of a capable executive; in small plants it may outline home study for those interested. Some means of instilling ambition and promoting action to acquire further knowledge of the equipment with which they work will be to the benefit of all concerned.

There are several organizations of operating and power engineers which are represented in larger cities and whose aims are purely educational. The educational and social functions of such groups are recommended highly.

Management should annually review the training available for the power plant personnel to note if the relationship to the rest of the plant is clearly understood. Many plant engineers or powerhouse superintendents simulate plant upsets in frequent in-plant drills. This is an excellent method of preparing for all types of contingencies.

Large boiler systems require special training programs because of the complexity of the controls, valving, instrumentation, and depth of hardware. Large boiler manufacturers sell training programs to utilities and large industrial customers. With government regulations on emissions standards affecting operations, training by personnel from the boiler manufacturer should help operators in taking over a new plant. Simulators are more and more used as the boiler plants become more complex.

Many state laws require full-time or part-time operator attendance on automatic-fired boilers. The meaning of *attendance* in many jurisdictions is as follows: Watertube boilers with capacities over 20,000 lb/hr and those fired with pulverized coal should have full-time operator attendance. High-pressure firetube boilers should have full-time or part-time operator attendance. *Full-time attendance* means the presence of a competent operator who never leaves the boiler room for more than 20 min. *Part-time attendance* means the presence of an operator who may leave the boiler room for more than 20 min, leaving the boiler operating unattended, but making periodic checks at least every 2 hr. A boiler is considered operated unattended when it is operated for more than 2 hr without a competent operator checking it for proper operation. State and city laws must be checked on operator-attendance requirements.

Where no legal requirements exist for operator attendance, it is good practice to train people to the following minimum know-how requirements depending on the size of the plant:

1. Trace and sketch a boiler-fuel system and associated valves, strainers, gauges, and controls.

2. Trace and sketch main and auxiliary steam systems and condensate-return and boiler-feed systems, including valves, gauges, controls, and interlocks.

3. Inspect and test boiler casings and settings.

4. Inspect fire sides and water sides for leakage, corrosion, cracking, bulging, blistering, and other conditions that weaken the boiler.

5. Clean, inspect, and test oil and gas burners and registers.

6. Check fuel tanks and, if necessary, clean strainers, lines, and valves.

7. Line up, recirculate, check for leaks, and light off a fuel-oil system. Check combustion safeguards.

8. Observe and record fuel-system pressures and temperatures. Check and adjust control settings.

9. Inspect and operate soot blowers properly.

10. Light off, fire, and bank fires in a coal-burning boiler (if installed).

11. Inspect and regulate stoker and pulverizer operation in a coal-fired boiler. Check and adjust controls and interlocks.

12. Start, regulate, and secure forced-draft and induced-draft fans. Check for proper operation, controls, and interlocks.

13. Inspect breechings, uptakes, and stack.

14. Determine draft, wind-box, and furnace pressures and furnace and stack temperatures. Adjust the draft where needed.

15. Take and analyze flue gases for CO_2, CO, and O_2. Interpret the results and make the necessary corrections.

16. Blow down a boiler, both bottom and surface blows.

17. Blow down gauge glasses and water columns.

18. Test high- and low-water alarms and low-water cutoff. Dismantle and clean alarms and low-water fuel cutoff.

19. Regulate feed pumps and change-over pumps and adjust feedwater governors.

20. Regulate boiler water level.

21. Cut in, adjust, and secure feedwater-level regulator.

22. Test safety valves.

23. Put a boiler on line.

24. Warm up and cut in steam lines.

25. Take a boiler off line, pull fires (for coal-burning boilers), and secure the boiler.

26. Conduct boiler-water analysis, interpret the results, treat the feedwater, and adjust continuous blow according to a water-treatment specialist's advice, if required.

27. Shift combustion control from manual to automatic and back again. Check the safety controls in doing this.

28. Prepare boiler logs and operating records.

29. Cut superheaters in and out properly.

30. Adjust feedwater heater pressures and temperatures.

31. Renew and repack gauge glasses.

32. Remove, regasket, and replace manhole and handhole plates.

33. Inspect, repair, and set safety valves within Code limits.

34. Clean the fire side and water side of the boiler.

35. Make refractory and other furnace repairs.

36. Conduct a hydrostatic test.

37. Lay up a boiler.

38. Know how to remove and replace tubes in the boiler, superheater, economizer, air heater, and feedwater heater.

39. Adjust soot blowers and lances.

40. Inspect, clean, and repair boiler gauges, instruments, and controls.

In addition to the above, an operator should continuously study any new developments in: boiler operation, maintenance procedures, repairs, legal requirements including safety and environmental laws affecting boiler plants, Code changes affecting boiler repairs, and inspection or installation requirements.

New plant familiarization. An efficient way to become familiar with a new plant is to trace every important piping system and make a sketch of all the valves and equipment in each system. Do the same with electrical systems. An operator should know the plant well enough to be able to go to key valves and controls in the dark during an emergency. All new operators should trace and sketch each system. A file of literature pertaining to machinery, instrumentation, and equipment in the plant should be maintained for study. Also, a filing system for blueprints of important equipment should be kept in case they are needed in a hurry. Set up a spare-parts inventory of critical components so they are always on hand or make a list of suppliers that can respond immediately to supply you with essential parts or repair services. This includes boiler tubes and components of equipment that would shut down the plant if they were not in stock and needed because of failure.

Emergency training. Training for emergencies of the operating personnel is indispensable to the boiler plant. In the small plant, individual instruction may suffice, but in larger plants, as with a fire drill, the operating force should be schooled for every conceivable emergency. Whether the emergency is low water in a boiler, loss of auxiliary power, failure of an interlocking system, or a rupture in the pressure system, each worker should have a definite station and specific duties to fulfill. Everyone should be drilled to understand the reason for these duties and to carry them out quickly with no confusion or overlapping in their scope.

Certain alarm systems are invaluable for warning of emergencies, but there is no alarm system that can replace the almost instinctive anticipation of many emergencies by a well-trained, alert engineer.

Flooding of boiler equipment has created emergencies that normally require power plant equipment to be reconditioned after the waters recede. The following precautions should be taken before a boiler that has experienced flooding is returned to service:

1. *Safety valves:* Clean, lubricate, inspect, and test the valves. Inspect the discharge pipe for mud obstructions.

2. *Low-water cutoffs:* Clean, rewire if necessary, renew probes and mercoid switches if necessary, inspect, and test.

3. *Limit controls:* Clean, inspect, and test. Renew if necessary.

4. *Flame safeguard devices:* Clean, inspect, and test. Renew if necessary.

5. *Electric motors, transformers, relays, and wiring:* Electric motors should be cleaned, baked out, and tested before energizing. Conduit, BX, wiring, and relays should be checked by a qualified electrician. If any is defective, renew. Failure to follow safe procedures regarding electrical facilities could lead to fatal electrocution and/or electrical failures including burnouts and fires.

6. *Refractory, brickwork, and insulation:* After inspection and repair, all these should be dried out as much as possible before firing. Initial firing should be light and in short intervals, in order to allow moisture to escape gradually. Rapid firing could build up steam pockets with resultant pushing out of refractory.

7. *Internal inspection:* Boilers should be thoroughly inspected internally, and any accumulation of slime and mud should be removed.

8. *Start-up:* As soon as the boiler is started, the following should be done: Test flame safeguard devices, low-water cutoffs, safety valves, and limit controls.

9. *Surveillance:* Maintain the boiler log after the boiler becomes operational. Diligently monitor the operation of controls and safety devices. Keep the boiler under strict surveillance for leaks, dirty gauge glass, muddy blowdown, and similar evidence that the effect of the flooding may require further cleaning and repairs.

Forced outages must always be considered a possibility. Most plants have no spare capacity or not enough capacity when the largest unit suffers an unexpected failure. Emergency planning should include an adequate inventory of spare parts, such as tubes, plugs, gaskets, and similar components, that will help in expediting a repair. Extensive repairs of a key unit may require considering renting a boiler while the damaged unit is being repaired. Much time can be gained if plant engineers preplan for such an emergency by having water, steam, and fuel piping installed for the rental boiler so that the rental unit can be quickly connected to the plant's system. A list of suppliers and reliable repair firms should be made available to all shift operators.

Control room operation. Large plants depend heavily on control-room-type operations. This practice has eliminated continuous area surveillance of large power plant sections, with heavy reliance placed on instrumentation and alarms to alert the control room to an abnormal situation. It is not possible to instrument an alarm for all possibilities of malfunction. Therefore, care is needed in avoiding too much reliance on control-room-type operation unless it is supplemented by area surveillance of operation at stipulated, safe intervals. As a result of several nuclear plant experiences, the following concepts in control room operation are being developed. Control rooms should display normal, marginal, and out-of-limit conditions on a system. Control room operators should be trained to identify when a safety system on the loop may be involved and what the implications to safe operation may be as a result. An upper-limit alarm system should force the operator to decide what preventive or corrective actions must be taken to avoid passing through a potentially dangerous condition. Critical information displayed in a control room should have means provided for rapid validation of readings. Alarms should ring on a priority system based on urgency for operator action whenever a plant upset jeopardizes equipment.

Safety in operation and maintenance. Safe and effective operation of pressure equipment is also important because of governmental regulations; however, a safe plant is also a means of reducing financial risk to an enterprise. Operators must assure themselves at all times that their procedures are technically sound, and will not violate environmental or safety regulations, such as those of the Occupational Safety and Health Administration or the Environmental Protection Agency. Improper operation or procedures that violate these regulations can impose severe liability costs or fines. Large plants sometimes employ a safety engineer whose duty is to see that all hazards are reduced to a minimum, to educate the personnel to be safety-minded, and to select educational material on safety for posting.

Obviously, the small plant cannot afford such a specialist. But the same ideas of safety can be developed among a smaller personnel at a proportionately lower cost. Such a program supplies another instance of how spending ten dollars may save hundreds.

No matter how small a plant may be, a bulletin board should be included in its equipment. If employee's compensation insurance is carried (as required in many states), the insurance companies usually furnish excellent safety bulletins free. These should be posted. A set of safety rules for the particular plant should be suitably framed and posted permanently in a conspicuous place. A set of these rules typical of those observed in many plants is shown in App. 3.

Safety laws involving power plant equipment and surroundings are covered by state statutes, Occupational Safety and Health Administration (OSHA) standards, and fire and casualty insurance company requirements. The modern approach to safety is to consider an entire system or loop and then determine the effect to the system if one of the components were to fail or if an operator did not respond to a situation properly. The extent of analysis needed will depend on the criticalness of the plant. Certainly in a nuclear plant, the malfunction of a valve may require more instrumentation, alarms, and redundancy than would be necessary in a small industrial plant. The term *plant interaction reviews,* in the analysis of the effect of a component failure, is gaining increased attention in reliability studies.

General safety rules that are applicable to boiler work include the following:

Use safe ladders with no defects as well as scaffolds, hoists, and cables in gaining access to boiler components.

When entering a boiler, make sure all valves, lines, and similar connections for steam, water, fuel, air, and flue gas are tightly closed or blanked off.

Make sure there is sufficient oxygen present in any drum, header, or shell that is to be entered per confined-space-entry requirements of OSHA. An instrument for measuring the presence of at least 21 percent oxygen should be used.

Make sure all asbestos that may have to be disposed of is by suitable containers, and that the work area is properly isolated to prevent the asbestos to spread in the boiler plant.

Use low-voltage lamps and extension cords with proper guards and insulation.

Make sure drums and furnaces are properly purged and vented before entering.

Wear hard hats in any area where head injury is possible from falling objects or from accidentally bumping into protruding parts, such as valves or elbows of piping.

Wear special clothing to avoid skin contact of potentially harmful contents in the boiler.

Observe and follow all other safety rules stipulated in company or legal statutes. Hydrostatic tests require consideration of the adequacy of support to handle the weight of water involved, test pressure to be imposed, relief valve gagging or removal, accuracy of test gauges, venting of vessel prior to filling with water, and suitability of connected apparatus to withstand the test pressure.

Shift schedules are another important function of boiler plant management. These must be arranged sometimes for a 24-hr day, and various schedules are followed in order to limit the workweek to a certain number of hours.

Many plants that employ a number of operating engineers find it necessary to have at least one spare worker, or "floating" operator. This person should be capable and trained to fill in any shift vacancy due to illness, absence, or vacation periods. During normal conditions, the floating operator may be absorbed in a maintenance or test schedule.

It should be a strict rule in shift work that no operator should leave the station to prepare to go home until the relief operator has reported at the post ready to carry on the duties. It is often during the few minutes' interim of neglect when one is preparing to leave the instant relief shows up that an accident happens.

Boiler Efficiency Monitoring

Boiler efficiency monitoring can reveal degradation that, if stopped and corrected, can improve boiler performance in technical and financial ways by perhaps reducing fuel costs. Two methods will be illustrated in calculating boiler efficiency for monitoring of boiler performance.

Direct method. The simplest and easiest to apply is the direct method in which

$$\text{Efficiency} = \frac{\text{output}}{\text{input}} \times 100$$

It is necessary to stabilize boiler conditions to agreed set points so that comparisons can be made to the same boiler operating conditions.

Example A 50,000 lb/hr boiler operating at 200 psi saturated steam has 47,000 lb/hr condensate return, 3500 lb/hr continuous blowdown. Makeup water temperature is 72°F, and condensate return is 175°F. Boiler uses natural gas with a heating value of 995 Btu/ft³. A check shows average natural gas flow for the 50,000 lb/hr steam flow is 69,800 ft³/hr. What is the efficiency of the boiler by the direct method?

solution

Makeup flow = 53,500 − 47,000 = 6500 lb/hr

Average temperature of entering feedwater is 72(6500) + 175(47,000) = 53,500(T_f), and solving, T_f = 162.5°F.

Enthalpy of steam at 215 psi absolute = 1199.2 Btu/lb

Enthalpy of water at 162.5°F = 120.4 Btu/lb

Enthalpy of liquid at 215 psia = 361.58 Btu/lb

Output of boiler = 50,000(1199.2 − 120.4) + 3500(361.58 − 120.4)

Output of boiler = 54,784,130 Btu/hr

Input = 69,800(995) = 69,451,000 Btu/hr. Substituting output/input,

Efficiency = 78.9%

This simplified method of tracking efficiency can be used to detect trends of performance by comparing previous efficiencies with current results. The best comparison is obtained with a benchmark reading when the boiler is new, or after a thorough cleaning and overhaul.

Indirect method. The indirect method is also called the input-loss method and closely follows the ASME method of tabulating where the losses are occurring. The efficiency is then determined by

$$E = \frac{\text{input} - \text{losses}}{\text{input}} \times 100$$

This method requires measuring the flue gas and also a fuel ultimate analysis. Its chief advantage is that it indicates where the losses are occurring, thus making it possible to improve the efficiency if the identifiable losses can be reduced. Its disadvantage is that much data and calculations are necessary, as the following example of the indirect method will show.

The following losses must be determined in the indirect method.

1. Loss due to *moisture in the fuel.* This is especially applicable to coal that is washed or stored outdoors. For other fuels, it is considered an inherent loss with little chance for improvement unless there is a fuel change.
2. Loss due to the *combustion of hydrogen* in the fuel that also forms *moisture* and goes up the stack as vapor. Gas and oil have the most hydrogen in them. This loss is also considered an inherent loss and not controllable unless the fuel specifications are changed.

3. Loss due to *moisture in the air* used for combustion. This is one of the reasons large-size boilers preheat the air; otherwise it is also considered an inherent loss originating from the use of ambient air in the combustion of the fuel.

4. Loss due to the heat carried up the stack by the *flue gas*. Factors that can influence this loss item include:

   a. High excess-air as revealed by flue gas analysis.

   b. Dirty or scaled water-side and fire-side heat-transfer surfaces.

   c. Poor water circulation in comparison to fireside gas flows.

   d. Faulty gas baffles that permit bypassing of heat-transfer surfaces.

   e. Gas velocity through boiler too high (draft) so that not enough time is available for efficient heat transfer.

5. Losses due to *incomplete combustion* from:

   a. Insufficient air supply.

   b. Cool furnace at low loads.

   c. Poor atomization or pulverization of fuel.

6. Losses due to *combustibles in ash,* especially applicable to solid fuel-burning boilers, caused by:

   a. Grate not large enough for complete combustion of the fuel.

   b. Too high solid fuel injection or overload on boiler.

   c. Too frequent dumping of ash before complete combustion of the fuel occurs.

7. *Miscellaneous losses* from radiation, convection and leaks, such as:

   a. Poor insulation around drums and setting.

   b. Furnace refractories in need of repair.

   c. Piping, joints, seal, and other leaks around boiler setting.

Example of Indirect Method   A coal-burning boiler uses 2000 lb/hr of coal, which produces 420 lb/hr ash. The unburned carbon in the ash is 18 percent. Air and fuel temperature at entrance to furnace is 74°F. Relative humidity of air is 70 percent. Barometric pressure is 29.92 in. Hg. Steam temperature is 360°F. Flue gas temperature leaving the boiler is 452°F. The following test results were obtained:

1. Ultimate *fuel analysis* for coal, percentage per lb coal:

Component	Percentage
Carbon	62.0
Hydrogen	4.0
Nitrogen	1.0
Oxygen	8.0
Sulfur	2.0
Moisture	8.0
Ash	18.0

Heating value of coal = 11,800 Btu/lb

2. Flue gas analysis by volume:

Component	Percentage
CO_2	13.0
CO	1.0
O_2	5.0
N_2	81.0

Find and tabulate the heat losses per lb of coal fired and determine boiler efficiency if the miscellaneous losses are assumed to be 4.11 percent of total losses.

solution   The ASME recommends the following in *heat balance calculations:*

1. Evaporation temperature of moisture in fuel and air = 100°F because of low vapor pressure in the furnace path.
2. Specific heats to be used:

$$\text{Water vapors in flue gas} = 0.47 \text{ Btu/lb/°F}$$

$$\text{Water} = 1 \text{ Btu/lb/°F}$$

$$\text{Flue gas} = 0.24 \text{ Btu/lb/°F}$$

1. *Heat loss due to water in fuel, H_w*

    Use

$$H_w = W_m[1105.2 + 0.47(t_g - 100) - 1(t_f - 32)]$$

and clearing

$$H_w = W_m(1090.2 + 0.47t_g - t_f)$$

where H_w = loss due to moisture in fuel per lb fuel fired

Enthalpy of steam at 100°F = 1105.2 Btu/lb

Heat loss due to water vapor = $0.47(t_g - 100)$

Heat gained by water from 32°F = $1(t_f - 32)$

t_g = temperature of flue gas leaving boiler = 452°F

t_f = temperature of fuel and air entering boiler = 74°F

W_m = weight of moisture per lb fuel fired = 8%

Substituting in above equation,

$$H_w = 0.08[1090.2 + 0.47(452) - 74]$$

$$H_w = 98.3 \text{ Btu/lb fuel fired}$$

2. *Heat loss due to moisture in combustion air, H_a*

    Use

$$H_a = W_v(0.47)(t_g - t_a)$$

where W_v = weight of water vapor required to saturate 1 lb dry air at 74°F × (relative humidity) × (weight of dry air used per lb fuel)

t_a = 74°F

t_g = 452°F as defined

Several calculations must be made to obtain W_v.

From properties of air tables, weight of water vapor required to saturate 1 lb dry air at 74°F = 0.01815 lb.

The weight of dry air required per lb of fuel fired is obtained from the ultimate analysis of the fuel as follows:

a. It is necessary to calculate the *weight of carbon*, C_1, appearing in the flue gas as part of a series of calculations to obtain the weight of dry air, W_{da}, used per lb of fuel. For determining C_1 use

$$C_1 = \frac{W_f C_f - W_r C_r}{W_f \times 100}$$

where W_f = weight of fuel fired = 2000 lb
C_f = percent carbon from ultimate analysis = 62.0%
W_r = weight of ash = 420 lb
C_r = percent carbon content of ash from ultimate analysis = 18.0%

Substituting,

$$C_1 = \frac{2000(62) - 420(18)}{2000(100)} = 0.5822 \text{ lb}$$

b. It is also necessary to calculate first the weight of dry chimney gases, W_{dg} lb/lb of fuel fired as follows:

$$W_{dg} = \left[\frac{4CO_2 + O_2 + 700}{3(CO_2 + CO)} \right] \left[\frac{W_f C_f - W_r C_r}{W_f(100)} \right]$$

and substituting,

$$W_{dg} = \left[\frac{4(13) + 5 + 700}{3(13 + 1)} \right](0.5822) = 10.493 \text{ lb/lb fuel}$$

c. The weight of dry air, W_{da}, lb/lb of fuel is determined as follows:

$$W_{da} = W_{dg} - C_1 + 8[H - O/8]$$

and substituting

$$W_{da} = 10.493 - 0.5822 + 8[0.04 - 0.08/8] = 10.15 \text{ lb/lb fuel}$$

Then, heat loss due to moisture in the air, H_a, with a relative humidity of 70 percent is:

$$H_a = 0.7(0.01815)(10.15)[0.47(452 - 74)] = 22.9 \text{ Btu/lb fuel fired.}$$

3. *Heat loss due to hydrogen in fuel forming water vapor*, H_h. Use

$$H_h = 9W_h[1090.2 + 0.47t_g - t_f]$$

where t_g = temperature of flue gas leaving boiler = 452°F
t_f = temperature of fuel and air entering boiler = 74°F
W_h = percentage hydrogen from ultimate analysis = 4.0%

Substituting,

$$H_h = 9(0.04)[1090.2 + 0.47(452) - 74] = 442 \text{ Btu/lb fuel}$$

4. *Heat loss due to dry chimney gas, H_{cg},* Btu/lb of fuel fired. Use

$$H_{cg} = W_{dg}(C_p)(t_g - t_f)$$

where W_{dg} = 10.493 lb from (2)
C_p = specific heat = 0.24 Btu/lb/°F
t_g = 452°F and t_f = 74°F
and substituting,

$$H_{cg} = 10.493(0.24)(452 - 74) = 951.9 \text{ Btu/lb fuel fired}$$

5. *Heat loss due to unburned combustibles, H_u,* caused by insufficient air or poor fuel-air mixing. Use percentages from flue gas analysis:

$$H_u = \frac{CO}{CO + CO_2} \times 10{,}160 \times C_1$$

Use C_1 from (2) = 0.5822 lb and substituting,

$$H_u = \left[\frac{1}{1 + 13}\right][10{,}160(0.5822)] = 422.5 \text{ Btu/lb fuel fired}$$

6. *Heat loss due to unconsumed combustibles in ash, H_r,* and substituting quantities as shown from previous calculations:

$$H_r = \frac{14{,}600(W_r C_r)}{W_f}$$

where W_r = 420 lb
C_r = 18%
W_f = 2000 lb

$$H_r = \frac{14{,}600(420)(0.18)}{2000} = 551.9 \text{ Btu/lb fuel fired}$$

7. *Miscellaneous losses, H_m,* given as 0.0411(11,800) = 485 Btu/lb of fuel fired. These are radiation, convection, and leakage losses.

Summation of losses (1) to (7) = 2,974.5 Btu/lb fuel fired

Output of boiler = 11,800 − 2,974.5 = 8,825.5 Btu/lb fuel fired

$$\text{Efficiency} = \frac{11{,}800 - 2{,}974.5}{11{,}800} \times 100 = 74.8\%$$

A heat balance would show the following heat absorption and losses for the 11,800 Btu in 1 lb of fuel:

Heat item	Btu	Percent of fuel Btu
Heat absorbed by boiler	8825.5	74.8
Heat loss, moisture in fuel	98.3	0.83
Heat loss, moisture in air	22.9	0.19
Heat loss, hydrogen in fuel	442.0	3.74
Heat loss, chimney flue gas	951.9	8.07
Heat loss, unburned combustibles	422.5	3.58
Heat loss, combustibles in ash	551.9	4.68
Heat loss, miscellaneous	485.0	4.11
Totals	11,800.00	100.00

Analysis of losses. The largest losses are in the heat going up the stack with the flue gas, and the unconsumed combustibles in the fuel and ash heat that was lost and did not contribute to the heat transfer. Many heat loss items are attributable to methods of operation or maintenance practices. These include:

1. Not maintaining proper air-to-fuel ratio to assure complete and safe fuel combustion, and wasting heat with too much air.

2. Scaling on water side from poor water treatment or poor blowdown procedures.

3. Fire-side fouling, such as ash accumulation from improper soot blowing.

4. Defective baffles in the gas passages that change heat-transfer flow fire side to water side.

5. Poor atomizing of oil or poor pulverization of solid fuels that cause incomplete combustion of fuel. Smoke is an indicator of incomplete fuel combustion.

6. Improper draft causing rapid flow of gases through boiler preventing proper mass flow convection heat transfer.

7. Overloading of boiler firing, contributing to unburned fuel going up the stack or into the ashes.

8. Poor maintenance, including heat leaks from poorly maintained insulation, steam and water leaks from tubes or boiler piping, and worn refractories.

Efficiency calculations and comparing the results to previous performance will assist operators to identify the possible sources of heat losses so that corrections can be made. As an example, calculations can be made on a monthly basis as shown by the following problem:

Problem   In the previous month, the direct method of boiler efficiency calculation showed that a boiler had a 78.9 percent efficiency based on monthly fuel input and monthly steam flow output. The operator noted the following for the current month: Steam output total for month = 6,200,000 lb at 250 psi and 500°F superheat. Average feedwater temperature was 202°F. Fuel oil consumed was 59,000 gal with an average heating value of 158,400 Btu/gal. Calculate this month's overall boiler efficiency by the direct method.

solution

$$\text{Efficiency} = \frac{\text{output}}{\text{input}}$$

$$= \frac{6,200,000(1261.5 - 180.6)}{59,100(158,400)} = 0.74.1 \text{ or } 74.1\%$$

The operator realized there was a 4.8 percent loss in efficiency and something would have to be done. The last internal inspection showed an increase of water-side scaling due to heavy makeup. He decided to have the boiler acid cleaned to improve the efficiency and thus save fuel costs. This decision was also influenced by observing a marked increase in stack temperature from the previous month. The following month showed that due to the acid cleaning, the efficiency had improved to 78.5 percent.

It is the usual practice in heat balance work to let the *enthalpy of the feedwater* to be that of the liquid enthalpy at the stated inlet feedwater temperature, unless the pressure of the feedwater is above about 600 psi.

Boiler tuneups and excess air.   Boiler tuneup has received increased attention as a means of improving combustion efficiency. The primary objective of the tuneup is to achieve the highest combustion efficiency by controlling the amount of excess air used in burning the boiler fuel and also to reduce the heat loss going up the stack. Some reduction in the power required for the induced and forced-draft fan is also possible.

Care is required in operating at *too low* an air-fuel ratio, because:

1. Incomplete combustion may occur, which can also cause a heat loss from combustibles going up the stack.

2. Smoke can be created, which may violate local environmental regulations.

3. Too rich a fuel mixture poses the hazard of unburned fuel in the gas passages igniting and causing a furnace explosion.

4. Soot deposits from incomplete combustion interferes in fire-side heat transfer, thus reducing efficiency.

Theoretical air required.   The theoretical air required to burn a fuel is determined by making an ultimate analysis of the fuel being burned, and then by using the decimal equivalents. The following equation is used to obtain the theoretical air, W_{ta}, required for burning the fuel under consideration:

$$W_{ta} = 11.52C + 34.56\left[H - \frac{O}{8}\right] + 4.32S$$

Example   From the heat balance problem previously described, the ultimate fuel analysis for coal showed, in percentages: carbon = 62.0; hydrogen = 4.0; nitrogen = 1.0; oxygen = 8.0; sulfur = 2.0. Substituting these in decimal form in the above equation,

$$W_{ta} = 11.52(0.62) + 34.56\left[0.04 - \frac{0.08}{8}\right] + 4.32(0.02)$$

W_{ta} = 8.27 lb theoretical air required per lb of fuel burned.

Actual air burned.   More than the theoretical amount of air is always needed in order to avoid the problems previously described. The actual weight of air used to burn 1 lb of fuel is obtained from the *fuel gas analysis,* and by using the ultimate analysis of the fuel. The principal constituents of the flue gas are taken by volume, but the molecular weights are used also to determine the actual weight of dry air, W_{aa}, lb/lb of fuel burned as follows:

$$W_{aa} = \left[\frac{28N_2}{12(CO_2 + CO)(0.769)}\right]\left[\frac{W_f C_f - W_r C_r}{100W_f}\right]$$

where the following is used from the previous heat-transfer problem. From the flue gas analysis: N_2 = 81.0; CO_2 = 13.0; CO = 1.0; O_2 = 5.0

$$W_f = \text{weight of fuel fired} = 2000 \text{ lb}$$

$$C_f = \text{percent carbon in fuel, from fuel analysis} = 62$$

$$W_r = \text{weight of ash formed} = 420 \text{ lb}$$

$$C_r = \text{percent carbon in the ash} = 18\%$$

Substituting,

$$W_{aa} = \left[\frac{28(81)}{12(13 + 1)(0.769)}\right]\left[\frac{2000(62) - 420(18)}{100(2000)}\right]$$

$$= 10.22 \text{ lb air per lb fuel burned}$$

Percentage excess air is then

$$\frac{W_{aa} - W_{ta}}{W_{ta}} = \frac{10.22 - 8.27}{8.27} \times 100$$

Percentage excess air = 23.6%

The excess air required varies with the fuel burned and type of burner. One boiler manufacturer, B&W, provides the following expected excess air guideline:

Fuel	Method of burning	Actual excess air (percent by weight)
Coal	Spreader stoker	30 to 60
	Traveling-grate stoker	15 to 50
	Pulverized coal	15 to 20
Fuel oil	Oil burner, register type	5 to 10
	Multifuel type	10 to 20
Natural gas	Register-type burner	5 to 10
Wood	Dutch-oven burning	20 to 25

Another *appropriate method* used by some operating engineers to determine excess air from the flue gas analysis is the following:

$$\text{Excess air, \%} = \left[\frac{O_2}{CO + CO_2} \right] \times 100$$

Example   Orsat analysis showed the following percentages by volume:

$$CO = 1; CO_2 = 13; O_2 = 5$$

$$\text{Excess air, \%} = \left[\frac{5}{1 + 13} \right] \times 100 = 35.7\%$$

New instruments.   Modern technology has developed handheld electronic flue gas analyzers that allow for reasonably accurate spot measurements of excess O_2 and flue gas exit temperatures. Such devices now can display the corresponding combustion efficiency. A portable fuel efficiency monitor made by Neotronics of North America, Inc., measures stack gas temperature and oxygen content, then calculates percentage combustion efficiency, and then displays all three readings in a clear digital readout. The portable combustion analyzer also measures stack carbon monoxide and stack soot density for oil-fired units. This permits a trained technician to observe changes in combustion patterns while simultaneously adjusting burner controls to obtain maximum efficiency.

Another model is constructed for *continuous monitoring* and analysis of the combustion efficiency, which will assist boiler operation and also make it possible to check performance on a continuous basis. Companies active in this field of combustion analysis instrumentation include: Neotronics of N.A., Inc., Jefferson, Georgia; Bacharach, Pittsburgh, Pennsylvania; Land Combustion, Bensalem, Pennsylvania; and Fisher-Rosemount, Pittsburgh, Pennsylvania.

Tracking efficiency. Operators must determine, from the boiler tuneup or the boiler/burner manufacturer, the minimum excess air permitted for the fuel fired. Periodic readings of O_2, CO, and CO_2 will help to determine if fuel-burning equipment adjustments may be necessary, such as: atomizing pressure and burner tips, fuel oil temperature to burner, flame direction, soot deposits, and similar degradation that occurs in operation. This will help in maintaining boiler performance efficiency and thus save fuel. On smaller boiler installations, on-off firing with purging prior to burner operation can be wasteful of fuel burning. The cost of installing a modulating burner system that varies firing rate by load demand can prove to be more economical than on-off firing. It also reduces the thermal stresses on combustion components of the boiler, such as furnace and tube sheet junctures of SM boilers.

It is also important to watch for any increases in flue gas temperature and high draft while maintaining excess air. Any significant increase is an indication of fouling of fire-side heat-transfer surfaces. High excess air sometimes is caused by excessive casing leaks into the furnace, especially when operating at negative furnace pressures. Increases in fuel consumption require also a review of water-side scaling possibilities. The importance of tracking performance and efficiency is to note the changes that may be occurring from previous set points or established desirable efficiencies in operating the boiler. It is the operator's skill that will determine what corrective actions are required to restore the boiler plant to expected performance criteria or efficiency.

There is an *economic benefit* in maintaining the highest boiler performance, immediately noticeable in fuel costs.

Example A 75,000 lb/hr boiler operates at 400 psi with 500°F superheat for an average of 16 hr/day, 246 days/yr. Natural gas is the fuel used at an average cost of $0.80 per therm. After one year of operation, tests show that the boiler efficiency dropped from 82.2 percent to an average of 75.6 percent. Continuous blowdown is at 2250 lb/hr. What extra fuel costs were incurred from the drop in average efficiency if feedwater temperature averaged 195°F?

solution Since enthalpy of steam = 1242.9 Btu/lb, enthalpy of feedwater = 152.9 Btu/lb, and enthalpy of blowdown water = 428 Btu/lb, then

Output = 75,000 (1242.9 − 152.9) + 2250 (428 − 152.9) = 82,368,975 Btu/hr

Since one therm = 100,000 Btus, this is 823.69 therms/hr.

Input at 82.2% efficiency = 823.69/0.822 = 1002.6 therms/hr

The cost of fuel at this efficiency for the time period is then

$$1002.06 \ (0.80)(16)(246) = \$3{,}155{,}286.50$$

Input at 75.6 percent efficiency = 823.69/0.756 = 1089.54 therms/hr. The cost of fuel at this efficiency for time period is then

$$1089.54(0.80)(16)(246) = \$3{,}430{,}743.50$$

The extra fuel cost incurred due to the drop in efficiency is thus

Extra fuel costs per year = $3,430,743.50 − 3,155,286.5 = $275,457.00

Economical evaluations. With the increase in available instruments and controls and the growth in environmental equipment that will assist boiler plants to comply with environmental regulations, boiler plant managers must evaluate many pieces of equipment for either repair options to restore performance or to replace it with new equipment. The engineering method of determining this type of problem is to compare cost of replacement with the cost of operating a retrofitted piece of equipment. However, even this may be influenced by such matters as environmental regulations, perhaps public relations with the community, and even tax laws on permissible depreciation. The following example is provided with certain assumptions made, and is one method of evaluating repairing old vs. buying new equipment from an economic standpoint.

Example Assume the same load conditions exist for the 75,000 lb/hr boiler reviewed in the previous example, but it is several years later. The tubes are 20 years old as are the combustion and other controls. The decision has to be made whether the boiler should be replaced with a new one or retubed with new controls installed. The following quotes are received:

1. Retubing and changing controls would cost $700,000 with an expected life of 15 years. The bank is willing to finance this purchase at 14 percent interest. The efficiency of the refurbished boiler is guaranteed at 80 percent. The bank loan would be for 15 years with an annual payment required.

2. A new boiler would cost $1,500,000 and have an efficiency of 85.5 percent. There would also be a $100,000 per year labor saving in operating the boiler. The expected life is 20 years. The bank is willing to finance the purchase at 12 percent interest with annual payments for 20 years.

Due to increase in production, it is planned to operate the new installation 16 hr/day for 250 days/yr. Steam output conditions would be the same, namely, 823.69 therms/hr as derived in the previous problem. Assume no salvage value at the end of expected life for the two options to be evaluated.

solution The economic considerations will be limited to comparing capital plus interest charges and operating costs on a yearly basis.

1. Option 1—Retubing with new controls
   a. Capital cost plus interest charges will be calculated as an annuity-type problem, where the $700,000 received is a present value, which is equated to 15 years of payments. From mathematics of finance, payments per year at 14 percent interest can be calculated as follows:

   P = Yearly payments; L = loan amount; i = interest;
   n = number of payments

   $$P = L\left[\frac{i(1 + i)^n}{(1 + i)^n - 1}\right]$$

   and substituting,

   $$P = 700,000\left[\frac{0.14(1.14)^{15}}{1.14^{15} - 1}\right]$$

   and solving,

   $$P = 700,000\left[\frac{0.14(7.146)}{7.146 - 1}\right]$$

   = $113,945.33 per year capital and interes costs.

   b. Operating costs = (823.69/0.80)(0.80)(16)(250) = $3,294,760.00.
   c. Total cost for Option 1 = $3,294,760.00 + 113,945.33 = $3,408,705.33.
2. Option 2—Buying a completely new boiler.
   a. Capital cost plus interest

   $$P = 1,500,000\left[\frac{0.12(1.12)^{20}}{1.12^{20} - 1}\right]$$

   and solving,

   $$P = 1,500,000\left[\frac{0.12(9.646)}{9.646 - 1}\right]$$

   = $200,818.86 per year capital and interest costs.

   b. Operating costs for fuel = 823.69/0.855 × 0.80 × 16 × 250 = $3,082,816.20. Subtracting labor saving, operating cost is then

   $3,082,816.20 - 100,000 = $2,982,816.20

   c. Total yearly cost with new boiler is then $2,982,816.20 + 200,818.86 = $3,183,635.

From this evaluation, yearly saving by buying a new boiler is $3,408,705.33 - 3,183,635.00 = $225,070.30.

The other advantage of the new boiler being selected is the speed of installation vs. the retubing and changing of controls on the older boiler. Parts for the new boiler and its controls will be more readily obtained in comparison to many obsolete parts on the old boiler controls.

Property conservation also affects operators of boiler plants, because of the inherent fire risk that exists when fuel is being burned. Insurance companies place great emphasis on the preservation of the boiler plant because of its potential fire and explosion hazard, and also for the plant to comply with legal inspection requirements. Quite often, the boiler plant being so vital in production, business interruption coverage is extended to the boiler plant for both a fire hazard and breakdown, as applied by a boiler and machinery policy. Operators will come in contact with insurance company representatives who will stress fire prevention, and licensed jurisdictional boiler inspectors who will stress ASME and jurisdictional Code requirements.

Fire prevention representatives. Thus, operators must become familiar with insurance companies' property conservation recommendations. Insurance companies' engineering representatives can provide information on property conservation, based on their company or national standard requirements or on incidents they have experienced with other similar properties. They generally specialize in identifying fire sources and what can be done to reduce the exposure. Many times these are based on NFPA standards and may have been incorporated into local legal requirements because these standards have received national recognition as good standards to follow. For copies of NFPA standards write to NFPA, Batterymarch Park, Quincy, MA 02269.

OSHA fire prevention plan. After several fires killed workers in an industrial plant, OSHA Labor Fact Sheet No. 91-41 was issued. This requires employers/owners of a facility to implement a *written fire prevention* plan with the following features:

1. The employer must provide proper and workable exits and fire-fighting equipment, have an emergency plan for contingencies that may arise based on the occupancy, and have an employee training program to prevent fire casualties in the workplace.

2. There must be housekeeping procedures for the storage and cleanup of flammable materials and waste. There is a recommendation to recycle flammable waste such as paper, but safe packaging and handling must be followed.

3. Workplace ignition sources such as smoking, welding, and burning-type operations must have written procedures to prevent igniting any flammable substances on the premises. Flammable material must be kept away from all sources of heat. This includes leaks and spills. All heat-producing equipment such as burners, boilers, heat exchangers, ovens, stoves, fryers, and similar equipment must be properly maintained so they function safely, and they must be kept clean and free of leaks and flammable residue.

4. The written plan must be available to employees for their review and input. The employees must be appraised of all potential fire hazards as they relate to their jobs and of the procedures to follow if a fire erupts, as specified in the employer's fire prevention plan. All new employees are to be similarly instructed. If the occupancy or conditions in the plant change, all employees must be so instructed.

Fire terms. Fire terms are briefly reviewed here to give plant operators an understanding of the quite complex subject of fire prevention engineering. *Fire* is generally defined as a chemical reaction that involves "oxidizing" a substance, solid, liquid, or gas; the visible evidence of this is a fire or flame and is accompanied by the release of heat. Combustion takes place in the presence of oxygen, a flammable substance, and a source of ignition. All three must be present to sustain it.

Fire protection engineers have classified fires as follows:

1. *Class A:* The burning of ordinary combustibles such as wood, paper, and cloth where water can easily lower the temperature of ignition and thus put out the fire.

2. *Class B:* Flammable liquid fires such as oil, gasoline, and grease where blanketing or smothering is needed to put out the fire (removing the oxygen).

3. *Class C:* Electrical equipment fires where an agent that does not conduct electricity is needed to fight the fire.

Fires in metals include the burning of magnesium, powdered aluminum, zinc, sodium, potassium, and similar hazardous metals in which ordinary extinguishing agents do not work to put out the fire. Special powdered or granular substances must be used to cover the burning metal and exclude oxygen from the fire.

A *flash point* is the temperature at which a substance (solid, liquid, or gas) gives off enough vapor to make an inflammable mixture with air. The *ignition, or burning, point* is that temperature at which a substance ignites or burns and continues to burn as long as flammable material and oxygen are present.

Flame luminosity is an indication of the intensity of combustion by the amount of heat liberated. It is related to flame temperatures listed in the table on the next page.

The *explosive range* is a range or percentage of volume mixture in air that limits the combustibility of the substance and rate of burning when mixed with air. In a lean mixture the particles of the substance are so widely separated that those set on fire by ignition will not set fire to the other particles nearby. When the particles are so close together that they exclude the proper mixing of air or oxygen with the particles, the mixture is called rich. The lean mixture's point of com-

Flame color, heat	Flame temperature, °F
Red	977
Dark red	1292
Cherry red	1562
Pale red	1742
Yellow	2012
Gray-white	2372
Full-white	2732

bustibility is called the lower explosion limit (lel), while the corresponding rich mixture's threshold limit is called the upper explosion limit (uel). For example, hydrogen has a lel of 4.1 percent and a uel of 74 percent.

Causes of fire. It may be well worth reviewing the many causes of fire because by knowing the potential sources, operators can identify the symptoms and take corrective action. The causes are:

1. Direct contact of combustibles with open flame or glowing materials.

2. Long-time exposure to low heat of combustibles until the released gas ignites and burns the surroundings. Hot steam pipes without sufficient insulation can ignite wood, dust, etc. that touch the steam pipes; this is an example of long-time heating effects.

3. Spontaneous heating or combustion. This is a slow oxidation at ordinary temperatures of material that is a poor conductor of heat. The heat of oxidation is not carried away, and as a result the temperature rises, with the reaction speeding up, until a temperature is reached at which a self-sustaining combustion, or burning, begins. Some heat is caused by fermentation such as hay or grass cuttings in a poorly ventilated pile.

4. Explosions or rapid propagations of flame. These result from rapid decomposition of a substance with large release of heat (exothermic reaction) and also rapid flame travel because of released flammable vapors or by rapid release of pressure.

5. Lightning igniting combustible material.

6. Dust fires and explosions. Dust can accumulate from such substances as lamp-black, grains, wood, flour, starch, sugar, wool, fibers and fluff from shearing, metal powders, resins, celluloids, and plastics.

7. Electric sparks. Electricity can be a source of intense heat, and if sparks develop from poor connections, failure of insulation, over-

load, or similar causes of electrical breakdowns, these sparks can ignite combustible vapors, liquids, or solids.

8. Chemical reactions. These can cause the release of flammable gases which in severe cases can form vapor clouds that when finally ignited cause very destructive explosions and fires.

9. Friction causes heat to develop, and this heat raises temperatures of the surroundings until combustion of any nearby flammable material may result.

10. Static electricity. This is produced by friction, or the rubbing together of two surfaces, which can produce a buildup of electric charge. The charge can reach a point where a spark-over occurs that can ignite combustible materials. Of great concern are flammable liquids flowing in pipes that generate static electricity and hold it if there is no ground drainage for it. It is necessary to ground any conducting-type discharge nozzle or tip so that the static is drawn off.

Autoignition temperature, as shown in Table 15.1, is the lowest temperature at which a solid, liquid, or gaseous substance will initiate a self-sustained combustion in the absence of a spark or flame. This temperature can vary considerably since it is influenced by the nature, size, and shape of the igniting surfaces.

Water supply and fire fighting plans. A good public water system is usually the most economical source of fire protection. This may have to be supplemented, depending on the occupancy or where the public supply is primarily for domestic water. However, all operators should be familiar with the *fire fighting plan* for the property. The following should receive attention:

1. Familiarity with the property's ground plan and the type of buildings on the property.

TABLE 15.1 Ignition Temperatures of Some Gases

Gas	Ignition, in oxygen, °F	Ignition, in air, °F	Autoignition, °F
Hydrogen	580–590	580–590	1076
Carbon monoxide	637–658	644–658	1204
Ethylene	500–519	542–547	842
Acetylene	416–440	406–440	635
Hydrogen sulfide	220–235	346–379	500
Methane	556–700	650–750	999
Ethane	520–630	520–630	959
Propane	490–570		871
Ammonia	700–860		1204

2. If one is not available, make a sketch of public water connections to the property including pipe and valve sizes and where they are located.

3. Also sketch check valves, hydrants, fire pumps, and hose connections.

4. Know where fire extinguishers and first aid equipment are located.

5. If there are sprinkler systems, know how they are activated and what valves must be open for them to be effective. This also applies to foam and fog systems.

6. Have readily available fire department alarms and phone numbers so fire brigades, police, and executives can be called in case of a fire or fire fighting impairment.

7. Know all exits and the procedure for occupants to follow to leave the building or boiler plant.

8. Become thoroughly familiar with the plant's emergency plan for disconnecting electrical equipment, shutting down process or service equipment in the boiler plant, and location of fire fighting equipment.

9. Determine what kind or class of fire exists so that proper extinguishing agents will be used in fighting the fire.

Impairment. *Impairment* is a term used by insurance fire protection engineers when fire protection sprinklers, fire pumps, detection devices, valves, alarms, water supply, gravity tanks, and similar equipment that is vital for fighting a fire are out of service either deliberately for maintenance or accidentally due to a mechanical or electrical failure or to building alterations or additions. During the impairment there is always the risk of a fire erupting. Whole properties have been destroyed, for example, because the water supply was shut off in the street to make pipe or valve repairs. *Concealed impairments* can be dangerous because they occur from some unknown action by an unauthorized person either on the property or on the water supply entering the plant. Regular inspections and tests of the fire protection system by plant personnel have the main goal of finding concealed impairments of the fire protection system so that corrective actions can be taken.

Planned impairment for repairs or alterations should be placed under the supervision of the plant's fire protection chief so that emergency plans are put in place for the actions to take during the impairment. Some actions to be taken if an impairment occurs are:

Notify the local fire department and key people involved with the plant's fire fighting plan of the impairment, its plant location, and its expected duration. Many fire insurance companies also want to be notified.

Establish a watch system in areas that contain combustibles or a flammable process, such as fuel tanks and fuel lines.

Provide extra portable fire extinguishers in the area where sprinklers are out of service.

Alert all employees where to take precautions when disposing of trash and combustibles.

Limit any source of ignition activity, such as cutting or welding, in the affected areas.

It is preferred that work on restoration be on an "around the clock" basis because of the risk involved during the impairment.

It is also preferred that hazardous process areas not be operated during the impairment.

It is essential to test the system for proper operation after the repairs are completed.

Notify plant management, fire departments, and fire brigades as well as the fire insurance company when the system is back to normal. This includes outside alarm agencies, such as ADT.

Insurance of boilers is an important item usually handled by the company executives, but one in which the plant engineering department quite often is involved, because boiler insurance is generally bought under a boiler and machinery contract that may cover other plant equipment besides boilers. Traditionally, the companies that sell boiler and machinery coverage also provide legal inspection service on boilers and pressure vessels as part of their effort to prevent accidents. Thus, this service serves a public good by protecting the public from potential dangerous accidents, and this is, of course, the main reason why legal authorities accept authorized insurance company inspections on boilers and pressure vessels.

Boiler and machinery coverage is twofold in scope. It is designed to provide financial reimbursement for any possible accidents, and it usually also provides a legal inspection service that will make it possible to reduce accidents to a minimum. This service helps point out any conditions in the boilers, overlooked by the plant personnel, that might be corrected to reduce operating or maintenance costs.

Two general types of boiler and machinery coverage may be written. One covers damages caused by accidental breakdown of boilers, pressure vessels, and machinery under comprehensive coverage plans, and the other reimburses the user for loss of production due to outage of equipment operated by the boiler in case of such an accident if such coverage is bought. The former is known as "breakdown or property coverage," and the latter is classed as "business-interruption coverage." There are also other options in coverages, such as contamination and consequential spoilage losses.

Many states require that boilers not under federal control be inspected by a state inspector (for whose services there is a charge) unless the boiler is insured by an authorized insurance company and inspected by one of its qualified inspectors. When a boiler is insured (usually by a company licensed to write boiler and machinery insurance in that state), the engineering department of the insurance company sends the state legal jurisdiction a notice that the boiler is insured. If the company has commissioned inspectors in their employ (and most companies writing boiler and machinery insurance do), the legal jurisdiction does not schedule an inspection on that boiler for which a notice was received. Why? Because it will be inspected and reported to them on a formal state report when the inspection comes due by the state-commissioned insurance company inspectors.

The owner must prepare the boiler for the internal inspection. On some low-pressure boilers, internals (inspections) may be scheduled only when it is deemed advisable, depending on the state law. In others, the boiler must be drained and opened for inspection whether it is of the high-pressure, low-pressure, steam, or hot-water-heating type.

The commissioned inspector makes the inspection, and if the conditions are satisfactory, files a report to the state, requesting the state to renew the operating certificate on the boiler. In some jurisdictions, the insurance company issues the certificate directly. If repairs are needed or if conditions need correction, certificates are withheld until the violation is removed. If this is not done, the inspector notifies the state of the violation and requests that no operating certificate be issued until the violation is removed. The state or legal jurisdiction can then use its police power to enforce the requirements.

When the insurance on a boiler is canceled or not renewed, the insurance company notifies the legal jurisdiction and gives the reasons for it. Dangerous conditions on a boiler can be reported to the state over the phone, if necessary, by the commissioned inspector. The owner will then have the full pressure of the legal jurisdiction to correct the condition or take the boiler out of service. Most insurance companies also have provisions for immediate suspension of insurance by the inspector under these adverse conditions.

A boiler operator can ensure that all possible steps to prevent boiler failure have been taken by doing the following: First, purchase the best equipment available for a given service. Second, see that the boiler is properly installed and equipped with all necessary Code-approved appurtenances and safety devices. Third, make sure the boiler is operated and maintained by licensed and/or competent operators and is inspected regularly by a commissioned inspector. Only then will all the legal requirements be covered. Keep a log-book check system and a set of preventive maintenance and testing procedures. See that such checks and testing procedures are followed and that the results are always recorded. Immediately correct any malfunctions found during any check or test. And never operate the boiler unattended until the proper repairs or replacements have been made.

Questions and Answers

1 On a packaged boiler, what would cause fire puffs when it is started up?

ANSWER: (1) Poor starting draft; (2) incorrect firebox; (3) wrong burner nozzle; (4) lean fire; (5) insufficient gas pilot or excessive gas pilot; and (6) water in the oil. The potential hazard of each is a furnace explosion.

2 The proximate analysis of a coal showed the following percentages: volatile matter, 35.67; fixed carbon, 52.26; ash, 7.05; and moisture, 5.02 percent. What would the readings be on a dry basis?

ANSWER: Divide the percentages by $(1 - 0.0502)$, except for moisture, to obtain the following readings on a dry basis: volatile matter, 37.55; fixed carbon, 55.03; ash, 7.42. This totals 100 percent.

3 What ASME stamps should be evident on the following components of a large field-assembled boiler: boiler manufacturer's nameplate; boiler installer or assembler in the field; steam piping from boiler to required stop valve; boiler and superheater safety valve; steam piping to and from the reheater.

ANSWER: "S" stamp; "A" stamp; "PP" stamp; "V" stamp; under B31.1 rule, not under Section I of Boiler Code; however, the reheater is considered part of the boiler requiring an "S" stamp.

4 A boiler operates at 400 psi with 500°F superheat. Output is 100,000 lb/hr with 90% condensate return at 190°F, and makeup is 14,500 lb/hr at 70°F. Fuel used has 19,000 Btu/lb heating value. Calculate the amount of fuel needed per hour if boiler efficiency by the direct method is 86.7 percent.

ANSWER: Steam enthalpy from superheat tables = 1243.1 Btu/lb. Average feedwater temperature is 173°F with liquid enthalpy of 140.9 Btu/lb. Blowdown = 90,000 + 14,500 − 100,000 = 4500 lb/hr. Enthalpy of blowdown at 415 psi absolute pressure = 441.2 Btu/lb. Then, output Btu by boiler is

$$100,000(1243.1 - 140.9) + 4500(441.2 - 140.9) = 109,868,650 \text{ Btu/hr.}$$

$$\text{Fuel needed per hour} = \frac{109,868,650}{0.867(19,000)} = 6669.6 \text{ lb/hr}$$

5 A fuel analysis shows the following percentages: carbon = 59; hydrogen = 2.0; nitrogen = 0.5; oxygen = 7.6; sulfur = 1.2. What is the theoretical air in lb/lb of fuel required to burn these constituents of the fuel?

ANSWER: Use

$$W_{ta} = 11.52C + 34.52\left[H - \frac{0}{8}\right] + 4.32S$$

and substituting

$$W_{ta} = 11.52(0.59) + 34.52\left[0.02 - \frac{0.076}{8}\right] + 4.32(0.012) = 7.21 \text{ lb air per lb fuel}$$

6 If a combustion analyzer for above fuel burning showed 30 percent excess air, what would be the actual amount of air used per lb of fuel?

ANSWER:

$$\text{Actual air used} = \frac{W_{ta}}{1 - E_a} = \frac{7.21}{1 - 0.3} = 10.3 \text{ lb actual air per lb fuel burned}$$

7 What is "tramp air"?

ANSWER: Air that leaks into the furnace, especially on negative pressure furnaces, through faulty seals and breaks in the casing. The tramp air may not be part of the combustion air, and thus will pick up heat and carry it up the stack. It will also affect the excess air readings and give a false picture of the actual combustion efficiency.

8 What is equivalent evaporation as applied to boiler output?

ANSWER: The pounds of water per hour that are evaporated from and at 212°F by the boiler heat, found by the following equation:

$$\text{Equiv. evap.} = \frac{W_s(h - h_f)}{970.3}$$

where W = weight of steam delivered by boiler, lb/hr
h = enthalpy of steam being delivered
h_f = enthalpy of feedwater at inlet temperature

For example, 215 psia steam delivers 10,000 lb/hr saturated steam with feedwater at 140°F.

$$\text{Equiv. evap.} = \frac{10,000(1199.2 - 107.89)}{970.3}$$

$$= 11,247 \text{ lb/hr of water being evaporated}$$

This term was used to calculate boiler hp by dividing by 34.5. Boiler hp for this boiler would then be 11,247/34.5 = 326 hp.

9 Describe the term "implosion."

ANSWER: Implosion refers to the inward buckling of large tube panels on large watertube boilers due to large negative pressures being created on the fire side of the boiler. Most large furnaces are not braced for the forces that a large negative pressure acting on large tube wall areas can create. Most incidents of implosions occurred where large volume ID fans are used. If the FD fan trips suddenly and the ID fan does not, large negative pressures can result.

10 What items require attention in efficiency-related maintenance programs for a boiler plant?

ANSWER: Efficiency-related maintenance for a boiler plant is directed toward the correction of conditions that result in increased fuel consumption when generating a standard established quantity of steam. These are conditions which increase from established set points and include: (1) stack gas temperature; (2) stack gas flow; (3) combustible content in flue gas or ash; (4) blowdown amount; and (5) miscellaneous losses from leaks in boiler casing and ductwork.

11 Does a Code inspector accept radiographic examinations performed by a parts manufacturer instead of the boiler manufacturer?

ANSWER: If the radiographic examinations meet Code acceptance criteria and are performed under the control of the parts manufacturer, the Code permits radiographic examination by a parts manufacturer, but a radiographic inspection report must be forwarded to the boiler manufacturer, and this must be available to the Authorized Code Inspector for review.

12 What is the maximum size of blowoff connection permitted by the Code?

ANSWER: Section I limits the size to 2½ in. maximum size.

13 If a high-pressure boiler has safety valves set at 750 psi, what is the minimum pressure rating required by the Code for the source of water to the boiler?

ANSWER: The water feeding source must be capable of producing a pressure 3 percent over the highest setting of any safety valve, and assuming 750 psi is the highest setting, the water feeding device must be capable of producing a minimum of 772.5 psi pressure.

14 What two conditions does the Code require for the qualification of an ASME approved inspector?

ANSWER: The two conditions are: (1) The inspector must be employed by an ASME accredited Authorized Inspection Agency, defined as the inspection agency of a municipality, state, or Canadian province that has adopted the ASME construction and installation requirements, or employed by an insurance company authorized to write boiler and pressure vessel insurance; and (2) the inspector must be qualified by written examination under the rules of any state of the United States or province of Canada which has adopted the ASME Code as jurisdictional requirements.

15   When are no telltale holes required on stay bolts?

ANSWER: When the stay bolt is attached by welding.

16   When may backing rings be permitted by the Code, and can they be left in place?

ANSWER: Backing rings are permitted in welding single-welded butt joints if complete penetration of the root cannot be accomplished; however, Code requirements must be met; for example, the material for backing rings must be compatible with the weld metal and base material so that no harmful alloying or contamination of the weld occurs. The backing rings may be left in place, but must be properly secured to prevent dislodgement. They must also be contoured on the inside to minimize any flow restriction and must have an inside diameter to permit the passage of a tube cleaner.

17   What are the ASME rules for at what stages of boiler fabrication that an authorized inspector should make inspections?

ANSWER: The ASME Code is very broad on this item, and basically states that the manufacturer must submit the boiler to inspection by the Authorized Inspector at such stages of fabrication as may be designated by the Inspector. The NB rules are followed by the Inspector and the procedure followed is found in the next question.

18   What does an inspector do in following the construction of a boiler through the shop?

ANSWER: Initially the inspector checks the chemical and physical properties of the steel from the mill test reports to see if they meet Code requirements. He or she then checks the melt and slab numbers on these reports with the numbers stamped on the plates to identify them. The shop's certificate of authorization for boiler construction as required by the ASME is checked to see if the shop is authorized to construct boilers for the state into which the boiler is to be installed. The plates are examined for any visible defects, such as scars, gouges, or laminations. The thickness of the plates is gauged, a tolerance up to 0.010 in. under that specified being allowed. The design of the proposed boiler is checked to see if it meets Code specifications. Subsequent visits are made to check methods and procedures used for welding, preparing welding edges, rivet holes, tube holes, welder qualifications, and assembling the boiler.

A visit is made on completion to view a hydrostatic test and examine the general work; if the boiler has been completed satisfactorily in accordance with Code specifications, the boiler is stamped and the manufacturer's data sheets are signed by the inspector.

19   What type of documentation must a boiler or pressure-vessel manufacturer maintain on file to be stamped as ASME quality?

ANSWER: The manufacturer is required to keep certain documentation. The type of documentation needed to satisfy Code requirements varies in different sections of the code. In general, the manufacturer must keep all radiographic film and must also prepare a manufacturer's data report, which must be signed by both the manufacturer's representative and the authorized inspector. The authorized inspector should not sign the data report until he or she has carefully checked it to make certain it properly describes the boiler or vessel to which it applies, that the boiler or vessel complies with the Code, and that the data report has already been signed by the manufacturer's representative. These data reports may be registered with the National Board. For nuclear vessels, more extensive documentation is required.

20 Does the ASME Code require that a boiler manufacturer have a formal quality-control program?

ANSWER: The manufacturer or assembler should have and maintain a quality-control system which will establish that all Code requirements, including material, design, fabrication, examination (by the manufacturer), and inspection (by authorized inspector), will be met. Provided that Code requirements are suitably identified, the system may include provisions for satisfying any requirement by the manufacturer or user which exceeds minimum Code requirements and may include provisions for quality control of non-Code work. In such systems, the manufacturer may make changes in parts of the system which do not affect the Code requirements without securing acceptance by the authorized inspector.

21 How are fires classified, and what substances are involved:

ANSWER: There are three classes: Class A involves ordinary combustibles such as wood, paper, cloth, which normally can be extinguished by water. Class B applies to oils, grease, gasoline, paint, and similar substances that cannot be extinguished by water. Class C applies to electrical equipment. Some jurisdictions classify burnable metals such as magnesium and soda as Class D.

22 What are the portable fire extinguishers that are only suitable for Class A fires?

ANSWER: Water pail, soda-acid, plain water with CO_2 cartridge, calcium chloride (antifreeze), and pump tank.

23 For what type of fire is the foam extinguisher used?

ANSWER: Class B for oil, grease, gasoline, and paint fires.

24 How would you define autoignition?

ANSWER: It is the lowest temperature that is needed to start combustion of a solid, liquid, or gas without requiring a spark or flame to get the combustion started.

25 What is the purpose of a standpipe?

ANSWER: Standpipes are used to get water to upper floors of tall buildings in order to fight a fire. Plant people use hoses that are located on each floor and connected to the standpipe. Standpipes also help the fire department get water quickly to the higher floors of the building.

26   Is there any classifications for standpipes?

ANSWER: This may depend on the jurisdiction. A common classification system is based on whether it is a wet pipe or dry system, classified as follows: Wet systems have an open supply valve and water under pressure at all times. This can create a water leakage hazard. A second system uses approved devices to admit water automatically by opening a hose valve. Another system admits water through the manual operation of approved, remote-control devices located at each hose station. Dry standpipes get water through the fire department pumper connection and are usually located outside the building for street connection to the pumper.

27   Where should hose outlets on standpipes be located?

ANSWER: Hose outlets should be near or in stairway enclosures or even on fire escapes on older buildings so that they are readily accessible to the local fire department.

28   What types of sprinkler systems are used to protect a property against the spread of fire?

ANSWER: Wet pipe, dry pipe, deluge, preaction, and firecycle.

29   Why are deluge systems used?

ANSWER: They are primarily used to inundate a large area at once, such as in hazardous areas. This will prevent the spread of fire due to fumes and gases that may be generated by the fire. Instead of only one sprinkler head opening, all sprinkler heads in the designated area are open at all times, and when a fire erupts, a deluge valve is opened by strategically placed sensors in the designated area that are activated by temperature or other criteria such as smoke, and this permits water to flow to all the open sprinkler heads in the designated area.

30   What types of sprinklers are used in unheated buildings?

ANSWER: Dry pipe systems are generally used. Wet pipe systems require antifreeze solutions and must be properly separated from the public water supply to prevent contamination. This would require an antifreeze solution storage tank that is properly sized for the exposure. Since deluge systems have open sprinkler heads and are activated by sensors that admit water if a fire is sensed, deluge systems that are properly designed on the water supply side to avoid freezing of the supply could also be used.

31   Do dry pipes need to be checked for possible freeze damage?

ANSWER: Before any freezing weather, it is wise to check all low-point drains for water to prevent freeze damage of the dry pipe.

32 How does a dry-chemical extinguisher work?

ANSWER: These are used for Class B and C fires. The extinguisher should be directed to a corner of the fire, and then the stream should be swept across the flames. The chemical releases a smothering gas on the fire, while a fog or dry chemical shields the operator from the heat. The spray range is 8 to 12 ft.

33 What does rusting do to a confined space that has been sitting idle, such as a boiler or tank?

ANSWER: Moist steel surfaces consume oxygen by rusting, and this can lower the oxygen content well below the 19 percent specified by OSHA as minimum oxygen content for a confined space.

34 What are important considerations for personnel who may have to rescue a person in a confined space?

ANSWER: Many people have been killed rushing in to rescue a person in a confined space. Before a rescuer enters the confined space, make sure there is sufficient oxygen, and if not, use an oxygen-breathing apparatus. Depending on the previous contents in the confined space, it may be necessary to first check for dangerous fumes or toxic gases by the use of gas-detecting instruments. It may be possible to blow fresh air into the enclosure. If not, make sure to wear the proper protective equipment. Finally, determine before entering the best method for getting the injured person out and what additional equipment or help may be needed to do this safely.

35 What precautions should be taken before entering a boiler shell?

ANSWER: If the blowdown enters a common line with other boilers in operation, it must be certain that all valves on the line to the open boiler are closed. If other boilers are operating on the same steam header, *both* stop valves must be closed and the drip valve between them must be open. Any other valves on lines under pressure leading to the boiler must be checked. The engineer in charge and the operator on duty must be told that someone is going inside. A responsible person should be stationed at the manhole or entrance doors while anyone is inside the boiler. The inside of the shell should be checked for oxygen content with an instrument per confined space OSHA regulation.

35 Explain the term "Life Safety Code."

ANSWER: High occupancy properties such as office buildings, schools, hospitals, nursing homes, and similar facilities where a large number of people could be affected immediately by smoke and heat from a fire before they can exit from the building require special fire protection considerations, and these special considerations are termed "Life Safety Code" requirements as they are addressed to people needs, and not property protection needs.

The NFPA has established a separate Life Safety Code for such occupancies, and many jurisdictions have included these NFPA requirements in local ordinances or laws. The objective is to minimize the effect of heat and smoke to the occupants of such facilities so that they can be safely removed from the premises without serious injury.

All plant service operators should become familiar with life safety systems. Among the items stressed in these codes are the following:

1. Have sufficient usable exits per floor, based on occupancy load, that can direct the occupants outside of the building safely. Stairways must be of fireproof construction.

2. Floors and buildings should be compartmentalized by installing approved fire walls and doors to limit the spread of fire, smoke, and heat.

3. All such facilities should have an emergency communication system that will make it possible to give instructions to the occupants on what to do in an emergency.

4. Automatic sprinkler protection is generally recommended in high human occupancy areas.

5. For theatres and auditorium-type occupancy, correct sizing guidelines are provided for aisle accessways.

6. For high-rise buildings, some of the items stressed are:

   a. Stairways must have battery backup for lighting.

   b. Above a certain height or square footage, emergency generators are required, sized for the critical emergency load of the building. Other occupancies such as hospitals are also required to have emergency generators.

   c. Elevators must be automatically directed to the first or ground floor in case of fire and be available only to fire fighting personnel.

   d. In some cities, high-rise buildings are required to have a designated and approved fire prevention engineer on site, who is responsible for all fire safety for the facility.

Plant operators may become involved in life fire safety code equipment maintenance and inspection and should become familiar with the NFPA Life Safety Code as well as local laws that govern human occupancy requirements for the facility in which they work.

1

Terminology and Definitions

absolute pressure The pressure above zero pressure, equal to gauge pressure plus atmospheric pressure.

acid cleaning The process of cleaning the interior surfaces of steam-generating units by filling the unit with a dilute acid accompanied by an inhibitor to prevent corrosion and by subsequently draining, washing, and neutralizing the acid by a further wash of alkaline water.

acidity Represents the amount of free carbon dioxide, mineral acids, and salts (especially sulfates or iron and aluminum) which hydrolyze to give hydrogen ions in water; is reported as milliequivalents per liter of acid, or ppm acidity as calcium carbonate, or pH, the measure of hydrogen ion concentration.

agglomeration Groups of fine dust particles clinging together to form a larger particle.

air-atomizing oil burner A burner for firing oil in which the oil is atomized by compressed air which is forced into and through one or more streams of oil, breaking the oil into a fine spray.

air-fuel ratio The ratio of the weight, or volume, of air to fuel.

air heater or air preheater Heat-transfer apparatus through which air is passed and heated by a medium of higher temperature, such as the products of combustion or steam.

1. *Regenerative air preheater.* An air heater in which heat is first stored up in the structure itself by the passage of the products of combustion, and which then gives up the heat so stored to the subsequent passage of air.

2. *Recuperative air heater.* An air heater in which the heat from products of combustion passes through a partition which separates the products from the air.

Sources: ASME Boiler Codes, American Society of Mechanical Engineers, New York; ABMA; and *Metals Handbook,* American Society for Metals.

673

a. Tubular air heater. An air heater containing a group of tubular elements through the walls of which heat is transferred from a flowing heating medium to an airstream.

b. Plate air heater. An air heater containing passages formed by spaced plates through which heat is transferred from a flowing heating medium to an airstream.

air purge   The removal of undesired matter by replacement with air.

air-swept pulverizers   A pulverizer through which air flows and from which pulverized fuel is removed by the stream of air.

air vent   A valved opening in the top of the highest drum of a boiler or pressure vessel for venting air.

alkalinity   The amount of carbonates, bicarbonates, hydroxides, and silicates or phosphates in the water; reported as grains per gallon, or parts per million as calcium carbonate.

allowable working pressure   The maximum pressure for which the boiler was designed and constructed; the maximum gauge pressure on a complete boiler; and the basis for the setting on the pressure-relieving devices protecting the boiler.

amplitude   In ultrasonic testing, the vertical pulse height of a signal, usually base to peak, when indicated by an A-scan presentation.

analysis, proximate   Analysis of a solid fuel determining moisture, volatile matter, fixed carbon, and ash; expressed as a percentage of the total weight of the sample.

analysis, ultimate   Chemical analysis of a solid fuel determining moisture, volatile matter, fixed carbon, and ash; expressed as a percentage of the total weight of the sample.

anthracite   ASTM coal classification by rank; Dry fixed carbon 92 percent or more and less than 98 percent; and dry volatile matter 8 percent or less and more than 2 percent on a mineral-matter-free basis.

approved   The word *approved* as used in a Code means acceptable to the authority having jurisdiction.

A-scan   In ultrasonic testing, a method of data presentation on a CRT (cathode ray tube) with the horizontal baseline indicating distance or time and vertical deflections from the baseline indicating amplitude of the ultrasonic reflection.

ash   The incombustible inorganic matter in the fuel.

ash sluice   A trench or channel used for transporting refuse from ash pits to a disposal point by means of water.

atomizing media   A supplementary medium, such as steam or air, which assists in breaking the fuel oil into a fine spray.

attemperator   Apparatus for reducing and controlling the temperature of a superheater vapor or of a fluid. See also **desuperheater**.

1. *Shell-and-tube type.* An attemperator consisting of a pressure vessel containing tubular elements through the walls of which heat is transferred.
2. *Spray type.* An attemperator in which a lower-temperature fluid is injected at relatively high velocity in an atomized state into the superheater vapor to reduce its temperature by direct contact with the atomized fluid.
3. *Submerged type.* An attemperator consisting of tubular elements located in the boiler circulation below the waterline.

authorized inspection agency   The inspection agency approved by the appropriate legal authority of a state or municipality of the United States or a province of Canada, which has adopted a section of the ASME Code, and which has also been accredited by the ASME as qualified.

automatic lighter or igniter   A means for starting ignition of fuel without manual intervention. Usually applied to liquid, gaseous, or pulverized fuel. See **igniter.**

available draft   The draft which may be utilized to cause the flow of air for combustion or the flow of products of combustion.

backing ring   A strip of thin plate used on the inner surfaces of the abutting ends of pipe, tubes, or plates which are to be butt-welded. Its purpose is to prevent irregularities at the base of the weld and to permit penetration at its root.

bag   A deep bulge in the bottom of the shell or furnace of a boiler.

bag filter   A device containing one or more cloth bags for recovering particles from the dust-laden gas or air which is blown through it.

balanced draft   The maintenance of a fixed value of draft in a furnace at all combustion rates by control of incoming air and outgoing products of combustion.

banking   Burning solid fuels on a grate at rates sufficient to maintain ignition only.

barrel   The cylindrical portion of a firetube-boiler shell that surrounds the tubes.

bituminous coal   ASTM coal classification by rank on a mineral-matter-free basis and with bed moisture only.

1. *Low volatile.* Dry fixed carbon 78 percent and less than 86 percent; dry volatile matter 22 percent or less and more than 14 percent.
2. *Medium volatile.* Dry fixed carbon 69 percent or more and less than 78 percent; dry volatile matter 22 percent or less and more than 31 percent.
3. *High volatile (A).* Dry fixed carbon less than 69 percent; dry volatile matter more than 31 percent. Btu value equal to or greater than 14,000 moist, mineral-matter-free basis.
4. *High volatile (B).* Btu value 13,000 or more and less than 14,000 moist, mineral-matter-free basis.
5. *High volatile (C).* Btu value 11,000 or more and less than 13,000 moist, mineral-free basis commonly agglomerating, or 8300 to 11,500 Btu agglomerating.

black light   In magnetic particle inspection, light in the near ultraviolet range of wavelengths, just shorter than visible light.

black liquor   Liquid by-product fuel extracted from wood in the alkaline pulp-manufacturing process and containing the chemical used to accomplish the extraction.

blowback   The number of pounds per square inch of pressure drop in a boiler from the point where the safety valve pops to the point where the safety valve reseats.

blowback ring   An adjustable ring in a safety valve, used to control the amount of blowback.

blowdown   The drain connection including the pipe and the valve at the lowest practical part of a boiler, or at the normal water level in the case of a surface blowdown. The amount of water that is blown down.

boiler   A closed vessel in which water is heated, steam is generated, steam is superheated, or any combination thereof, under pressure or vacuum by the application of heat from combustible fuels, electricity, or nuclear energy. The term does not include such facilities of an integral part of a continuous processing unit but does include fired units of heating or vaporizing liquids other than water where these units are separate from processing systems and are complete within themselves.

boiler assembler   Means a corporation, company, partnership, or individual who assembles a boiler which has been delivered knocked down in multiple pieces by bolting, threading, welding, or other methods of fastening to produce a finished pressure vessel. A boiler assembler may also be a boiler installer.

boiler, automatically fired   A boiler which cycles automatically in response to a control system.

boiler header (box)   A pressure part of a boiler consisting of a flat tube sheet into which the ends of the water tubes are rolled. In a parallel plane is a tube cap or handhole sheet. The two sheets are spaced about 4 to 8 in. or more apart. The top and bottom and both ends are flanged together and riveted or may be closed by a narrow flanged strip of plate riveted to each sheet. Circulating nipples connect the top of the header and drum, or the header may be flanged and riveted directly to the drum. Welding would be used today instead of rivets.

boiler, high-pressure, steam or vapor   A boiler in which steam or vapor is generated at a pressure exceeding 15 psig.

boiler, hot-water-heating   A boiler in which no steam is generated and from which hot water is circulated for heating purposes and then returned to the boiler.

boiler, hot-water-supply   A boiler functioning as a water heater.

boiler, low-pressure, steam or vapor   A boiler in which steam or vapor is generated at a pressure not exceeding 15 psig.

boiler manufacturer   A corporation or company which manufactures complete pressure parts for boilers or whose shop assembles parts into completed boilers.

boiling out   The boiling of a highly alkaline water in boiler pressure parts for the removal of oils, greases, etc. prior to normal operation or after major repairs.

bourdon tube   A hollow, metallic tube, bent semicircular, which forms the actuating medium of a pressure gauge.

breeching   A duct for the transport of the products of combustion between parts of a steam-generating unit or to the stack.

bridgewall   A wall in the furnace over which the products of combustion pass.

Brinell test   A hardness test performed by pressing a steel ball of standard hardness into a surface by a standard pressure.

British thermal unit   The mean British thermal unit (Btu) is $\frac{1}{180}$ of the heat required to raise the temperature of 1 lb of water from 32 to 212°F at a constant atmospheric pressure. It is about equal to the quantity of heat required to raise 1 lb of water 1°F [251.9957 cal or 1054.35 joules (J)].

brittle   A metal is brittle when it permits little or no plastic deformation prior to fracture.

brittle crack propagation   A very sudden propagation of a crack with no energy absorption except that stored elastically in the body of the material or metal.

B-scan   In ultrasonic testing, a means of data presentation which gives a cross-sectional ultrasonic view of the test piece.

buckstay   A structural member placed against a furnace or boiler wall to limit the motion of the wall against furnace pressure.

bulge   A local distortion or swelling outward caused by internal pressure on a tube wall or boiler shell due to overheating. Also applied to similar distortion of a cylindrical furnace due to external pressure when overheated, provided the distortion is of a degree that can be driven back.

bunker C oil   Residual fuel oil (no. 6 fuel oil) of high viscosity commonly used in marine and stationary steam power plants.

burner   A device for the introduction of fuel and air properly mixed in correct proportions to the combustion zone.

burner assembly   A burner that is factory-built as a single assembly or as two or more subassemblies which include all parts necessary for its normal function when installed as intended.

burner, atmospheric   A gas burner in which all air for combustion is supplied by natural draft, the inspirating force being created by gas velocity.

burner, automatically lighted   A burner in which fuel to the main burner is normally turned on and ignited automatically.

burner, natural-draft type   A burner which depends primarily on the natural draft created in the flue to induce the air required for combustion into the burner.

burner, power   A burner in which all air for combustion is supplied by a power-driven fan that overcomes the resistance through the burner to deliver the quantity of air required for combustion.

burner windbox   A plenum chamber around a burner in which an air pressure is maintained to ensure proper distribution and discharge of secondary air.

bypass temperature control   Control of vapor or air temperature by diverting part of or all the heating medium from passing over the heat-absorbing surfaces, usually by means of a bypass damper.

caking   Property of certain coals to become plastic when heated and form large masses of coke.

calcium   A scale-forming element found in some boiler feedwaters.

calorie   The mean calorie is $\frac{1}{100}$ of the heat required to raise the temperature of 1 g of water from 0 to 100°C at a constant atmospheric pressure. It is about equal to the quantity of heat required to raise 1 g of water 1°C (4.184 J).

carryover   The moisture and entrained solids forming the film of steam bubbles; a result of foaming in a boiler. Carryover is caused by a faulty boiler-water condition. See also **foaming.**

casing   A covering of sheets of metal or other material such as fire-resistant composition board used to enclose all or a portion of a steam-generating unit.

caulk   To make the contacting surfaces of a seam tight against leakage by upsetting or forcing (by distortion) the edge or abutment of the plate into the surface of the adjoining plate. Also, to close any pinhole or fissure in a metal plate, by virtue of the ductility of boiler plate, by distorting its surface to close a slight opening. A blunt tool is used in caulking.

caustic cracking   Also called caustic embrittlement cracking, usually occurring in carbon steels or iron-chromium nickel alloys that are exposed to concentrated hydroxide solutions at temperatures of 400 to 480°F.

chain grate stoker   A stoker which has a moving endless chain as a grate surface, onto which coal is fed directly from a hopper.

Charpy test   An impact test in which a V-notched, keyhole-notched, or U-notched specimen, supported at both ends like a beam, is struck behind the notch by a pendulum slung weight. The energy that is absorbed when the specimen fractures is calculated and compared to the energy or height that the pendulum would have risen had there been no notched specimen. This test is used to check on the brittle characteristics of the material.

checker work   An arrangement of alternately spaced brick in a furnace with openings through which air or gas flows.

check valve   A valve designed to prevent reversal of flow. Flow in one direction only is permitted.

chevron pattern   Also called herringbone patterns, found in fractographs of failures. These patterns are typically found on brittle fracture surfaces. The points of the chevron can be traced back to the origin of the fracture.

cinder   Particles of partially burned fuel from which volatile gases have been driven off, which are carried from the furnace by the products of combustion.

circulating tube   A boiler tube used to connect the water spaces of two drums or the pressure parts of a boiler.

closed feedwater heater   An indirect-contact feedwater heater; that is, one in which the steam and water are separated by tubes or coils.

closing-in-line   The sealing by plastic refractory between a boiler shell or head and the firebrick wall; used to prevent hot gases from contacting the boiler above the lowest safe waterline.

colloid   A finely divided organic substance which tends to inhibit the formation of dense scale and results in the deposition of sludge, or causes it to remain in suspension, so that it may be blown from the boiler.

combined feeder cutoff   A device that regulates makeup water to a boiler in combination with low-water fuel cutoff.

combustion   Chemical combination of the combustible (that part which will burn) in a fuel with oxygen in the air supplied for the process. Temperatures may range from 1850 to over 3000°F.

combustion (flame) safeguard   A system for sensing the presence or absence of flame and indicating, alarming, or initiating control action.

condensate   Condensed water resulting from the removal of latent heat from steam.

conduction   The transmission of heat through and by means of matter unaccompanied by any obvious motion of the matter.

conductivity   The amount of heat (Btu) transmitted in 1 hr through 1 ft^2 of a homogeneous material 1 in. thick for a difference in temperature of 1°F between the two surfaces of the material.

constant ignition   Usually a gas pilot that remains lighted at full volume whether the main burner is in operation or not.

control   A device designed to regulate the fuel, air, water, steam, or electrical supply to the controlled equipment. It may be automatic, semiautomatic, or manual.

control, limit   An automatic safety control responsive to changes in liquid level, pressure, or temperature; normally set beyond the operating range for limiting the operation of the controlled equipment.

control, operating   A control, other than a safety control or interlock, to start or regulate input according to demand and to stop or regulate input according to demand and to stop or regulate input on satisfaction of demand. Operating controls may also actuate auxiliary equipment.

control, primary safety   A control responsive directly to flame properties, sensing the presence of flame and, in the event of ignition failure or unintentional flame extinguishment, causing safety shutdown.

control, safety   Automatic controls and interlocks (including relays, switches, and other auxiliary equipment used in conjunction to form a safety control system) which are intended to prevent unsafe operation of the controlled equipment.

convection   The transmission of heat by the circulation of a liquid or a gas such as air. Convection may be natural or forced.

corner firing   A method of firing liquid, gaseous, or pulverized fuel in which the burners are located in the corners of the furnace. See also **tangential firing.**

corrosion   The wasting away of metals as a result of chemical action. In a boiler, usually caused by the presence of O_2, CO_2, or an acid.

corrosion fatigue   Cracks produced by the combined action of repeated or fluctuating stress and a corrosive environment, which produces the cracking at lower stress levels or fewer cycles of stress than would be the case if no corrosive environment were present.

course   A circumferential section of a boiler shell or drum. With usual diameters, the number of courses will equal the number of plates forming the shell or drum.

creep   The time-dependent stretching or strain, heavily influenced by temperature, of a material under stress.

creep-rupture strength   The stress that will cause fracture in a creep test at so much given time for it to occur and in a constant environment, usually temperature. This is also called stress-rupture strength.

crimping tool   A tool used to reduce the diameter of the end of a boiler tube preparatory to its removal from a boiler.

critical pressure and critical temperature   That point at which the difference between the liquid and vapor states for water completely disappears.

cross box   A boxlike structure to the longitudinal drum of a sectional header boiler for connecting circulating tubes.

crowfoot   The end of a brace in a boiler, split in two directions for riveting to the plate.

crown sheet   The plate forming the roof of an internally fired furnace or a combustion chamber.

C-scan   In ultrasonic testing, a means of data presentation to show a plan view of the material, and of any discontinuities therein.

damper   A device for introducing a variable resistance of regulating the volumetric flow of gas or air.
1. *Butterfly type:*   A single-blade damper pivoted about its center.
2. *Curtain type:*   A damper consisting of one or more blades, each pivoted about one edge.
3. *Flap type:*   A damper consisting of one or more blades, each pivoted about one edge.
4. *Louvre type:*   A damper consisting of several blades, each pivoted about its center and linked together for simultaneous operation.
5. *Slide type:*   A damper consisting of a single blade which moves substantially normal to the flow.

deaerating heater   A type of feedwater heater operating with water and steam in direct contact. It is designed to heat the water and to drive off oxygen.

design pressure   The pressure used in the design of a boiler for the purpose of determining the minimum permissible thickness or physical characteristics of the different parts of the boiler.

desuperheater   Apparatus for reducing and controlling the temperature of a superheated vapor. See also **attemperator.**

   1. *Shell-and-tube type:* A desuperheater consisting of a pressure vessel containing tubular elements through the walls of which heat is transferred.
   2. *Spray type:* A desuperheater in which a lower-temperature fluid is injected at relatively high velocity in an atomized state into the superheated vapor to reduce its temperature by direct contact with the atomized fluid.
   3. *Submerged type:* A desuperheater consisting of tubular elements located in the boiler circulation below the waterline.

developer   Used with liquid penetrant NDT, usually a white powder applied to the material being tested after the liquid penetrant application and the removal of excess surface penetrant. The white developer accentuates the bleed-out process from the fault, thus intensifying the discernibility of the flaw indication.

diagonal stay   A brace used in firetube boilers between a flathead or tube sheet and the shell.

distillate oil   Light fraction of oil which has been separated from crude oil by fractional distillation. See **fuel oil.**

downcomer   A tube or pipe in a boiler or waterwall circulating system through which fluid flows downward between headers.

dowtherm   An organic chemical with an exceedingly high boiling point, sometimes used in special types of boilers for high-temperature service. It is composed of diphenyl and diphenyloxide.

draft   The difference between atmospheric pressure and some lower pressure existing in the furnace or gas passages of the steam-generating unit.

draft control, barometric   A device that controls draft by means of a balanced damper which bleeds air into the breeching on changes of pressure to maintain a steady draft.

draft differential   The difference in static pressure between two points in a system.

drip leg   The container placed at a low point in a system of piping to collect condensate and from which it may be removed.

drum   A cylindrical shell closed at both ends, designed to withstand internal pressure.

dry bottom furnace   A pulverized-fuel furnace in which the ash particles are deposited on the furnace bottom in a dry, nonadherent condition.

dry pipe   A perforated or slotted pipe or box inside the drum and connected to the steam outlet.

dry steam   Steam containing no moisture. Commercially dry steam containing not more than 0.5 percent moisture.

ductile crack propagation   Slow crack propagation accompanied by noticeable plastic deformation, but which requires energy to be supplied from outside the material or body to continue the propagation.

ductility   A plastic property of metal to withstand deformation without failure.

dump grate stoker   One equipped with movable ash trays, or grates, by means of which the ash can be discharged at any desirable interval.

dutchman   A wedge or tapered plug used in butt-and-double-strap longitudinal seams of some boilers to fill the space between the abutting edges of the plate from the end of the inside butt strap to the edge of the adjoining course.

dutch oven   An extended furnace, external to the main setting of a boiler, used to increase the volume of an existing furnace.

economizer   A series of tubes located in the path of flue gases. Feedwater is pumped through these tubes on its way to the boiler in order to absorb waste heat from the flue gas.

eddy current testing   Applies to NDT in which eddy current flow is induced in the object under test. The changes in the flow caused by variations in the object's part under test are reflected into a coil, which relays the flow variation into a suitable instrument for analysis of variations that may be due to a defect.

efficiency   *Of boiler operation:* Output in heat units divided by input in heat units. The number of Btus contained in all steam evaporated is the useful output. The number of Btus contained in all fuel supplied to the boiler is the input. *Of a riveted seam:* Ratio of the strength of a unit length of a riveted seam to the same unit length of the seamless plate.

ejector   A device which utilizes the kinetic energy in a jet of water or other fluid to remove a fluid or fluent material from tanks or hoppers.

elasticity   The property of a material deformed under stress which allows return to its original shape when the stress is removed.

elastic limit   The maximum tensile load to which a metal may be subjected without becoming permanently deformed upon cessation of the load.

electric boiler   A boiler converting electric energy to heat energy.

electric furnace   A furnace used for the refinement of high-grade steel.

electromagnetic testing   An NDT method that uses electromagnetic energy at frequencies less than those of visible light to detect faults in the material. Magnetic particle inspection is an electromagnetic testing method.

electrostatic precipitator   A device for collecting dust, mist, or fume from a gas stream, by placing an electric charge on the particle and removing that particle onto a collecting electrode.

embrittlement   An intercrystalline corrosion of boiler plate occurring in highly stressed zones. Cracking may result.

endurance limit   The maximum stress below which a material is assumed to be able to withstand an infinite number of stress cycles.

enthalphy A thermal property of a fluid which is a function of state and is defined as the sum of stored mechanical potential energy and internal energy. It is generally expressed in Btu per pound of fluid (joules per kilogram).

entrainment The conveying of particles of water or solids from the boiler water by the steam.

equalizing tube A boiler tube used to connect the steam spaces of two steam drums, or pressure parts of a boiler.

erosion The wearing away of refractory or of metal parts by the action of slag or fly ash.

evaporation rate The number of pounds of water evaporated in a unit of time.

evaporator A pressure vessel used to evaporate raw water by means of a steam coil. The steam is condensed by means of cooling water coils, and this distilled water is used as makeup boiler feed.

evaporator condenser That section of an evaporator installation which condenses the vapor.

excess air Air supplied for combustion in excess of that theoretically required for complete oxidation.

expanded joint The pressuretight joint formed by enlarging a tube end in a tube seat.

explosion door A door in a furnace or boiler setting designed to be opened by a predetermined gas pressure.

explosion fireside Combustion which proceeds so rapidly that a high pressure is generated suddenly, also called a furnace explosion.

extended surface Heating surface in the form of fins, rings, or studs, added to heat-absorbing elements.

external corrosion A chemical deterioration of the metal on the fireside of boiler heating surfaces.

extraction feedwater heater A closed feedwater heater supplied with steam extracted or bled from a stage of a steam turbine. See also **feedwater heater.**

factor of safety The ratio between that stress which will cause failure and the working stress. This ratio often applies to pressures instead of stresses.

fan performance A measure of fan operation in terms of volume, total pressures, static pressures, speed, power input, and mechanical and static efficiency, at a stated air density.

fan performance curves The graphical presentation of total pressure, static pressure, power input, and mechanical and static efficiency as ordinates and the range of volumes as abscissas, all at constant speed and air density.

fatigue The phenomenon of a material experiencing fracture under repeated or fluctuating stresses having a maximum value less than the ultimate strength of the material.

fatigue crack growth rate, da/dN   The rate of crack extension caused by constant-amplitude fatigue loading, expressed in terms of crack extension per cycle of load application.

fatigue limit   A measure of the ability of a material to withstand repeated stress reversals without fracture or damage to the crystalline structure. A piece of soft iron wire may be broken easily by hand when it is bent back and forth a few times. Its fatigue limit is low. Conversely, a piece of spring steel may be flexed many thousands of times without showing any indication of distress. In this case, the fatigue limit is high. This property is of special value in steam-boiler construction.

fatigue notch factor   The ratio of the fatigue strength of an unnotched specimen to the fatigue strength of a notched specimen of the same material and under the same test conditions, including same number of stress cycles.

feed through   A trough or pan from which feedwater overflows in the drum.

feedwater heater   A device used to heat feedwater with steam. See also **extraction feedwater heater.**

feedwater regulator   A device for admitting feedwater to a boiler automatically on demand. Practically a constant water level should result.

ferromagnetic   Applied to materials that can be magnetized or strongly attracted by a magnetic field, such as in magnetic particle inspection.

ferrule   A short, metallic ring rolled into a tube hole to decrease diameter. Also a short, metallic ring rolled inside a rolled tube end. Also, a short, metallic ring for making up handhole joints.

fin   Usually a strip of steel welded longitudinally or circumferentially to a tube for the purpose of increasing heat transfer,

fin tube   A tube with one or more fins.

fire crack   A crack starting on the heated side of a tube, shell, or header resulting from excessive temperature stresses.

fire tube   A tube in a boiler having water on the outside and carrying the products of combustion on the inside.

firing rate control   A pressure or temperature flow controller which controls the firing rate of a burner according to the deviation from pressure or temperature set point. The system may be arranged to operate the burner on-off, high-low, or in proportion to load demand.

fixed carbon   The carbonaceous residue less the ash remaining in the test container after the volatile matter has been driven off in making the proximate analysis of a solid fuel.

flame detector   A device which indicates if fuel, such as liquid, gaseous, or pulverized, is burning or if ignition has been lost. The indication may be transmitted to a signal or to a control system.

flange   A circular metal plate threaded or otherwise fastened to an end of a pipe for connection with a companion flange on an adjoining pipe. Also that part of a boiler head (dished or flat) which is fabricated to a shape suitable for riveted or welded attachment to a drum or shell.

flange   To fabricate the flange in a head or similar plate.

flareback   A burst of flame from a furnace in a direction opposed to the normal flow, usually caused by the ignition of an accumulation of combustible gases.

flared tube end   The projecting end of a rolled tube which is expanded or rolled to a conical shape.

flue gas   The gaseous products of combustion in the flue to the stack.

fluorescent liquid penetrant   A highly penetrating liquid used in liquid penetrant testing, characterized by its ability to fluoresce under black light.

fluorescent magnetic particle inspection   A magnetic particle inspection method that coats the area under investigation with a powdered ferromagnetic material that fluoresces when activated by light of suitable wavelength.

fly ash   Suspended ash particles carried in the flue gas.

foaming   Formation of steam bubbles on the surface of boiler water due to high surface tension of the water. See also **carryover.**

forced-draft fan   A fan supplying air under pressure to the fuel burning equipment.

forge-weld   The welding together of metals by raising the temperature to the plastic point and by applying pressure or blows.

free-blow   A pipe open and free to blow to atmosphere.

fuel oil   A petroleum product, requiring comparatively minor refinement, used as a combustible for steam boilers. The following terms are used to describe its properties:

1. *Flash point:* The flash point of a fuel oil is an indication of the maximum temperature at which it can be stored and handled without serious fire hazard.
2. *Pour point:* The pour point is an indication of the lowest temperature at which a fuel oil can be stored and still be capable of flowing under very low forces.
3. *Viscosity:* The viscosity of an oil is a measure of its resistance to flow. In fuel oil it is highly significant since it indicates both the relative ease with which the oil will flow or may be pumped and the ease of atomization. See also **viscosity.**

fuel-oil heater   A tank and coil-type heater using steam as a heating medium to reduce heavy low-priced fuel oil to the proper viscosity for good atomization and combustion. Also an electric-coil heater making use of an electric-resistance coil because the heating medium is used sometimes where steam is not available for starting up a "cold" boiler plant.

furnace explosion   A violent combustion of dust or gas accumulations in a furnace or combustion chamber of a boiler.

furnace release rate   Furnace release rate is the heat available per square foot of heat-absorbing surface in the furnace. That surface is the projected area of tubes and extended metallic surfaces on the furnace side including walls, floor, roof, partition walls, and platens and the area of the plane of the furnace exit which is defined as the entrance to the convection tube bank.

fusible plug   A brass bushing having a tapered core composed of 99 percent pure tin and a melting temperature of 400 to 500°F and installed at the lowest safe water level of a boiler. The large end of the tapered core is exposed to boiler pressure; the other end is exposed to products of combustion. The core of fusible plug is designed to melt if the boiler water level approaches a dangerously low level. When the core melts, escaping steam will sound warning.

fusion weld   To weld the edges or surfaces of metal by raising the temperature to the fusion point and adding a "filler" metal (of the same characteristics as the metal being welded) at the same temperature.

gag   A clamp designed to prevent a safety valve from lifting. Used in applying a hydrostatic test at higher pressure than the safety-valve setting.

galvanic corrosion   Accelerated corrosion of a metal in contact with a more noble metal or nonmetallic conductor when in a corrosive electrolyte.

gas recirculation   The reintroduction of part of the combustion gas at a point upstream of the removal point, in the lower furnace for the purpose of controlling steam temperature.

gate valve   A stop valve using the wedge-and-double-seat principle. It may be used to control fluids containing some solids, for when wide open, it operates on a straight-through flow. There is little likelihood of its becoming obstructed.

gauge glass   A glass-enclosed visible indicator of the water level in a boiler. Many gauge glasses are tubular, but modern high-pressure practice and railroad locomotives use two thick, flat strips of glass bolted between flanged plates, with the water and steam between the glass strips.

gauge pressure   The pressure above that of the atmosphere, 14.7 psi at sea level; absolute pressure minus 14.7 at sea level.

generating tube   A boiler tube used for evaporation.

girth seam   A roundabout, or circumferential, seam connecting two courses of a boiler shell or drum.

globe valve   A stop valve using the round-disk-and-seat principle. Used where the fluid controlled is comparatively clean.

grain   Metallurgically, an individual crystal in a polycrystalline metal or alloy.

grain boundary   Metallurgically, an interface separating two grains at which the orientation of the lattice structure changes from that of one grain to the other.

grains per cubic foot   The term for expressing dust loading in weight per unit of gas volume (7000 grains equals 1 lb).

granular fracture   When a metal is broken, the surface has a rough, grainlike appearance, rather than a smooth or fibrous one. Sometimes called "crystalline fracture."

grate   The surface on which fuel is supported and burned and through which air is passed for combustion.

grindability   Grindability is the characteristic of coal representing its ease of pulverizing and is one of the factors used in determining the capacity of a pulverizer. The index is relative, with the large values, such as 100, representing coals easy to pulverize such as Pocahontas and smaller values such as 40 representing coals difficult to pulverize.

grooved tube seat   A tube seat having one or more shallow grooves into which the tube may be forced by the expander.

ground   A conducting connection, whether intentional or accidental, between an electric circuit or equipment and either the earth or a conducting body which serves in place of the earth.

grounded   Connected to earth or to some conducting body which serves in place of the earth.

grounded conductor   A system or circuit conductor which is intentionally grounded.

grounding conductor, equipment   The conductor used to connect noncurrent-carrying metal parts of equipment, raceways, and other enclosures to the system-grounded conductor at the service and/or the grounding electrode conductor.

guarded   Covered, shielded, fenced, enclosed, or otherwise protected by means of suitable covers, casings, barriers, rails, screens, mats, or platforms to remove the likelihood of contact by persons or objects.

gun   (1) A pneumatic riveter. (2) A gun-type oil burner, of the kind having a long-shaped flame. (3) An injector, in railroad terminology.

handhole   An inspection, a sight, or a cleanout opening in a boiler; often elliptical and closed by a handhole plate.

hand lance   A manually manipulated length of pipe carrying air, steam, or water for blowing ash and slag accumulations from heat-absorbing surfaces.

hardness   A measure of the amount of calcium and magnesium salts in a boiler water. Usually expressed as grains per gallon or parts per million as $CaCO_2$.

hard patch   A riveted patch made pressuretight by caulking.

hard water   Water which contains calcium or magnesium in an amount which requires an excessive amount of soap to form a lather.

header   A distribution pipe supplying a number of smaller lines tapped off of it. A main receiving pipe supplying one or more main pipe lines and receiving a number of supply lines tapped into it. Typical is the boiler *header and superheater header.*

heat-affected zone   That portion of the base metal juncture with weld metal that was not properly fused together, causing the microstructure and mechanical properties of the welded joint to be altered by the heat of welding. Abbreviated HAZ.

heating surface   That surface which is exposed to the heating medium for absorption and transfer of heat to the heat medium per American Boiler Manufacturers Association (ABMA) rules as follows:

1. *Boiler and Waterwall Heating Surface:* This surface consists of all the apparatus in contact on one side with the water or wet steam being heated and on the other side with gas or refractory being cooled in which the fluid being heated forms part of the circulating system; this surface is measured on the side receiving heat.

    Waterwall heating surface in the furnace, including walls, floor, roof, partition walls, and platens, consisting of bare or covered tubes, is measured as the sum of the projected areas of the tubes and the extended metallic surface on the furnace side.

    Continuation of furnace tubes beyond the furnace gas outlet is included as boiler heating surface, and this surface is measured on that portion on the circumferential and the extended metallic surface receiving heat.

    All other boiler surfaces, including furnace screen tubes, are measured on that portion of the circumferential and the extended metallic surface receiving heat. The surface is not included in more than one category.

2. *Superheater and Reheater Surface:* This heating surface consists of all the heat-transfer apparatus in contact on one side with steam being heated and on the other side with gas or refractory being cooled; this surface is measured on the side receiving heat.

    The radiant superheating or radiant reheating surface in the furnace, including walls, floor, roof, partition walls, and platens, is measured as the sum of the projected areas of the tubes' extended metallic surface on the furnace side.

    Continuation of superheater tube beyond the furnace gas outlet is included as convection superheater surface, and this surface is measured on that portion of the circumferential and the extended metallic surface receiving heat.

    All other superheater and reheater surface, including screen tubes, is measured on the basis of the circumferential and the extended metallic surface receiving heat.

heat release   The total quantity of thermal energy above a fixed datum introduced into a furnace by the fuel, considered to be the product of the hourly fuel rate and its high heat value, expressed in Btu per hour per cubic foot of furnace volume or square foot of heating surface.

high-cycle fatigue   Failure from repetitive stress or strain that occurs as a result of a large number of cycles, usually taken as above 10,000 to 100,000 cycles depending on the material and on the magnitude of the stress or strain being imposed.

high fire   The input rate of a burner at or near maximum.

high-heat value or higher heating value   The total heat obtained from the combustion of a specified amount of fuel which is at 60°F before the quantity of heat released is measured.

hopper bottom furnace   A furnace bottom with one or more inclined sides forming a hopper for the collection of ash and for the easy removal of same.

hot-short   Brittle when hot.

hot well   A tank used to receive condensate from various sources on its passage back to a boiler through the feedwater system. It usually is vented to atmosphere.

huddling chamber   A space provided under the valve disks of many safety valves, permitting the steam pressure in the boiler to act on an increased area when the valve disk lifts, to permit the valve to pop open rather than to rise gradually.

hydrogen damage   Temporary reduction in ductility of steel without significant reduction in tensile strength as a result of absorption of hydrogen by the steel.

hydrogen embrittlement   A condition of low ductility or hydrogen-induced cracking in metals that results from the absorption of hydrogen into the metal.

hydrostatic test   A pressure test by water at room temperature applied to a boiler to determine its safety, as a check on repairs or to trace suspected leakage.

igniter   A burner smaller than the main burner, which is ignited by a spark or other independent and stable ignition source and which provides proven ignition energy required to immediately light off the main burner.

ignition   A system in which the fuel to a main burner or gas or oil pilot is ignited directly either by an automatically energized spark or glow coil or by a gas or oil pilot.

ignition temperature   Lowest temperature of a fuel at which combustion becomes self-sustaining.

impeller   The rotating wheel of a centrifugal pump.

impingement   The striking of moving flame against boiler parts, causing local overheating.

incomplete combustion   The partial oxidation of the combustible constituents of a fuel.

inhibitor   A substance which selectively retards a chemical action. An example in boiler work is the use of an inhibitor, when using acid to remove scale, to prevent the acid from attacking the boiler metal.

injector   A device for feeding water into a boiler, making use of the high-velocity-momentum principle to feed water back against boiler pressure by use of steam at the same pressure.

input rating   The fuel-burning capacity of a burner at sea level in Btu per hour (watts) as specified by the manufacturer.

integral economizer   A segregated portion of a watertube boiler in which the feedwater is preheated before its admixture with the circulating boiler water.

interbank superheater   A superheater located in a space between the tube banks of a bent-tube boiler.

interdeck superheater   A superheater located in a space between the tube banks of a straight-tube boiler.

intergranular cracking   Cracking that occurs between the grains or crystals in a polycrystalline aggregate.

interlock A device to prove the physical state of a required condition and to furnish that proof to the primary safety control circuit.

intermittent firing A method of firing by which fuel and air are introduced and burned in a furnace for a short period after which the flow is stopped, this succession occurring in a sequence of frequent cycles.

intermittent ignition An igniter which burns during light-off and while the main burner is firing and which is shut off with the main burner.

internally fired boiler A firetube boiler having an internal furnace such as a scotch, locomotive firebox, vertical tubular, or other type have a water-cooled plate-type furnace.

intertube economizer An economizer, the elements of which are located between tubes of a boiler convection bank.

intertube superheater A superheater, the elements of which are located between tubes of a boiler convection bank.

ion A charge atom or radical which may be positive or negative.

ion exchange A reversible process by which ions are interchanged between solids and a liquid. These ions exist throughout the solution and act almost independently.

izod test Impact test in which a V-notched specimen, mounted vertically, is subjected to a sudden blow delivered by the weight at the end of a pendulum arm. The energy required to break off the free end is a measure of the impact strength, or toughness, of the material.

lagging Blocks of asbestos or magnesia insulation wrapped on the outside of a boiler shell or steam piping.

lamination As applied to boiler plate, a slag stratum or inclusion rolled into a piece of steel plate during rolling-mill operation.

lazy bar A bar fitting across the latches of firing doors of hand-fired boilers; used as a balance and rest for long, heavy firing tools.

ligament A series of holes in one or more rows.

lining The material used on the furnace side of a furnace wall. It is usually high-grade refractory tile or brick or plastic refractory material.

listed Equipment or materials included in a list published by a nationally recognized testing laboratory that maintains periodic inspections of production of listed equipment or materials. Listing indicates compliance with nationally recognized standards.

live steam Steam which has not performed any of the work for which it was generated.

longitudinal seam A riveted or welded seam along the longitudinal axis of a boiler shell or drum.

low-cycle fatigue Fatigue that occurs at a relatively small number of cycles, taken as below 10,000 cycles of repeated stress or strain on the material. Low-cycle fatigue may be accompanied by plastic or permanent deformation.

low-heat value The high heating value minus the latent heat of vaporization of the water formed by burning the hydrogen in the fuel.

low-oil-temperature switch A cold-oil switch; a control to prevent burner operation if the temperature of the oil is too low.

low-water cutoff A device to stop the burner on unsafe water conditions in the boiler.

lug As applied to boiler suspension, a steel eyepiece fitted and riveted or welded to the curvature of a boiler shell or drum and connected by a steel U-bolt or sling rod to overhead steel structure; used to support the weight of the boiler.

magnesium A scale-forming element found in some boiler feedwaters.

magnetic particle inspection NDT method for detecting cracks and other discontinuities at or near the surface in ferromagnetic materials. Finely divided magnetic particles are applied to the surface of a part that has been suitably magnetized. The particles are attracted to regions of magnetic nonuniformity associated with defects or discontinuities, thus producing indications that can be seen visually. In the *wet method,* the magnetic particle inspection employs ferromagnetic particles suspended in a liquid form instead of a powder.

makeup water The amount of raw water necessary to compensate for the amount of condensate that is not returned in the feedwater supply to the boiler.

manhole An access opening to the interior of a boiler, elliptical and 11 in. by 15 in. or larger or circular 15-in. diameter or larger.

manual-reset device A component of a control which requires resetting by hand to restart the burner after safe operating conditions have been restored.

mechanical-atomizing oil burner A burner which uses the pressure of the oil for atomizing.

mechanical stoker A device consisting of a mechanically operated fuel-feeding mechanism and a grate; is used for the purpose of feeding solid fuel into a furnace, distributing it over the grate, admitting air to the fuel for the purpose of combustion, and providing a means for removal or discharge of refuse.

micrometer One millionth of a meter, or 0.00039 in. (1/25400 in.); formerly called a micron. The diameter of dust particles is often expressed in micrometers.

microstructure Metallurgically, the structure of metals and alloys as revealed after polishing and etching the specimen and then observing the crystal structure at magnification greater than 25×.

mill scale An iron oxide scale formed on the surface of a steel plate by cooling and exposing the plate to air just after it has been rolled at high temperatures.

mill test report An affidavit from a steel mill testifying as to the physical and chemical properties of the steel referred to by the report.

miniature boiler   A boiler, the dimensions and working pressure of which do not exceed the following limits: diameter, 16 in.; working pressure, 100 psig; gross volume, 5 ft^3; or heating surface, 20 ft^2.

mud or lower drum   A pressure chamber of a drum or header type located at the lower extremity of a water-tube boiler convection bank which is normally provided with a blowoff valve for periodic blowing off of sediment collecting in the bottom of the drum.

multifuel burner   A burner by means of which more than one fuel can be burned either separately or simultaneously, such as pulverized fuel, oil, or gas.

multiple-retort stoker   An underfeed stoker consisting of two or more retorts, parallel and adjacent to one another, but separated by a line of tuyères and arranged so that the refuse is discharged at the ends of the retorts.

natural circulation   The circulation of water in a boiler caused by differences in density; also referred to as thermal or thermally induced circulation.

neutron embrittlement   Embrittlement of a metal resulting from bombardment with neutrons as usually encountered by metals that have been exposed to a neutron flux in a reactor core. In steels, neutron embrittlement is evidenced by a rise in the ductile-to-brittle transition temperature. Also called *radiation damage.*

nipple   A short length of pipe or tubing.

nonreturn trap   A trap designed to discharge its condensate at atmospheric pressure or at considerably lower pressure than at its inlet.

notch brittleness   Susceptibility of a material to brittle fracture at points of stress concentration. If in a notch tensile test, the material fails at less than the tensile strength of an unnotched specimen, the material is considered notch brittle, otherwise it is said to be notch ductile.

notch sensitivity   A measure of the reduction in strength of a metal caused by the presence of stress concentration.

notch tensile strength   The maximum load on a notched tension-test specimen divided by the root area of the notch.

nozzle   A short flanged or welded neck connection on a drum or shell for the outlet or inlet of fluids; also a projecting spout for the outlet or inlet of fluids; also a projecting spout through which a fluid flows.

ogee flange   A flange in the form of a reverse curve, used to connect the edges of two concentric shells.

oil burner   A burner that atomizes fuel oil and blows it into the combustion chamber in the form of a fine mist or vapor. Steam or mechanical motion plus air may be used as the operating medium.

once-through boiler   A steam-generating unit usually operated above the critical pressure in which there is no recirculation of the working fluid in any part of the unit. In the case of a supercritical steam generator, there is a constant increase in temperature and enthalphy from inlet to outlet.

operating control   A control to start and stop the burner; it must be set below high-limit control.

orsat   An instrument for determining the chemical analysis of flue gas.

oscillogram   Term applied to photographs showing data as displayed on a CRT screen as used in ultrasonic testing.

oxygen attack   Corrosion or pitting in a boiler caused by oxygen.

packaged steam generator   A boiler equipped and shipped complete with fuel-burning equipment, mechanical draft equipment, automatic controls and accessories; usually shipped in one or more major sections.

palm   The end of a brace in a boiler, forged flat and riveted or welded to the shell plate; used to stay flat surfaces.

patch   A piece of boiler plate used to replace a defection section cut out of a boiler.

pendant-tube superheater   An arrangement of heat-absorbing elements which are substantially vertical and suspended from above.

penetrameter   In radiography NDT, a device employed to obtain evidence on a radiograph that the technique used was satisfactory. It is not intended for use in judging the size of the discontinuities or for establishing acceptance limits for materials or products, but rather to establish that the technique of radiography was satisfactory.

penetrant   Applied to liquid penetrant NDT, a liquid that has unique properties rendering it highly capable of entering small openings, a characteristic or property that makes the liquid especially suitable for use in the detection of surface discontinuities in materials.

perfect or stoichiometric combustion   The complete oxidation of all the combustible constituents of a fuel, utilizing all the oxygen supplied.

pH   The hydrogen ion concentration of a water to denote acidity or alkalinity. A pH of 7 is neutral. A pH above 7 denotes alkalinity while one below 7 denotes acidity. This pH number is the negative exponent of 10 representing hydrogen ion concentration in grams per liter. For instance, a pH of 7 represents 10^{-7} g/L.

pilot   A small burner which is used to light off the main burner.

pilot, constant   A pilot that burns without turndown throughout the entire time the boiler is in service.

pilot flame establishing period   The length of time fuel is permitted to be delivered to a proved pilot before the flame-sensing device is required to detect pilot flame.

pilot, proved   A pilot flame which has been proved by flame-failure controls.

pit   Corrosion localized in a small spot.

pitch   The unit spacing of a series of holes, tube holes, or other holes in a plate.

plastic deformation   The permanent distortion of a metal under applied stress to the extent it strains the metal beyond its elastic limit.

platen   A plane surface receiving heat from both sides and constructed with a width of one tube and a depth of two or more tubes bare or with extended surface.

platen superheater   A superheater made up of close back-spaced tubes forming plane elements located so as to absorb heat primarily by radiation.

plenum   An enclosure through which gas or air passes at relatively low velocities.

porcupine boiler   A boiler consisting of a vertical shell from which project a number of dead end tubes.

postpurge   A period after the fuel valves close during which the burner motor or fan continues to run, to supply air to the combustion chamber.

power-actuated relief valve   A safety or relief valve, actuated by a separate power source, usually electrical or pneumatic; set to operate slightly below the spring-loaded pressure-actuated valve. This valve is for the express purpose of saving wear and tear on the Code valve and may be installed with an isolating valve to permit maintenance and repair without the necessity of shutting down the boiler. The relieving capacity of this valve is not to be included in the relieving capacity in calculating total Code required.

prepurge period   A period on each start-up during which air is introduced into the combustion chamber and associated flue passages in volume and manner as to completely replace the air or fuel-air mixture contained therein prior to an attempt to initiate ignition.

pressure   As applied to boilers, the force exerted by a liquid or gas on a unit area. Three pressures may be involved: gauge pressure, the unit pressure above the atmospheric pressure; absolute pressure, gauge pressure plus the atmospheric pressure; vacuum pressure, the pressure below atmospheric pressure, usually expressed in inches of Hg.

primary air   Air introduced with the fuel at the burners. In direct-fired systems this may be the same as pulverizer air bypassed around the pulverizer or bled in at the exhauster suction.

priming   An induction of boiler water caused by the steam flow into the steam line. The water may be in the form of a spray or a solid body.

projected grate area   The horizontal projected area of the stoker grate.

proportional control   A mode of control in which there is a continuous linear relation between value of the controller variable and position of the final control element (modulating control).

pulse-echo method   In ultrasonic testing, the presence and position of a reflection on the CRT screen indicated by the echo amplitude and time along the horizontal axis.

pulverized fuel   Solid fuel reduced to a fine size, such as 70 percent through a 200-mesh screen.

pulverizer   A machine which reduces a solid fuel to a fineness suitable for burning in suspension. Types used are:

1. *High speed:* (over 800 r/min)

   a. *Impact pulverizer:* A machine in which the major portion of the reduction in particle size of the fuel to be pulverized is effected by fracture of larger sizes by sudden shock, impingement, or collision of the fuel with rotating members and casing.

   b. *Attrition pulverizer:* A machine in which the major portion of the reduction on particle size is by abrasion, either by pulverizer parts on coal or by coal on coal.

2. *Medium speed:* (between 70 and 300 r/min)

   a. *Roller pulverizer:* A machine having grinding elements consisting of conical or cylindrical rolls and a bowl, bull-ring mating rings, or table, any of which may be the rotating member, the fuel to be pulverized being reduced in size by crushing and attrition between the rolls and the rings.

   b. *Ball pulverizer:* A machine in which the grinding elements consist of one or more circular rows of metal balls arranged in suitable raceways, wherein the fuel to be pulverized is reduced in size by crushing and attrition between the balls and raceways.

3. *Low speed:* (under 70 r/min)

   a. *Ball or tube pulverizer:* A machine having a rotating cylindrical or conical casing charged with metal ball or slugs and the fuel to be pulverized, with reduction in particle size being effected by crushing and attrition resulting from continuous relative movement of the charge on rotation of the casing.

quenching crack   A crack resulting from the thermal stresses produced by rapid cooling from a high temperature.

radiant   As applied to heat, having the property that permits heat to be transmitted by rays similar to those of light. To absorb radiant heat, an object must be in the "light" of the fire.

radiant superheater   A superheater exposed to the direct radiant heat (light) of the fire.

radiograph   A permanent, visible image placed on a recording medium, such as a film, that was produced by penetrating radiation that passed through the material being tested or examined for defects or discontinuities.

ram   A form of plunger used in connection with underfeed stokers to introduce fuel into retorts; a form of pusher.

rated capacity   The manufacturer's stated capacity rating for mechanical equipment, for instance, the maximum continuous capacity in pounds of steam per hour for which a boiler is designed.

raw water   Untreated feedwater.

recycle   The process of sequencing a normal burner start-up following shut-down.

red liquor   An acid-water mixture of organic material (wood residue) and spent inorganic pulping chemicals, generally associated with a sulfite pulping process in the paper mill industry.

refractory   A heat-insulating material, such as firebrick or plastic fire clay, used for such purposes as lining combustion chambers.

reheater   A device using highly superheated steam or high-temperature flue gases as a medium serving to restore superheat to partly expanded steam; used often between high- and low-pressure turbines.

relay   A device that is operative by a variation in the conditions of one electric circuit to start the operation of other devices in the same or another electric circuit (such as pressure or temperature relay).

residual stress   Stress remaining in a material after external forces or thermal gradients are removed.

resonance method   In ultrasonic testing, a technique that varies the frequency of continuous ultrasonic waves in order to obtain excitation for a maximum amplitude of vibration in the material under test. This technique is especially used for thickness measurement.

retarder   A straight or helical strip inserted in a fire tube primarily to increase the turbulence.

retractable blower   A soot blower in which the blowing element can be mechanically extended into the boiler and retracted or pulled back.

return trap   A trap designed to discharge its condensate against boiler pressure and feed to the boiler without additional mechanical equipment.

Ringlemann chart   A series of four rectangular grids of black lines of varying widths printed on a white background, used as a criterion of blackness for determining smoke density from chimneys.

riser tube   A tube through which steam and water pass from an upper waterwall header to a drum.

rolled joint   A joint made by expanding a tube into a hole by a roller expander.

rotary oil burner   A burner in which atomization is accomplished by feeding oil to the inside of a rapidly rotating cup.

safe-end   To replace a deteriorated end of a fire tube by cutting off the end and welding on a short length of new tube.

safety valve   A valve that automatically opens when pressure attains the valve setting which is adjustable; used to prevent excessive pressure from building up in a boiler.

safety-valve drain   A hole of at least ⅜-in. diameter required through the body below the valve-seat level in safety valves larger than 2-in. diameter; used to prevent condensate from collecting at this point.

safety-valve escape A pipe conducting steam discharged from a safety valve to a safe location.

safety-valve lifting lever A lever by which a safety valve may be lifted from its seat.

safety-valve muffler A silencer designed so that it will not cause appreciable restriction to steam flow.

safety-valve nozzle A flanged nozzle by which a safety valve is connected to a boiler shell or drum.

scale A deposit of medium to extreme hardness occurring on water heating surfaces of a boiler because of an undesirable condition in the boiler water.

scrubber An apparatus for the removal of solids from gases by entrainment in water.

seal weld A weld used primarily to obtain tightness and prevent leakage.

secondary combustion Combustion which occurs as a result of ignition at a point beyond the furnace. See also **delayed combustion.**

secondary treatment Treatment of boiler feedwater or internal treatment of boiler water after primary treatment.

segregation A nonuniform distribution of alloying elements, impurities or phases of a metal.

sensing head In electromagnetic testing, a probe unit containing a coil, magnet, or magnetic circuit from which a test signal is obtained.

sensitization Applies to austenitic (nonmagnetic) stainless steels, where the precipitation of chromium carbides, usually at grain boundaries, on exposure to temperatures of about 1000 to 1500°F, leaves the grain boundaries depleted of chromium, and therefore the stainless steel is susceptible to preferential attack by a corroding or oxidizing medium.

separator A tank-type pressure vessel installed in a steam pipe to collect condensate to be trapped off and thus providing comparatively dry steam to connected machinery.

shore scleroscope A device to test the hardness of a material, performed by dropping a diamond-pointed hammer from a standard height.

shrinkage crack A crack, usually starting at elevated temperature, that forms because of the internal shrinkage stresses that develop while the metal cools and solidifies. Also termed *hot crack.*

silica A scale-forming element found in some boiler feedwaters.

single-retort stoker An underfeed stoker using only one retort in the assembly of a complete stoker. A single furnace may contain one or more single-retort stokers.

sinuous header A header of a sectional header-type boiler in which the sides are curved back and forth to suit the stagger of the boiler tubes connected to the header faces.

siphon A pigtail-shaped pipe or a drop leg in the pipe leading to a steam pressure gauge, serving to trap water in the gauge and prevent its overheating from direct contact with steam.

slag A residue deposited by ash particles that have attained their softening temperature (1900 to 2700°F) depending on their composition. Slag may be plastic and viscous when hot. It hardens and is rather porous and brittle when cool.

slag-tap furnace. A pulverized-fuel-fired furnace in which the ash particles are deposited and retained on the floor in molten condition, and from which molten ash is removed by tapping either continuously or intermittently.

slicer A slicing bar; a long steel bar used for breaking up a fuel bed in coked or caked condition.

slug A solid body of boiler water passed into the steam flow by priming or picked up from a pocket of condensate in the steam line.

smoke Small gas-borne particles of carbon or soot, less than 1 μm in size, resulting from incomplete combustion of carbonaceous materials and of sufficient number to be observable.

S-N curve Usually applied in calculating the endurance limit, a plot of stress (S) against the number of cycles to failure (N). Also known as the S-N diagram.

softening The act of reducing scale-forming calcium and magnesium impurities from water.

soft patch A patch applied with tap bolts, with a gasket under the patch plate to prevent leakage.

soot blower A tube from which jets of steam or compressed air are blown for cleaning the fireside of tubes or other parts of a boiler.

spalling The breaking off of the surface refractory material as a result of internal stresses.

specific gravity The ratio of the weight of a unit volume of a material to the weight of the same unit volume of water.

specific heat The quantity of heat, expressed in Btu (joule) required to raise the temperature of 1 lb (kilogram) of a substance 1°F (°C).

spontaneous combustion Ignition of combustible material following slow oxidation without the application of high temperature from an external source.

spray nozzle A nozzle from which a liquid fuel is discharged in the form of a spray.

spud A flange nut wrench, open at one end and pointed at the other, as a drift pin. The pointed end is used for aligning boltholes of pipe flanges.

stack A steel "chimney."

standard reference In NDT work, a reference used as a basis for comparison or to make calibration of the testing procedure.

stay bolt A stay threaded and riveted over at each end, used to connect two flat or curved pressure parts of a boiler.

steam Water vapor produced by evaporation. Dry saturated steam contains no moisture and is at a specific temperature for every pressure; it is colorless. The white appearance of escaping steam is due to condensation at the lowered temperature; it is the water vapor that shows white.

steam-and-water drum A pressure chamber located at the upper extremity of a boiler circulatory system in which the steam generated in the boiler is separated from the water and from which steam is discharged at a position above a water level maintained there.

steam-atomizing oil burner A burner for firing oil which is atomized by steam. It may be of the inside or outside mixing type.

steam binding A restriction in circulation due to a steam pocket or a rapid steam formation.

steam-generating unit A unit to which water, fuel, and air or waste heat are supplied and in which steam is generated. It can consist of a boiler furnace and fuel-burning equipment and may include as component parts waterwalls, superheater, reheater, economizer, air heater, or any combinations.

steam quality The percentage by weight of vapor in a steam-and-water mixture.

strain hardening An increase in hardness and strength caused by plastic deformation at temperatures below the recrystallization range. Also called *work hardening*.

stress The internal resistance of a material to an external force changing, or tending to change, the shape or position of the material. See also **total stress** or **unit stress.**

stress concentration An abrupt change in the contour of a material or a discontinuity that magnifies or increases the stress under load above the normal expected stress.

stress-corrosion cracking A cracking process that requires the simultaneous action of a corrodent and sustained tensile stress.

stress-relieve To dissipate pent-up stresses caused by welding, by means of heat treatment.

strongback A heavy steel bar bolted to tube sheets of firetube boilers during construction, while braces are being installed, to prevent the tube sheet from buckling before installation of the tubes.

stud A projecting pin serving as a support or means of attachment.

stud tube A tube having short studs welded to it.

subpunch To drive a pilot hole through a plate preparatory to drilling a larger hole.

subsurface discontinuity In magnetic particle inspection, any defect which does not open onto the surface of the part in which it exists.

superheated steam   Steam heated to a temperature higher than that corresponding to the temperature equivalent to the pressure.

superheater   A series of tubes exposed to high-temperature gases or to radiant heat. Steam from the boiler passes through these tubes to attain a higher temperature than would be possible otherwise. This superheated steam ensures dryness. See also **radiant superheater.**

superheater header   A large-diameter (about 4- to 8-ft) thick-walled shell or drum into which a row of superheater tubes is rolled.

surface blowoff   Removal of water, foam, etc. from the surface at the water level in a boiler; the equipment for such removal.

surge   The sudden displacement or movement of water in a closed vessel or drum.

suspended solids   Undissolved solid in boiler water.

switch, air-flow-proving   A device installed in an airstream which senses air flow or loss thereof and electrically transmits the resulting impulses to the flame-failure circuit.

switch, high-pressure   A device to monitor liquid, steam, or gas pressure and arranged to open and/or close contacts when the pressure value is exceeded.

switch, low-pressure   A device to monitor liquid, steam, or gas pressure and arranged to open and/or close contacts when pressure drops below the set value.

switch, oil-temperature-limit   A device to monitor the temperature of oil between preset limits and arranged to open and/or close contacts should improper oil temperature be detected.

tack   To hold edges of plate in correct position for riveting by a few scattered bolts, known as "tack bolts," placed through rivet holes or by small, scattered spot welds known as "tack welds" or "stitch welds."

tangential firing   A method of firing by which a number of fuel nozzles are located in the furnace walls so that the centerlines of the nozzles are tangential to a horizontal circle. Corner firing is usually included in this type.

tangent tube wall or tube-to-tube wall   A waterwall in which the tubes are substantially tangent to one another with practically no space between the tubes.

tap hole   An opening for the removal of slag from a slag tap furnace.

telltale hole   A hole drilled into the ends of a staybolt. The hole extends at least ½ in. inside the inner surface of the stayed sheets; or if the staybolt is reduced in diameter at its middle portion, the hole extends ½ in. inside the point of diameter reduction. The purpose is to show leakage, through the telltale hole, if the staybolt breaks or cracks.

tensile strength (ultimate)   That stress which causes breaking in tension.

tertiary air   Air for combustion supplied to the furnace to supplement the primary and secondary air.

theoretical air   The quantity of air required for perfect combustion.

therm   A unit of heat applied especially to gas. One therm equals 100,000 Btu.

thermal fatigue   Temperature gradients that vary with time in such a manner as to produce cyclic stresses that result in a crack or fracture.

thermal shock   The sudden development of a steep temperature gradient that produces high stresses within a material from the rapid expansion or contraction.

thermal sleeve   A spaced internal sleeve lining of a connection for introducing a fluid of one temperature into a vessel containing fluid at a substantially different temperature; used to avoid abnormal stresses.

thermal stress   Stresses in a material resulting from nonuniform temperature distribution.

thermostatic trap   A nonreturn trap using a thermostatic expansion and contraction principle as its actuating medium.

through-stay   A brace used in firetube boilers between the heads or tube sheets.

tie bar   A structural member designed to maintain the spacing of furnace waterwall tubes.

tie rod   A tension member between buckstays or tie plates.

tile   A preformed, burned refractory, usually applied to shapes other than standard brick.

time delay   A deliberate delay of a predetermined time in the action of a safety device or control.

titration   A chemical process used in analyzing feedwater.

titration point   That point at which a solution changes color when an indicating chemical is introduced drop by drop.

total stress   The total resistance of a material to an external force on its entire cross-sectional area in a plane perpendicular to the direction of the force. See also **stress.**

transducer   Used in ultrasonic testing, an electro-acoustical device for converting electrical energy into acoustical energy or the reverse, acoustical to electrical.

transgranular cracking   Cracking or fracturing that occurs through or across a crystal or grain, sometimes called transcrystalline cracking.

trap   A device designed to remove condensate from steam automatically, with negligible loss of steam. See **nonreturn trap, return trap,** and **thermostatic trap.**

traveling-grate stoker   A stoker similar to a chain-grate stoker with the exception that the grate is separate from but is supported on and driven by chains. Only enough chain strands are used as may be required to support and drive the grate.

trial for ignition   That period of time during which the programming flame-failure controls permit the burner fuel valves to be open before the flame-sensing device is required to detect the flame.

trial for main-flame ignition   A timed interval when, with the ignition means proved, the main valve is permitted to remain open. If the main burner is not ignited during this period, the main valve and ignition means are cut off. A safety-switch lockout follows.

trial for pilot ignition   A timed interval when the pilot valve is held open and an attempt is made to ignite and prove it. If the presence of the pilot is proved at the termination of the interval, the main valve is energized; if not, the pilot and ignition are cut off, followed by a safety lockout.

try cock   One of three valves mounted on a boiler or water column within the visible range of the gauge glass and used to check the water level.

tube cap   An elliptical or a circular handhole plate used opposite the end of a watertube in a header of a watertube boiler; used for inspection, cleaning, or tube removal.

tube rattler   A vibrating tool designed to be passed through fire tubes to crack scale loose from the tube as a result of vibration.

tube sheet   A flat head of a boiler or that part of a boiler drum into which boiler tubes are rolled.

tube turbine   A rotating tool used with water or compressed air pressure, designed to be passed through watertubes to remove scale.

tubular-type collector   A dust collector utilizing a number of essentially straight-walled cyclone tubes in parallel.

turbidity   The optical obstruction to the passing of a ray of light through a body of water, caused by finely suspended matter; used to check feedwater.

turbulent burner   A burner in which fuel and air are mixed and discharged into the furnace in such a manner as to produce turbulent flow from the burner.

tuyères   Forms of grates, located adjacent to a retort, through which air is introduced.

ultrasonic   As applied to ultrasonic testing, mechanical vibrations having a frequency greater than approximately 20,000 cycles per second or 20,000 Hz.

unit stress   A value expressed in pounds per square inch and found by dividing the total stress or force by the cross-sectional area stressed. See **stress, total stress.**

upset   To enlarge or increase the cross-sectional area of any part of a metal by forging it back to a shorter length.

valve   See **check valve, gate valve, globe valve, safety valve.**

valve, manual-reset safety shutoff   A manually opened, electrically latched, electrically operated safety shutoff valve designed to automatically shut off fuel when de-energized.

valve, safety shutoff A valve automatically closed by the safety control system or by an emergency device to completely shut off fuel supply to the burner.

vane A fixed or adjustable plate inserted in a gas or airstream used to change the direction of flow.

vane control A set of movable vanes in the inlet of a fan to provide regulation of airflow.

vane guide A set of stationary vanes to govern direction, velocity, and distribution of air or gas flow.

vapor generator A container of liquid, other than water, which is vaporized by the absorption of heat.

vent An opening in a vessel or other enclosed space for the removal of gas or vapor.

vent valve (gas burner) A normally open, power closed valve piped between the two safety shutoff valves, vented to a safe location.

viscosity Measure of the internal friction of a fluid or its resistance to flow.

visible dye penetrant Applies to liquid penetrant inspection, an intensely colored (usually red) highly penetrating liquid that provides maximum contrast with the white developer when used for detecting surface flaws.

vortex eliminator Baffles, screens, or plates at the entrance to a large downcomer designed to prevent the formation of a free vortex.

washout plug An inspection, sight, and cleanout opening, circular, threaded, and fitted with a threaded pipe plug, and not to be used for any pipe connection.

waste fuel Any by-product fuel that is waste from a manufacturing process.

water column A vertical, hollow chamber located between a boiler and the gauge glass for the purpose of steadying the water level in the glass through the reservoir capacity of the column. Also, the column may eliminate the obstruction of small-diameter gauge-glass connections by serving as a sediment chamber.

water hammer A sudden increase in pressure of water due to an instantaneous conversion of momentum to pressure.

water leg That space which is full of boiler water between two parallel plates. It usually forms one or more sides of internally fired furnaces.

water screen A screen formed by one or more rows of water tubes spaced above the bottom of a pulverized-fuel furnace.

water tube A boiler tube through which the fluid under pressure flows. The products of combustion surround the tube.

watertube boiler A boiler in which the water or other fluid flows through the tubes and the products of combustion surround the tubes.

waterwall A row of watertubes lining a furnace or combustion chamber, exposed to the radiant heat of the fire; used to protect refractory and to increase capacity of the boiler.

weld   To join two edges or surfaces of metal by the application of heat. Also **forged-weld, fusion-weld.**

welded wall   A furnace closure wall made up of closely spaced waterwall tubes welded together or to an intermediate fin to form a continuous airtight structure.

windbox   A chamber below the grate or surrounding a burner, through which air under pressure is supplied for combustion of the fuel.

wire drawing   A cutting of surfaces caused by the abrasive action of high-velocity flow under restricted outlet.

wrapper sheet   The outside plate enclosing the firebox in a firebox or locomotive boiler. Also the thinner sheet in the shell of a two-thickness boiler drum.

yield point   The point at which a metal, under a mounting tensile load, exceeds its elastic limit. At the yield point the metal becomes permanently deformed and will not return to its original shape or position upon cessation of the load.

yoke magnetization   In magnetic particle inspection, a longitudinal magnetic field induced in a part, or in an area of a part, by means of an external electromagnet shaped like a yoke.

Appendix

2

Water Treatment
Tables

TABLE A2.1   Terms Used in Descriptive Literature on Water Treatment

Alkalinity   The protection of boiler plate from acidity and also the exposure to caustic embrittlement is indicated by the alkalinity report. Water containing sodium carbonate or bicarbonate tends to be alkaline when heated in the boiler; they change into sodium hydroxide.

Chlorides   The chloride concentration is used as an indication of the total amount of dissolved and suspended solids in the boiler water, such as chemicals used in treatment to dissolve scale. A yardstick for using blowdown.

Phosphates   Scale formation by sulfates, chlorides, and nitrates of calcium and magnesium [$CaSO_4$, $CaCl_2$, $Ca(NO_3)_2$, $MgSO_4$, $MgCl_2$, $Mg(NO_3)_2$] do not influence the alkalinity of water, and form a hard scale in boilers when water is evaporated. These are usually treated with a combination of Na_2HPO_4 (disodium phosphate) and Na_2CO_3 (sodium carbonate). The purpose being to precipitate out $Ca_3(PO_4)_2$ (calcium phosphate) while the Na_2SO_4 (sodium sulfate), being highly soluble, does not deposit scale. Phosphates are used to help prevent scale formation. The use of the two chemicals Na_2HPO_4 and Na_2CO_3 in conjunction are not only more thorough than either by itself but also form a mixed sludge which is more easily removed during blowdown.

Salt   A salt is any compound produced when all or part of the hydrogen of an acid is replaced by an electropositive radical or a metal. $Ca(HCO_3)_2$ heated near 212°F yields $CaCO_3$ plus CO_2 plus H_2O, H_2CO_3 being the highly soluble carbonic acid found in common soda water.

Acid   An acid has a predominant strength of hydrogen (positive) ions, H_2.

Alkali   An alkali has a predominant strength of hydroxyl (negative) ions, OH.

Hardness   Water containing lime and/or magnesia is considered hard to the degree these salts are present. When fatty acid found in soap such as stearic ($C_{17}H_{35}COOH$) of a given quantity and water of a given quantity are mixed, all the metal of the salts (that is Ca and Mg) combines with the stearic solution used in the test before a permanent lather is obtained; at this point the hardness is determined.

TABLE A2.2   Effect of Impurities in Boiler Water

Common name	Chemical term	Manifestation (when excessive)	Symbol
Lime	Calcium bicarbonate	Soft scale (dissolved in carbonic acid)	$Ca(HCO_3)_2$
	Calcium carbonate	Soft scale	$CaCO_3$
Magnesia	Magnesium carbonate	Chalky scale	$MgCO_3$
Silica	Silicon dioxide	Brittle, light, hard scale	SiO_2
Gypsum (plaster of paris)	Calcium sulfate and water in crystal form	Hard smooth scale	$CaSO_4 + H_2O$
Magnesium chloride	Same as common name	Forms hydrochloric acid with water	$MgCl_2$
Epsom salts	Magnesium sulfate and water	Corrosive	$MgSO_4 + H_2O$
Table salt	Sodium chloride	Causes foaming	NaCl
Glauber's salts	Sodium sulfate	Causes foaming	Na_2SO_4
Soda ash	Sodium carbonate	Causes foaming	Na_2CO_3
Baking soda	Sodium bicarbonate	Causes foaming	$NaHCO_3$
Gases			
Oxygen	Same as common name	Accelerates electrolytic corrosion	O_2
Carbon dioxide	Same as common name	Forms acid in water	CO_2
Chlorine	Same as common name	Forms acid in water	Cl
Organic Substances Such as Leaf Mold, also Mud			
		Foaming and deposits	
Acids			
Sulfuric acid	Same as common name	Corrosive	H_2SO_4
Hydrochloric acid	Muriatic acid	Corrosive	HCl
Alkalis			
Sodium hydroxide	Caustic soda	Caustic removal of protective oxides from metal	NaOH
Magnesium hydroxide	Same as common name	Caustic removal of protective oxides from boiler metal	$Mg(OH)_2$

TABLE A2.3   Some Chemicals and Their Purpose in Water Treatment

Chemical	Purpose	Comment
Sodium hydroxide NaOH (caustic soda)	Increase alkalinity, raise pH, precipitate magnesium	Contains no carbonate, so doesn't promote CO_2 formation in steam
Sodium carbonate Na_2CO_3 (soda ash)	Increase alkalinity, raise pH, precipitate calcium sulfate as the carbonate	Lower cost, more easily handled than caustic soda. But some carbonate breaks down to release CO_2 with steam
Sodium phosphates NaH_2PO_4, Na_2HPO_4, Na_3PO_4, $NaPO_3$	Precipitate calcium as hydroxyapatite $[Ca_{10}(OH)_2(PO_4)_6]$	Alkalinity and resulting pH must be kept high enough for this reaction to take place (pH usually above 10.5)
Sodium aluminate $NaAl_2O_4$	Precipitate calcium, magnesium	Forms a flocculent sludge
Sodium sulfite Na_2SO_3	Prevent oxygen corrosion	Used to neutralize residual oxygen by forming sodium sulfate. At high temperatures and pressures, excess may form H_2S in steam
Hydrazine hydrate $N_2H_4 \cdot H_2O$ (35% solution)	Prevent oxygen corrosion	Removes residual oxygen to form nitrogen and water. One part of oxygen reacts with three parts of hydrazine (35% solution)
Filming amines Octadecylamine, etc.	Control return-line corrosion by forming a protective film on the metal surfaces	Protects against both oxygen and carbon dioxide attack. Small amounts of continuous feed will maintain the film
Neutral amines Morpholine, cyclohexylamine, benzylamine	Control return-line corrosion by neutralizing CO_2 and adjusting pH of condensate	About 2 ppm of amine is needed for each ppm of carbon dioxide in steam. Keep pH in range of 7.0 to 7.4 or higher
Sodium nitrate $NaNO_3$	Inhibit caustic embrittlement	Used where the water may have embrittling characteristics
Tannins, starches, glucose and lignin derivatives	Prevent feed line deposits, coat scale crystals to produce fluid sludge that won't adhere as readily to boiler heating surfaces	These organics, often called protective colloids, are used with soda ash, phosphate. Also distort scale crystal growth, help inhibit caustic embrittlement
Seaweed derivatives (Sodium alginate, sodium mannuronate)	Provide a more fluid sludge and minimize carryover	Organics often classed as reactive colloids since they react with calcium, magnesium and absorb scale crystals
Antifoams (Polyamides, etc.)	Reduce foaming tendency of highly concentrated boiler water	Usually added with other chemicals for scale control and sludge dispersion

TABLE A2.4   Valence, Ionic, Molecular, and Equivalent Weights of Chemical Compounds Used in Water Treatment[1]

	Ion formula	Ionic weight	Equivalent weight
Cations			
Aluminum	Al^{+++}	27.0	9.0
Ammonium	NH_4^+	18.0	18.0
Calcium	Ca^{++}	40.1	20.0
Hydrogen	H^+	1.0	1.0
Ferrous Iron	Fe^{++}	55.8	27.9
Ferric Iron	Fe^{+++}	55.8	18.6
Magnesium	Mg^{++}	24.3	12.2
Manganese	Mn^{++}	54.9	27.5
Potassium	K^+	39.1	39.1
Sodium	Na^+	23.0	23.0
Anions			
Bicarbonate	HCO_3^-	61.0	61.0
Carbonate	CO_3^{--}	60.0	30.0
Chloride	Cl^-	35.5	35.5
Fluoride	F^-	19.0	19.0
Nitrate	NO_3^-	62.0	62.0
Hydroxide	OH^-	17.0	17.0
Phosphate (tribasic)	PO_4^{---}	95.0	31.7
Phosphate (dibasic)	HPO_4^{--}	96.0	48.0
Phosphate (monobasic)	$H_2PO_4^-$	97.0	97.0
Sulfate	SO_4^{--}	96.1	48.0
Sulfite	SO_3^{--}	80.1	40.0

	Formula	Molecular weight	Equivalent weight
Compounds			
Aluminum hydroxide	$Al(OH)_3$	78.0	26.0
Aluminum sulfate	$Al_2(SO_4)_3$	342.1	57.0
Alumina	Al_2O_3	102.0	17.0
Sodium aluminate	$Na_2Al_2O_4$	164.0	27.3
Calcium bicarbonate	$Ca(HCO_3)_2$	162.1	81.1
Calcium carbonate	$CaCO_3$	100.1	50.1
Calcium chloride	$CaCl_2$	111.0	55.5
Calcium hydroxide (pure)	$Ca(OH)_2$	74.1	37.1
Calcium hydroxide (90%)	$Ca(OH)_2$	—	41.1
Calcium sulfate (anhydrous)	$CaSO_4$	136.2	68.1
Calcium sulfate (gypsum)	$CaSO_4 \cdot 2H_2O$	172.2	86.1
Calcium phosphate	$Ca_3(PO_4)_2$	310.3	51.7
Disodium phosphate	$Na_2HPO_4 \cdot 12H_2O$	358.2	119.4
Disodium phosphate (anhydrous)	Na_2HPO_4	142.0	47.3
Ferric oxide	Fe_2O_3	159.6	26.6

TABLE A2.4   Valence, Ionic, Molecular, and Equivalent Weights of Chemical Compounds Used in Water Treatment[1](Continued)

	Formula	Molecular weight	Equivalent weight
Compounds			
Iron oxide (magnetic)	Fe_3O_4	321.4	—
Ferrous sulfate (copperas)	$FeSO_4 \cdot 7H_2O$	278.0	139.0
Magnesium bicarbonate	$Mg(HCO_3)_2$	146.3	73.2
Magnesium carbonate	$MgCO_3$	84.3	42.2
Magnesium chloride	$MgCl_2$	95.2	47.6
Magnesium hydroxide	$Mg(OH)_2$	58.3	29.2
Magnesium phosphate	$Mg_3(PO_4)_2$	263.0	43.8
Magnesium sulfate	$MgSO_4$	120.4	60.2
Monosodium phosphate	$NaH_2PO_4 \cdot H_2O$	138.1	46.0
Monosodium phosphate (anhydrous)	NaH_2PO_4	120.1	40.0
Metaphosphate	$NaPO_3$	102.0	34.0
Sodium bicarbonate	$NaHCO_3$	84.0	84.0
Sodium carbonate	Na_2CO_3	106.0	53.0

[1]Note: The valences are shown as + and − in the Ion Formula column. (*Courtesy Nalco Chemical Co.*)

3

Observing Boiler Safety Rules

The rules that follow are brief reminders of the possible consequences in a boiler plant of inappropriate operator response or questionable actions in maintaining the boiler plant. In all cases, follow the published or oral safety rules of your employer, jurisdiction, and the Federal OSHA safety regulations to avoid possible disciplinary actions. In addition, study the safety guidelines of the manufacturer of your boiler.

NEVER	ALWAYS
NEVER fail to anticipate emergencies. Do not wait until something happens to start thinking.	ALWAYS study every conceivable emergency and know exactly what moves to make.
NEVER start work in a strange plant without tracing every pipeline and learning the location and purpose of every valve. Know your job.	ALWAYS proceed to proper valves or switches rapidly but without confusion in time of emergency. You can think better walking than running.
NEVER allow sediment to accumulate in gauge-glass or water-column connections. A false water level may fool you and make you sorry.	ALWAYS blow out each gauge-glass and water-column connection at least once each day. Forming good habits may mean longer life for you.
NEVER give verbal orders for important operations or report such operations verbally with no record. Have something to back you up when needed.	ALWAYS accompany orders for important operations with a written memorandum. Use a log book to record every important fact or unusual occurrence.
NEVER light a fire under a boiler without a double check on the water level. Many boilers have been ruined and many jobs lost this way.	ALWAYS have at least one gauge of water before lighting off. The level should be checked by the gauge cocks. You will not be fired for being too careful.

NEVER

NEVER light a fire under a boiler without checking all valves. Why take a chance?

NEVER open a valve under pressure quickly. The sudden change in pressure, or resulting water hammer, may cause piping failure.

NEVER cut a boiler in on the line unless its pressure is within a few pounds of header pressure. Sudden stressing of a boiler under pressure is dangerous.

NEVER bring a boiler up to pressure without trying the safety valve. A boiler with its safety valve stuck is nearly as safe as playing with dynamite.

NEVER take it for granted that the safety valves are in proper condition. The power plant is no place for guesswork.

NEVER increase the setting of a safety valve without authority. Serious accidents have occurred from failure to observe this rule.

NEVER change adjustments of a safety valve more than 10 percent. Proper operation depends on the proper spring.

NEVER tighten a nut, bolt, or pipe thread under steam or air pressure. Many have died doing this.

NEVER strike any object under steam or air pressure. This is another sure path to the undertaker.

NEVER allow unauthorized persons to tamper with any steam-plant equipment. If they do not injure themselves, they may cause injury to you.

ALWAYS

ALWAYS be sure blowdown valves are closed and proper vents, water-column valves, and pressure-gauge cock are open.

ALWAYS use the bypass if one is provided. Crack the valve from its seat slightly and await pressure equalization. Then open it slowly.

ALWAYS watch the steam gauge closely and be prepared to cut the boiler in, opening the stop valve only when the pressures are nearly equal.

ALWAYS lift the valve from its seat by the hand lever when the pressure reaches about three-quarters of popping pressure.

ALWAYS raise the valve from its seat with the lifting lever periodically while the boiler is under pressure. Test by raising to popping pressure at least once per year.

ALWAYS consult an authorized boiler inspector and accept his or her recommendations before increasing the safety-valve pressure setting.

ALWAYS have the valve fitted with a new spring and restamped by the manufacturer for changes over 10 percent.

ALWAYS play safe on this rule. The one that is going to break does not have a special warning sign.

ALWAYS play safe on this rule. You cannot tell which straw might break the camel's back.

ALWAYS keep out loiterers and place plant operation in the hands of proper persons. A boiler room is not a place for a club meeting.

NEVER leave an open blowdown valve unattended when a boiler is under pressure or has a fire in it. Play safe; memory can fail.

ALWAYS check the water level before blowing down and have a second person watch the water gauge level while you blowdown the boiler. Close the blowdown valve, and then recheck the water level. You will avoid dry-firing the boiler this way.

NEVER allow anyone to enter a drum of a boiler without following OSHA rules for entering a confined space.

ALWAYS make sure the boiler is cool to enter, has enough oxygen per OSHA rules, has a sign by the entrance stating "Worker Inside," has an emergency person at the entrance, and that all valves going to and from the boiler are locked closed.

NEVER allow major repairs to a boiler without authorization. If you do not break a law, you may break your neck.

ALWAYS consult an authorized boiler inspector before proceeding with boiler repairs.

NEVER try to light a second burner from the flames of the first on-line burner. You might be inviting a serious puffback.

ALWAYS follow the starting sequence of the manufacturer on multiburner boilers, including ignition and main flame proving by installed burner controls, and you will avoid a furnace explosion.

NEVER attempt to light a burner without venting the furnace until clear. Burns are painful.

ALWAYS allow draft to clear furnace of gas and dust for prescribed purge period. Change draft conditions slowly.

NEVER fail to report unusual behavior of a boiler or other equipment. It may be a warning of danger.

ALWAYS consult someone in authority. Two heads are better than one.

Selected Bibliography

ASME Boiler and Pressure Vessel Codes, Sections I through VI, IX, and XI, American Society of Mechanical Engineers, New York.
Elonka, S. M.: *Standard Plant Operator's Manual,* McGraw-Hill, New York, 1975.
_____, and A. L. Kohan: *Standard Heating and Power Boiler Plant Questions and Answers,* McGraw-Hill, New York, 1984.
Fundamentals of Welding, American Welding Society, Miami, Fla.
National Board Inspection Code, National Board of Boiler and Pressure Vessel Inspectors, Columbus, Ohio.
National Fire Protection Codes, National Fire Protection Association, Boston.
Power Piping Code, ANSI B31.1, American National Standards Institute, New York.
Recommended Practices for NDT Personnel Qualifications and Certification, American Society for Nondestructive Testing, Evanston, Ill.
State, County, and City Synopsis of Boiler and Pressure Vessel Laws on Design, Installation, and Reinspection Requirements, Uniform Boiler and Pressure Vessel Laws Society, Louisville, Kentucky.
Kohan, A. L.: *Pressure Vessel Systems,* McGraw-Hill, New York, 1987.
Kohan, A. L.: *Plant Services and Operations Handbook,* McGraw-Hill, New York, 1995.

Index

ABOUT THE AUTHOR

Anthony Kohan is a consultant with more than 30 years of experience as a technician, tester, and inspector of boilers, pressure vessels, and machinery. He was formerly the manager of the boiler and machinery technical specialists at the Royal Insurance Co. Mr. Kohan is the author or coauthor of *Standard Boiler Room Questions and Answers, Standard Heating and Power Boiler Plant Questions and Answers, Pressure Vessel Systems,* and *Plant Services and Operations Handbook,* all published by McGraw-Hill.